QUANTUM MECHANICS FOR APPLIED PHYSICS AND ENGINEERING

by Albert Thomas Fromhold, Jr.
DEPARTMENT OF PHYSICS
AUBURN UNIVERSITY
AUBURN, ALABAMA

Dover Publications, Inc., New York

Published in Canada by General Publishing Company, Ltd., 30 Lesmill Road, Don Mills, Toronto, Ontario.
Published in the United Kingdom by Constable and Company, Ltd., 3 The Lanchesters, 162–164 Fulham Palace Road, London W6 9ER.

This Dover edition, first published in 1991, is an unabridged, corrected republication of the work first published by Academic Press, N.Y., 1981.

Manufactured in the United States of America
Dover Publications, Inc., 31 East 2nd Street, Mineola, N.Y. 11501

Library of Congress Cataloging-in-Publication Data

Fromhold, A. T.
Quantum mechanics for applied physics and engineering / by Albert Thomas Fromhold, Jr.
p. cm.
Reprint. Originally published: New York : Academic Press, 1981.
Includes bibliographical references and index.
ISBN 0-486-66741-3 (pbk.)
1. Quantum theory. 2. Engineering. I. Title.
QC174.12.F76 1991
530.1'2—dc20 91-8712
CIP

To my sons
Thomas William and Matthew Albert

May each be inspired
to lead a life governed by
lightheartedness, logic, and self-discipline

CONTENTS

PART II QUANTUM STATISTICS OF MANY-PARTICLE SYSTEMS; FORMULATION OF THE FREE-ELECTRON MODEL FOR METALS

PART III APPROXIMATION TECHNIQUES FOR THE SCHRÖDINGER EQUATION

PREFACE

To those interested principally in the application of quantum physics to the development of new technology, it may seem eminently pragmatic to accept experimental phenomena following from the wavelike behavior of particles as concrete facts without troubling to consider the quantum leap in the intellect required to explain the phenomena. To those who experience the joy of formulating and solving equations of motion resulting from the application of Newton's second and third laws to macroscopic particles, however, the dichotomy of the reasoning processes required to view an entity simultaneously as a wave and as a particle is very real. One must marvel at the theoretical physicists such as Werner Heisenberg and Erwin Schrödinger, who developed the conceptual basis and the theoretical structure required for quantum calculations. As one probes more deeply into the origins of this abstruse and highly mathematical discipline, however, it becomes quite clear that these esoteric theorists were in fact hardheaded realists, driven to develop the new approach only because of the failure of the esteemed "classical" physics to provide an accurate description of nature at the electronic and nuclear levels. The acceptance of this failure and the development of a successful alternative is in the spirit of the highest principles of applied science; in this respect, the architects of quantum theory are excellent models for those aspiring to become outstanding engineers and applied physicists. After all, P. A. M. Dirac, who developed the well-known relativistic wave equation, received his early training in electrical engineering, and the physics Nobel laureate Ivar Giaever, famous for his work on electron tunneling in superconductors, was initially a successful mechanical engineer.

The predictive powers of quantum theory are so great that those aspiring to develop new technology certainly wish to be able to use the theory. Quantum mechanics is our only method of understanding and predicting the behavior of modern atomic and solid-state devices, including masers and lasers, atomic clocks, superconducting magnets, Esaki diodes, and Josephson tunnel junctions. Many of these are already employed in industry, and current technical advances are leading to the development and use of even more such quantum devices. Therefore the need is urgent for applied scientists to develop a functional knowledge of quantum concepts and methodology. It can be said that the Schrödinger equation is to quantum devices what Newton's laws of motion are to the space shuttle. However convincing these arguments may be, the question of the practicality of obtaining such a working knowledge without working through the equivalent of earning a physics Ph.D. remains. Pragmatic individuals have often concluded that it is beyond the realm of possibility for them to clarify their muddled concept of the semiconductor energy gap, and that quantum tunneling of a particle through an energy barrier is a theoretical exercise involving talents second only to those required to be an international chess champion. This text was conceived to enable such individuals to "tunnel" through "barriers" of ignorance so that they may participate in the enlightened spirit of the quantum way of viewing nature. It is intended to lead them into developing a mental facility for doing practical quantum calculations. To sense nature's beautiful harmony at work as electrons flow through a metal wire is as delightful as listening to a Mozart piano concerto. Moreover, it is vastly elevating as well as pragmatic to have full assurance that one is correctly applying a mathematical formula to a given situation, such confidence stemming from an understanding of the derivation of the formula.

This book evolved from practical experience in training engineers and applied physicists. Initially a set of notes was written with the objective of providing an elementary and self-contained development of the fundamentals of quantum mechanics. Emphasis was placed on those aspects of quantum mechanics and quantum statistics that are essential to an understanding of solid-state theory. The notes were kept honest in the sense that all developments were carefully worked out, complete details were given for each of the derivations, and successive derivations were developed on a firm basis provided by the preceding material. This approach was adopted to minimize frustration for the serious reader. The notes were so successful in their objective of grounding students thoroughly in the quantum method that the author was prompted to submit them as the basis for a full text. Encouragement was offered by the editorial staff of Academic Press to develop the material into its present form. The changes and additions were field-tested on upper-level undergraduates and on graduate students at Auburn University as they were incorporated, and simultaneously, problems and projects were developed for the text.

Basic quantum mechanics, using the Schrödinger equation, is completely developed from first principles in Chap. 1. Quantum statistics (Chap. 2) is developed as a prelude to the important free-electron theory of metals

(Chap. 3). Perturbation theory (Chap. 5) is developed and employed to evaluate modifications in free-electron theory to accommodate the effects of ion cores in the solid. The WKB approximation (Chap. 4) is employed to deduce the transmission coefficient for electron tunneling in solids. The theory of electronic energy bands (Chap. 7) is developed by applying the Schrödinger equation to the problem of the periodic potential of a crystalline solid (Chap. 6). Throughout the text, examples from solid-state physics are employed to illustrate specific applications and to demonstrate the principal results that can be deduced by means of quantum theory. This serves to bridge the artificial gap between quantum mechanics and solid-state physics, thereby circumventing the cardinal difficulties encountered by applied physicists and engineers in learning solid-state physics from conventional monographs. Because of the strong emphasis on the rapid development of the background needed to understand energy bands and electron transport in crystals, a somewhat lesser emphasis has been placed on the mathematical details of the hydrogen atom and harmonic oscillator problems than is traditional. Even though the advisability of this sacrifice might be questioned by some, it can be argued that so many excellent developments of these two problems are already available that it is somewhat difficult to justify still another.

Several tactics are used to increase the effectiveness of the material as a learning aid. First, a determined effort has been made to avoid requiring from the reader a heavy mathematical background. The usual courses in calculus and differential equations are presumed, but very little beyond this is required. Second, the physics background required of the student is minimal. In fact, the usual two-semester sequence in calculus-based elementary physics will be found to be adequate for most students. Rigorous detailed derivations are sometimes preceded by simpler derivations containing the essential features. Parallel developments leading to the same important final result are occasionally given. Previews are given whenever needed. (For example, the energy band picture for an insulator is introduced in Chap. 4 for metal–insulator–metal tunneling; this provides an overview that anticipates the illustration of energy gaps using perturbation theory in Chap. 5 and the formal development of energy bands using Bloch functions in Chap. 7.) Expanded treatments of some topics are given because of high current interest in specific areas of applied research, electron tunneling being one case in point.

This is sufficient material for a four-semester sequence for upper-level undergraduates and beginning graduate students in applied physics and engineering. The four parts of the book (Elementary Quantum Theory; Quantum Statistics and the Free-Electron Model; WKB Approximation and Perturbation Theory; and Energy Bands in Crystals) each require a semester. In an alternative usage mode, Parts I, II, and III can be covered in a two-semester course in basic quantum mechanics for physics majors. In a third mode, Chaps. 3, 4, 6, and 7 have been used frequently for an elementary course in solid-state physics. In a fourth mode, the author has on occasion used Chap. 2 for a special-topics course in the quantum mechanics of many-particle systems and the development of quantum statistics.

Exercises for the reader provide motivation to work out simple details of the central developments and in some cases to extend these developments. Problems have been designed to encourage practical numerical computations with hand-held calculators or computer-terminal facilities. A series of projects, frequently requiring outside study and consultation of technical literature, will be found to be enjoyable and profitable by many. Such projects are intended to provide enough in the way of important extended developments to enable the reader to expand upon the "bare bones" of the textual material and develop competence and style in open-ended applied research problems. Apart from the goals of motivating the reader to understand and develop a working knowledge of the material, are not the really important objectives those of training the reader to develop ideas logically and to synthesize diverse concepts? Is this not one of the better ways to develop latent research capabilities? These projects will often require other textbooks to broaden the reader's basic knowledge in physics, to develop his capability in relevant mathematical techniques, or to acquaint him with the published experimental data on a topic. They may be pursued as an individual or a group effort. The number and variety of these study aids are thus sufficient to allow choices in accord with individual tastes and abilities. One can view the text as a "variable-difficulty" learning aid, with the "difficulty index" determined by the degree of incorporation of these materials. The major objective should, of course, be initially the mastery of the text material. Then, even a careful reading of the problems and projects will broaden scientific horizons and will aid in correlating the material with that in more advanced works.

ACKNOWLEDGMENTS

Dr. W. Beall Fowler, Jr., and Dr. Frank J. Feigl of Lehigh University carefully read the entire first draft of the manuscript and made numerous helpful comments and detailed suggestions; the author acknowledges with sincere thanks their labors, which led to a much higher quality end product. The editorial staff of Academic Press offered continuous encouragement to the author, yet remained firm in its exhortation to render the manuscript into the best possible form before going to press. The author is much indebted to his former department head, Dr. Howard E. Carr, who offered him the opportunity to teach so many courses in modern physics, quantum mechanics, and solid-state physics at Auburn University. Dr. Charles H. Holmes made several constructive suggestions for improvement of the work, based on his extensive experience in working with engineering students and practicing engineers. Never to be forgotten, of course, are the students who represent the raison d'être for the typed class notes, and who were always courteous and encouraging when questioned regarding the value of the notes as a learning aid. The author and these students are much indebted to a number of departmental secretaries who worked at one time or another over the years on the class notes, painstakingly typing and retyping the several versions which culminated in this text. Mrs. Gail Pressnell and Mrs. Patricia Ray, in particular, spent much time on this effort; Mrs. Sara Watkins, Mrs. Carol Henderson, Mrs. Karen Sollie, Mrs. Mary Childers, and Mrs. Paula Howell also worked diligently on the notes during their tenures, while Mrs. Sherry Walton and Mrs. Jo Hawkins graciously assisted whenever they could spare the time from their regular duties. The author is likewise very grateful to the hardworking and talented student draftsmen, Mr. Danny Creamer and Mr. Don Cooper, who labored diligently to create professional-quality drawings out of sometimes crude and abstruse sketches. This list would not be complete without the author's acknowledgment of the invaluable technical assistance of his wife, Regina, who aided throughout with editorial suggestions, typing of preliminary drafts, and proofreading. Even more important were the words of encouragement she offered when his spirits were low and the patience she exhibited during his preoccupation with this long-term effort.

PART I

Elementary Quantum Theory

CHAPTER 1

AN INTRODUCTION TO QUANTUM MECHANICS

Today we know that no approach which is founded on classical mechanics and electrodynamics can yield a useful radiation formula. A. Einstein (1917)

1 Wave–Particle Duality

1.1 Domain of Quantum Mechanics

Quantum mechanics is a theory that can be used to correlate and predict the behavior of atomic and subatomic systems. These systems constitute the microscopic domain in nature where the predictions of "classical" physics (e.g., Newton's three laws of motion) are not always in accord with experimental results. Quantum mechanics not only predicts correctly the results of physical observations in the microscopic domain, it also continues to predict correctly in the macroscopic domain where classical physics is applicable.

1.2 Particle and Wave Properties

Matter and electromagnetic radiation individually have both particlelike properties and wavelike properties. By "particlelike" we mean localized and acting in some sense as individual entities. By "wavelike" we mean nonlocalized and periodic, with the capability of interacting constructively or destructively with similar entities. The coexistence of wavelike and particlelike aspects in a single physical entity is known as "wave–particle duality."

1.3 Quantization and Discreteness

Certain properties of matter and certain properties of radiation are found to be quantized. Matter in the atomic and nuclear domain consists of a variety of particles of electronic and nuclear scale (e.g., electrons, protons, neutrons) and combinations of such; each is found to have a discrete set of values for the various physical properties such as mass, charge, spin angular momentum, magnetic moment, and electric quadrupole moment. For example, a continuous range of values of the charge on an electron is not found, but instead, all electrons have the same fixed charge. In this sense the charge is quantized. Similarly, the mass, spin angular momentum, and magnetic moment of the electron are quantized.

The quantization of radiation, on the other hand, is manifested by the fact that it is frequently observed to interact with matter as if it were an ensemble of discrete entities, each such entity having a fixed amount of energy and momentum. An example of this behavior is the photoelectric effect, which is the

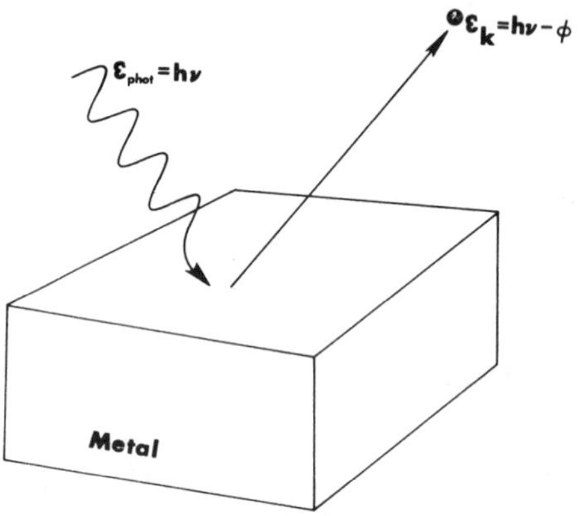

Fig. 1.1 Photoelectric effect. (A photon of frequency ν and energy $\mathscr{E}_{\text{phot}} = h\nu$ ejects an electron from a metal surface with a work function ϕ, the outgoing electron having a maximum kinetic energy $\mathscr{E}_{\text{K}} = h\nu - \phi$.)

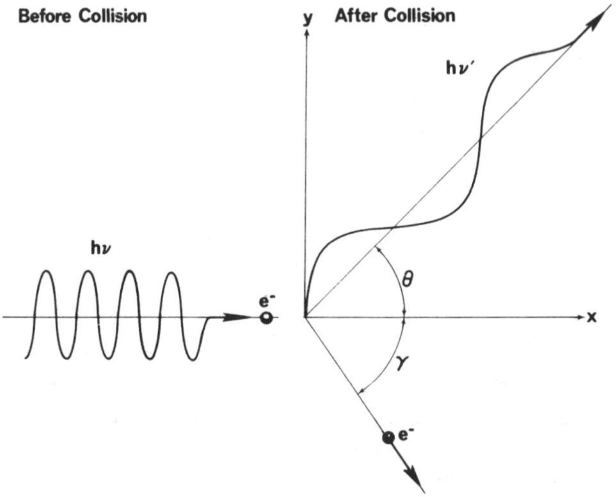

Fig. 1.2 Compton effect. (An incoming photon of frequency ν and energy $h\nu$ is scattered at an angle θ by a free electron e^-, thereby undergoing a decrease in frequency to ν' and a decrease in energy to $h\nu'$, with the difference in energy $h(\nu - \nu')$ appearing as an increase in the kinetic energy of the free electron scattered at an angle γ.)

ejection of electrons from a metal by a particlelike interaction of the incident electromagnetic radiation with the conduction electrons in the metal. This is illustrated in Fig. 1.1. Another example is the Compton effect, as illustrated in Fig. 1.2, which is the observation of an increase in the wavelength of electromagnetic radiation due to particlelike collisions of photons (such as x rays) with unbound electrons in a solid. These characteristic properties of matter and radiation may be viewed as "particlelike" properties rather than specifically "quantum" properties, since the word "quantization" as it is used in quantum mechanics often has the connotation of certain specific sets of allowed and disallowed values.

The discreteness of physical properties persists when elementary particles combine to form atoms and nuclei. In particular, experimental measurements of spectral lines and particle–atom collision processes lead directly to the viewpoint that atoms possess discrete energy levels. This observation provides the basis for the semiclassical atomic model (Bohr atom) in which the electron is visualized as undergoing planetary-type motion in certain "allowed orbits" around a parent nucleus (see Fig. 1.3). The corresponding quantum-mechanical model is one in

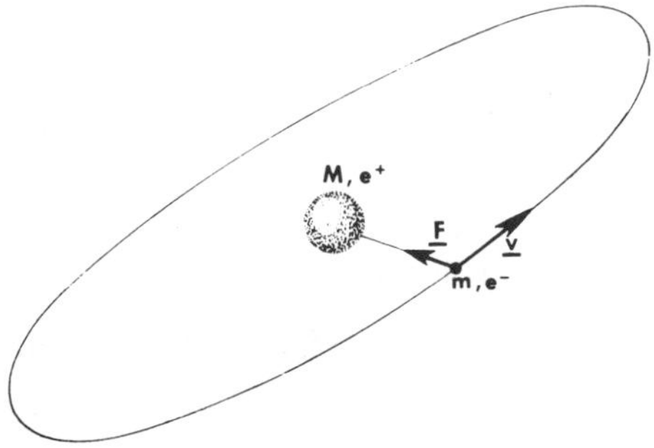

Fig. 1.3 Classical picture (Bohr model) of the atom. (An electron of mass m and charge e^- is visualized as undergoing planetary-type motion at a speed v in certain "allowed" orbits around an oppositely charged nucleus having a much greater mass M.)

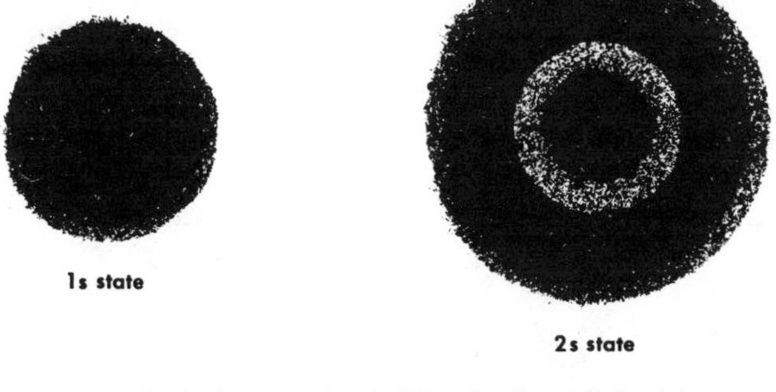

Fig. 1.4 Quantum-mechanical picture (probability density model) of the atom. (An electron is visualized in terms of "patterns" of probability density about the nucleus, the pattern being characteristic of the electronic state.)

which the electron is described at a given time by certain "allowed patterns" of probability density (see Fig. 1.4). The allowed "orbits" or "patterns" are designated simply as *electronic states*. These discrete electronic states can be shown to be characterized by specific values for the energy and angular momentum of the electron. The atomic absorption of electromagnetic radiation or the collision of the atom with a bombarding particle promotes the electron in the atom to one of the higher energy states. The emission of electromagnetic radiation is then considered to follow from a transition of the electron to one of the lower energy states. The lowest energy state of the electron for the system in question is called the *ground state*; this state is stable with respect to decay by radiation.

1.4 Nature of Radiation

The picture of transitions between discrete electronic states through absorption and emission of electromagnetic radiation is consistent with energy conservation. The amount of radiation absorbed or emitted in a given transition represents a relatively fixed amount of energy, the amount being equal to the difference in energy between the initial and final states (cf. Fig. 1.5). Although this in itself does not give us any details concerning the various possible forms in which the electromagnetic radiation might be emitted, we deduce from experimental optical spectral lines that the energy $\Delta\mathscr{E}$ emitted in the transition of an electron from a given higher energy (or excited) state to a given lower energy state occurs in the form of radiation which has a characteristic spectral frequency ν. The frequency ν is linearly related to $\Delta\mathscr{E}$,

$$\Delta\mathscr{E} = h\nu \qquad \textbf{(energy–frequency relation)}. \tag{1.1}$$

The proportionality factor h, known as **Planck's constant**, has the value

$$h = 6.626 \times 10^{-34} \quad \text{J sec} = 6.626 \times 10^{-27} \quad \text{erg sec}. \tag{1.2}$$

This restriction of the emission or absorption of radiation to discrete bundles (or "packets") of energy $\Delta\mathscr{E} = h\nu$ having a specific spectral frequency ν leads to the postulate that the radiation emitted by natural atomic radiators occurs in the form of a particlelike quantity (known as the *photon*) with energy $h\nu$. A radiation "field" can then be established by a radiation source consisting of a large

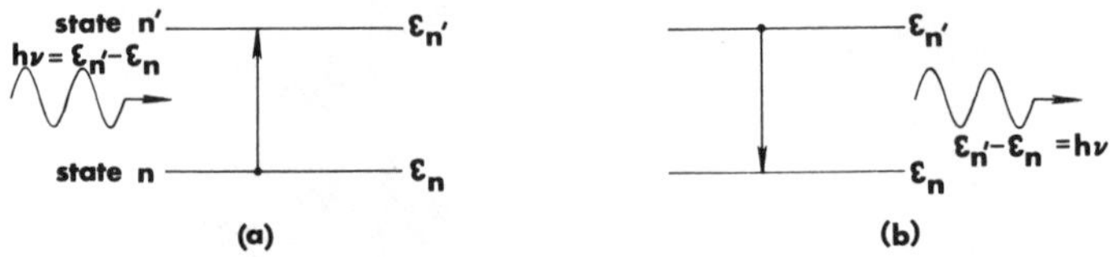

Fig. 1.5 Energy absorption (a) and emission (b) in electronic transitions between discrete energy levels involving electromagnetic radiation of frequency ν in quantized increments $h\nu$. (The amount of radiation absorbed or emitted in an electronic transition represents a relatively fixed amount of energy, the amount being equal to the difference in energy between the initial and final electronic states.)

number of atoms emitting photons simultaneously. The electromagnetic field is known to have wavelike properties; nevertheless, the wavelike field (associated, for example, with a propagating light wave) can ultimately be resolved into discrete packets of radiation called photons which are emitted by the discrete atoms of the source. This in itself can lead one to speculate that light (or more generally, electromagnetic radiation) has a *dual* nature, since it is endowed with both wavelike and particlelike properties.

1.5 Photoelectric Effect

Einstein applied the concept of particlelike properties to light to explain the photoelectric effect. This experimental effect, associated with the ejection of electrons from metal surfaces when the surfaces are illuminated with electromagnetic radiation (see Fig. 1.1), cannot be understood on the basis of purely classical physics [Gamow (1966)]. For example, if the illumination were perfectly uniform, it would take a very long time for any single electron in the metal to receive enough energy from direct radiation to surmount the metal–vacuum work-function barrier because an electron is extremely small, having a radius of the order of 2.8×10^{-13} cm. In the experimental observation of electron ejection, however, there is no minimum time for electrons to be ejected after illumination begins, even in the limit of very low illumination intensities. This is in accordance with the concept that each quantum of radiation $h\nu$ is absorbed by a single electron and serves to eject the electron immediately whenever $h\nu$ is larger than some minimum energy $h\nu_0$ necessary to remove the electron from the metal. This minimum energy $h\nu_0$ is called the photoelectric work function ϕ for the metal in question. If $h\nu$ is less than ϕ, ejection is not generally found to occur even if the metal is illuminated for very long periods of time at very high intensities. It therefore appears that the ejection of electrons from metals by electromagnetic radiation is not a process in which energy is absorbed uniformly by the metal with subsequent electron ejection; rather it is a process in which individual photons of energy $h\nu$ interact with the metal to eject individual electrons from the metal. This conclusion is especially reinforced by the above-mentioned experimental observation of the cutoff in electron ejection as the photon energy $h\nu$ is decreased below ϕ. (The absorption of a single photon with energy *less* than ϕ does not provide the electron with enough energy to surmount the work function barrier, and the probability for the simultaneous absorption of two or more photons to provide sufficient energy for surmounting the work function barrier is too small under normal experimental conditions for this process to be observed.) In addition, the kinetic energy of the individual ejected electrons is found to increase with the frequency of the light as $h(\nu - \nu_0)$ but it is found to be independent of the *intensity* of the radiation. The *rate* of electron ejection for fixed frequency, however, is related linearly to the intensity of the radiation provided $\nu > \nu_0$ (see Figs. 1.6a and 1.6b). These facts are in accord with the concept of an interaction of individual photons of energy $\mathscr{E} = h\nu$ with the unbound electrons in the metal, with the density of such photons being proportional to the radiation intensity. In summary, the photoelectric effect can be said to be an experimental observation that emphasizes the "corpuscular" or "quantum" properties of light.

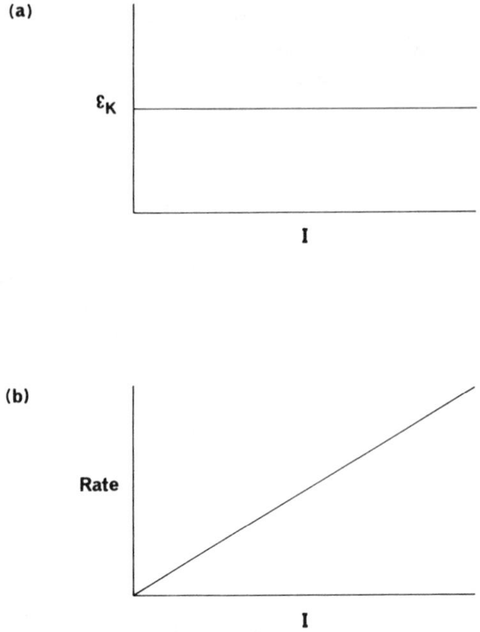

Fig. 1.6 Photoelectric emission of electrons from a metal surface having workfunction ϕ by radiation of a given frequency $\nu > \phi/h$. (a) The maximum kinetic energy $\mathcal{E}_K$ of the ejected electrons is independent of radiation intensity I. (b) The rate of electron ejection increases linearly with radiation intensity I.

1.6 Gedanken Experiments with Light

Can one reconcile the well-known phenomena of the interference and diffraction of light with the photoelectric effect? Let us investigate a typical interference experiment, in which light passing through the slits of a grating (see Fig. 1.7) is analyzed. As a detector we could use a photographic plate, which on detailed examination would reveal a *multitude* of spots with a density given by the classical wave theory (see Fig. 1.8). Each individual spot is actually the result of a photochemical reaction that is triggered by a single photon. This is shown by reducing the beam intensity to a point where, on the average, only one photon is passing through the apparatus at a time. Thus only one chemical reaction in the detector is triggered at a time. This result is in complete disagreement with the classical theory, which predicts that the *continuous* interference pattern should remain entirely unaltered, regardless of the degree to which the total intensity is reduced. In the one-photon-at-a-time experiment, we see only one spot at a time. In this sense it can be said that an individual photon does not interfere with itself. If we make a long exposure so that many photons pass through the apparatus, such that we fail to resolve the different spots on the photographic plate, we regain the classical wave pattern illustrated in Fig. 1.7. Thus in a statistical sense the photon does experience an interference effect in that its trajectory is determined by its interaction with the entire grating. In other words, the wave theory gives the specific *probability* of a light quantum striking the detector at a given point. Thus we have a quantum-statistical process that is based on the

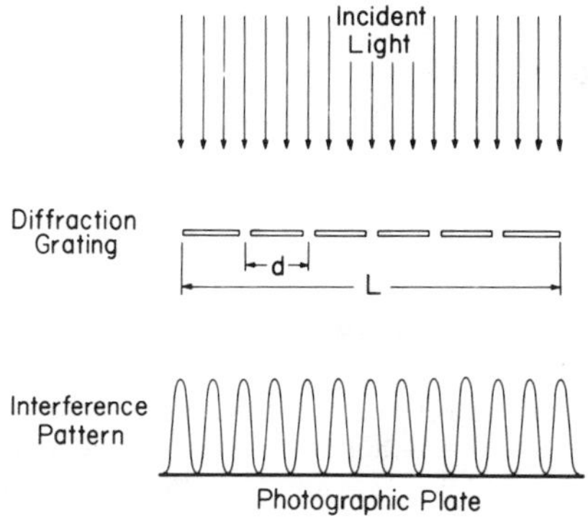

Fig. 1.7 Schematic illustration of the interference of light passed through the slits of a grating. (An interference pattern characteristic of the wavelength λ of the light, the spacing d between slits, and the length L of the transmission grating is observed in the transmitted light by means of a suitable detector such as a photographic plate. The pattern illustrated is characteristic of that expected for two very narrow slits, for which the intensity I in terms of the maximum intensity I_0 is $I = I_0 \cos^2\beta$, where $\beta = (\pi d/\lambda)\sin\theta$. The angle θ is determined by $\tan\theta = y/D$, where D is the perpendicular distance between the grating and the photographic plate and y is the distance along the photographic plate. Thus for small values of y/D, $\beta \simeq \pi(d/\lambda)(y/D)$, and I is a cosine-squared function of y.)

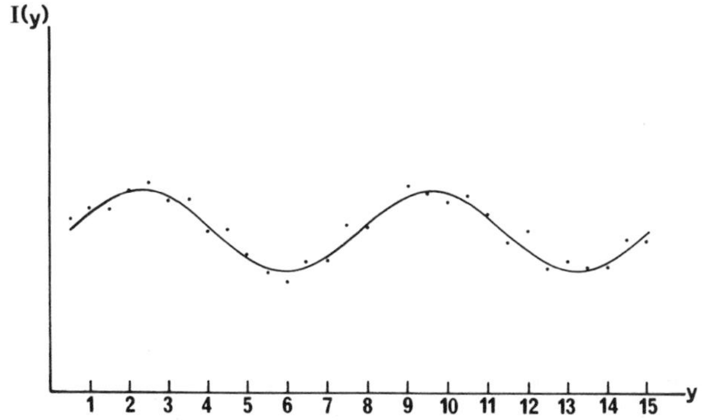

Fig. 1.8 Intensity I versus position y along the photographic plate. (In the limit of high intensities, the smooth continuous curve for intensity versus position along the photographic plate is obtained, in accordance with a wave theory of light. In the limit of low intensities, however, the intensity must be obtained by counting the areal density of discrete spots at incremental positions along the photographic plate. The seemingly random spots accrue at a given position with a probability governed by the wave theory.)

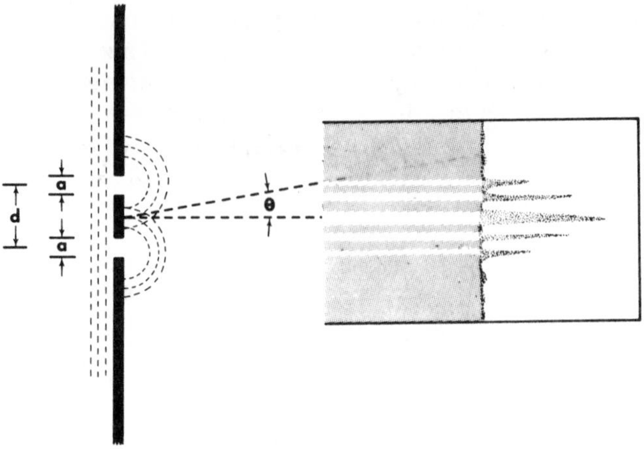

Fig. 1.9 Interference pattern produced by light passing through two slits of finite width a separated by a distance d. (The light bands indicate exposure on the photographic plate, and the peaks to the right indicate the variation of the intensity of the transmitted light with position given by the scattering angle θ. The ang'ɜ θ determines the distance y along the photographic plate in accordance with the relation, $\tan\theta = y/D$, where D is the perpendicular distance between the transmission grating and the photographic plate.)

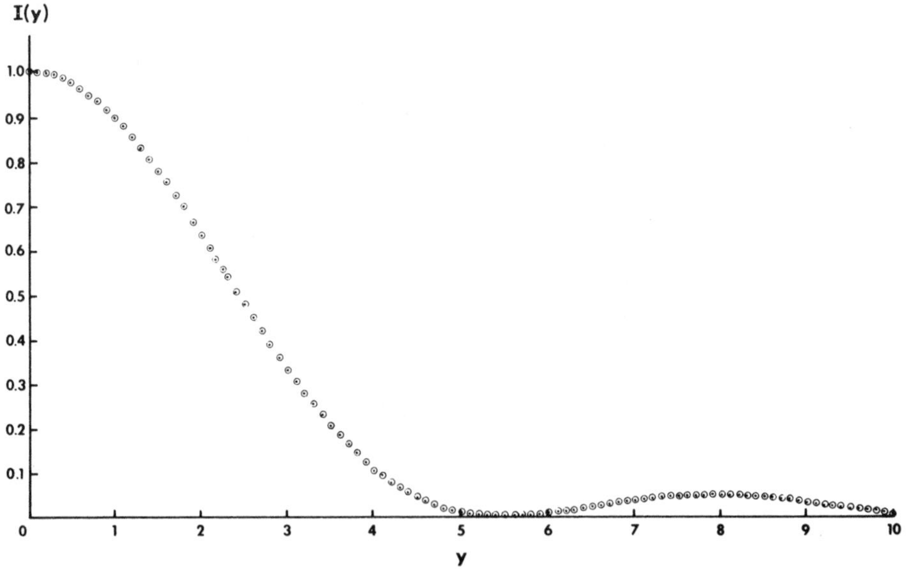

Fig. 1.10 Single-slit diffraction pattern produced by light of wavelength λ passing through one slit of width a. (The expression for the intensity I in terms of the maximum intensity I_0 and as a function of the angle θ, the wavelength λ, and the slit width a is given by $I = I_0\alpha^{-2}\sin^2\alpha$, with $\alpha = (\pi a/\lambda)\sin\theta$, as derived by considering interference between all components originating at different points on the wave front at the slit [see Halliday and Resnick (1974)]. This expression was numerically evaluated by means of a calculator, choosing $a = 0.01$ mm and $\lambda = 5461$ Å, with distance $D = 1$ m to the detector screen. The results of this calculation for the single slit is to be compared to the results for the double slit illustrated in Fig. 1.11.)

innate properties of radiation (and also matter) separate from the thermodynamic temperature that underlies classical statistical considerations.

It is of interest to ask whether a specific trajectory can be ascribed to the photons that reach the detector (cf. Fig. 1.9). That is, can the photon be localized not only in the act of absorption by the detector, but also throughout its course in the apparatus? Let us presume that initially the entire grating in Fig. 1.7 is illuminated with photons uniformly, and the resulting pattern observed. Then let us form a second pattern by illuminating the grating in sections $L' \ll L$ for a fixed time increment. (For example, L' could be chosen to include only two or three slits.) At any given moment the trajectories of the photons striking the photographic plate in this second case are thus restricted to a given portion L' of the grating. If different sections L' of the grating are then exposed sequentially until the entire grating has been exposed to the same intensity as in the first case, we might, in our ignorance, presume that the resultant pattern formed in the second experiment would be much the same as the pattern formed in the first experiment. This is not found to be the case. The patterns obtained in the second experiment are each characteristic of a grating of length L', whereas the pattern

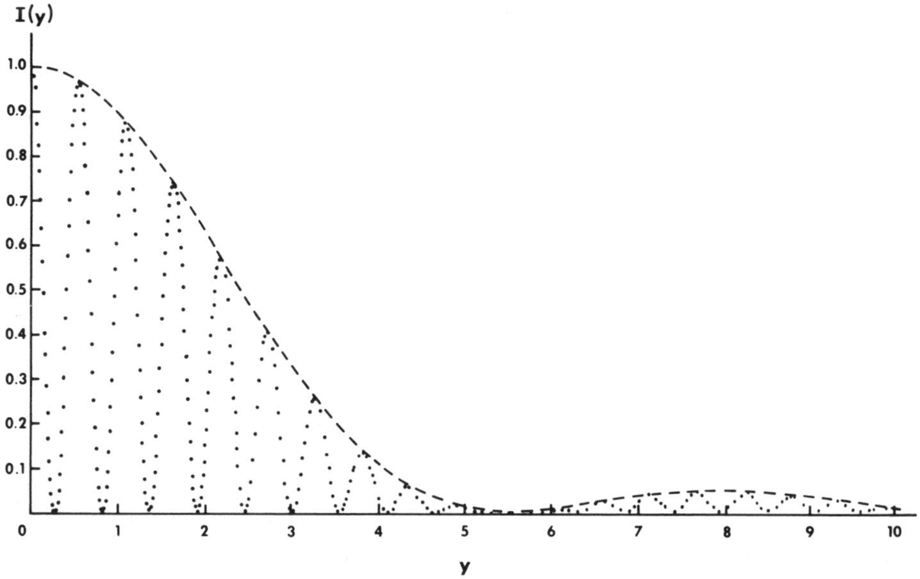

Fig. 1.11 Double-slit diffraction pattern produced by light of wavelength λ passing through two parallel slits, each of width a, separated by a distance d. (The expression for the intensity I in terms of the maximum intensity I_0 and as a function of the angle θ, the wavelength λ, and the slit width a is given by $I = I_0\alpha^{-2}\sin^2\alpha\cos^2\beta$, with $\alpha = (\pi a/\lambda)\sin\theta$, and $\beta = (\pi d/\lambda)\sin\theta$, as derived by modulating the single-slit pattern illustrated in Fig. 1.10 by the $\cos^2\beta$ factor characteristic of the two slits [see Halliday and Resnick (1974)]. This expression was numerically evaluated by means of a calculator, choosing $a = 0.01$ mm, $d = 0.10$ mm, and $\lambda = 5461$ Å, with distance $D = 1$ m to the detector screen. The parameter y was defined in the caption of Fig. 1.9; its units in Figs. 1.10 and 1.11 are centimeters.)

obtained in the first experiment is characteristic of a grating of length L. In fact, if we cover all the slits but one, on the assumption that the photon has atomic dimensions and goes through only one slit, then we completely lose the multislit interference pattern. This is well illustrated by Figs. 1.10 and 1.11 (see also Fig. 1.12). The act of localizing the trajectory of the photon in the manner just described does affect the resulting pattern. Therefore it appears that each individual photon in some way interacts with the entire apparatus, since its trajectory is influenced by the total number and arrangement of slits. It is in this sense, and in this sense only, that it can be said that an individual photon interferes with itself.

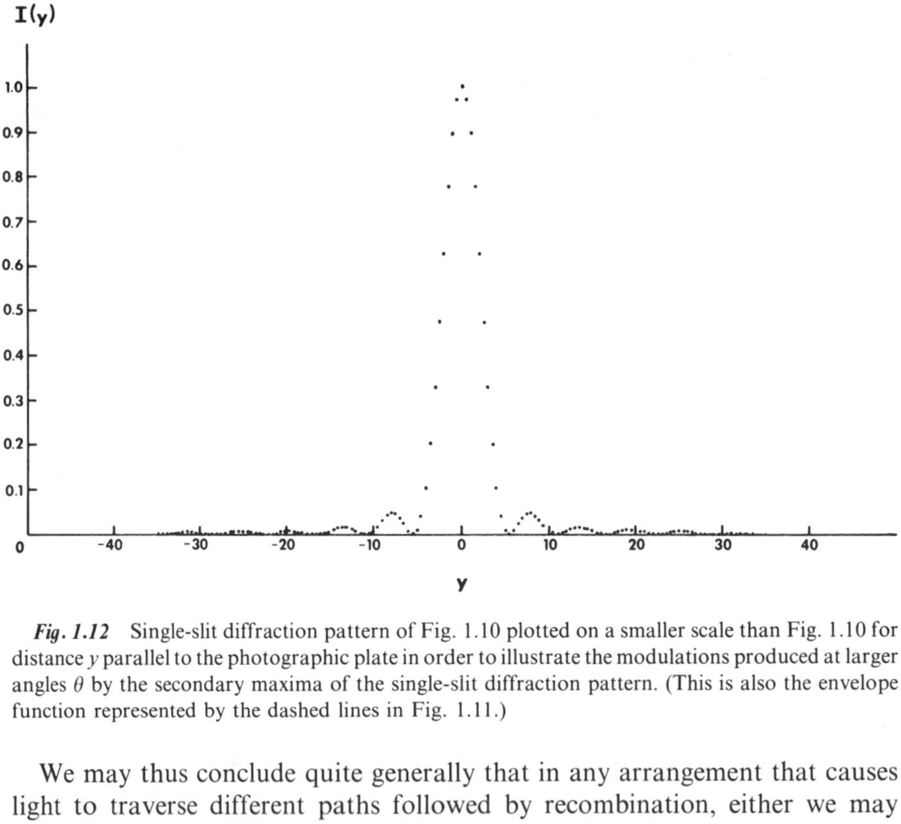

Fig. 1.12 Single-slit diffraction pattern of Fig. 1.10 plotted on a smaller scale than Fig. 1.10 for distance y parallel to the photographic plate in order to illustrate the modulations produced at larger angles θ by the secondary maxima of the single-slit diffraction pattern. (This is also the envelope function represented by the dashed lines in Fig. 1.11.)

We may thus conclude quite generally that in any arrangement that causes light to traverse different paths followed by recombination, either we may observe an interference pattern and remain ignorant of the photon's path, or we may experimentally determine which path the photon followed but thereby destroy the interference pattern. This is the enigma with which we are forced to live; experimentally it is found to be an innate property of particles as well as of radiation. It is so fundamental that it has been enshrined in the so-called "*complementarity principle*" of Bohr: *An experimental arrangement designed to manifest one of the classical attributes* (e.g., wavelike or particlelike aspects) *precludes the possibility of observing at least some of the other classical attributes.*

1.7 Wave Properties of Matter

Since light, considered generally to be wavelike, can therefore be shown in certain experiments to exhibit particlelike properties, it is certainly a reasonable extrapolation to speculate that particles of matter may have certain wavelike properties that could be emphasized under the proper experimental conditions. A man possessing the imagination for such speculation was de Broglie, who proposed that the wave–particle duality may be characteristic not only of light, but may be a universal characteristic of nature. De Broglie thus postulated the *wave nature of matter*. If particles do exhibit a wavelike character, they should give rise to interference and diffraction phenomena such as those commonly observed for light. Successful experiments designed to test this hypothesis are discussed in §5.

If the reader initially wonders how a wavelike description of particles could ever be in accord with our visual observations of the existence and motion of macroscopic objects, it may be recalled that the wave equations of electromagnetic theory have wavelike solutions that, in the limit of short wavelengths, yield results equivalent to the rays (or straight-line paths) of geometrical optics. An appropriate wave equation for matter would have wavelike solutions that in the appropriate limit of short wavelengths yield results equivalent to the predictions of classical physics such as Newton's equations of motion for particles.

At this point it proves helpful to delineate more clearly just what we mean by the concepts of particle and wave. *Particle traditionally means to us an object with a definite position in space. Wave means a periodically repeated pattern in space with no particular emphasis on any one crest or valley; it is characteristic of a wave that it does not define a location or position sharply.* Although these concepts at first seem to be mutually exclusive, some thought will show that a synthesis of discreteness and wave motion is sometimes evident in classical macroscopic physics, one prime example being the discrete frequencies of vibrating bodies of finite extension such as strings, membranes, and air columns. These systems have natural standing-wave modes representing sets of discrete wavelengths and frequencies. Mathematically the discreteness arises from boundary conditions imposed on the solutions from physical considerations such as zero-amplitude displacements at the fixed ends of a vibrating string. These discrete waves can be superimposed to yield various wave shapes; this is closely analogous to the superposition of sine and cosine functions to form arbitrary functions in a Fourier series or Fourier integral (see §4). This is the really important point, namely, that classical sinusoidal waves, each representing a nonlocalized disturbance, can be superimposed to obtain a localized disturbance. This is effected by constructive and destructive interference of the various wavelength components at different points in space. These statements are further explained and justified in the following sections dealing with classical waves and Fourier series.

2 Classical Wave Motion

2.1 Solutions to the Classical Wave Equation

Let us consider some of the mathematical details of wave motion, as would be generally suitable for a description of electromagnetic waves or sound waves. Solutions $\psi(x, t)$ of the classical wave equation

$$\frac{\partial^2 \psi}{\partial x^2} = \frac{1}{c^2} \frac{\partial^2 \psi}{\partial t^2} \qquad \textbf{(one-dimensional classical wave equation)} \tag{1.3}$$

give a representation of the amplitude of some wavelike disturbance which happens to be a function of position in one direction x and time t. The function ψ could represent, for example, the displacement of a uniform stretched string or the electromagnetic field in free space. Consider a function f which is arbitrary except for the requirement that its argument be $x + ct$ or $x - ct$. Both of these cases can be indicated by writing $x \pm ct$. Although f is a function of the two variables x and t, we have restricted the manner in which these two variables appear. If we designate $x \pm ct$ by η and further assume that the first and second derivatives of f with respect to η exist, then we obtain the results

$$\frac{\partial f}{\partial x} = \left(\frac{\partial f}{\partial \eta}\right)\left(\frac{\partial \eta}{\partial x}\right) = \left(\frac{\partial f}{\partial \eta}\right)\left(\frac{\partial (x \pm ct)}{\partial x}\right) = \frac{\partial f}{\partial \eta}, \tag{1.4}$$

$$\frac{\partial^2 f}{\partial x^2} = \frac{\partial}{\partial x}\left(\frac{\partial f}{\partial x}\right) = \frac{\partial}{\partial x}\left(\frac{\partial f}{\partial \eta}\right) = \left(\frac{\partial \eta}{\partial x}\right)\frac{\partial}{\partial \eta}\left(\frac{\partial f}{\partial \eta}\right) = \frac{\partial^2 f}{\partial \eta^2}, \tag{1.5}$$

$$\frac{\partial f}{\partial t} = \left(\frac{\partial f}{\partial \eta}\right)\left(\frac{\partial \eta}{\partial t}\right) = \left(\frac{\partial f}{\partial \eta}\right)\left(\frac{\partial (x \pm ct)}{\partial t}\right) = \pm\, c\left(\frac{\partial f}{\partial \eta}\right), \tag{1.6}$$

$$\frac{\partial^2 f}{\partial t^2} = \frac{\partial}{\partial t}\left(\frac{\partial f}{\partial t}\right) = \pm\, c\frac{\partial}{\partial t}\left(\frac{\partial f}{\partial \eta}\right) = \pm\, c\left(\frac{\partial \eta}{\partial t}\right)\frac{\partial}{\partial \eta}\left(\frac{\partial f}{\partial \eta}\right) = (\pm\, c)^2\left(\frac{\partial^2 f}{\partial \eta^2}\right). \tag{1.7}$$

Equating expressions for $\partial^2 f/\partial \eta^2$ obtained from Eqs. (1.5) and (1.7) yields an equation identical to Eq. (1.3) except that f replaces ψ. Therefore we have shown that any arbitrary properly differentiable function f satisfies the wave equation and thus represents one possible solution $\psi(x, t)$ provided only that the argument of f is $x + ct$ or $x - ct$.

Consider $f(x \pm ct)$ evaluated at $t = 0$ to be a curve represented by $f(x)$, and choose some time-dependent point x_p on the x axis which is denoted at $t = 0$ by x_p^0 with the corresponding amplitude of f given by $f(x_p^0)$. As t increases, the value of $f(x_p \pm ct)$ will change unless we permit x_p to be time dependent in such a way that $x_p \pm ct$ remains constant. If we define $x_p(t)$ as the value of x_p that satisfies

$$x_p(t) \pm ct = x_p^0, \tag{1.8}$$

then differentiation gives the velocity v_p of the point $x_p(t)$,

$$v_p = dx_p(t)/dt = -(\pm)c. \tag{1.9}$$

Thus the point of constant amplitude moves with a constant velocity, independent of the initial choice of x_p^0. Since the point represented by x_p was initially arbitrary, it follows that the solutions $f(x + ct)$ and $f(x - ct)$ represent displacements $f(x)$ which move without deformation with velocities c in the negative and positive x directions, respectively. These solutions are linearly independent since it is clear that one of these solutions is not a constant multiple of the other. Moreover, f itself as originally defined was arbitrary, so that it can represent any number of different functional forms at $t = 0$. This class of solutions of the wave equation therefore has the property that the amplitude and shape of the wave remain undeformed in time and the wave translates along the direction x with a velocity of $\pm c$. Since the wave equation (1.3) is linear, any linear combination of such solutions will also be a solution, as can be shown immediately by substitution of the linear combination into Eq. (1.3).

EXERCISE Deduce the classical wave equation by applying Newton's second law to small transverse displacements of a stretched wire.

EXERCISE Deduce the classical wave equation by combining Maxwell's equations for time-dependent electric and magnetic fields in free space.

2.2 Elementary Properties and Superposition of Waves

Fourier series and Fourier integrals (see §4) constitute powerful mathematical tools for representing arbitrary functions [such as the function $f(x)$ introduced above] in terms of a linear combination of sine and cosine terms. If each of the individual components of the Fourier series (or Fourier integral) representing $f(x)$ is given the argument $x + ct$ (or alternatively, each is given the argument $x - ct$), as discussed above, then the resultant superposition of these components in the form of a linear combination will represent the propagating solutions $f(x \pm ct)$ deduced above for the wave equation. This will become more apparent as we continue our development.

Let us now examine in more detail the particular function

$$g_k(x \pm ct) = \sin[k(x \pm ct)], \tag{1.10}$$

which can represent the individual terms making up a "Fourier representation" of $f(x \pm ct)$. [Alternatively we could examine the function $h_k(x \pm ct) = \cos[k(x \pm ct)]$.] The parameter k is at this point simply an arbitrary constant. Clearly $g_k(x \pm ct)$ satisfies Eq. (1.3), in accordance with our above discussion of solutions with argument $x \pm ct$, or as can be verified immediately by direct substitution into Eq. (1.3). The function $g_k(x \pm ct)$ has the following elementary properties:

(a) A change of phase by any multiple of 2π has no effect on the value of the function at any point in space or at any point in time. That is

$$\sin[k(x \pm ct) + 2m\pi] = \sin[k(x \pm ct)], \tag{1.11}$$

where m is an arbitrary positive or negative integer.

(b) It follows, therefore, that at a given time t the function $g_k(x \pm ct)$ repeats itself spatially in basic intervals $\Delta x = 2\pi/k$. This basic interval is called the *wavelength* λ,

$$\lambda = 2\pi/k, \tag{1.12}$$

and it is illustrated in Fig. 1.13.

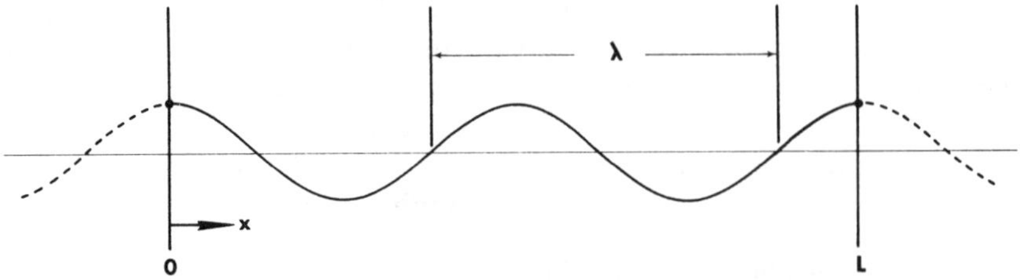

Fig. 1.13 Diagrammatic representation of a sinusoidal wave with wavelength λ within the spatial interval $0 \leqslant x \leqslant L$.

(c) It likewise follows that at a given position x the function $g_k(x \pm ct)$ repeats itself in time in basic intervals $\Delta t = 2\pi/kc$. This basic interval is called the *period* T,

$$T = 2\pi/kc. \tag{1.13}$$

Thus $g_k(x \pm ct)$ is periodic in space and in time.

Since the *temporal frequency* ν is the number of repetitions of the periodic function in unit time, and T is the time for one repetition, then

$$\nu = 1/T. \tag{1.14}$$

For example, if the time required for each repetition is $T = 0.1$ sec, then the frequency ν is 10 cycles/sec. Thus from Eqs. (1.13) and (1.14) we obtain $\nu = kc/2\pi$, which yields upon substitution of $\lambda = 2\pi/k$ as deduced in Eq. (1.12), the relation

$$\lambda\nu = c. \tag{1.15}$$

For those who are knowledgeable in elementary wave motion, this is immediately recognized as the familiar relation between wavelength, frequency, and wave-propagation velocity. Since the *angular frequency* ω is $2\pi\nu$, Eq. (1.15) can be written

$$\omega = ck, \tag{1.16}$$

which is called the *dispersion relation* for the wave motion under consideration. The velocity $c = \omega/k$ is known as the *phase velocity* of the sinusoidal wave under consideration, and $k = 2\pi/\lambda$ is a measure of the *wave number*, since at a given time it gives the number of spatial oscillations of the function in a spatial interval of length 2π. Some of these elementary wave relationships can be noted in Fig. 1.14, where $g_k(x - ct)$ is plotted as a function of time. If the velocity c is to be the same for any function $g_k(x \pm ct)$ satisfying the wave equation, it is clear that $\lambda\nu = \omega/k = c$ must be independent of the frequency ν and wavelength λ of the wave under consideration. In this case, where each Fourier component $g_k(x \pm ct)$ of an arbitrary wave $f(x \pm ct)$ has the same phase velocity, the velocity of the *superimposed* Fourier components (namely, the velocity of the wave represented by f) will also be c. The velocity with which $f(x \pm ct)$ moves is known as the *group velocity* because $f(x \pm ct)$ is represented by the superposition of the *group* of Fourier component waves. We have already shown independently that the velocity of the wave represented by $f(x \pm ct)$ is c along the x axis. Thus the group velocity is equal to the phase velocity for the classical wave motion under consideration. In physical situations where the phase velocity depends on the frequency ν of the component wave, the velocity of a group of such waves can no longer be obtained by such simple considerations. The superposition of a group of waves differing from each other in wavelength yields what is commonly known as a *wave packet*. Wave packets can have various shapes, depending upon the distribution of wavelengths and the relative amplitudes of the superimposed waves in the packet. A wave packet is the physical analog of the mathematical superposition of waves in a Fourier integral (see §4). A general treatment of the group velocity is presented in §7.4.

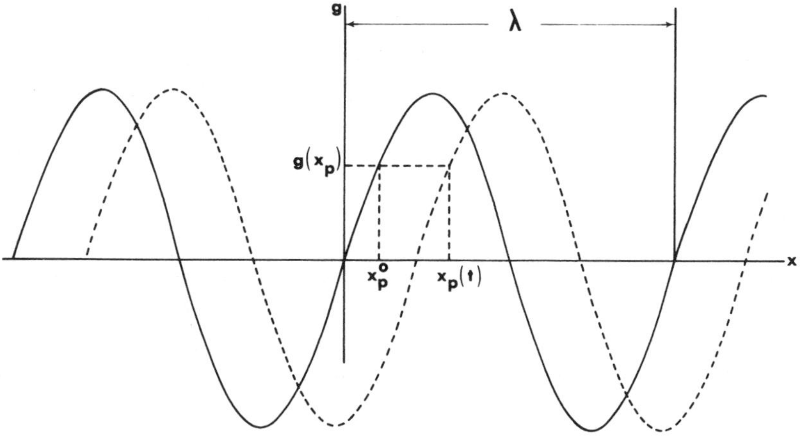

Fig. 1.14 Traveling wave $g_k(x - ct) = \sin[k(x - ct)]$ illustrated at $t = 0$ (solid curve) and at a later time t (dashed curve); during time t, the point on the wave denoted by the displacement $g(x_p)$ moves parallel to the x axis from the initial point x_p^0 at $t = 0$ to the new point $x_p(t)$.

2.3 Standing Waves

Consider now the specific linear superposition of two of the above solutions $g_k(x \pm ct)$ defined by

$$G_k(x, t) = \tfrac{1}{2}[g_k(x + ct) + g_k(x - ct)] = \tfrac{1}{2}[\sin(kx + \omega t) + \sin(kx - \omega t)], \quad (1.17)$$

which is the same as

$$G_k(x, t) = \sin kx \cos \omega t. \tag{1.18}$$

Equation (1.18) follows simply from application of the Euler formulas to Eq. (1.17), or more directly from the trigonometric identity

$$\sin x + \sin y = 2 \sin[\tfrac{1}{2}(x + y)] \cos[\tfrac{1}{2}(x - y)]. \tag{1.19}$$

The "packet" or "group" $G_k(x, t)$ of two sinusoidal waves of equal amplitude traveling with speed c in opposite directions represents a *standing wave* in space which oscillates periodically in time. The spatial points where $G_k(x, t) = 0$ are entirely stationary with passing time. These are the *nodal points* or "*nodes*" of the function. A little thought (or mathematical manipulation) shows quickly that there is in general no continuous function $x_p(t)$ that can be employed to trace the time dependence of the position of an arbitrary (but fixed) value for G_k on the wave. That is, it is simply impossible to pick an arbitrary $G_k(x_p^0, 0)$ and maintain $G_k(x_p, t) = G_k(x_p^0, 0)$ with continuously progressing time. This is in contrast to the case discussed previously for the solutions $f(x \pm ct)$. We thus conclude that the group velocity of this particular group or packet composed of two equal amplitude waves traveling in opposite directions has no meaning within the context of our previous discussion of group velocity. The nodes of the function $G_k(x, t)$ remain stationary in time, so from this standpoint we can say that the group velocity for $G_k(x, t)$ is zero, thus justifying the terminology "*standing wave*." This example shows quite clearly that the superposition of traveling-wave solutions of the type $g_k(x + ct)$ with those of the type $g_k(x - ct)$ can yield resultant solutions that are not traveling waves. The standing-wave solutions formally satisfy the wave equation as well as the traveling-wave solutions, as can be seen immediately by the substitution of $G_k(x, t)$ into Eq. (1.3); the resulting condition $\omega^2 = c^2k^2$ allows each of the two component waves to be considered as retaining its individual phase velocity $\pm c$ in the negative or positive x direction. A general solution to the wave equation will require a linear combination (i.e., a *superposition*) of the two types of solution denoted by $f(x + ct)$ and $f(x - ct)$, and the *group velocity, i.e., the velocity of a group or packet of such superimposed waves*, can in general be quite different from the "*phase*" *velocity, i.e., the velocity* $\pm c$ *of the individual periodic component waves.*

2.4 Fixed Boundary Conditions

If we apply the wave equation to describe the displacement of a vibrating string as a function of the time and the position along the string, the boundary conditions of the physical problem will serve to restrict the range of solutions to the wave equation. For example, the quantity k employed in our function $g_k(x \pm ct)$ is an arbitrary constant, but it may prove to be somewhat restricted in its allowable values because of requirements of the boundary conditions. For example, suppose that the ends of a string of length L are fixed such that the displacement is required to be zero at the two end points $x = 0$ and $x = L$. This is an example of *fixed boundary conditions*. A little thought convinces us that individual traveling waves such as $g_k(x + ct)$ or $g_k(x - ct)$ as defined by Eq. (1.10) will not be zero at $x = 0$ or at $x = L$ for arbitrary times; however, a superposition of these functions such as is given by the quantity $G_k(x, t)$ in Eq. (1.18) will indeed satisfy these conditions for *all* time t if it satisfies the conditions at $t = 0$. That is, the conditions

$$0 = G_k(0, t) = \{\sin[(k)(0)]\} \cos \omega t, \tag{1.20}$$

$$0 = G_k(L, t) = \{\sin[(k)(L)]\} \cos \omega t \tag{1.21}$$

can be satisfied for arbitrary t. The first condition is of course satisfied for any arbitrary finite value of k, while the second equation requires that $kL = m\pi$, with m being any integer. We can designate these allowed values of k by the symbol k_m. Therefore the set of functions

$$G_k(x, t) = [\sin(m\pi x/L)] \cos \omega t \tag{1.22}$$

satisfy both the wave equation and the imposed boundary conditions. One such function is illustrated in Fig. 1.15. Any wave packet constructed by arbitrary linear combinations of the $G_k(x, t)$ will also satisfy both the wave equation and the boundary conditions for this problem. Such linear combinations at a given time t take on the appearance of a Fourier series (see §4) for an arbitrary function of position.

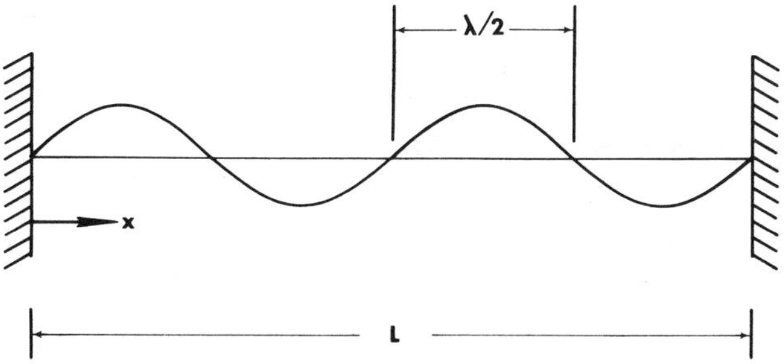

Fig. 1.15 Sinusoidal wave satisfying *fixed boundary conditions* of zero displacement at positions $x = 0$ and $x = L$. (This condition of zero displacement at $x = 0$ and $x = L$ requires an *integral* number of *half-wavelengths* within the interval $0 \leqslant x \leqslant L$. The wave illustrated is the one given by Eq. (1.22) for $m = 4$; at a given time t the accumulated phase over the spatial region $0 \leqslant x \leqslant L$ is 4π, which is equivalent to two wavelengths. The time factor $\cos \omega t$ causes the displacement at any given point x to oscillate periodically with time, taking on both positive and negative values.)

It is an important point that the restriction in the values of k corresponds to a restriction in the values of λ, since $k = 2\pi/\lambda$ [Eq. (1.12)]. Thus for *fixed boundary conditions*,

$$\tfrac{1}{2}\lambda_m = \tfrac{1}{2}(2\pi/k_m) = L/m \qquad (m = \text{integer}), \tag{1.23}$$

which shows that each wave in the discrete set of allowed wavelengths satisfies the condition that an integral number of half wavelengths must equal the length L. Furthermore, this restricts the values of the allowed frequencies. The relation $\omega = ck$ [Eq. (1.16)] leads to the set of frequencies

$$\nu_m = \omega_m/2\pi = ck_m/2\pi = m(c/2L). \tag{1.24}$$

Waves having other wavelengths than the discrete set given by $\lambda_m = 2L/m$ [see Eq. (1.23)] are excluded. To elaborate on this, a linear combination of terms such as [$\sin(kx) \cos(\omega t)$] for $k \neq m\pi/L$ which do not individually meet the boundary conditions for arbitrary t, but with coefficients chosen for mutual cancellation effects in order to meet the boundary conditions at some time t', would in general no longer meet the boundary conditions at a time $t' + \Delta t$, since the oscillation frequencies ν of the different waves are different.

Standing waves varying as $\cos kx$ instead of $\sin kx$ follow directly from the superposition of $g_k(x + ct)$ and $-g_k(x - ct)$, quite analogous to the construction of the sinusoidal standing waves given by Eq. (1.18). The cosine standing waves, however, cannot individually meet the given boundary conditions for the above example at $x = 0$, since $\cos(0) \neq 0$. For this reason, the Fourier representation of an arbitrary function which satisfies the wave equation and the boundary conditions currently under consideration need contain no terms other than those constructed by linear superposition of the above sine functions $G_k(x, t)$. On the other hand, different boundary conditions, or even a different choice of coordinate system for the same problem, could well require the cosine terms. This would be the case, for example, for a string extending from $x = -\frac{1}{2}L$ to $x = \frac{1}{2}L$. These points will be clarified by the detailed treatment of Fourier series and Fourier integrals given in §4. First, it is helpful to consider complex (as contrasted with real) solutions to the wave equation and give some consideration to an alternate type of boundary condition.

EXERCISE For a string extending from $x = -\frac{1}{2}L$ to $x = \frac{1}{2}L$, what functions are required in the Fourier-series representation of an arbitrary transverse displacement?

3 Periodic Boundary Conditions and Complex Fourier Components

3.1 Complex Basis Functions

For a general solution of the wave equation, we would of course need to consider the inclusion of terms such as

$$h_k(x \pm ct) = \cos[k(x \pm ct)]. \tag{1.25}$$

Complex linear combinations of the traveling waves $g_k(x + ct)$ [see Eq. (1.10)] and $h_k(x + ct)$, or alternatively of $g(x - ct)$ and $h(x - ct)$, are frequently very useful. Consider, for example, the functions $H_k(x \pm ct)$ defined by

$$\begin{aligned} H_k(x \pm ct) &= h_k(x \pm ct) + ig_k(x \pm ct) \\ &= \cos[k(x \pm ct) + i \sin[k(x \pm ct)] = \exp[ik(x \pm ct)]. \end{aligned} \tag{1.26}$$

These complex functions satisfy the conditions necessary to be solutions to the wave equation, and any linear combination of such complex functions will also be a solution. The coefficients in arbitrary linear combinations of these complex functions should be allowed to be complex; otherwise the superposition would not be capable in general of representing real functions. The choice of complex coefficients for linear combinations of the H_k provides us with an alternate description of any linear combination of the real functions $h_k(x \pm ct)$ and $g_k(x \pm ct)$ containing real coefficients. The $H_k(x \pm ct)$ can be considered to be *basis functions* for a *complex-number representation* of an arbitrary solution to the wave equation. (This is elaborated upon in §4 on Fourier series and Fourier integrals.) Although the linear combination of complex basis states can represent a physically meaningful solution to the wave equation (such as the displacement of a vibrating string as a function of position and time), there is certainly no reason to expect all such linear combinations to represent physically meaningful solutions. For example, any solution that is complex would not in itself be physically meaningful for the vibrating string. Since the basis vectors $H_k(x \pm ct)$ are themselves complex, it follows that one of these functions considered alone does not represent anything that is physically meaningful for a vibrating string.

3.2 Use of Complex Numbers for Real Physical Problems

There are some generally accepted practices using complex numbers, apart from simple superposition of complex solutions to effect real solutions, which sometimes lead to confusion in this matter. One practice crops up when phase differences exist for energy storage in different elements of a mechanical or electrical system with an attendant continuous time-dependent energy transfer between elements. One example is the impedance diagrams in ac circuits, where impedances are plotted in a complex plane; the angles between vectors in the plane denote the phase differences between the voltages across the various elements or currents through the various elements. This procedure works for series ac circuits because there are 90° phase differences between voltages across capacitive, resistive, and inductive elements carrying a common ac current, as can be shown directly by solving the relevant differential equation for the circuit. The maximum applied ac voltage is thus given in terms of a right-triangle relation between the total voltage (hypotenuse), the resistive component (which is plotted on the real axis), and the net value of the reactive components (which are plotted on the imaginary axis). This type of procedure works also for parallel ac circuits because there are 90° phase differences between the currents through parallel capacitive, resistive, and inductive elements having a common impressed ac voltage, as also can be shown directly by solving the relevant differential equation for the circuit. The applied ac current is thus given in terms of a right-triangle relation between the total applied current (hypotenuse), the current through the resistive component (plotted on the real axis), and the net current through the reactive components (plotted on the imaginary axis). Analogous practices exist for mechanical systems. For example, the kinetic and potential energies of a simple harmonic oscillator (such as a mass attached to a fixed spring with the mass set in horizontal motion on a frictionless plane) are 90° out of phase, as one finds in §12.3 by direct solution of the relevant differential equation for the classical behavior of this mechanical system. A right-triangle relation can thus be set up between a velocity-dependent quantity and a position-dependent quantity, with the hypotenuse denoting the constant total energy of the mechanical system.

Another confusing practice is to write the complex functions $H_k(x \pm ct)$ with the understanding that only the real (or occasionally the imaginary) part is the physically meaningful portion. This is often done for propagating electromagnetic fields in free space. This practice also arises often in ac circuits, with the *real* parts of several complex rotating phasors used to give the instantaneous values of the time-dependent voltages and currents associated with the various discrete resistors, capacitors, and inductors in the circuit.

In addition, an extension of the practice of writing complex functions with the understanding that only the real part is physically meaningful constitutes a frequently employed trick for simplifying mathematical manipulations when solving nonhomogeneous differential equations involving sine and cosine functions. Namely, real driving forces varying sinusoidally in time as $\cos \omega t$ or $\sin \omega t$ are denoted by considering either the real or the imaginary part of $e^{\pm i\omega t}$ to be the physically meaningful portion, and trial solutions of the form of linear combinations of $e^{\pm i\omega t}$ with real or complex coefficients are employed. The corresponding real or imaginary part of the resulting complex solution is then taken as the physically meaningful solution for the problem. This method is valid for linear equations in the real-number domain, since linearity assures that there will be no mixing of the real and imaginary parts in the homogeneous part of the equation.

3.3 Physical Implications of Basis States

The conclusion stands that unless complex solutions are qualified in some manner, such as outlined above, or else are combined in such a way as to yield a real result, then they are not physically meaningful for real quantities such as amplitudes. This provides evidence for a more far-reaching conclusion, namely, *basis states are not necessarily either physically meaningful or physically realizable.* This conclusion it supported by the fact that the choice of basis states is to a large extent arbitrary, and frequently there exist transformations that can generate new sets of basis states from any given set. [Note, for example, that the $H_k(x \pm ct)$ in Eq. (1.26) were formed from a linear combination of $g_k(x \pm ct)$ and $h_k(x \pm ct)$.] These conclusions will be referred to in Chap. 7 when the so-called "Bloch functions" are used to describe electronic states in solids.

3.4 Periodic Boundary Conditions

The complex functions $H_k(x \pm ct)$ defined by Eq. (1.26) are generally used as basis states whenever the *periodic boundary condition concept*, as explained below, is used.

Periodic boundary conditions require simply that each of the component waves represent a function that is periodic with respect to some spatial unit of periodicity derived from physical considerations. For example, the total length L of a vibrating string is a natural choice, in which case we require the displacement and phase at $x = L$ to be the same as at $x = 0$. Periodic boundary conditions in one sense are not as stringent as fixed boundary conditions, since each wave can have an arbitrary phase at $x = 0$. However, the requirement that the wave be periodic over the length L imposes the condition that the phase at $x = L$ be the same as the phase at $x = 0$, even though this phase may vary periodically in time, as it would in the case of running waves. There must therefore be an integral number of *whole wavelengths* between $x = 0$ and $x = L$, as can be noted in Fig. 1.16. This is similar to, but not quite the same as, the requirement for the case of *fixed boundary conditions* [cf. Eq. (1.23)] based on the assumption that the displacement is zero at fixed positions 0 and L. This requires that an integral number of *half wavelengths* equal the length L.

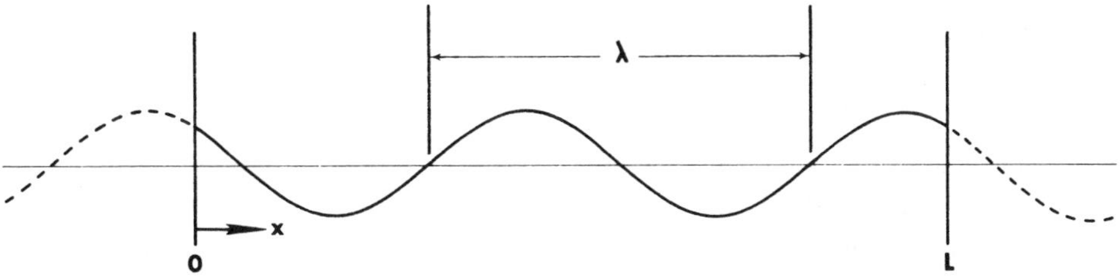

Fig. 1.16 Sinusoidal wave satisfying *periodic boundary conditions* requiring the *phase* of the wave to be the same at positions $x = 0$ and $x = L$. (The requirement of the same phase can be met only if there are an *integral* number of *whole* wavelengths within the interval $0 \leqslant x \leqslant L$. The phase at $x = 0$ is unspecified, and thus it can be *time dependent* as required for a traveling wave such as $g_k(x - ct) = \sin[k(x - ct)]$.)

3.5 Alternative Method of Solving the Classical Wave Equation

It is interesting to see how naturally the $H_k(x \pm ct)$ arise when we employ an elementary trial solution in the complex number domain for the wave equation. Suppose, for example, that we attempt a trial solution having the separated form

$$\Psi(x, t) = \eta(x) \exp(\pm i\omega t). \tag{1.27}$$

Substitution into the classical wave equation [Eq. (1.3)] yields

$$d^2\eta(x)/dx^2 + k^2\eta = 0, \tag{1.28}$$

where we have used the relation $\omega = ck$. Attempting a second trial solution in the complex number domain, we find that the function

$$\eta(x) = \exp(\pm ikx) \tag{1.29}$$

satisfies Eq. (1.28). Thus we obtain product solutions given by Eq. (1.27) which are of the form

$$\Psi_k(x, t) = A_k \exp(\pm ikx \pm i\omega t), \tag{1.30}$$

where any combination of signs in $\pm k$ and $\pm\omega$ is allowable, and A_k is an arbitrary constant. These solutions for the case of the positive sign for k are simply the solutions given by our functions $H_k(x \pm ct)$. If k were chosen to be imaginary or complex instead of real, the above mathematical formalism would remain unchanged; however, the relation $\omega = ck$ [Eq. (1.16)] would then impose the condition that either ω or c (or else both) would have to be imaginary or complex. Formally Eq. (1.30) would still represent a valid solution to Eq. (1.28). Physically, however, such a solution would be exponentially damped or exponentially increasing with x, since ik would then have a real component. For the propagation of electromagnetic radiation in free space, we know that the velocity c and the angular frequency ω are real, so that k for this situation is real, corresponding to an unattenuated propagating wave.

4 Fourier Series and Fourier Integrals

4.1 Basic Concepts

The concept of resolution of arbitrary solutions $f(x \pm ct)$ to the wave equation into a superposition of individual Fourier components such as $g_k(x \pm ct) = \sin[k(x \pm ct)]$ and $h_k(x \pm ct) = \cos[k(x \pm ct)]$ was introduced in §2, and we discussed the wavelike properties exhibited by the individual Fourier components. The pertinent mathematical formulas that govern this resolution into Fourier components are summarized in this section.

First we consider only functions that are real (i.e., not complex) and are periodic in the mathematical sense. This is not as restrictive from a physical standpoint as it would appear to be, since nonperiodic functions which happen to be physically meaningful only over a finite interval L in some specific problem, such as a string of length L, can be described completely by real periodic functions with period L or period L/m, where m is a positive integer. Since all solids are bounded in extent, all wave amplitudes in solids can thus be described by appropriate Fourier series. Therefore, for wave propagation in solids it is generally unnecessary to consider the strictly aperiodic (nonperiodic) limit of unbounded media in which the Fourier series must be replaced by the corresponding Fourier integrals. We must remember, however, to restrict our consideration of the final results to the region that is physically meaningful.

The Fourier series for the spatially periodic function $f(x)$,

$$f(x + \Lambda) = f(x), \tag{1.31}$$

having fundamental period Λ is given by

$$f(x) = \sum_{n=0}^{\infty}\left[A_n \cos\left(\frac{2\pi nx}{\Lambda}\right) + B_n \sin\left(\frac{2\pi nx}{\Lambda}\right)\right] \quad \textbf{(real Fourier series)}, \tag{1.32}$$

where the coefficients A_n and B_n are real. The Fourier components are seen to be individually periodic with period $\lambda_n \equiv \Lambda/n$. In analogy with the relation $k = 2\pi/\lambda$ given in §2 for a wave of wavelength λ, we define k_n

$$k_n \equiv 2\pi/\lambda_n = 2\pi/(\Lambda/n) = 2\pi n/\Lambda \tag{1.33}$$

as the corresponding quantity for each component wave of our Fourier series. The coefficients are easily determined in principle: simply multiply both sides of Eq. (1.32) by one of the Fourier components, such as $\cos(2\pi n'x/\Lambda)$ with $n' = 0, 1, 2, \ldots$, and integrate over any interval of length Λ, such as from x_0 to $x_0 + \Lambda$. The integral appearing on the left-hand side of the resulting expression can be performed graphically or numerically, if not analytically. The series of integrals on the right-hand side obtained from term-by-term integration are zero for $n' \neq n$ due to orthogonality of the trigonometric functions in the series. Likewise the integral for $n' = n$ which involves the product $\cos(2\pi n'x/\Lambda) \times \sin(2\pi n'x/\Lambda)$ is also zero due to orthogonality of the two functions (see exercise). The only nonzero integral is the one multiplying the coefficient $A_{n'}$, thus yielding the evaluation of $A_{n'}$. In a similar manner, multiplying by $\sin(2\pi n'x/\Lambda)$ instead of $\cos(2\pi n'x/\Lambda)$ and integrating yields the coefficient $B_{n'}$. The final results can be written as follows. For $n = 0$,

$$A_0 = \frac{1}{\Lambda}\int_{x_0}^{x_0+\Lambda} f(x)\, dx, \tag{1.34}$$

$$B_0 = 0. \tag{1.35}$$

For $n \neq 0$,

$$A_n = \frac{2}{\Lambda}\int_{x_0}^{x_0+\Lambda} f(x) \cos\left(\frac{2\pi nx}{\Lambda}\right) dx, \tag{1.36}$$

$$B_n = \frac{2}{\Lambda}\int_{x_0}^{x_0+\Lambda} f(x) \sin\left(\frac{2\pi nx}{\Lambda}\right) dx. \tag{1.37}$$

Considering the specific choice $x_0 = -\frac{1}{2}\Lambda$, it can be seen by a change in variable $x' = -x$ in the integrals that if $f(x)$ is an *even function*, namely,

$$f(-x) = f(x), \tag{1.38}$$

then the coefficients B_n are zero. On the other hand, if $f(x)$ is an *odd function*, namely,

$$f(-x) = -f(x), \tag{1.39}$$

then the coefficients A_n are zero (see exercise).

The above results are based on the assumptions that the series can be integrated legitimately term by term and that the products of $f(x)$ with the functions $\sin(2\pi n'x/\Lambda)$ and $\cos(2\pi n'x/\Lambda)$ are integrable. This restricts the range of possibilities for the function $f(x)$ somewhat. The convergence of the Fourier series to $f(x)$ is expected over regions where $f(x)$ is continuous, since the set of sine and cosine functions is *complete*. The class of functions that is generally suitable for Fourier expansion and the convergence at points where the functions are discontinuous are matters that are delineated by the *Dirichlet theorem*, which can be stated as follows: *If $f(x)$ is a bounded periodic function with at most a finite number of maxima and minima and a finite number of discontinuities in any one period, then the Fourier series of $f(x)$ converges to $f(x)$ at all points where $f(x)$ is continuous, and converges to the average of the right- and left-hand limits of $f(x)$ at each point where $f(x)$ is discontinuous.* The conditions on $f(x)$ embodied in the Dirichlet theorem are called the *Dirichlet conditions.* Functions that satisfy the Dirichlet conditions have a number of important properties. For example, the *integral* of any periodic function $f(x)$ satisfying the Dirichlet conditions can be found by *termwise integration* of the Fourier series representing the function. If, in addition, the function is continuous everywhere and has a derivative $df(x)/dx$ which satisfies the Dirichlet conditions, then the derivative $df(x)/dx$ can be found anywhere it exists by *termwise differentiation* of the Fourier series for $f(x)$. It is also found that for sufficiently large n, the Fourier coefficients of a function satisfying the Dirichlet conditions always decrease in magnitude at least as rapidly as $1/n$. If the function $f(x)$ has one or more discontinuities, the coefficients can decrease no more rapidly with n than $1/n$. If the function is continuous everywhere but has one or more points where its derivative is discontinuous, the Fourier coefficients decrease as $1/n^2$. If a function and its various derivatives all satisfy the Dirichlet conditions and if the lth derivative is the first which is not continuous everywhere, then for sufficiently large n the Fourier coefficients of the function approach zero as $1/n^{l+1}$.

Since the Dirichlet conditions are not particularly stringent, periodic functions which represent physically meaningful quantities do generally meet the conditions necessary for expansion in a Fourier series. Furthermore, the smoother the function, the more rapid the convergence of the series. The fact that the function can be differentiated and integrated easily by termwise differentiation and integration of the sine and cosine functions in the series makes it very useful for many applications. Thus Fourier series constitute very powerful tools in solid state physics and in all branches of engineering.

EXERCISE Prove (by direct integration) the statements in this section concerning the zero values of the integrals.

EXERCISE Prove the statements in this section relating to even and odd functions by carrying out the suggested variable change in the integrals.

4.2 Fourier Series in the Complex-Number Domain

The complex Fourier series, containing terms such as the $H_k(x \pm ct) = \exp[ik(x \pm ct)]$ discussed in §3, can be obtained readily [Wylie (1951)] by substitution of the exponential equivalents of the sine and cosine terms into the real form of the Fourier series. The result (see exercise) is

$$f(x) = \sum_{n=-\infty}^{\infty} C_n \exp(i2\pi nx/\Lambda) \qquad \textbf{(complex Fourier series),} \qquad (1.40)$$

with

$$C_n = \frac{1}{\Lambda}\int_{x_0}^{x_0+\Lambda} f(x)\exp\left(-\frac{i2\pi nx}{\Lambda}\right)dx \qquad (n = 0, \pm 1, \pm 2, \ldots). \quad (1.41)$$

EXERCISE Evaluate the coefficients C_n given in Eq. (1.40), obtaining as a result Eq. (1.41).

EXERCISE Using the Euler identities, show that the complex Fourier series representation [Eqs. (1.40) and (1.41)] reduces to the real Fourier series representation [Eqs. (1.32)–(1.37)] for cases in which the periodic function $f(x)$ is real.

4.3 Fourier Integrals

We now proceed from the *Fourier series* representation of a *periodic function* to the *Fourier-integral* representation of an *aperiodic function.* Suppose we let $x_0 = -\frac{1}{2}\Lambda$ specifically and change the dummy variable from x to x' in the integral for C_n. Substituting the result into $f(x)$ yields

$$f(x) = \sum_{n=-\infty}^{\infty}\left[\frac{1}{\Lambda}\int_{-\Lambda/2}^{\Lambda/2} f(x')\exp\left(-\frac{i2\pi nx'}{\Lambda}\right)dx'\right]\exp\left(\frac{i2\pi nx}{\Lambda}\right). \quad (1.42)$$

The fundamental unit of periodicity of $f(x)$ is Λ, and we have previously defined quantities $k_n = 2\pi/\lambda_n = 2\pi n/\Lambda$. The difference Δk between successive values of k_n is simply

$$\Delta k = k_{n+1} - k_n = 2\pi/\Lambda. \quad (1.43)$$

Using these definitions, the above expression for $f(x)$ can be written in the form

$$f(x) = \sum_{n=-\infty}^{\infty}\left[\frac{1}{2\pi}\int_{-\Lambda/2}^{\Lambda/2} f(x')\exp(-ik_nx')\,dx'\right]\exp(ik_nx)\,\Delta k. \quad (1.44)$$

If now we consider the limiting process in which Λ becomes larger and larger, then the range of the integral becomes greater and greater and $\Delta k = 2\pi/\Lambda$ becomes smaller and smaller. The parameter $k_n = n\,\Delta k$, with $n = 0, 1, 2, \ldots$, takes on the properties of a continuous variable which we can call k,

$$k_n = n\,\Delta k \to k, \quad (1.45)$$

$$k_{n+1} - k_n = \Delta k \to dk. \quad (1.46)$$

If the limiting process of letting $\frac{1}{2}\Lambda \to \infty$ is carried out in a mathematically proper fashion, the limit of the sum is a definite integral, so that $f(x)$ becomes

$$f(x) = \int_{-\infty}^{\infty}\left[\frac{1}{2\pi}\int_{-\infty}^{\infty} f(x')\exp(-ikx')\,dx'\right]\exp(ikx)\,dk. \quad (1.47)$$

In this limit for which $\frac{1}{2}\Lambda \to \infty$, the function is aperiodic instead of periodic, since the period has become infinite in length. Defining $g(k)$ as

$$g(k) = \frac{1}{2\pi}\int_{-\infty}^{\infty} f(x')\exp(-ikx')\,dx', \quad (1.48)$$

then $f(x)$ becomes

$$f(x) = \int_{-\infty}^{\infty} g(k) \exp(ikx)\, dk. \tag{1.49}$$

This pair of expressions $g(k)$ and $f(x)$ constitutes a *Fourier transform pair*, and the integral expression for $f(x)$ is called a *Fourier integral.*

The Fourier integral is a valid representation of $f(x)$ provided that in every finite interval the function $f(x)$ satisfies the Dirichlet conditions, and the integral

$$\int_{-\infty}^{\infty} |f(x)|\, dx \tag{1.50}$$

exists. The Fourier integral converges to $f(x)$ at all points where $f(x)$ is continuous, and it converges to the average of the right- and left-hand limits at all points where $f(x)$ is discontinuous.

4.4 Application to Solutions to the Classical Wave Equation

The above consideration of Fourier series and Fourier integrals involves only one variable x which can be considered to be the position variable in the arbitrary solution $f(x \pm ct)$ to the wave equation discussed in §2. Thus $f(x, 0)$, representing $f(x \pm ct)$ evaluated at $t = 0$, can be expanded in a Fourier series or Fourier integral with the formulas of the present section. In accordance with the conclusions arrived at in §2, then, the arbitrary solution $f(x \pm ct)$ is obtained by substituting the argument $x \pm ct$ in place of x in each of the Fourier components. Thus, for the complex Fourier series representation we have

$$f(x \pm ct) = \sum_{n=-\infty}^{\infty} C_n \exp\left[\frac{i2\pi n(x \pm ct)}{\Lambda}\right] \tag{1.51}$$

with

$$C_n = \frac{1}{\Lambda}\int_{x_0}^{x_0+\Lambda} f(x, 0) \exp\left(-\frac{i2\pi nx}{\Lambda}\right) dx. \tag{1.52}$$

For the Fourier-integral representation we have

$$f(x \pm ct) = \int_{-\infty}^{\infty} g(k) \exp[ik(x \pm ct)]\, dk \tag{1.53}$$

with

$$g(k) = \frac{1}{2\pi}\int_{-\infty}^{\infty} f(x', 0) \exp(-ikx')\, dx'. \tag{1.54}$$

If we define a new function $G(k)$ as

$$G(k) = (2\pi)^{1/2} g(k), \tag{1.55}$$

then Eq. (1.49) for $f(x)$ can be modified in appearance (though not in value) by substituting $(2\pi)^{-1/2}G(k)$ for $g(k)$. Let us refer to this modified form of $f(x)$ as $F(x)$. Then we have as an alternative to Eqs. (1.48) and (1.49) the Fourier transform pair $F(x)$ and $G(k)$ given by the **Fourier integrals**

$$F(x) = \frac{1}{(2\pi)^{1/2}} \int_{-\infty}^{\infty} G(k) \exp(ikx)\, dk, \tag{1.56}$$

$$G(k) = \frac{1}{(2\pi)^{1/2}} \int_{-\infty}^{\infty} F(x) \exp(-ikx)\, dx. \tag{1.57}$$

This pair has greater symmetry than the pair $f(x)$ and $g(k)$; it is frequently employed in wavepacket descriptions because it has the property of preserving normalization of $F(x)$ and $G(k)$. Nonsymmetrical Fourier transform pairs are converted immediately to the symmetrical form by replacing the symbols f and g by F and G while simultaneously changing the factors of unity and $1/2\pi$ in front of the integrals for f and g, respectively, to $(1/2\pi)^{1/2}$.

If F happened to be a function of several independent variables $x_1, x_2, \ldots, x_N$ instead of a single variable x, then from a mathematical standpoint we could consider F to be a function of one variable x_j at a time and obtain a function $G(k_j)$ defined analogously to $G(k)$ above. Carrying out this mathematical procedure for each of the N independent variables and considering the factor of $(2\pi)^{-1/2}$ arising each time, we obtain the result

$$F(x_1, x_2, \ldots, x_N) = \left(\frac{1}{2\pi}\right)^{N/2} \int_{-\infty}^{\infty} dk_1 \int_{-\infty}^{\infty} dk_2 \cdots \int_{-\infty}^{\infty} dk_N\, G(k_1, k_2, \ldots, k_N) \times \exp[i(k_1x_1 + k_2x_2 + \cdots + k_Nx_N)], \tag{1.58}$$

$$G(k_1, k_2, \ldots, k_N) = \left(\frac{1}{2\pi}\right)^{N/2} \int_{-\infty}^{\infty} dx_1 \int_{-\infty}^{\infty} dx_2 \cdots \int_{-\infty}^{\infty} dx_N\, F(x_1, x_2, \ldots, x_N) \times \exp[-i(k_1x_1 + k_2x_2 + \cdots + k_Nx_N)]. \tag{1.59}$$

If we consider

$$\mathbf{x} = (x_1, x_2, \ldots, x_N), \tag{1.60}$$

$$\mathbf{k} = (k_1, k_2, \ldots, k_N) \tag{1.61}$$

to be two vectors in an N-dimensional orthogonal *abstract* linear vector space, then the arguments of the exponentials contain the simple dot product,

$$\mathbf{k} \cdot \mathbf{x} = k_1x_1 + k_2x_2 + \cdots + k_Nx_N. \tag{1.62}$$

In particular, at a specific time (such as $t = 0$) the function $F(\mathbf{x})$ given by Eq. (1.58) can represent the spatial part of a solution to the **three-dimensional classical wave equation**

$$\frac{\partial^2\psi}{\partial x^2} + \frac{\partial^2\psi}{\partial y^2} + \frac{\partial^2\psi}{\partial z^2} = \frac{1}{c^2}\frac{\partial^2\psi}{\partial t^2}. \tag{1.63}$$

The vector **r**,

$$\mathbf{r} = x\,\hat{\mathbf{x}} + y\,\hat{\mathbf{y}} + z\,\hat{\mathbf{z}}, \tag{1.64}$$

represents the position vector in real space with projections x, y, and z along the unit vectors $\hat{\mathbf{x}}, \hat{\mathbf{y}}$, and $\hat{\mathbf{z}}$ in a Cartesian coordinate system. Then for real three-dimensional space,

$$F(\mathbf{r}) = \left(\frac{1}{2\pi}\right)^{3/2} \int_{\Omega_\mathbf{k}} G(\mathbf{k}) \exp(i\mathbf{k}\cdot\mathbf{r})\, d\mathbf{k}, \tag{1.65}$$

$$G(\mathbf{k}) = \left(\frac{1}{2\pi}\right)^{3/2} \int_{\Omega_\mathbf{r}} F(\mathbf{r}) \exp(-\, i\mathbf{k}\cdot\mathbf{r})\, d\mathbf{r}. \tag{1.66}$$

The vector **k** is given by

$$\mathbf{k} = k_x\hat{\mathbf{x}} + k_y\hat{\mathbf{y}} + k_z\hat{\mathbf{z}}. \tag{1.67}$$

The unit vector

$$\hat{\mathbf{k}} = \mathbf{k}/|\mathbf{k}| \tag{1.68}$$

characterizes the direction of propagation of the particular plane wave $\exp[i(\mathbf{k}\cdot\mathbf{r} - \omega t)]$. The symbols $d\mathbf{r}$ and $d\mathbf{k}$ represent the volume elements $dx\,dy\,dz$ and $dk_x\,dk_y\,dk_z$, respectively, in "real space" and in "**k** space." The symbols $\Omega_\mathbf{r}$ and $\Omega_\mathbf{k}$ associated with the integrals mean that the three-dimensional integrals are to be carried out from $-\infty$ to ∞ in each of the three orthogonal directions in "real space" and in "**k** space," respectively. The term "real space" refers to the domain of all position vectors $\mathbf{r} = x\hat{\mathbf{x}} + y\hat{\mathbf{y}} + z\hat{\mathbf{z}}$. The term "**k** space" refers to the domain of all vectors $\mathbf{k} = k_x\hat{\mathbf{x}} + k_y\hat{\mathbf{y}} + k_z\hat{\mathbf{z}}$. Since $\mathbf{k}\cdot\mathbf{r}$ must be dimensionless, occurring as it does in Eqs. (1.65) and (1.66) as the argument of an exponential function, the dimensions of k_x, k_y, k_z must be reciprocal to the dimensions of x, y, z. If x, y, and z are measured in meters, then k_x, k_y, and k_z are measured in units of $(\text{meters})^{-1}$. Thus the **k** vectors could be referred to as "reciprocal vectors," and the domain mapped out by such vectors could be referred to as "reciprocal space." These latter terms, however, will be reserved for the *subset* of such **k** vectors which are sufficient for a Fourier series representation of periodic functions in a solid. (See Chap. 6 for further details.)

In the three-dimensional case, as in the one-dimensional case [see Eq. (1.12)], the magnitude of **k** is $2\pi/\lambda$. This is required if we are to have self-consistency between the one-dimensional and three-dimensional cases whenever one of the axes of the three-dimensional system is chosen to coincide with the direction of propagation of the plane wave, thus effectively reducing the three-dimensional case to the one-dimensional case. Let us now interpret physically the **k**-vector components k_x, k_y, k_z of a plane wave in terms of the wavelength as measured in each of the three component directions. In terms of the spherical polar coordinates r, θ, ϕ, the components k_x, k_y, and k_z of the vector **k** are

$$k_x = |\mathbf{k}| \sin\theta \cos\phi = (2\pi/\lambda) \sin\theta \cos\phi, \tag{1.69}$$

$$k_y = |\mathbf{k}| \sin\theta \sin\phi = (2\pi/\lambda) \sin\theta \sin\phi, \tag{1.70}$$

$$k_z = |\mathbf{k}| \cos\theta = (2\pi/\lambda) \cos\theta. \tag{1.71}$$

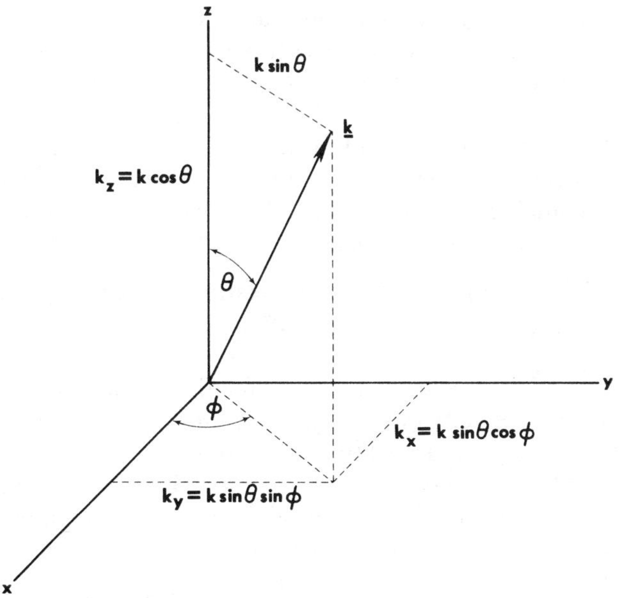

Fig. 1.17 Resolution of the wave propagation vector **k** into components k_x, k_y, and k_z in a spherical polar coordinate system.

The resolution of **k** into these components is illustrated in Fig. 1.17. Defining "component wavelengths" λ_x, λ_y, and λ_z as $2\pi/k_x$, $2\pi/k_y$, and $2\pi/k_z$, respectively, we thus obtain

$$\lambda = \lambda_x \sin\theta \cos\phi, \tag{1.72}$$

$$\lambda = \lambda_y \sin\theta \sin\phi, \tag{1.73}$$

$$\lambda = \lambda_z \cos\theta. \tag{1.74}$$

Therefore each of the component wavelengths is *larger* than the wavelength

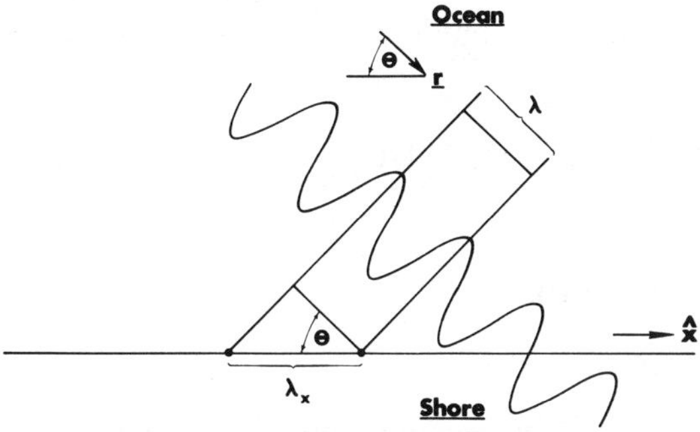

Fig. 1.18 Illustration showing that the x component k_x of the wave vector **k** represents a longer wavelength $\lambda_x = 2\pi/k_x$ than the **k** vector itself, for which $\lambda = 2\pi/|\mathbf{k}|$.

along the propagation direction, as one can readily visualize from a physical picture such as that given by ocean waves traveling at some angle with respect to the two rectangular coordinates located in a plane parallel to the surface of the ocean (see Fig. 1.18).

To convert the time-independent Fourier integral $F(\mathbf{r})$ to an arbitrary solution $F(\mathbf{r}, t)$ of the three-dimensional wave equation (1.63), we must add to $\mathbf{k} \cdot \mathbf{r}$ a time-dependent term analogous to the term $\pm kct$ added to kx in Eq. (1.53) for the one-dimensional case. We thus replace $\mathbf{k} \cdot \mathbf{r}$ by $\mathbf{k} \cdot \mathbf{r} \pm \omega t$, where the quantity

$$\omega = |\mathbf{k}|c = \mathbf{k} \cdot \mathbf{c} = \mathbf{c} \cdot \mathbf{k} \qquad \textbf{(dispersion relation for classical wave motion)} \tag{1.75}$$

is the three-dimensional classical dispersion relation analogous to Eq. (1.16) for one dimension. The velocity of propagation is in this three-dimensional case a vector quantity **c** which is parallel to **k**. In writing Eq. (1.75), it is assumed that the frequency ω depends only on the magnitude of **k** and not on the direction of **k**. This is equivalent to assuming that the wave propagation is taking place in an isotropic medium such as free space. Equation (1.75) leads to

$$\mathbf{k} \cdot \mathbf{r} \pm \omega t = \mathbf{k} \cdot \mathbf{r} \pm c|\mathbf{k}|t = \mathbf{k} \cdot (\mathbf{r} \pm c\hat{\mathbf{k}}t), \tag{1.76}$$

where $\hat{\mathbf{k}} = \mathbf{k}/|\mathbf{k}|$ is a unit vector parallel to **k**. Thus we obtain

$$F(\mathbf{r}, t) = \left(\frac{1}{2\pi}\right)^{3/2} \int_{\Omega_\mathbf{k}} G(\mathbf{k}) \exp[i\mathbf{k} \cdot (\mathbf{r} \pm c\hat{\mathbf{k}}t] \, d\mathbf{k}, \tag{1.77}$$

$$G(\mathbf{k}) = \left(\frac{1}{2\pi}\right)^{3/2} \int_{\Omega_\mathbf{r}} F(\mathbf{r}, 0) \exp(-i\mathbf{k} \cdot \mathbf{r}) \, d\mathbf{r}, \tag{1.78}$$

where $F(\mathbf{r}, 0)$ is the solution $F(\mathbf{r}, t)$ evaluated at $t = 0$. Since **k** represents the propagation direction for any plane wave $\exp[i(\mathbf{k} \cdot \mathbf{r} \pm \omega t)]$ in question, and the integration is over the domain of **k**, the integral expression (1.77) can be interpreted as a superposition of plane waves of various wavelengths propagating in various directions. Changing the dummy variable in the integral expression (1.78) for $G(\mathbf{k})$ from **r** to **r**′ and substituting $G(\mathbf{k})$ into $F(\mathbf{r}, t)$ yields the integral form for a general solution to the classical three-dimensional wave equation (1.63):

$$F(\mathbf{r}, t) = \left(\frac{1}{8\pi^3}\right) \int_{\Omega_\mathbf{k}} \left[\int_{\Omega_\mathbf{r}} F(\mathbf{r}', 0) \exp[i\mathbf{k} \cdot (\mathbf{r} - \mathbf{r}')] \, d\mathbf{r}' \right] \exp(\pm ic|\mathbf{k}|t) \, d\mathbf{k}. \tag{1.79}$$

5 Wave Nature of Particles

5.1 Diffraction of Waves

We have previously discussed the wondrous fact that light exhibits both wavelike and particlelike properties, and have enunciated the startling hypothesis put forth by de Broglie that such wave–particle duality may also be a property of matter (see §1). Then we treated the fascinating subject of classical wave motion in some detail to establish the background for a wave description of particles (see §2). One such wave description of particles is given by the seemingly omnipotent Schrödinger equation. Before carrying out our own development of the Schrödinger equation, however, we should ask whether there exists unambiguous experimental evidence which supports the de Broglie hypothesis that matter can exhibit wavelike properties. Some of the most important experiments which convince us that matter does indeed have wavelike properties are provided by experimental observations of electron diffraction and neutron diffraction. These experiments using particles give results that are entirely analogous to results obtained using x rays.

First of all, let us consider the essentials of x-ray diffraction in solids. The x-ray diffraction results obtained from crystals are readily understood from a simple one-dimensional model in which a monochromatic x-ray beam of wavelength λ impinges at angle θ with respect to a given set of atomic planes of spacing d in a crystal, as illustrated schematically in Fig. 1.19. Constructive interference is obtained whenever the waves reflected from the various planes in the crystal happen to be in phase. This requires that the difference in path length between the waves reflected from the different planes be an integral number n of wavelengths. The dashed line $\overline{\mathrm{BP}}$ in Fig. 1.19 is drawn perpendicular to the direction of propagation of the incident waves. The difference in path length between the waves reflected from adjacent planes is $\overline{\mathrm{BA}} - \overline{\mathrm{PA}}$, so the condition for constructive interference of the waves is

$$n\lambda = \overline{\mathrm{BA}} - \overline{\mathrm{PA}}. \tag{1.80}$$

If this condition is satisfied, then reflected waves from all such parallel planes will be in phase. We now deduce the expression for $\overline{\mathrm{BA}} - \overline{\mathrm{PA}}$ in terms of the angle θ and the lattice spacing d. The two angles designated ϕ in Fig. 1.19 are

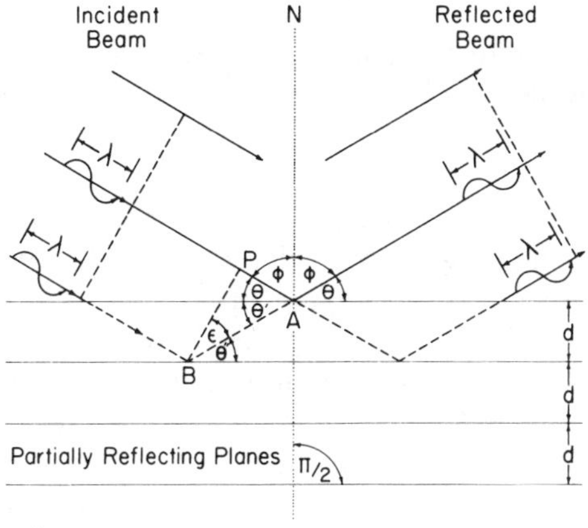

Fig. 1.19 Reflection of incident waves of wavelength λ by a sequence of partially transparent parallel planes of equal spacing d.

equal because the angle of reflection is equal to the angle of incidence, a well-known law in geometrical optics. It can be shown geometrically that this is the requirement that all parts on a plane-wave front incident at angle ϕ to a reflecting surface travel equal optical path lengths to reconstitute a plane-wave front after reflection. The two angles designated θ in Fig. 1.19 are therefore equal.

From elementary geometry, it can be seen that the angle θ' in Fig. 1.19 is equal to θ, so the angle labeled ε is given by

$$\varepsilon = \tfrac{1}{2}\pi - 2\theta. \tag{1.81}$$

Since $\theta = \theta' = \theta''$, it follows from the figure that

$$\sin\theta = \sin\theta'' = d/\overline{\mathrm{BA}}. \tag{1.82}$$

Note further that

$$\sin\varepsilon = \overline{\mathrm{PA}}/\overline{\mathrm{BA}}. \tag{1.83}$$

We thus can use Eqs. (1.82) and (1.83) to write

$$\overline{\mathrm{BA}} = d/(\sin\theta), \tag{1.84}$$

$$\overline{\mathrm{PA}} = \overline{\mathrm{BA}}\sin\varepsilon = (d\sin\varepsilon)/(\sin\theta). \tag{1.85}$$

Therefore

$$\overline{\mathrm{BA}} - \overline{\mathrm{PA}} = d(1 - \sin\varepsilon)/(\sin\theta). \tag{1.86}$$

Employing Eq. (1.81), we see from trigonometry that

$$\sin\varepsilon = \sin(\tfrac{1}{2}\pi - 2\theta) = \cos 2\theta = \cos^2\theta - \sin^2\theta. \tag{1.87}$$

Substituting into Eq. (1.86) gives

$$\overline{\mathrm{BA}} - \overline{\mathrm{PA}} = d[(1 - \cos^2\theta) + \sin^2\theta]/\sin\theta = 2d\sin\theta. \tag{1.88}$$

Substituting this result into the condition (1.80) for constructive interference gives the **Bragg condition**

$$n\lambda = 2d\sin\theta. \tag{1.89}$$

This derivation is rigorous in the sense that it considers the proper superposition of reflected waves originating from all the various reflecting planes, and thus correctly describes the net wave impinging on a detector. It is therefore preferred over the much simpler derivation using the condition that the phases of the reflected components arising from the points of intersection of the perpendicular N to the reflecting planes (cf. Fig. 1.19) differ by integral multiples of 2π. Although this is a necessary requirement because the reflected wave from the first plane is a plane wave, it is more straightforward to consider explicitly the superposition of all wave contributions arising from different points in the various reflecting planes. To express this matter in somewhat different terminology, we have considered the conditions for waves to interfere (constructively or destructively) with one another at given positions in space instead of imposing the usual condition of constant phase for selected points on the wavefront of an outgoing plane wave. That the phase is uniform over a plane wavefront when the Bragg condition is satisfied is interesting in itself since it implies that the outgoing wave is indeed a plane wave, but this is not so important with regard to the detector reading for these reflected waves.

An alternate derivation of the Bragg condition (1.89) is illustrated in Fig. 1.20, with the approach outlined in the caption. It is suggested that the reader carry through this derivation with the details.

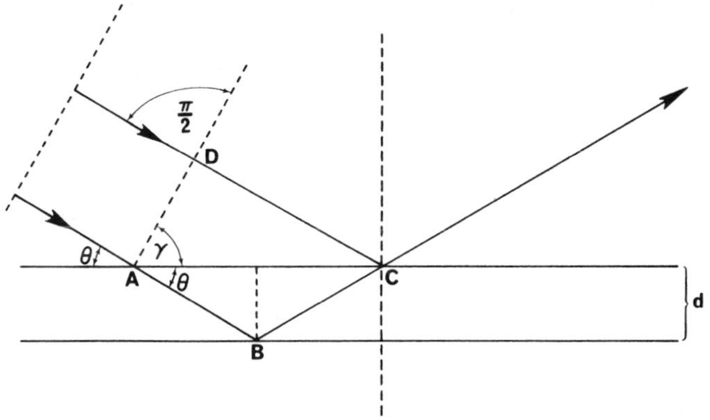

Fig. 1.20 Alternate geometrical proof of the Bragg condition (1.89) for constructive interference between reflected wave components. (Note that $\theta + \gamma = 90°$, so that $\sin \gamma = \cos \theta$. The path difference between the two superimposed reflected rays is $(\overline{AB} + \overline{BC}) - \overline{DC}$. However, $\overline{BC} = \overline{AB} = d/\sin \theta$, and $\overline{DC} = 2\overline{AB} \cos \theta \sin \gamma$. The path difference is therefore $2(d/\sin \theta) \cdot [1 - \cos^2\theta] = 2d \sin \theta$, which in turn must equal an integer multiple of the wavelength λ for constructive interference between the superimposed reflected waves.)

EXERCISE Derive the Bragg condition (1.89) by means of Fig. 1.20.

5.2 The Wave Behavior of Particles

Clearly the explanation of diffraction which has just been presented is based on the consideration that light is wavelike; no particlelike properties have been invoked in the explanation. Now if matter exhibits wavelike properties also, in accordance with the mind-boggling hypothesis of de Broglie, then a beam of monoenergetic particles should have associated with it some wavelength, and this beam should also be diffracted by the crystal whenever the Bragg condition (1.89) is met. Since the wavelength of the x-ray beam is related to the x-ray photon energy $\mathscr{E}_p$ by

$$\mathscr{E}_p = h\nu = hc/\lambda, \tag{1.90}$$

the wavelength associated with the beam of particles should also presumably be energy dependent. The experiment could be carried out by varying the energy of the incident particle beam. This is easily done for an electron beam by changing the voltage on an accelerating grid electrode. A change in energy is somewhat harder to achieve in the case of a beam of neutrons, but the experiment can be executed by filtering out a narrow band of energies from a broader energy spectrum of neutrons with the aid of a mechanical chopper (see Fig. 1.21). Putting aside the complexities of the experimental apparatus and the technical difficulties of the experiments, however, the results can be summarized briefly by saying that diffraction peaks are indeed observed! The diffraction peaks are analogous to those observed in x-ray diffraction, which we explained above by invoking the wave properties of electromagnetic radiation. The results appear unambiguous in verifying the hypothesis of de Broglie that matter has wavelike properties; furthermore, the wavelength associated with the beam of particles can be deduced from the Bragg condition. In this way, a correlation has been established between the particle energy and the wavelength associated with the particle beam. That is, if it is assumed that θ and d are determined from the analogous x-ray experiment, then the Bragg condition gives $n\lambda$ for a given diffraction peak. If a series of diffraction peaks are then measured for several of the sets of crystal planes differing in d spacing from one another, the value of n can be established for each of the peaks. The correlation that is found between wavelength and particle energy is expressed most simply in terms of the momentum p of the individual particles in the beam,

$$\lambda = h/p \qquad \textbf{(de Broglie relation)}, \tag{1.91}$$

where h is Planck's constant. This is called the de Broglie relation, since it was first postulated by de Broglie from theoretical considerations based on wave–particle duality. The experimental verification of this hypothesis has

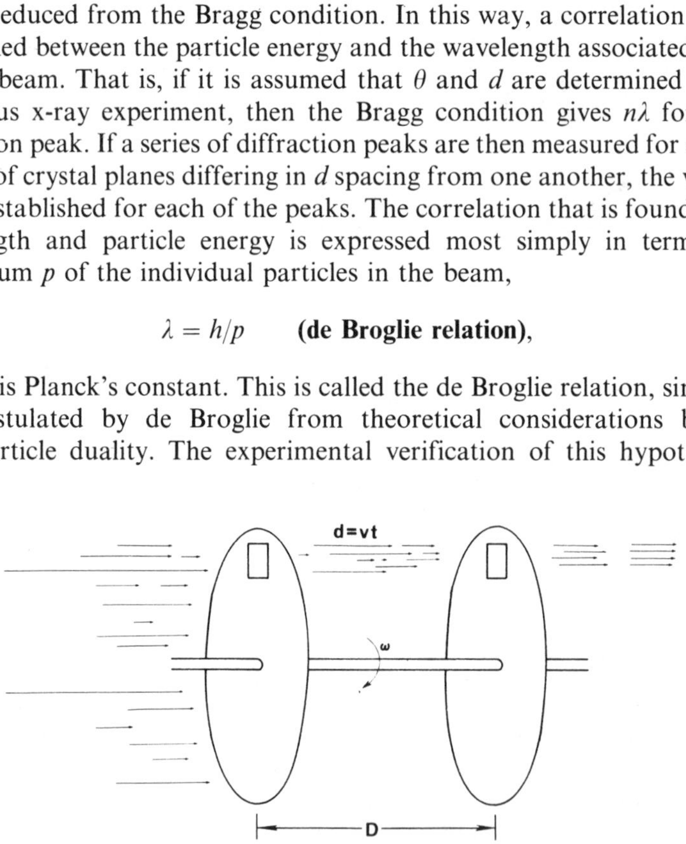

Fig. 1.21 A mechanical chopper consisting of two slotted wheels rotating at a common angular velocity ω can serve as a velocity selector for particles in a beam. (During the period $T = 1/\nu = 2\pi/\omega$ of one revolution, a particle having speed v passing through the slot in the first wheel will travel a distance $d = vT$; it will thus pass through the slot in the second wheel only if the separation distance D between the two wheels is equal to d or some integer multiple of d, namely, $D = nd = nvT$ $(n = 1, 2, 3, \ldots)$. The discrete velocities v_n selected are thus given by $v_n = D/nT = \omega D/2\pi n$. In practice, there will be some spread about each of these values due to the finite width of the slots.)

far-reaching consequences since it provides a sound basis for the development of a wave equation for matter.

It is perhaps worthwhile to emphasize that the wavelength as given by the above expression (1.91) involves the particle mass but is independent of the particle charge; the lack of dependence on charge is also evident from the fact that the same equation is obeyed in both electron diffraction and neutron diffraction. It is also very significant that it is the momentum of the individual particles of the beam instead of the beam intensity which determines the wavelength.

In many respects, the results of particle diffraction are similar to the description given earlier of the experiment involving photons passing through an array of slits and individually triggering a photochemical reaction on a photographic plate behind the slits. The distribution of photons impinging individually on the photographic plate is in statistical accord with the predictions of classical wave theory. Particles can be identified individually in modern detection devices such as counters, cloud chambers, bubble chambers, and photographic plates [see Leighton (1959)]. It has been established unequivocally with such detectors that the observed diffraction peaks are simply the statistical result of an experiment involving a very large number of independent particles, each particle retaining its discrete individuality and traversing its individual path. The distribution of a statistical group of such particles is found to be in accordance with the predictions of a wave picture. That

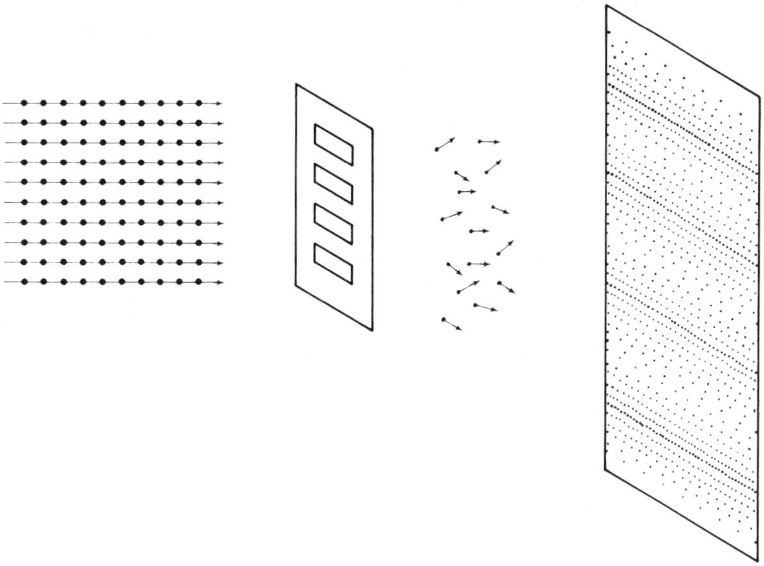

Fig. 1.22 Particles in a beam passing through wave interference slits form individual spots on a photographic plate detector, with the density of spots statistically distributed in accordance with wave predictions. (Corresponding intensity measurements are illustrated schematically as in the case of photons by Fig. 1.8.)

particles can be accelerated and detected individually allows us to produce a diffraction pattern in a stepwise fashion, a particle at a time (see Fig. 1.22). The individual particles seemingly traverse random paths, but when many particles have been detected in this way, a regular diffraction pattern is formed just as in the case of high beam intensities and correspondingly shorter exposure times. The conclusion seems inevitable that the behavior of each individual particle is governed by statistical laws which are in accord with the predictions of a wave picture. This is emphasized by the fact that individual particles even traverse paths which would be completely inaccessible from the viewpoint of classical mechanics where the particles follow straight-line trajectories. In a sense, quantum particles act in much the same way as larger particles observed with an optical microscope that undergo Brownian motion when in thermal equilibrium with a heat bath; the seemingly random movement of the larger particles, however, is in fact produced by collisions with much smaller molecules unobserved in the optical microscope. In both cases the prediction of the exact trajectory and the time dependence of the momentum of each observed particle is impossible, albeit for different reasons, but statistical techniques predict very accurately the behavior of a large ensemble of such particles in both cases.

It is certainly a paradox that each individual localized particle seemingly obeys statistical laws in accord with a wave picture in which interference effects are explained by an interaction of the wave with the entire diffracting structure. Apparently the particle interacts with the diffracting crystal in somewhat the same way as if it were some wave packet capable of being resolved into Fourier-component waves, with each Fourier-component wave being able to interact with the entire diffracting crystal since the component waves are not spatially localized. The difficulty with this picture is that the conditions for reflection for each component wave would be different, so that the packet of Fourier component waves might be expected to be decomposed by the diffraction process. We know, however, that the particle which we have represented by the packet of Fourier component waves is not decomposed by diffraction. Thus it must be concluded that internal forces maintaining the integrity of the particle take precedence over the tendency for the diffraction process to decompose the packet. The overall effect is that the particle can seemingly traverse any path which is consistent with the diffraction pattern obtained from the superposition of Fourier component waves, the probability for selection of a given path being determined by the relative intensity of the diffraction pattern in that direction. To keep this view of the diffraction process realistic, we must remember that the wavelength of each of the component waves depends on the momentum of the component wave in accordance with the de Broglie relation, and in fact, a particle with a well-defined momentum can be described in terms of a *single* component wave.

Thus we can say that single particles are subjected to wave interference effects in the sense that the particle can follow a path which would be a forbidden path according to classical mechanics. If we perform a *Gedanken experiment* (i.e., a thought experiment) in which we pass a beam of particles one by one through a two-slit structure similar to the grating previously discussed (see §1.6) for the case of light interference, then some of the particles will strike the detection screen behind the slits at locations which they could not reach if they moved along straight-line paths through either slit. The appearance of the interference fringes depends on the passage of the wave through both slits at once, just as a propagating electromagnetic plane wave is described as passing through both slits in the analogous optical interference experiment. If the wave describes the behavior of a single particle, then it follows that we cannot decide through which one of the two slits the particle has gone. (If we try to avoid this consequence by determining experimentally with some monitoring device through which slit the particle has passed, we shall by the very action of the monitoring mechanism localize the particle and thereby change its wave packet drastically and destroy the plane-wave interference pattern. A single particle would then go definitely through one slit or the other, and the accumulation of a large number of particles on the screen would result in two well-separated traces. Exactly the same traces would be obtained by closing one slit at a time, thereby predetermining the path of each particle. We are forced to conclude that the conditions necessary for producing the interference pattern forbid a determination of the slit through which the particle passes.)

The simultaneous appearance of wave and particle aspects compels us to be resigned to some degree of inevitable indeterminism in the results of an experiment. Wave aspects and particle aspects in one and the same object are compatible only if we do not ask certain questions of nature which are not really meaningful from an experimental standpoint, such as whether or not we can see the interference fringes produced by particles whose paths through an arrangement of slits we have determined. This line of thought could very quickly get us into an interesting discussion of the interaction between the object being observed and the experimenter (or the experimental apparatus).

On the other hand, the *probability doctrine* of quantum mechanics accepts as a basic hypothesis that such *indeterminism* is a property inherent in nature. If this is a valid hypothesis, then we can never expect to explain this seeming indeterminism by a future theory which is either better or more complete. However, the probability doctrine is purely philosophical speculation, and it therefore has at best a precarious place in physics. The important point is that we can construct a coherent theory for predicting experimental observations on the basis of the wave picture of matter, even though it does involve use of some radically new thought patterns on our part as compared to those developed in working physics problems with Newton's equations of motion or with other classical approaches.

The experimental manifestation of the dual nature of matter led Bohr to the formulation of what is called the *principle of complementarity*, in which the wave nature and the particle nature are considered to be complementary aspects of matter. Both of these classical aspects are equally essential for a full description of matter; although they may appear to be mutually inconsistent, they are assumed to be capable of coexistence. One important aspect of the complementarity principle is that an experimental arrangement designed to manifest one of the classical attributes (e.g., a wavelike or a particlelike aspect) precludes the possibility of observing at least some other classical attribute. This was emphasized in our earlier description of the interference and diffraction of light as compared to the photoelectric effect.

PROJECT 1.1 Wavelengths: Electromagnetic and Particle Waves

1. Compute the range of λ for the following: (a) γ rays, (b) x rays, (c) green light, (d) microwaves, (e) radio waves. (Tabulate results; give equations.)
2. Compute the wavelength λ for the following particles at speeds of 10^{-2}, 1, 100, 10,000, and 10^6 m/sec: (a) electron, (b) proton, (c) neutron, (d) silver atom, (e) macromolecule consisting of approximately 10^6 carbon atoms and 10^6 oxygen atoms, (f) 0.1 g speck of copper, (g) basketball. Express results in tabular form, and give equations used.)

PROJECT 1.2 Photon Production and Electron Ejection

1. What are your concepts of energy absorption and energy emission for a system in which the angular momentum is quantized?
2. Give a qualitative explanation of the production of light in a mercury arc source in terms of electronic energy levels.
3. Describe the effect of an interference filter (with peak transmission at 5461Å) on photons emerging from such a source in terms of the wave properties of light. (How does this differ from the results that would be expected if photons were purely corpuscular?)
4. Compute the energy and momentum of photons emerging from the source-filter system in part 3.
5. What would be the results of an experiment in which the photons in part 3 with a beam intensity of one W/m^2 impinge on a metal with a vacuum work function of 1 eV? (Consider both energy and momentum transfer.)
6. How would part 5 be modified if photons were purely wavelike?
7. Compute the wavelength of the de Broglie wave associated with a photoelectron (if any are ejected in part 5) if all the energy of a single photon is absorbed by a single electron.
8. Would the situation described in part 7 be allowable, assuming the conservation of both energy and momentum?

PROJECT 1.3 Diffraction of Particles

For a crystal lattice constant $d = 4$Å and an incidence angle $\theta = 45°$, compute λ, p, $\mathscr{E}$, and v for the first 5 diffraction peaks for electron and neutron diffraction. Present the results in tabular form so that a comparison between the electrons and neutrons can be made.

5.3 The Bohr Atom and Energy Quantization

Accepting that individual particles have wavelike properties, then, let us ask what this implies qualitatively regarding the behavior of a particle in the neighborhood of a potential energy minimum. Consider an electron bound by the attractive *Coulomb potential* of a fixed nucleus. The **Coulomb potential energy** $\mathscr{U}(\mathbf{r})$ [see Eq. (5.221)] between two point charges q_1 and q_2 in free space decreases inversely with increasing separation $r = |\mathbf{r}|$ between the two charges, $\mathscr{U}(\mathbf{r}) = q_1 q_2 / 4\pi\varepsilon_0 r$, where ε_0 is the permittivity of free space. From the standpoint of classical mechanics, we can picture the specific case of a circular orbit for which the momentum in a direction tangent to the orbit (see Fig. 1.3) has a fixed magnitude independent of position on the orbit. Let us digress a bit in order to summarize the approach to the hydrogen atom as viewed from classical Newtonian physics, and note where the results obtained are incomplete with regard to providing a formula which correctly predicts the experimentally observed optical spectra.

The **Coulomb force** between the two charges follows by taking the negative gradient of the Coulomb potential energy,

$$\begin{aligned}
\mathbf{F} = -\nabla\mathscr{U}(\mathbf{r}) &= -\left(\hat{\mathbf{x}}\frac{\partial}{\partial x} + \hat{\mathbf{y}}\frac{\partial}{\partial y} + \hat{\mathbf{z}}\frac{\partial}{\partial z}\right)\frac{q_1q_2}{4\pi\varepsilon_0 r} \\
&= -\frac{q_1q_2}{4\pi\varepsilon_0}\left(\hat{\mathbf{x}}\frac{\partial}{\partial x} + \hat{\mathbf{y}}\frac{\partial}{\partial y} + \hat{\mathbf{z}}\frac{\partial}{\partial z}\right)(x^2+y^2+z^2)^{-1/2} \\
&= -\frac{q_1q_2}{4\pi\varepsilon_0}(-\tfrac{1}{2})(x^2+y^2+z^2)^{-3/2}[\hat{\mathbf{x}}(2x) + \hat{\mathbf{y}}(2y) + \hat{\mathbf{z}}(2z)] \\
&= \frac{q_1q_2}{4\pi\varepsilon_0 r^3}\mathbf{r} = \frac{q_1q_2}{4\pi\varepsilon_0 r^2}\hat{\mathbf{r}},
\end{aligned}$$

where $\hat{\mathbf{r}} \equiv \mathbf{r}/r$ is a positive unit vector directed outward along the line of centers between the two charges. This has a positive sign for a repulsive force (q_1 and q_2 of same sign); it has a negative sign for an attractive force (q_1 and q_2 of opposite sign). The force can be noted to have the characteristic inverse-square dependence on separation distance r.

In the hydrogen atom, an electron of charge $q_2 = -e$ may be viewed from the standpoint of classical mechanics as circulating around an essentially stationary nucleus of opposite charge $q_1 = Ze$, in which case the product q_1q_2 is negative and is given by $q_1q_2 = -Ze^2 = -|q_1q_2|$. The potential energy $\mathscr{U}(\mathbf{r})$ and the attractive Coulomb force are both negative. In this case the orbit for a bound state may be elliptical or circular [see Goldstein (1956)]. The dynamical equilibrium requirement of classical mechanics must be satisfied at each point on the orbit. This requirement is simply that the centripetal force provided by the attractive Coulomb force must have the magnitude $mv_\perp^2/r$ at each point on the orbit, where $v_\perp$ is the component of electron velocity $\mathbf{v}$ perpendicular to the vector $\mathbf{r}$ giving the instantaneous position of the electron with respect to the nucleus. The parameter m is the electron mass, which is quite small relative to the nuclear mass. We neglect correction terms of the order of the ratio of electron mass to nuclear mass. For a circular orbit, $v_\perp = v$ and the dynamical equilibrium condition becomes

$$mv_0^2/r_0 = |q_1q_2/4\pi\varepsilon_0 r_0^2| = Ze^2/4\pi\varepsilon_0 r_0^2,$$

where the subscript 0 denotes specifically a circular orbit. The kinetic energy $\mathscr{E}_\mathrm{K} = \tfrac{1}{2}mv_0^2$ is therefore seen to be

$$\mathscr{E}_\mathrm{K} = \tfrac{1}{2}mv_0^2 = \tfrac{1}{2}Ze^2/4\pi\varepsilon_0 r_0 = -\tfrac{1}{2}\mathscr{U}(\mathbf{r}),$$

and the total energy $\mathscr{E}$ of the electron given by the sum of the kinetic and potential energies is simply

$$\mathscr{E} = \mathscr{E}_K + \mathscr{U}(\mathbf{r}) = -\tfrac{1}{2}\mathscr{U}(\mathbf{r}) + \mathscr{U}(\mathbf{r}) = +\tfrac{1}{2}\mathscr{U}(\mathbf{r}) = -\mathscr{E}_K.$$

The angular momentum L for a circular orbit is given by $L = r_0 p_0 = m v_0 r_0$, where the subscript 0 again denotes the circular orbit case. Substituting the value of $v_0 = (Ze^2/4\pi\varepsilon_0 m r_0)^{1/2}$ obtained from the above dynamical equilibrium condition gives $L = (mZe^2 r_0/4\pi\varepsilon_0)^{1/2}$, or equivalently, the radius r_0 increases as the square of the angular momentum, $r_0 = (4\pi\varepsilon_0/mZe^2)L^2$. The total energy can thus be written in terms of L^2,

$$\begin{aligned}\mathscr{E} &= \tfrac{1}{2}\mathscr{U}(\mathbf{r}) = -\tfrac{1}{2}Ze^2/4\pi\varepsilon_0 r_0 = -\tfrac{1}{2}(Ze^2/4\pi\varepsilon_0)(mZe^2/4\pi\varepsilon_0)L^{-2} \\ &= -mZ^2e^4/(32\pi^2\varepsilon_0^2 L^2).\end{aligned}$$

Classical mechanics imposes no restriction upon the orbit radius r_0, in which case the total energy $\mathscr{E}$ and the angular momentum L can take on any values between 0 and $-\infty$ and 0 and ∞, respectively. This classical mechanics result for the hydrogen atom ($Z = 1$) problem is in stark contrast to the quantum mechanical treatment which yields the following set of discrete (quantized) values for the total energy and the angular momentum,

$$\mathscr{E}_n = -me^4/2(4\pi\varepsilon_0 n\hbar)^2, \qquad L_n = n\hbar \qquad (n = 1, 2, 3, \ldots, \infty),$$

where $\hbar$ is Planck's constant h divided by 2π, and the integer n is called the *principal quantum number*. The "allowed angular momentum values" are thus integer multiples of a basic unit given by Planck's constant, and the "allowed energy values" $\mathscr{E}_n$ (in units of electron volts) are $\mathscr{E}_n = -(13.6/n^2)$ eV. Optical spectra involving absorption and emission of radiation are in good experimental agreement with energy differences obtained using the energy formula, and this constitutes strong experimental evidence for quantization. The quantization is brought about by appropriate boundary conditions applied to the "stationary-state" wavelike solutions obtained for the quantum problem. For example, the probability of finding the bound electron at a given point in space in the neighborhood of the nucleus is assumed to be a single-valued function of

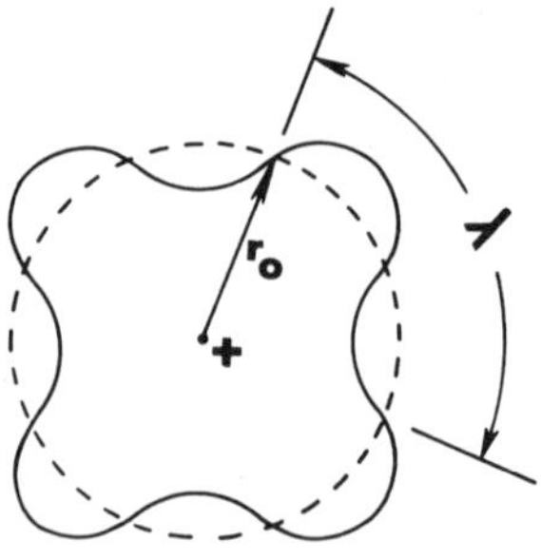

Fig. 1.23 Concept of a single-valued wave of wavelength $\lambda = h/p$ existing around the circular orbit of radius r_0 for classical planetary motion of an electron about an oppositely charged nucleus.

position which approaches zero at infinite separation. We expect from the de Broglie relation (1.91) that such an electron would have a fixed value for the wavelength λ, independent of position on a circular orbit (see Fig, 1.3). If, in addition, at each given instant we consider the wave as existing on the entire ring of the *classical orbit* (see Fig. 1.23) of the electron around the nucleus, then from the physical consideration that wave value must be a single-valued function of position we deduce that there should be an integral number of wavelengths around the orbit. That is, mathematically we impose periodic boundary conditions (cf. §3.4) as might be expected from the wave properties of matter, even though it is not completely clear to us what this means physically for a particle. We simply speculate that there is some single-valued stationary or traveling wave associated with an electron bound to a nucleus which is somehow related to the confinement of the particle to an orbit. By means of the de Broglie relation $\lambda = h/p$, the criterion of an integral number of wavelengths around a circular orbit of radius r_0 yields the result

$$2\pi r_0 = n\lambda = nh/p_0 \qquad (n = 1, 2, \ldots), \tag{1.92}$$

where p_0 is the fixed value of the magnitude of the momentum tangent to the orbit. Since the product $r_0 p_0$ is the angular momentum L of a particle in a circular orbit, this result can be stated as

$$L = r_0 p_0 = nh/2\pi = n\hbar \qquad (n = 1, 2, 3, \ldots, \infty)$$

$$\textbf{(quantized angular momentum values)}; \tag{1.93}$$

that is, the angular momentum must be the product of some integer and Planck's constant divided by 2π. Therefore we have shown by means of de Broglie waves that the angular momentum for the circular orbits is quantized! The basic quantization unit (cf. §1) for the angular momentum can be noted to be Planck's constant divided by 2π.

The semiclassical approach, based on combining the classical formulas with the de Broglie relation $\lambda = h/p$ for the wavelength of the electron, has thus led us to the concept of quantization, and we shall show that it yields the correct expression for the quantized *energy* values. The argument has been simple enough if one accepts the above-mentioned idea that the de Broglie wave is in a "stationary state" such that it appears as a smooth continuous wave around the classical orbit. In order for the wave to close on itself around a circular orbit without a discontinuity in value and slope, an integral number of wavelengths $\lambda = h/p_0$ must be contained within the circumference $2\pi r_0$. This gives condition (1.92), which leads immediately to the quantized values L_n for the angular momentum given by Eq. (1.93). However, the classical dynamical equilibrium condition of a balance between centripetal and centrifugal forces was shown to lead to the relation $r_0 = (4\pi\varepsilon_0/Ze^2 m)L^2$ between the radius r_0 of the circular orbit and the angular momentum L of the electron in its classical mechanical orbit. In contrast to the situation in classical mechanics where there is no restriction on the value of the angular momentum of the electron in its orbit, the de Broglie wave concept imposes the restriction contained in Eq. (1.93) on the values of the angular momentum consistent with single-valuedness of the wave on the orbit. Substituting the semiclassical quantization condition $L_n = n\hbar$ given by Eq. (1.93) into this expression for the radius r_0 of the circular orbit gives a set of *quantized values* for the radius of the semiclassical circular "orbits,"

$$r_0 = (4\pi\varepsilon_0/Ze^2m)n^2\hbar^2 = n^2a_0/Z \qquad (n = 1, 2, 3, \ldots, \infty), \tag{1.94}$$

where the parameter

$$a_0 \equiv 4\pi\varepsilon_0\hbar^2/me^2 \tag{1.95}$$

is called the **Bohr radius**. These "allowed" orbits are illustrated in Fig. 1.24. The value of a_0 is approximately 0.529 Å. Substituting the expression for r_0 into the above expression for the total energy, $\mathscr{E} = \frac{1}{2}\mathscr{U}(\mathbf{r})$, then yields the following quantized values for the total energy,

$$\begin{aligned}\mathscr{E}_n &= \tfrac{1}{2}\mathscr{U}(\mathbf{r}) = \tfrac{1}{2}(q_1q_2/4\pi\varepsilon_0)[(n^2\hbar^2)^{-1}(Ze^2m/4\pi\varepsilon_0)] \\ &= -mZ^2e^4/32n^2\pi^2\varepsilon_0^2\hbar^2 = -(13.6\ Z^2/n^2)\ \ \text{eV}\end{aligned}$$

(quantized energy levels for Bohr atom). (1.96)

This result agrees with the exact quantum result for the hydrogen atom, assuming the nucleus to be a stationary proton with $Z = 1$. The derivation is referred to as the *Bohr theory of the hydrogen atom.* It was presented to the scientific world by Niels Bohr in 1913, a decade or so before the advent of quantum mechanics.

If we consider the more general case of elliptical orbits as deduced in classical mechanics for the motion of an electron about a fixed nucleus, then the radius of

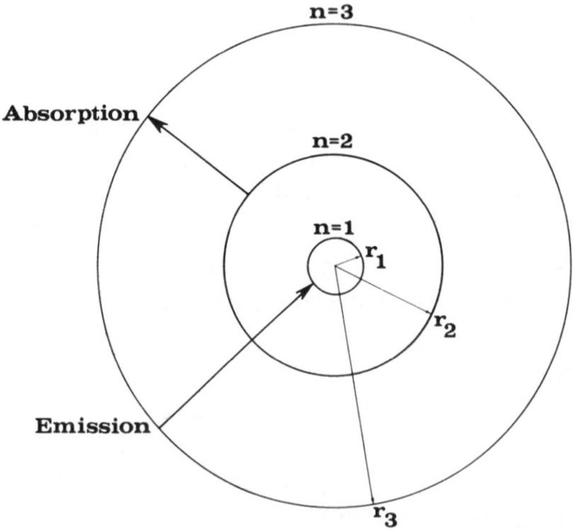

Fig. 1.24 Schematic diagram of discrete circular semi-classical orbits. [These are predicted by Eq. (1.94), which is derived on the basis of the de Broglie relation $\lambda = h/p$ and the wave concept illustrated in Fig. 1.23.]

the classical orbit and the magnitude of the momentum p tangent to the orbit are functions of position on the orbit. The de Broglie relation would then imply that the wavelength λ varies around the orbit, as illustrated schematically in Fig. 1.25. However, the physical considerations of continuity and single-valuedness again lead to the conclusion that there must be an integral number of wavelengths around the orbit.

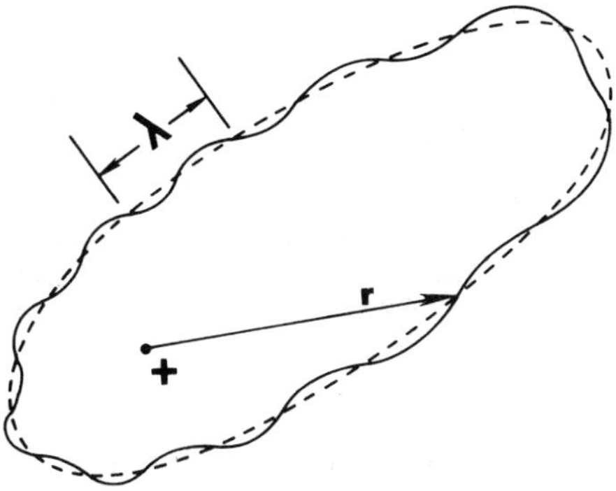

Fig. 1.25 Concept of a single-valued wave with local wavelength $\lambda = h/p$ varying with position around an elliptical orbit for classical planetary motion of an electron about the oppositely charged nucleus.

The integral number n of wavelengths around the orbit is given by $n = \oint ds/\lambda$, where ds is an increment of length on the orbit, and the circle through the integral sign means, by convention, that the integral is a line integral over the entire closed orbit. Employing the de Broglie relation $\lambda = h/p$, where in the present instance p represents the magnitude of the momentum parallel to the orbit at any given position on the orbit, we obtain $\oint p\, ds = nh$ $(n = 1, 2, \dots)$. This integral has the dimensions of angular momentum. However, angular momentum has the same dimensions as the product of energy and time, such products being called *action*. Planck's constant likewise has the units of action. The above integral expression is analogous to the *semiempirical quantum conditions* employed in the early stages of the development of quantum mechanics: *The classical action* (*or phase*) *integrals for periodic motion were required to be quantized* according to $\oint p_i\, dq_i = nh$, where the quantum number n is an integer and the integral is over the complete period of the *generalized coordinate* q_i. The momentum p_i is that which is *canonically conjugate* to q_i. Canonically conjugate variables are defined and utilized in advanced treatments of classical mechanics [see Goldstein (1956)]. Although Bohr succeeded in calculating the discrete energy levels of the hydrogen atom using this condition, he was unsuccessful in extending the calculation to two-electron systems, such as the helium atom, and a fortiori to higher electron systems.

For large quantum numbers, $n\hbar$ represents a large quantity of angular momentum, in which case the total angular momentum is much larger than the separation $\hbar$ between adjacent quantized values. The discreteness of the angular momentum values in nature can be ignored in such cases, and the quantity can be considered almost as if it were a continuous variable. In this limit, classical mechanics provides an adequate description of nature. This is a manifestation of what is called the *correspondence principle*, which states that in the limit of large quantum numbers ($n \to \infty$), the quantum results become identical with the predictions of classical mechanics. Alternatively, it is sometimes said that *the correspondence principle requires that quantum theory be consistent with classical physics in the limit of large quantum numbers.*

PROJECT 1.4 The Bohr Theory and Line Spectra

Refer to the original papers written by Bohr to find precisely the wording he used to describe the two fundamental assumptions that he made to formulate his theory to explain experimental line spectra. Also determine exactly how he formulated what is today known as the Bohr theory of the hydrogen atom. [*Hint*: See van der Waerden (1967) for references to the original literature.]

5.4 The Short Wavelength Limit

From the de Broglie relation (1.91), we see that $\lambda \to 0$ as the momentum p becomes larger and larger. In the limit that $\lambda \to 0$, waves cease to be diffracted; instead, they follow a straight rectilinear path. A particle of very large momentum thus tends to obey the laws of classical mechanics. This constitutes the short wavelength limit of wave mechanics in the same way as geometrical optics constitutes the short wavelength limit of wave optics. The same result is obtained by considering h to approach zero, since $\lambda = h/p$ again approaches zero. Because $p = mv$ (nonrelativistic approximation), where m is the particle mass and v is the particle velocity, large mass particles (such as macroscopic bodies) and fast particles tend toward classical (i.e., nonquantum) behavior. Thus classical mechanics must be contained in quantum mechanics. That classical mechanics must be the limit obtained from the quantum formulation as $h \to 0$ is one aspect of the correspondence principle.

If we attempt to extend the de Broglie relation $\lambda = h/p$ as discussed above for matter to the case of light, which also manifests wave–particle duality, we reach the conclusion that photons have a momentum $p = h/\lambda$ associated with them. If this is true (and indeed it has been experimentally verified), then a classical electromagnetic wave consisting of many quanta can carry (or possess) a significant amount of momentum, even though the photon rest mass is zero. Since the photon energy $\mathscr{E}_p$ is simply $h\nu$, and $\lambda\nu = c$,

$$p = h/\lambda = h\nu/c = \mathscr{E}_p/c \quad \textbf{(energy–momentum relation for photons).} \tag{1.97}$$

Thus, for photons, $\mathscr{E}_p = cp$. This relation also follows directly from special relativity. Particle energy $\mathscr{E}$ is related to the rest mass energy $\mathscr{E}_0$ and the momentum p through $\mathscr{E}^2 = \mathscr{E}_0^2 + p^2c^2$, so that zero rest mass particles such as photons and neutrinos obey the relation $\mathscr{E}^2 = p^2c^2$, consistent with $\mathscr{E} = pc$. In collisions between particles and photons, it is found experimentally that both energy and momentum must be conserved.

6 Development of the Time-Dependent and Time-Independent Schrödinger Wave Equations

6.1 The Wave Function for Free Particles

As shown from diffraction experiments and discussed in the preceding section, it is remarkable that a monochromatic wave of wavelength $\lambda = h/p$ is to be associated with a beam of electrons, neutrons, or other particles traveling with a definite momentum of magnitude p. Let us consider the waves to be represented mathematically by a function $\psi(x, y, z, t)$, called the *wave function*. We postulate that the scalar ψ is some measure of the presence of the particle. In order to have interference, we must allow both positive and negative values for ψ. However, the probability that a particle can be found somewhere is always positive (or zero); it cannot be negative. Thus if we measure probability in terms of a *probability density function*, then this function must vary somehow as the *magnitude* of ψ in order that it always be positive. In physical optics, interference patterns are produced by the superposition of waves **E** characterizing the electric field, but the intensity of the fringes is measured by the scalar product of $\mathbf{E}(x, y, z, t)$ with itself,

$$\mathbf{E} \cdot \mathbf{E} = E^2. \tag{1.98}$$

The same is true with the magnetic field $\mathbf{H}(x, y, z, t)$: Superposition of the components of **H** gives rise to interference effects, but the intensity of the field varies as $H^2 = \mathbf{H} \cdot \mathbf{H}$. Similarly, the energy density of an electromagnetic field is a positive-definite quantity involving the sum of E^2 and H^2 contributions.

In analogy to this situation, we assume that the positive quantity

$$|\psi(x, y, z, t)|^2 = \psi^*(x, y, z, t)\psi(x, y, z, t)$$

$$\textbf{(particle probability density)} \tag{1.99}$$

measures the probability of finding a particle at x, y, z at time t. The absolute value allows for the possibility that ψ may be complex. The complex conjugate ψ^* of a function ψ is obtained by replacing the quantity i, defined as $(-1)^{1/2}$, by the quantity $-i$. We choose ψ to be a scalar quantity for the present work, although it must be replaced by a vectorlike quantity when the intrinsic *spin angular momentum* of the particles is also under consideration, in somewhat the same way as the electric and magnetic fields **E** and **H** are vector quantities. That is, *spin* corresponds to a *polarization* of the matter waves.

We consider the scalar ψ to be capable of representing in a statistical fashion the behavior of each particle in the beam; it will then likewise contain a description of the statistical effects of the superposition of a large number of particles of the same energy making up a particle beam. Let us consider a particle beam to be monoenergetic and of uniform intensity, and choose a coordinate system so that the beam is traveling in the x direction. If this beam is to be described by a plane wave (in analogy with electromagnetic wave propagation) such that $|\psi(x, t)|^2$ can be interpreted as a uniform density of electrons, then

$$\psi(x, t) = A \exp\{i[(2\pi/\lambda)x - \omega t]\}. \tag{1.100}$$

We have no dependence on y and z since we assume the wavefront to be infinite in extent. Equation (1.12) states that $k = 2\pi/\lambda$; this can be used to replace the parameter λ by the parameter k if desired. The complex form (1.100) chosen for $\psi(x, t)$ has the property that the spatial density of electrons $|\psi|^2 = A^2$ is uniform; this would not be the case if we chose the real function $\psi(x, t) = A \cos[(2\pi/\lambda)x - \omega t]$, for example.

We must now face the question of what to use for ω in (1.100). Equation (1.16) is not immediately applicable because c represents the velocity of light instead of a velocity characteristic of matter waves. In analogy with the properties of the light quantum (photon), we may postulate that the circular frequency ω of the wave is related to the energy $\mathscr{E}$ of each particle in the beam by

$$\mathscr{E} = h\nu = \hbar\omega, \tag{1.101}$$

where again h is Planck's constant given by Eq. (1.2) and $\hbar = h/2\pi = 1.054 \times 10^{-34}$ J sec. This relationship between particle energy and frequency unites particle and wave concepts in much the same way as the de Broglie relation $p = h/\lambda$. Substituting $\lambda = h/p$ and $\omega = \mathscr{E}/\hbar$ into Eq. (1.100) gives

$$\psi(x, t) = A \exp[(i/\hbar)(px - \mathscr{E}t)]. \tag{1.102}$$

6.2 Development of the Time-Dependent Schrödinger Wave Equation

Suppose we differentiate ψ with respect to t and x,

$$\partial\psi/\partial t = -(i/\hbar)\mathscr{E}\psi, \tag{1.103}$$

$$\partial\psi/\partial x = (i/\hbar)p\psi, \tag{1.104}$$

$$\partial^2\psi/\partial x^2 = -(p^2/\hbar^2)\psi. \tag{1.105}$$

Solving for p^2 and $\mathscr{E}$ and substituting into the classical relation $p^2/2m = \mathscr{E}$ gives

$$-\frac{\hbar^2}{2m}\frac{1}{\psi}\frac{\partial^2\psi}{\partial x^2} = -\frac{\hbar}{i}\frac{1}{\psi}\frac{\partial\psi}{\partial t}, \tag{1.106}$$

or equivalently,

$$-\frac{\hbar^2}{2m}\frac{\partial^2\psi}{\partial x^2} = i\hbar\frac{\partial\psi}{\partial t}. \tag{1.107}$$

This is the *one-dimensional time-dependent Schrödinger equation satisfied by a flux of free particles.*

Now if the particles were not free, but instead were moving in a region of varying potential such that the potential energy of the particles depends on position in the manner $\mathscr{V} = \mathscr{V}(x)$, then if $\mathscr{E}$ is still considered to be the total energy of the particles, we have the usual classical relation

$$(p^2/2m) + \mathscr{V}(x) = \mathscr{E}. \tag{1.108}$$

Assuming that $p^2/2m$ is still represented by the left-hand side of Eq. (1.107) and the total energy $\mathscr{E}$ is still represented by the right-hand side of Eq. (1.107), we obtain

$$-\frac{\hbar^2}{2m}\frac{\partial^2\psi}{\partial x^2} + \mathscr{V}\psi = i\hbar\frac{\partial\psi}{\partial t}. \tag{1.109}$$

This important equation is known as the **one-dimensional time-dependent Schrödinger equation.**

For regions in which the potential energy $\mathscr{V}$ is not a function of position or time, namely, $\mathscr{V} \neq \mathscr{V}(x)$ and $\mathscr{V} \neq \mathscr{V}(t)$ so that $\mathscr{V} = \text{const}$, direct substitution of the plane wave (1.102) into the Schrödinger equation (1.109) shows that any plane wave of this form is a solution. Because of linearity of the Schrödinger equation, any linear combination of waves of this form also represents a formal mathematical solution, assuming only that $\mathscr{E} = (p^2/2m) + \mathscr{V}$ for each $\mathscr{E}$ and p in question. The solution of the Schrödinger equation for cases in which $\mathscr{V}$ is position dependent, however, yields solutions which are not of the plane-wave form.

Although it appears initially that the plane wave (1.102) formally satisfies the differential equation (1.109) and the energy–momentum relation (1.108) for particles even when $\mathscr{V} = \mathscr{V}(x)$, the dependence of $\mathscr{V}$ on x with a fixed total energy $\mathscr{E}$ requires p to be position dependent, namely $p = p(x)$, in which case Eq. (1.102) does not represent a plane wave and, more important,

$$-\frac{\hbar^2}{2m}\frac{\partial^2}{\partial x^2}\left\{A\exp\left[\left(\frac{i}{\hbar}\right)(px - \mathscr{E}t)\right]\right\} \neq \left(\frac{p^2}{2m}\right)\left\{A\exp\left[\left(\frac{i}{\hbar}\right)(px - \mathscr{E}t)\right]\right\}.$$

This may cause us to question the merits of our deduction of this equation on the basis of plane waves. We do not concern ourselves with this matter; suffice it to state that experiment has shown this equation to have a far greater degree of validity than the above simple derivation might indicate. In general, all characteristic solutions are of the nature of a complete set of *basis states* from which a general solution must be constructed by linear superposition, so in this sense the type of basis state used is not of fundamental significance. This is similar to our discussion of Fourier series and Fourier integral solutions of the classical wave equation presented in §§ 3 and 4.

Suppose now that we consider three dimensions instead of one dimension. The position x must be replaced by the position vector $\mathbf{r} = x\hat{\mathbf{x}} + y\hat{\mathbf{y}} + z\hat{\mathbf{z}}$, where $\hat{\mathbf{x}}$, $\hat{\mathbf{y}}$, and $\hat{\mathbf{z}}$ are unit vectors in a Cartesian coordinate system, and the momentum p must be replaced by $\mathbf{p} = p_x\hat{\mathbf{x}} + p_y\hat{\mathbf{y}} + p_z\hat{\mathbf{z}}$. The corresponding plane wave is of the form

$$\psi(\mathbf{r}, t) = A\exp[i(\mathbf{k}\cdot\mathbf{r} - \omega t)], \tag{1.110}$$

where $\mathbf{k}$ is a vector of magnitude $2\pi/\lambda$ that points in the direction of propagation of the plane wave. (This is in accordance with the discussion in §4). Thus $\mathbf{k}$ is parallel to $\mathbf{p}$. The de Broglie relation for the case in which $\mathbf{p}$ is considered to be a vector quantity becomes $\lambda = h/|p|$. Thus

$$\lambda = 2\pi/|\mathbf{k}| = h/|\mathbf{p}|, \tag{1.111}$$

so

$$|\mathbf{k}| = 2\pi|\mathbf{p}|/h = |\mathbf{p}|/\hbar. \tag{1.112}$$

Combining this with the fact that **k** and **p** are parallel yields the important result

$$\mathbf{p} = \hbar\mathbf{k}. \tag{1.113}$$

Substituting this into $\psi(\mathbf{r}, t)$, given by Eq. (1.110) together with the relation $\mathscr{E} = \hbar\omega$, gives

$$\psi(\mathbf{r}, t) = A \exp[(i/\hbar)(\mathbf{p} \cdot \mathbf{r} - \mathscr{E}t)] \qquad \textbf{(wave function for free particles)} \tag{1.114}$$

for the wave function. Since $\mathbf{p} \cdot \mathbf{r} = p_x x + p_y y + p_z z$, we can differentiate this wave function with respect to t, x, y, and z to obtain

$$\partial\psi/\partial t = -(i/\hbar)\mathscr{E}\psi, \tag{1.115}$$

$$\partial\psi/\partial x = (i/\hbar)p_x\psi, \tag{1.116}$$

$$\partial^2\psi/\partial x^2 = -(p_x^2/\hbar^2)\psi, \tag{1.117}$$

$$\partial^2\psi/\partial y^2 = -(p_y^2/\hbar^2)\psi, \tag{1.118}$$

$$\partial^2\psi/\partial z^2 = -(p_z^2/\hbar^2)\psi. \tag{1.119}$$

Solving for p_x^2, p_y^2, p_z^2, and $\mathscr{E}$ and substituting into the classical nonrelativistic relation

$$(p_x^2 + p_y^2 + p_z^2)/2m = \mathscr{E} \tag{1.120}$$

gives

$$-\frac{\hbar^2}{2m}\frac{1}{\psi}\left(\frac{\partial^2\psi}{\partial x^2} + \frac{\partial^2\psi}{\partial y^2} + \frac{\partial^2\psi}{\partial z^2}\right) = -\frac{\hbar}{i}\frac{1}{\psi}\frac{\partial\psi}{\partial t}. \tag{1.121}$$

Using the fact that the Laplacian operator ∇^2 is simply $\partial^2/\partial x^2 + \partial^2/\partial y^2 + \partial^2/\partial z^2$, we obtain

$$-(\hbar^2/2m)\nabla^2\psi = i\hbar\,\partial\psi/\partial t. \tag{1.122}$$

This is called the three-dimensional time-dependent Schrödinger equation for *free particles*. If we consider the particles as moving in a region of varying potential such that the potential energy is $\mathscr{V}(x, y, z) = \mathscr{V}(\mathbf{r})$, then the total energy $\mathscr{E}$ is given classically by

$$\mathscr{E} = \mathscr{V}(\mathbf{r}) + [(p_x^2 + p_y^2 + p_z^2)/2m]. \tag{1.123}$$

In the same way that we deduced Eq. (1.109) from Eqs. (1.107) and (1.108), we deduce Eq. (1.124) from Eqs. (1.122) and (1.123),

$$-(\hbar^2/2m)\nabla^2\psi + \mathscr{V}(\mathbf{r})\psi = i\hbar\,\partial\psi/\partial t. \tag{1.124}$$

This important result is known as the **three-dimensional time-dependent Schrödinger equation.** It is most remarkable that the solutions to this equation predict quite accurately the probability densities and energy levels of single-particle systems. This equation is truly the "workhorse" of present-day quantum mechanics.

EXERCISE Show that the function $\psi(x, t) = \int_{-\infty}^{\infty} A(k)e^{i(kx-\omega t)}dk$ satisfies the Schrödinger equation in which $V(r) = 0$.

EXERCISE Write the Schrödinger equation for a particle of mass m in a uniform gravitational field $\mathbf{g} = g\hat{\mathbf{x}}$.

PROJECT 1.5 Schrödinger Equation with Applied Magnetic Field

1. Deduce the total momentum for a charged particle in an electromagnetic field.
2. Use this result to formulate the Schrödinger equation for situations involving magnetic fields.

6.3 Solution by the Separation of Variables Technique

In the above justification of the Schrödinger equation for free particles, we assumed a function ψ of the plane-wave form, formed the various first and second derivatives, and by substituting into the classical relation $\mathscr{E} = (p_x^2 + p_y^2 + p_z^2)/2m$ we obtained the famous Schrödinger equation. It is clear that given the Schrödinger equation for free particles [i.e., $\mathscr{V}(\mathbf{r}) = 0$], we should be able to solve this partial differential equation to regain the function $\psi(\mathbf{r}, t)$. The solution can be obtained easily if we employ the technique known as the *separation of variables* by which we assume a solution in the form of a product $\psi(\mathbf{r}, t) = X(x)Y(y)Z(z)\theta(t)$, in which X is a function only of x, Y a function only of y, Z a function only of z, and θ a function only of t. To illustrate the procedure and to derive the important *time-independent* Schrödinger equation, we perform a partial separation of variables in the time-dependent Schrödinger equation for the case in which $\mathscr{V}(\mathbf{r})$ is an arbitrary function of position. Substituting the assumed product solution

$$\psi(x, y, z, t) = \phi(x, y, z)\theta(t) = \phi(\mathbf{r})\theta(t) \qquad (1.125)$$

into the time-dependent Schrödinger equation (1.124) and dividing by the product $\phi\theta$ gives

$$-(\hbar^2/2m\phi)\nabla^2\phi + \mathscr{V}(\mathbf{r}) = (i\hbar/\theta)\, d\theta/dt. \qquad (1.126)$$

The left-hand side does not involve t, so that the right-hand side cannot be a function of t. Since θ is a function of t only, as assumed when we write the separated product $\phi(x, y, z)\theta(t)$ for ψ, the right-hand side cannot be a function of x, y, or z. Since the right-hand side is not a function of x, y, z, or t, it can only be a constant which we can denote by α^2. Then

$$(i\hbar/\theta)\, d\theta/dt = \alpha^2 \qquad (1.127)$$

or

$$d\theta/dt + (i/\hbar)\alpha^2\theta = 0. \qquad (1.128)$$

From our previous consideration of the plane-wave form of $\psi(x, y, z, t)$, we expect the time-dependence of $\theta(t)$ to be of the form

$$\theta(t) = \theta_0 \exp[-(i/\hbar)\mathscr{E}t], \tag{1.129}$$

where θ_0 is an arbitrary constant. We expect this to have the same functional form whether or not $\mathscr{V}(\mathbf{r})$ is zero, since the right-hand side of the separated equation is independent of $\mathscr{V}(\mathbf{r})$. Substituting $\theta(t)$ as a trial solution into the above equation shows that it indeed is a good solution provided we identify the separation constant α^2 as the total energy $\mathscr{E}$ of the particle in question, i.e., $\alpha^2 = \mathscr{E}$. Substituting this result for $\theta(t)$ into the separated form of the partial differential equation (1.126) and multiplying both sides by ϕ yields

$$-(\hbar^2/2m)\nabla^2\phi(\mathbf{r}) + \mathscr{V}(\mathbf{r})\phi(\mathbf{r}) = \mathscr{E}\phi(\mathbf{r}). \tag{1.130}$$

This important equation is known as the **three-dimensional time-independent Schrödinger equation**; the solutions $\phi(\mathbf{r})$ are called stationary-state solutions for the potential $\mathscr{V}(\mathbf{r})$ in question. It is clear that the time-dependent solution corresponding to any given stationary-state solution $\phi(\mathbf{r})$ is the product of $\phi(\mathbf{r})$ and the corresponding $\theta(t) = \theta_0 \exp[-(i/\hbar)\mathscr{E}t]$. Any linear combination of such time-dependent solutions is also a solution, since the time-dependent Schrödinger equation is linear. Therefore it is possible to superimpose these solutions to form *wave packets*, similar to the wave packets discussed in §4. Such wave packets for matter waves are found to change with time, in contrast to packets of electromagnetic waves propagating in free space.

In the case of free particles, for which $\mathscr{V}(\mathbf{r}) = 0$, the separation process can be continued by assuming the trial form

$$\phi(\mathbf{r}) = X(x)Y(y)Z(z), \tag{1.131}$$

where X, Y, and Z have been defined previously. Symmetry tells us that X, Y, and Z will be of the same functional form. Substitution of the above trial solution into the three-dimensional time-independent equation for free particles [Eq. (1.130) with $\mathscr{V}(\mathbf{r}) = 0$] gives

$$-\frac{\hbar^2}{2m}\left(\frac{1}{X}\frac{d^2X}{dx^2} + \frac{1}{Y}\frac{d^2Y}{dy^2} + \frac{1}{Z}\frac{d^2Z}{dz^2}\right) = \mathscr{E}. \tag{1.132}$$

Setting

$$(1/X)\, d^2X/dx^2 = -\alpha_x^2, \tag{1.133}$$

and similarly defining α_y^2 and α_z^2, we have

$$(\hbar^2/2m)(\alpha_x^2 + \alpha_y^2 + \alpha_z^2) = \mathscr{E}. \tag{1.134}$$

Clearly α_x^2, α_y^2, and α_z^2 are constants; this is deduced by the same type of argument used to show that our previous separation constant α^2 was a constant.

The trial solution

$$X(x) = X_0 \exp(i\alpha_x x) \tag{1.135}$$

is a valid solution to the above differential equation; similar trial solutions will be valid for $Y(y)$ and $Z(z)$. The quantities X_0, Y_0, and Z_0 are arbitrary constants. If we identify α_x as $p_x/\hbar$, α_y as $p_y/\hbar$, and α_z as $p_z/\hbar$, then the above relation between α_x, α_y, α_z, and $\mathscr{E}$ satisfies the classical relation $\mathscr{E} = (p_x^2 + p_y^2 + p_z^2)/2m$, and the product solution

$$\psi(x, y, z, t) = X(x)Y(y)Z(z)\theta(t) \tag{1.136}$$

is nothing more than the plane wave

$$\psi(\mathbf{r}, t) = A \exp[(i/\hbar)(\mathbf{p} \cdot \mathbf{r} - \mathscr{E}t)] \tag{1.137}$$

previously considered [cf. Eq. (1.114)], with $A = \theta_0 X_0 Y_0 Z_0$. If there is no restriction on the separation constant $\mathscr{E}$, as in the case of free particles, then we have a *continuous spectrum* of energies available. In other cases, the imposition of boundary conditions arising from the requirement that the solution be physically meaningful for some particular problem puts restrictions on the possible values of $\mathscr{E}$, so that an unbounded continuum of energies is unacceptable for many physical problems.

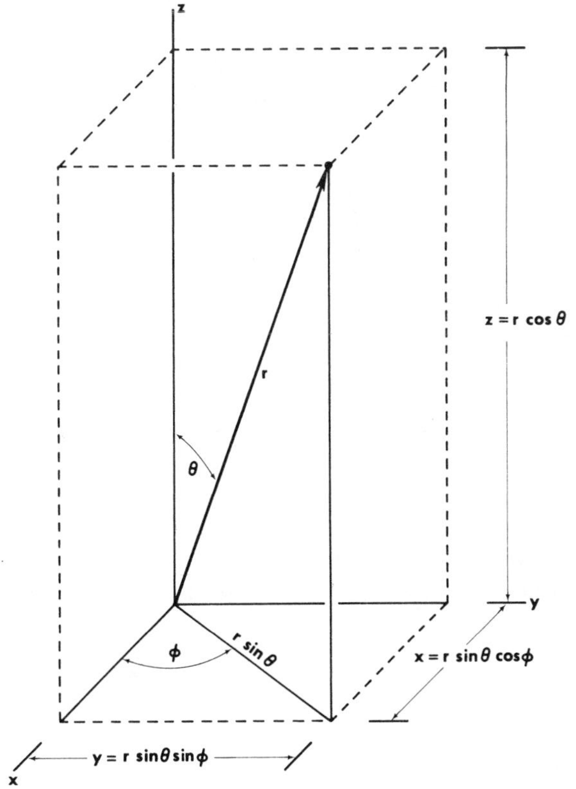

Fig. 1.26 Spherical polar coordinates r, θ, ϕ for locating a point in a three-dimensional space. (The coordinates x, y, z of the corresponding point in rectangular Cartesian coordinates follow from elementary trigonometry, as indicated in the diagram.)

The technique of separation of variables can be used even if $\mathscr{V}(\mathbf{r}) \neq 0$ provided $\mathscr{V}(\mathbf{r})$ itself is a function of only one independent variable in some orthogonal coordinate system. Spherical polar coordinates (Fig. 1.26) are ideal for central forces, for which $\mathscr{V}$ is a function only of $|\mathbf{r}|$ and is independent of θ and ϕ. The separated differential equations are still frequently difficult to solve for arbitrary potentials, so that a good knowledge of ordinary linear differential equations and their various solutions is required. The solutions to the better known potentials, such as the simple Coulomb potential and the harmonic oscillator potential, have been developed in detail and can be found in textbooks on quantum mechanics and modern physics such as those by Schiff (1968), Merzbacher (1970), Leighton (1959), Bohm (1951), Pauling and Wilson (1935), and others.

6.4 Quantum Operators

It is worthwhile to note that differentiation of the plane-wave solution (1.137) with respect to x, y, z, and t yields the product of important physical quantities with ψ:

$$\partial\psi/\partial x = (i/\hbar)p_x\psi, \tag{1.138}$$

$$\partial^2\psi/\partial x^2 = -(p_x^2/\hbar^2)\psi, \tag{1.139}$$

$$\partial\psi/\partial y = (i/\hbar)p_y\psi, \tag{1.140}$$

$$\partial^2\psi/\partial y^2 = -(p_y^2/\hbar^2)\psi, \tag{1.141}$$

$$\partial\psi/\partial z = (i/\hbar)p_z\psi, \tag{1.142}$$

$$\partial^2\psi/\partial z^2 = -(p_z^2/\hbar^2)\psi, \tag{1.143}$$

$$\partial\psi/\partial t = -(i/\hbar)\mathscr{E}\psi. \tag{1.144}$$

Therefore if we define the differential operators

$$p_x^{\text{op}} = -i\hbar\ \partial/\partial x, \tag{1.145}$$

$$p_y^{\text{op}} = -i\hbar\ \partial/\partial y, \tag{1.146}$$

$$p_z^{\text{op}} = -i\hbar\ \partial/\partial z, \tag{1.147}$$

$$\mathscr{E}^{\text{op}} = i\hbar\ \partial/\partial t, \tag{1.148}$$

then we can say that the differential operator acts on the wave function ψ to give the product of the value of the physical observable corresponding to the operator and ψ:

$$p_x^{\text{op}}\psi = p_x\psi, \tag{1.149}$$

$$p_y^{\text{op}}\psi = p_y\psi, \tag{1.150}$$

$$p_z^{\text{op}}\psi = p_z\psi, \tag{1.151}$$

$$\mathscr{E}^{\text{op}}\psi = \mathscr{E}\psi. \tag{1.152}$$

We therefore say that the plane wave solution is an eigenfunction of operators for the x, y, and z components of the momentum and the energy. These differential equations are known as the *eigenvalue equations* for the operators in question. Furthermore, Eqs. (1.138), (1.140), and (1.142) lead to the relation

$$\nabla\psi = (i/\hbar)\mathbf{p}\psi. \tag{1.153}$$

Thus

$$\mathbf{p}^{\text{op}} = -i\hbar\,\nabla \tag{1.154}$$

is the **linear momentum operator**, since

$$\mathbf{p}^{\text{op}}\psi = -i\hbar\,\nabla\psi = \mathbf{p}\psi. \tag{1.155}$$

From Eqs. (1.139), (1.141), and (1.143), we obtain the result

$$-\hbar^2\left(\frac{\partial^2}{\partial x^2}+\frac{\partial^2}{\partial y^2}+\frac{\partial^2}{\partial z^2}\right)\psi = (p_x^2+p_y^2+p_z^2)\psi = p^2\psi = \mathbf{p}\cdot\mathbf{p}\,\psi \tag{1.156}$$

so that the p^2 operator, $(p^2)^{\text{op}}$, can be considered to be given by

$$(p^2)^{\text{op}} = -\hbar^2\left(\frac{\partial^2}{\partial x^2}+\frac{\partial^2}{\partial y^2}+\frac{\partial^2}{\partial z^2}\right) = -\hbar^2\,\nabla^2. \tag{1.157}$$

From the equations of vector analysis [see Wylie (1951)] and Eq. (1.155), however,

$$\begin{aligned}\mathbf{p}^{\text{op}}\cdot\mathbf{p}^{\text{op}}\psi &= -i\hbar\,\nabla\cdot(-i\hbar\nabla\psi) = -i\hbar\,\nabla\cdot(\mathbf{p}\psi) = -i\hbar[(\nabla\cdot\mathbf{p})\psi + \mathbf{p}\cdot\nabla\psi]\\ &= -i\hbar[0+\mathbf{p}\cdot(i/\hbar)\mathbf{p}\psi] = \mathbf{p}\cdot\mathbf{p}\,\psi = p^2\psi.\end{aligned} \tag{1.158}$$

Therefore

$$(p^2)^{\text{op}} = \mathbf{p}^{\text{op}}\cdot\mathbf{p}^{\text{op}}. \tag{1.159}$$

Referring back to the time-independent Schrödinger equation (1.130), we see from Eqs. (1.157) and (1.159) that it can be written in the form

$$[(1/2m)\mathbf{p}^{\text{op}}\cdot\mathbf{p}^{\text{op}} + \mathscr{V}(\mathbf{r})]\phi(\mathbf{r}) = \mathscr{E}\phi(\mathbf{r}), \tag{1.160}$$

so that we can consider $\mathscr{H}$, defined as

$$\mathscr{H} = (1/2m)\mathbf{p}^{\text{op}}\cdot\mathbf{p}^{\text{op}} + \mathscr{V}(\mathbf{r}) = -(\hbar^2/2m)\,\nabla^2 + \mathscr{V}(\mathbf{r}), \tag{1.161}$$

to be the energy operator $\mathscr{E}^{\text{op}}$. The operator $\mathscr{H}$ is called the **Hamiltonian operator** in quantum mechanics, and the result of operating on the stationary-state wavefunction $\phi(\mathbf{r})$ with $\mathscr{H}$ is to generate the product of $\{[(p_x^2+p_y^2+p_z^2)/2m]+\mathscr{V}(\mathbf{r})\}$ and $\phi(\mathbf{r})$. The quantity

$$(p_x^2+p_y^2+p_z^2)/2m + \mathscr{V}(\mathbf{r}) = \mathscr{E} \tag{1.162}$$

is nothing more than the total energy of the particle, which is designated as the *classical Hamiltonian* H_c for the one-particle system in question. The total energy is called the "Hamiltonian" in classical mechanics, since it plays a central role in the formulation known as "Hamilton's equations of motion." The analogy between quantum mechanics and classical mechanics is perhaps best seen by means of Hamilton's equations [see, e.g., Ikenberry (1962)].

Using Eq. (1.161), we can write the *time-dependent Schrödinger equation* as

$$\mathscr{H}\psi = i\hbar\, \partial\psi/\partial t, \tag{1.163}$$

and write the *time-independent Schrödinger equation* as

$$\mathscr{H}\phi = \mathscr{E}\phi. \tag{1.164}$$

The *Hamiltonian operator*

$$\mathscr{H} = (1/2m)(\mathbf{p}^{\text{op}} \cdot \mathbf{p}^{\text{op}}) + \mathscr{V}(\mathbf{r}) \tag{1.165}$$

can be considered to be obtained from the *classical Hamiltonian* H_c,

$$H_c = (1/2m)(\mathbf{p} \cdot \mathbf{p}) + \mathscr{V}(\mathbf{r}), \tag{1.166}$$

by replacing the *classical momentum* $\mathbf{p}$ by the *momentum operator* $\mathbf{p}^{\text{op}}$ given by Eq. (1.154), which in turn corresponds to the replacement of the classical scalar quantity $p^2 = \mathbf{p} \cdot \mathbf{p}$ by the operator $-\hbar^2 \nabla^2$. Since the position coordinate $\mathbf{r}$ remains unchanged in the transformation, we can say that $\mathbf{r}$ and $\mathbf{r}^{\text{op}}$ are the same in the classical and operator forms of the Hamiltonian for the equations given above. That is,

$$\mathbf{r}^{\text{op}} = \mathbf{r} \qquad \textbf{(position operator)}. \tag{1.167}$$

The operators given in this section are said to be those appropriate for the *position representation*. Because of the symmetry of the plane-wave function (1.137) in position $\mathbf{r}$ and momentum $\mathbf{p}$, an alternate formulation (known as the *momentum representation*) is possible in which the momentum operator is a multiplicative factor and the position operator is a differential operator involving derivatives with respect to p_x, p_y, and p_z [see, e.g., Ikenberry (1962)]. The following prescription is therefore given for transforming the classical Hamiltonian into the operator form of the Hamiltonian needed to formulate the time-dependent and time-independent Schrödinger equations (1.163) and (1.164): *Replace the momentum and position coordinates in the classical Hamiltonian by the operator equivalents.*

PROJECT 1.6 Angular Momentum Operators

1. Using the definition

$$\mathbf{L} = \mathbf{r} \times \mathbf{p} = \begin{vmatrix} \hat{\mathbf{x}} & \hat{\mathbf{y}} & \hat{\mathbf{z}} \\ x & y & z \\ p_x & p_y & p_z \end{vmatrix}$$

$$= \hat{\mathbf{x}}(yp_z - zp_y) + \hat{\mathbf{y}}(zp_x - xp_z) + \hat{\mathbf{z}}(xp_y - yp_x) = \hat{\mathbf{x}}L_x + \hat{\mathbf{y}}L_y + \hat{\mathbf{z}}L_z$$

for the angular momentum $\mathbf{L}$ and its vector components L_x, L_y, L_z in classical mechanics, construct quantum mechanical operators $\mathscr{L}^{\text{op}}$, $\mathscr{L}_x^{\text{op}}$, $\mathscr{L}_y^{\text{op}}$, $\mathscr{L}_z^{\text{op}}$ for these physical observables. (*Hint*: Use the prescription of substituting operator forms such as x^{op} and p_x^{op} for the corresponding classical quantities x and p_x.)

2. Can you construct a quantum mechanical operator for the square of the total angular momentum, $(\mathscr{L}^{\text{op}})^2$? (*Hint*: Classically, $L^2 = L_x^2 + L_y^2 + L_z^2$.)

7 Wave-Packet Solutions and the Uncertainty Relation

7.1 Linearity and Superposition

In our treatment of the wave equation (1.3) for the propagation of electromagnetic waves in free space, we discussed the fact that an arbitrary superposition of solutions corresponding to waves of various frequencies and wavelengths also represents a solution because of the property of linearity of the wave equation. The Schrödinger equation is likewise linear, so that the superposition principle holds for this equation also. We know from experience in solving differential equations that we can expect to obtain complete sets of eigenfunctions as solutions to the time-independent Schrödinger equation in regions where $\mathscr{V}(\mathbf{r}) \neq$ const. in the same way in which the complete set of plane waves constitute solutions to the Schrödinger equation whenever $\mathscr{V} \neq \mathscr{V}(\mathbf{r})$. Each of these eigenfunctions, when multiplied by the appropriate time factor $\exp[-(i/\hbar)\mathscr{E}t]$, represents a perfectly valid solution $\psi(\mathbf{r}, t)$ to the Schrödinger equation. Furthermore, each such solution will have a probability density $\psi^*\psi$ which is time independent. As an alternative, we know from our earlier treatment of Fourier series and Fourier integrals (see §4) that any function satisfying the Dirichlet conditions can be expanded in a Fourier series or a Fourier integral, so we expect that linear combinations of plane waves can provide a representation of each of these solutions (or any linear combination of these solutions) at any given instant. Thus, in accordance with Eqs. (1.65) and (1.66) we can write at $t = 0$,

$$\psi(\mathbf{r}, 0) = \left(\frac{1}{2\pi}\right)^{3/2} \int \chi(\mathbf{k})\, e^{i\mathbf{k}\cdot\mathbf{r}}\, d\mathbf{k}, \tag{1.168}$$

$$\chi(\mathbf{k}) = \left(\frac{1}{2\pi}\right)^{3/2} \int \psi(\mathbf{r}, 0)\, e^{-i\mathbf{k}\cdot\mathbf{r}}\, d\mathbf{r}. \tag{1.169}$$

7.2 Example of Plane-Wave Superposition and the Uncertainty Relation

Let us consider the simple one-dimensional example sketched in Fig. 1.27 for which $\chi(k)$, called the probability amplitude in k space, has values $[2(\delta k)]^{-1/2}$ over the domain $k_0 - \delta k \leqslant k \leqslant k_0 + \delta k$ and is zero outside this interval. The factor $[2(\delta k)]^{-1/2}$ is chosen to give normalization of $\chi(k)$ over the chosen interval, as can be verified by direct integration of $|\chi|^2$ over this interval. The domain in k corresponds to a spread in wave vector of $2(\delta k)$ and a corresponding spread (or uncertainty) in momentum of $2\hbar(\delta k)$. Then using Eq. (1.56) we obtain

$$\begin{aligned}\psi(x, 0) &= (2\pi)^{-1/2} \int_{-\infty}^{\infty} \chi(k)\, e^{ikx}\, dk = [4\pi(\delta k)]^{-1/2} \int_{k_0 - \delta k}^{k_0 + \delta k} e^{ikx}\, dk \\ &= [4\pi(\delta k)]^{-1/2} (ix)^{-1} \left[e^{i(k_0 + \delta k)x} - e^{i(k_0 - \delta k)x}\right] \\ &= [\pi(\delta k)]^{-1/2} x^{-1} e^{ik_0 x} \sin[(\delta k)x].\end{aligned} \tag{1.170}$$

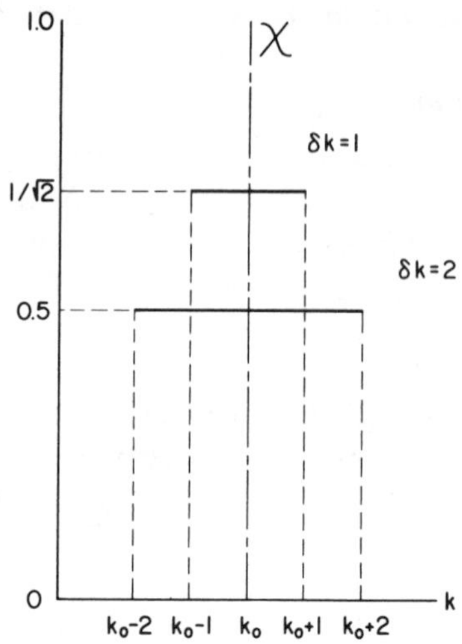

Fig. 1.27 Wave packets as a function of wave vector k which can be represented by step functions in wave-vector space and in momentum space $p = \hbar k$.

It is especially noteworthy that a specification of $\chi(k)$, constituting the extent of our *statistical knowledge* of the wavelength of the particle through the general wave relation $k = 2\pi/\lambda$, allows through the Fourier integral formulation a deduction of $\psi(x, 0)$ containing our entire *statistical knowledge* of the position of the particle. Because the wavelength λ is intimately related to the momentum p through the de Broglie relation $\lambda = h/p$, we conclude that the complete specification of our knowledge of *either* the particle's momentum *or* the particle's position is entirely sufficient to enable us to obtain the unspecified complementary member of this pair of physical observables.

The *probability density* corresponding to the wave function (1.170) is

$$\psi(x, 0)^* \psi(x, 0) = [\pi(\delta k)]^{-1} x^{-2} \sin^2[(\delta k)x]. \tag{1.171}$$

Both the value of the constant factor for χ and the symmetrical form of the Fourier transform were chosen to effect proper normalization of $\chi(k)$ and $\psi(x, 0)$: The probability of finding the particle somewhere in all of space with some zero or nonzero momentum value is unity.

The probability density (1.171) in real space is plotted in Fig. 1.28 for the case in which $\delta k = 1$; it is plotted in Fig. 1.29 for the case in which $\delta k = 2$. It can be noted that the *width* of the central maximum in $\psi(x, 0)$ is roughly twice as large for $\delta k = 1$ as it is for $\delta k = 2$, but the central maximum is only one-half as *high* for the $\delta k = 1$ case. Furthermore, if we estimate the distance to half the peak value in Fig. 1.28 as approximately 2 units in distance, where $\delta k = 1$ in corresponding reciprocal distance units, then for this example we can say that $\delta x \; \delta k \approx 2$. Essentially the same estimate holds for Fig. 1.29 because of the inverse relationship already noted above between δx and δk. This is the really remarkable thing to note, namely, that to narrow the distribution in δx we are

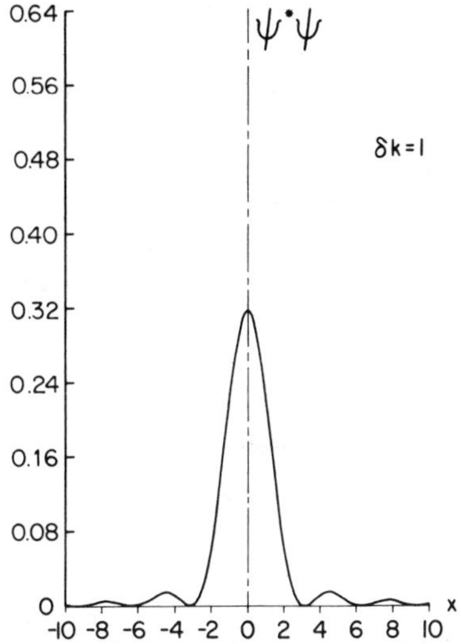

Fig. 1.28 Wave packet in real space corresponding to the narrow high ($\delta k = 1$) step-function wave packet in wave-vector space illustrated in Fig. 1.27. [This result is obtained by direct numerical evaluation of Eq. (1.170).]

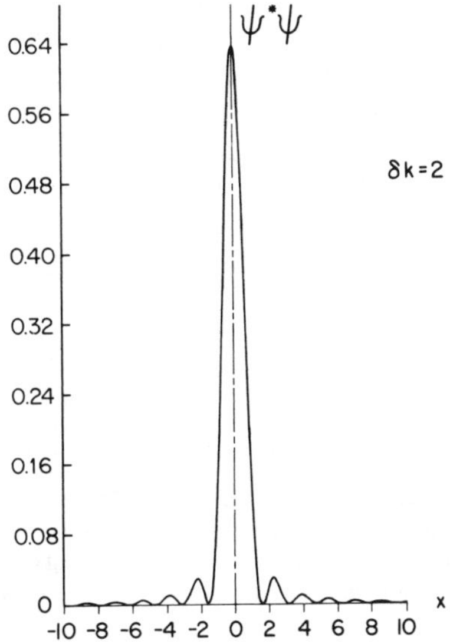

Fig. 1.29 Wave packet in real space corresponding to the broad low ($\delta k = 2$) step-function wave packet in wave-vector space illustrated in Fig. 1.27. [This result, which is likewise obtained by a direct numerical evaluation of Eq. (1.170), is to be compared with Fig. 1.28.]

required to broaden the distribution in δk, and vice versa. For this example, at least, it is conservative to conclude that

$$\delta x \, \delta k \gtrsim 1, \tag{1.172}$$

where δx and δk represent the widths of the distributions in position and k value, respectively. Although Eq. (1.172) has been deduced for a particular distribution $\chi(k)$, it is found to be generally true for all functional forms of $\chi(k)$ which satisfy the Dirichlet conditions. Since the quantity δk corresponds to a spread in momentum for the component waves of amount $\hbar\,\delta k$ on either side of the average value $p_0 = \hbar k_0$ (corresponding to a spread or uncertainty Δp in our *knowledge* of the particle's momentum), and since δx represents a spatial width of the packet (constituting an *uncertainty* Δx in our *knowledge* of the particle's position), then it can be concluded that

$$\Delta x\ \Delta p = \hbar\ \Delta x\ \Delta k \gtrsim \hbar, \tag{1.173}$$

which is the essence of the *position–momentum form of the Heisenberg uncertainty relation.* We have employed conventional notation from the standpoint that we have used the symbol Δ to express uncertainty. The symbol δ was employed to denote the *spread* in the values of x or k for the plane waves making up the packet. In summary, it can be concluded that an increased spread in *momentum* for the packet allows a greater spatial localization of the packet within the immediate neighborhood of the position of the maximum value of the wavepacket. Because $|\psi(x, 0)|^2$ represents the real-space probability density for the particle, the greater localization of the central maximum in the neighborhood of the origin corresponds to an increase in the probability that the particle will be found near the point of the maximum. This of course is a qualitative statement of the content of the Heisenberg uncertainty relation (1.173), namely, that *an increase in our knowledge of the position of the particle requires a corresponding decrease in our knowledge of the momentum of the particle, and vice versa.* For an alternate form of the uncertainty relation and its development, see §7.8 of this chapter. Also given there are alternate developments of the two forms (viz, position–momentum and energy–time forms) of the uncertainty relation together with clear statements and examples of the operational meaning and use of the relations.

PROJECT 1.7 The Uncertainty Relation: When Is It Important?

Plot a family of curves for objects of different mass (including electrons, protons, helium atoms, uranium atoms, baseballs, and elephants, among others); each curve being a log–log plot of the minimum uncertainty in position of the object versus the uncertainty in velocity of the object. Then translate this information into common-sense conclusions by comparing the uncertainties to typical sizes and speeds of these objects.

7.3 Philosophical Implications of Quantum-Mechanical Indeterminism

One may dislike the quantum mechanical approach for its lack of determinism, but nevertheless it remains our most accurate description of the way in which particles act in the limit of extremely small mass. That we do not understand why particles should act in a wavelike fashion clouds our understanding; the cloudiness is enhanced because particles large enough to be observed with our eyes have masses so large that quantum effects do not influence appreciably their location or future trajectories. This, in fact, is the critical impediment to our understanding, namely, that physical entities which have extremely small (but nonzero) masses do not always behave according to our mental definition and mental image of a particle, which is derived from our sense perception of the motion and interaction of large masses. The discipline of quantum mechanics does not appear to provide a complete description of nature because we do not like to admit that our ultimate knowledge concerning the position, momentum, and future trajectory of a given particle is only of a statistical nature, and therefore incomplete insofar as the exact motion of any particular particle in question is concerned. For these reasons it is just as well to consider quantum mechanics to be merely an elegant and accurate computational tool. It gives us excellent statistical information but does not provide us with a deterministic understanding of the universe.

The nodes and oscillations (wiggles) in the probability function in real space illustrated in Figs. 1.28 and 1.29 may also be found to be intellectually bothersome, since we generally prefer to consider a particle as an entity. These oscillations are due to the sharp cutoff of the function of k (cf. Fig. 1.27) which we happened to choose. Analogous oscillations are found in optical diffraction patterns, and indeed such oscillations represent the most characteristic feature of such patterns. An optical diffraction pattern is produced by passing a wave through an aperture which provides a sharp spatial cutoff of a portion of the beam wavefront, thereby producing spatial oscillations in the electric field intensity in the region behind the aperture. In the present case, we may eliminate the oscillations in the wave packet by considering a smoother function in k space; for example, the choice of a one-dimensional Gaussian function $\exp(-\Lambda^2k^2)$ of width Λ with an appropriate normalization factor leads to a Gaussian function in real space, so that there are no oscillations at all. The function looks something like the central maximum in Fig. 1.28. There are no nodes, but instead the function approaches zero rapidly in an asymptotic manner as $x \to \pm\infty$. Whether or not we have real-space oscillations will therefore depend upon the momentum (or **k**-vector) distribution for our physical system.

7.4 Time Dependence of Wave Packets in a Homogeneous Medium

If we now return to Eq. (1.168) and add the usual $\exp(-i\omega t)$ time dependence to each component wave in this general three-dimensional relation, and furthermore assume (perhaps naively) that each component wave propagates independently of the other component waves, we obtain

$$\psi(\mathbf{r}, t) = \left(\frac{1}{2\pi}\right)^{3/2} \int \chi(\mathbf{k})\, e^{i(\mathbf{k}\cdot\mathbf{r} - \omega t)}\, d\mathbf{k} \qquad (1.174)$$

for the time-dependence of the linear combination of plane waves. We must, of course, be careful to use the ω value appropriate for each k vector in question, since $\omega = \omega(\mathbf{k})$. Such linear superpositions of plane waves are called *wave packets*. The function $\psi(\mathbf{r}, t)$ is characterized by a degree of localization in space achieved at the expense of introducing components with a range (or *spectrum*) of $\mathbf{k}$ values; the packet given by Eq. (1.174) satisfies the time-dependent Schrödinger equation (1.122) for free particles, as can be seen by direct substitution. The phase velocity v_p of any individual component of the packet is $\omega/|\mathbf{k}|$, in accordance with Eq. (1.75) and our general discussion of plane waves in §2. For the case of free particles, with energy $\mathscr{E}$ and momentum $\mathbf{p}$,

$$\hbar\omega = \mathscr{E} = p^2/2m = \hbar^2 k^2/2m, \tag{1.175}$$

so that

$$\omega = \hbar k^2/2m. \tag{1.176}$$

This leads to

$$v_p = \omega/|\mathbf{k}| = \hbar|\mathbf{k}|/2m \qquad \textbf{(phase velocity for matter waves)}. \tag{1.177}$$

Therefore the phase velocity for the component waves depends on the k value, which results in some *dispersion* of the wave packet with time. That is, the probability density $\psi^*\psi$ for this superposition of plane waves will be *time dependent*. In addition, the frequencies of the component waves will differ from one another since $\omega = \omega(\mathbf{k})$, so in general the frequency of any given component cannot be exactly $\mathscr{E}/\hbar$ if $\mathscr{E}$ is the total energy of the single particle under consideration. Thus we are forced to conclude that even though a superposition of plane waves can represent a general solution to the Schrödinger equation at any given instant, this wave packet will not be a *stationary-state solution* since its probability density will change in appearance as time progresses. This, of course, is different from the situation for the group of electromagnetic waves in free space which we treated in §2. The wave-packet description of matter waves is nevertheless found to be useful in many problems as long as we consider packets which do not spread (or disperse) appreciably over the time period under consideration. In this manner one can even treat the problem of a localized particle colliding with a potential barrier or a potential well [Messiah (1965)].

The relation $v_p = \hbar|\mathbf{k}|/2m$ leads to another interesting conclusion by considering the momentum $\mathbf{p}$ and velocity $\mathbf{v}$,

$$\mathbf{p} = m\mathbf{v} = \hbar\mathbf{k}^{(\text{peak})}, \tag{1.178}$$

so

$$v_p = \hbar|\mathbf{k}|/2m = |\mathbf{p}|/2m = \tfrac{1}{2}|\mathbf{v}|. \tag{1.179}$$

Thus the phase velocity of the wave at the peak value in the packet associated with a particle moving with velocity $\mathbf{v}$ is only one-half the velocity of the particle. If the wave packet is to provide any representation of the motion of the particle whatsoever, in accordance with the correspondence principle, then the group velocity $\mathbf{v}_g$ of the packet must be equal to the particle velocity $\mathbf{v}$. We thus see that matter waves associated with free particles will have a phase velocity of the order of one-half of the group velocity of the packet. This means that the individual waves move more slowly than the packet and therefore pass back through the packet as it advances. This is not an uncommon phenomenon in wave motion. For example, the individual wavelets on an ocean wave can be noted to move with a velocity different from that of the ocean wave itself.

Because a packet includes plane waves with various ω and $\mathbf{k}$ values, each traveling with its characteristic phase velocity $v_p = \omega/|\mathbf{k}| = \hbar|\mathbf{k}|/2m$, we must have some scheme for deducing the average velocity of the packet. This average velocity which we call the group velocity must correspond to the classical velocity of the free particle, as mentioned previously. (The velocity $\mathbf{p}/m = \hbar|\mathbf{k}|/m$ used above for the group velocity is of course perfectly acceptable if the packet consists of only a single plane wave of wave vector $\mathbf{k}$ and frequency ω, corresponding to the momentum $\mathbf{p} = \hbar\mathbf{k}$ and energy $\mathscr{E} = \hbar\omega$.)

Let us assume that we know the dispersion relation $\omega(k)$ for some particular one-dimensional case of wave motion, or else have measured the properties of the physical system in question sufficiently to be able to write the first terms in a Taylor series approximation for $\omega(k)$. Our treatment will not be restricted to particles, but will apply also, for example, to the case of electromagnetic waves propagating through a homogeneous dispersive or nondispersive medium. Let us assume, however, that the values of k are restricted to a narrow band centered about k_0, as in Fig. 1.27. Then we can make a Taylor series expansion for $\omega(k)$,

$$\omega(k) \simeq \omega(k_0) + \left(\frac{d\omega(k)}{dk}\right)_{k=k_0}(k-k_0) + \frac{1}{2!}\left(\frac{d^2\omega(k)}{dk^2}\right)_{k=k_0}(k-k_0)^2 + \cdots, \tag{1.180}$$

keeping only the lowest order terms. The validity of this approximation depends, as usual, upon the size of the higher order terms relative to the linear term in $(k - k_0)$, and the approximation improves with decreasing magnitude of $(k - k_0)$. The value of k at any point in the packet can be expressed as

$$k = k_0 + (k - k_0), \tag{1.181}$$

and we restrict $k - k_0$ to small values relative to k_0. We do this out of necessity so that our Taylor series expansion will be valid. In practice, this does not restrict the limits on the integral in Eq. (1.182) since we choose $\chi(k)$ to be large only in the region of k around $k = k_0$. What is the physical meaning of expanding $\omega(k)$ about the point k_0? We recognize that $\omega(k)$ may vary markedly over the complete domain $-\infty < k < \infty$ such that values of $\omega(k)$ for k values far removed from k_0 may be quite different from the value $\omega(k_0)$; furthermore there are situations in which $\omega(k)$ can vary in peculiar ways in the neighborhood of particular k values. The rationale behind the Taylor series expansion is the assumption that the multiplicative factor $\chi(k)$ in the integral of Eq. (1.174) will be small over most of the domain of k, such that the only values of $\omega(k)$ which affect the value of the integral appreciably are those corresponding to the limited domain of k where $\chi(k)$ is relatively large. In cases where $\chi(k)$ is appreciable only over a very limited range of k values, there is a reasonable hope that a Taylor series can approximate $\omega(k)$ satisfactorily over the domain of k for which $\chi(k)$ is relatively large. The use of a Taylor series approximation for $\omega(k)$ in the region around k_0 also involves the assumption that $\omega(k)$ is reasonably well behaved around k_0, thus precluding regions of k near discontinuities, singularities, and other critical points where small changes in k lead to exceptionally large variations in $\omega(k)$.

Substituting the first two terms of the relation (1.180) into the packet

$$\psi(x, t) = \left(\frac{1}{2\pi}\right)^{1/2} \int_{-\infty}^{\infty} \chi(k) e^{i(kx - \omega t)}\, dk \tag{1.182}$$

gives

$$\psi(x, t) \simeq \left(\frac{1}{2\pi}\right)^{1/2} \exp\{i[\omega'(k_0)k_0 - \omega(k_0)]t\}$$
$$\times \int_{-\infty}^{\infty} \chi(k) \exp\left\{ik\left[x - \left(\frac{d\omega(k)}{dk}\right)_{k=k_0} t\right]\right\} dk \quad \textbf{(wavepacket)} \tag{1.183}$$

where $\omega'(k_0)$ is merely an abbreviation for $(d\omega/dk)_{k=k_0}$. At $t = 0$ the integral is a relatively slowly varying function of x whenever $k - k_0$ is restricted to small values (cf. Figs. 1.27–1.29). The exponential factor in front of the integral introduces rapid temporal modulations of the integral which, however, do not contribute in any way to the value of $\psi^*(x, t)\psi(x, t)$. (For example, $e^{i\theta} \times e^{-i\theta} = 1$ for any real value of θ, and in our example ω and k are real.) We assume that $\chi \geqslant 0$, since χ represents the amplitudes of the various component waves in our wave packet. Considering the integral as a function of x, it will be peaked in the neighborhood of the point

$$x_p = (d\omega(k)/dk)_{k=k_0} t \equiv \omega'(k_0)t, \tag{1.184}$$

because for this value the slowly oscillating exponential function of k in the integrand has a very long period in k so that the integral is essentially the value of the entire area under the curve given by the positive function $\chi(k)$. For other values of x, the oscillations of the real and imaginary parts of the exponential function in the integrand will introduce a partial cancellation so that the integral will be less than the area under the curve $\chi(k)$. Considering the velocity of the wave packet to be governed by the velocity of the peak, then the time derivative of x_p can be identified with the *group velocity* v_g of the packet. Hence

$$v_g = dx_p/dt = (d\omega(k)/dk)_{k=k_0}. \tag{1.185}$$

As a check on the validity of the relation (1.185), let us apply it to particles in free space for which the dispersion relation is $\omega(k) = \hbar k^2/2m$. We obtain immediately $v_g = \hbar k_0/m = p_0/m$, which is the classical velocity of a particle with a definite momentum p_0, in accord with our previous conclusions. As a second check, let us apply this relation to electromagnetic waves propagating in free space, for which the dispersion relation is $\omega(k) = ck$. We obtain immediately $v_g = c$, which is in complete accord with our previous conclusions in §2 that a packet of electromagnetic waves moves with a group velocity equal to the universal phase velocity c.

7.5 Group Velocity in Three Dimensions

The above relation for the group velocity is quite generally valid for one-dimensional systems; it is therefore worthwhile to extend it to the three-dimensional case. The group velocity must in this case be a vector quantity $\mathbf{v}_g$, with components given by the derivative of $\omega(\mathbf{k})$ with respect to k_x, k_y, and k_z, respectively. Thus

$$\mathbf{v}_g = [\nabla_{\mathbf{k}}\, \omega(\mathbf{k})]_{\mathbf{k}=\mathbf{k}_0} \qquad \textbf{(group velocity of a wave packet).} \qquad (1.186)$$

The relation $\mathscr{E} = \hbar\omega$ between total energy and frequency is valid for any dispersion relation $\omega(\mathbf{k})$, so an alternate expression for the group velocity is

$$\mathbf{v}_g = \hbar^{-1}(\nabla_{\mathbf{k}}\, \mathscr{E}(\mathbf{k}))_{\mathbf{k}=\mathbf{k}_0} \qquad \textbf{(group velocity of a particle).} \qquad (1.187)$$

7.6 Effects of Successive Terms in the Taylor Series Expansion of $\omega(k)$

We should ask ourselves what effects are introduced if higher derivatives in the Taylor series expansion for $\omega(k)$ are important. For free particles, $\omega = \hbar k^2/2m$, so that use of only the first two terms in the Taylor series approximation (1.180) results in the neglect of the second term

$$(1/2!)(d^2\omega/dk^2)|_{k=k_0}(k - k_0)^2 = \hbar(k - k_0)^2/2m \qquad (1.188)$$

with respect to the first term

$$(d\omega/dk)|_{k=k_0}(k - k_0) = \hbar k_0(k - k_0)/m. \qquad (1.189)$$

The ratio of the second term to the first term is $(k - k_0)/2k_0$, and thus it can be quite small if the spread in the values of k over the packet is small relative to the value of k at the peak of $\chi(k)$ for the packet, namely k_0. By referring to the general expression (1.182) for the one-dimensional wave packet, we see that addition of this second term leads to the exponential factor $\exp[-i\hbar(k - k_0)^2 t/2m]$ in the integrand. As time increases, this gives rise to a time-dependent broadening in the packet, which represents a type of dispersion of the wave packet as it travels through free space. Similar effects are found for electromagnetic waves traveling through solids or other *dispersive* media [Stratton (1941); Jackson (1962)] for which the refractive index n and therefore phase velocity c/n depend on the frequency.

If $\omega(k)$ contains terms which are higher order in $(k - k_0)$, this can cause a change in shape of the packet of matter waves with time as it travels through the medium in question. Of course, for free particles, $\omega = \hbar k^2/2m$, so the Taylor series expansion terminates with the second-order term. Our treatment starting with Eq. (1.180) has been quite general, since it is applicable for any dispersion relation. We attempt to extend our understanding of wave packets at this point by returning to the specific case of particles in free space, since we have a simple exact dispersion relation for this case. If we successively keep terms through zero, first, and second order in the Taylor series expansion of $\omega(k) = \hbar k^2/2m$, we obtain the approximate free-particle dispersion relations,

$$\omega(k) \simeq \hbar k_0^2/2m, \tag{1.190}$$

$$\omega(k) \simeq \hbar k_0^2/2m + (\hbar k_0/m)(k - k_0), \tag{1.191}$$

$$\omega(k) \simeq \hbar k_0^2/2m + (\hbar k_0/m)(k - k_0) + (\hbar/2m)(k - k_0)^2. \tag{1.192}$$

This last approximation is in fact the exact free-particle dispersion relation

$$\omega(k) = \hbar k^2/2m, \tag{1.193}$$

as can be seen readily by combining like terms. Let us examine the results of using each of these relations in the specific time-dependent wavepacket (1.174) illustrated in Figs. 1.27–1.29. We must be careful to choose a scale factor for χ to retain normalization of $\psi(\mathbf{r}, t)$. For the one-dimensional case, with the choice χ independent of k over $k_0 - \delta k \leqslant k \leqslant k_0 + \delta k$ and zero outside this range, as in Fig. 1.27,

$$\psi(x, t) = \left(\frac{1}{2\pi}\right)^{1/2} \chi(k_0) \int_{k_0 - \delta k}^{k_0 + \delta k} e^{i(kx - \omega t)}\, dk. \tag{1.194}$$

Using the zeroth-order dispersion relation (1.190), we obtain

$$\psi(x, t) = \left(\frac{1}{2\pi}\right)^{1/2} \chi(k_0)\, e^{-i\hbar k_0^2 t/2m} \int_{k_0 - \delta k}^{k_0 + \delta k} e^{ikx}\, dk. \tag{1.195}$$

Except for the modulating time-dependent phase factor $\exp(-i\hbar k_0^2 t/2m)$, which contributes nothing to $|\psi(x, t)|^2$, we have the same expression as Eq. (1.170). Therefore the value of $\chi(k_0)$ can again be chosen to have the value $[2(\delta k)]^{-1/2}$ for proper normalization. The position of the maximum of this packet in real space, as for Figs. 1.28 and 1.29, occurs at $x = 0$ independent of the time t. In addition, $|\psi(x, t)|^2$ is time independent. *By ignoring terms higher than zero order in the Taylor series expansion for the dispersion relation, we have thus neglected motion of the wave packet as well as any broadening of the wave packet.*

Considering next the first-order dispersion relation (1.191), we obtain

$$\psi(x, t) = \left(\frac{1}{2\pi}\right)^{1/2} \chi(k_0)\, e^{i\hbar k_0^2 t/2m} \int_{k_0 - \delta k}^{k_0 + \delta k} e^{ik[x - (\hbar k_0 t/m)]}\, dk. \tag{1.196}$$

Again we have a rapidly modulating time-dependent exponential factor in front of the integral which does not contribute to $|\psi(x, t)|^2$. In accordance with our previous discussion involving group velocity, we conclude that this packet will have a peak at the position x_p which changes with time according to the relation

$$x_p = \hbar k_0 t/m, \tag{1.197}$$

so the velocity v_g of the packet is

$$v_g = dx_p/dt = \hbar k_0/m. \tag{1.198}$$

This is satisfying because it agrees with our previous conclusions (1.178) and (1.179) regarding the particle velocity, and it also agrees with the general relation (1.185) as applied to our specific case of free particles. Using $v_g = \hbar k_0/m$ in Eq. (1.196) for $\psi(x, t)$ together with a comparison of the integral in Eq. (1.170) shows us that we have formally the same problem if we substitute $x - v_g t$ for x in Eq. (1.170). On this basis, our normalization factor $\chi(k_0)$ will again be $[2(\delta k)]^{-1/2}$, thus giving for this case

$$\psi(x, t) = \{[\pi(\delta k)]^{1/2}(x - v_g t)\}^{-1} e^{ik_0(x - v_p t)} \sin[(\delta k)(x - v_g t)], \tag{1.199}$$

where

$$v_p = \hbar k_0/2m \tag{1.200}$$

is the phase velocity given by Eq. (1.177) for this case. This leads to

$$|\psi(x, t)|^2 = \sin^2[(\delta k)(x - v_g t)]/[\pi(\delta k)(x - v_g t)^2]. \tag{1.201}$$

The packet moves with velocity v_g, and it has the same functional dependence on x around the point $x_p = \hbar k_0 t/m$ as the packets in Figs. 1.28 and 1.29 have around $x = 0$. *Within the limits of this first-order approximation, then, the packet propagates but does not broaden or change its shape as time progresses.*

We now investigate the effects of going to our second-order Taylor series approximation (1.192) for the dispersion relation, which is in fact the exact dispersion relation (1.193) for free particles. We obtain

$$\psi(x, t) = \left(\frac{1}{2\pi}\right)^{1/2} \chi(k_0, t) \int_{k_0 - \delta k}^{k_0 + \delta k} e^{i[kx - (\hbar k^2 t/2m)]}\, dk. \tag{1.202}$$

The argument of the exponential function is thus quadratic in k. Unfortunately the integration cannot be carried out immediately, so we are not able to examine an exact final result. It is intuitively clear that since each of the component waves over the interval $k_0 - \delta k \leqslant k \leqslant k_0 + \delta k$ has its individual phase velocity $v_p = \hbar k/2m$, the group of waves will tend to disperse as time elapses. This dispersion increases the width of the wave packet in real space. If a numerical integration is performed on the above expression for specific values of the parameters and for various times t, *a time-dependent broadening* should indeed be found to take place.

EXERCISE Carry out a numerical evaluation of the integral in Eq. (1.202) for a series of different times t in order to achieve an understanding of the time dependence of $\psi(x, t)$ and its magnitude.

7.7 Gaussian Wave Packets

One way to study the type of behavior expected with the exact dispersion relation is to replace the constant function $\chi(k_0)$ by a Gaussian function

$$\chi(k) \propto \exp[-\Lambda^2(k - k_0)^2], \tag{1.203}$$

where the width of the Gaussian determined by Λ is of the order of the width of the constant function $\chi(k_0)$, and some constant preexponential factor is chosen to maintain normalization. This functional form permits the range of integration to be extended over the interval $-\infty \leqslant k \leqslant \infty$ without any resulting divergences, since the Gaussian function falls rapidly to zero for $|k - k_0| > \Lambda^{-1}$. The argument of the resulting exponential function in the integrand can be expressed as a perfect square and a phase factor, with the result that we have a definite integral with the evaluation given in standard integral tables. We can thus obtain an explicit expression for $\psi(x, t)$ for this case. It is found to be a Gaussian function in position, and it travels with the group velocity $\hbar k/m$ and undergoes a broadening with time. We do not pursue the solution in detail here, since Gaussian wave packets are treated at length in many textbooks, so much so that we are sometimes led to believe that there is something more fundamental about a Gaussian function in quantum mechanics than is actually warranted by either basic theoretical considerations or experimental evidence. Instead we leave the mathematical treatment of the Gaussian wave packet as an exercise for the reader. The merits of the Gaussian wave packet are its smooth functional form (no nodes), its good convergence properties for integration over an infinite domain, and the relative simplicity with which the resulting definite integral can be obtained. It also has the unique property that its Fourier transform is likewise a Gaussian function, as mentioned above. However, the major conclusion of the treatment of the Gaussian wavepacket which is pertinent for our present development is that the quadratic term in k^2 (which is present in the exact dispersion relation for free particles) does indeed result in a broadening of the wave packet with time.

PROJECT 1.8 Gaussian Wave Packet

Work out the mathematical details of the Gaussian wave packet. (*Hint*: Read the qualitative description given in §7.7 and then carry out the details in the same manner as the extended treatment given in §7.2 for the alternate wave packet. In particular, obtain expressions for the probability density as a function of position and probability density as a function of momentum. Plot these quantities after evaluating the probability densities for a range of x values and k values, using reasonable values for the fixed parameters and a programmable calculator. Evaluate the uncertainty product $\Delta p\, \Delta x$ at $t = 0$ and for $t > 0$.)

7.8 Energy–Time Form of the Uncertainty Relation and Alternative Approaches

Now that we have discussed in some detail the width of wave packets in real space and the momentum spread due to the range of k values, the uncertainty relation $\Delta x\, \Delta p \gtrsim \hbar$ given by Eq. (1.173) takes on added meaning. The particle has a high statistical probability of being found at any point where $|\psi(x, t)|^2$ is large, so the position of the particle is uncertain roughly to the extent of Δx. Similarly, the momentum of the particle is uncertain to the extent Δp. When we also include the fact that the wave packet is traveling with velocity v_g, the time at which the particle described by the packet goes past a given point in space is uncertain by an amount Δt, where

$$\Delta t \simeq \Delta x / v_g. \tag{1.204}$$

Since $\mathscr{E} = p^2/2m$ for free particles, an uncertainty in momentum requires a corresponding uncertainty in energy,

$$\Delta\mathscr{E} = \Delta(p^2/2m) \simeq p(\Delta p)/m = v_g(\Delta p). \tag{1.205}$$

Solving for Δx and Δp in these last two relations and substituting into Eq. (1.173) yields the **energy–time form of the Heisenberg uncertainty relation**,

$$\Delta\mathscr{E}\ \Delta t \gtrsim \hbar. \tag{1.206}$$

Although we have deduced the relation (1.206) in only one way, there are in fact many other ways to obtain it. One of the most interesting approaches is to return to the basic consideration of waves and wave motion. To measure the temporal frequency ν (or corresponding angular frequency $\omega = 2\pi\nu$) of a sinusoidal wave we can perform an experiment with a fine stopwatch in which we count the number of intensity maxima which occur over a given time interval. Since the first maximum may occur just before, just after, or at some intermediate time point relative to the starting point ($t = 0$), and similarly for the last maximum relative to the stopping point ($t = t_s$), the number N of maxima corresponding to the time interval t_s will be (assuming the best possible measurement) equal to the number n which have been counted to within an uncertainty of 1 or 2 maxima. Thus the measured frequency $\nu_s = n/t_s$ will constitute an accurate representation of the true frequency $\nu = N/t_s$ to within some corresponding uncertainty $\Delta\nu$,

$$\nu = N/t_s = (n \pm 1)/t_s = \nu_s \pm (1/t_s) = \nu_s \pm \Delta\nu, \qquad \text{where} \quad \Delta\nu = 1/t_s.$$

The uncertainty in measured frequency thus decreases with an increase in the measuring time t_s. However, if we are to associate the probability of particle location with this particular wave, then the energy $\mathscr{E}$ of the particle is $h\nu = \hbar\omega$, and the time t_s of the measurement corresponds to the time interval during which we attempt to observe the particle. Since the amplitude of a pure sinusoidal wave is time independent, the probability of observing the particle over any fixed time interval containing many periods ($T = 1/\nu$) is essentially independent of time over the interval $(0, t_s)$. Thus an uncertainty in time Δt exists for particle observation which is equal to the time interval t_s chosen. Therefore $\Delta\nu \simeq 1/(\Delta t)$, or equivalently, $\Delta\nu\ \Delta t \simeq 1$. Recognizing that the uncertainty in frequency may indeed be somewhat greater than the minimum (e.g., suppose n has been miscounted), then we can write $\Delta\nu\ \Delta t \gtrsim 1$. Multiplying through by h gives once again an *energy–time form* of the *Heisenberg uncertainty relation* consistent with Eq. (1.206), $\Delta\mathscr{E}\ \Delta t = (h\,\Delta\nu)\Delta t \gtrsim h$. The time uncertainty can be given physical meaning by considering a measurement to determine the quantum state of a particle. If the average lifetime of the particle state is t_l, for example, then the particle's energy (and thus frequency) measurement can be carried out over this time interval, at least from an averaged or statistical viewpoint. However, the particle may well undergo a transition to another state at any instant during this time interval, thus corresponding to an uncertainty $\Delta t \simeq t_l$. If the energy of the state is $\mathscr{E}$, there will be an uncertainty in energy $\Delta\mathscr{E}$ in accordance with Eq. (1.206), where Δt can be replaced by the average lifetime t_l of the state in question.

A variation of the above-described counting approach can be used to deduce the position–momentum form of the uncertainty relation. If the wavelength λ is to be determined by counting the crests within a spatial interval L, then the experimental number n gives an experimental value λ_{exp}, $\lambda_{\text{exp}} = L/n$, which approximates the true wavelength λ to the extent to which n is a good measure of the number N of wavelengths associated with the length L. It is readily seen from the argument previously used that $n = N \pm 1$, so that the momentum $p = h/\lambda$ is measured (at best) to some limiting degree of precision,

$$p_{\text{exp}} = h/\lambda_{\text{exp}} = h(n/L) = h[(N \pm 1)/L] = (h/\lambda) \pm (h/L) = p \pm \Delta p$$

where $\Delta p \gtrsim h/L$. However, the particle may have any position over the distance L, so that the uncertainty in position Δx is L, and we obtain $\Delta x\, \Delta p \gtrsim h$, which is certainly consistent with the inequality (1.173).

An alternate approach to the position–momentum form of the uncertainty relation can be deduced directly from our Fourier integral development. Let us consider a uniform amplitude wave $y = A \sin kx$ over the spatial interval L. Expansion of this truncated wave (i.e., it has zero amplitude outside of the interval in question) leads to a distribution of Fourier components, with the width of the distribution $g(k)$ increasing proportionally to the reciprocal of the spatial interval L such that $L\, \Delta k \gtrsim 1$. Assigning the wave the particular role of describing the probability amplitude of the particle then leads again to the uncertainty Δx in particle position being of the order of L. Employing the de Broglie relation $p = h/\lambda = h(2\pi/k)^{-1} = \hbar k$ leads to $\Delta k = \Delta p/\hbar$, so that the above relation is converted to the position–momentum form of the uncertainty relation, $\Delta x\, \Delta p \gtrsim \hbar$. Alternatively, a periodic function in time can be Fourier expanded over a finite time interval t_{M}, and it is found that the width $\Delta\omega$ of the frequency distribution $g(\omega)$ needed to represent the truncated wave is inversely related to the time interval t_{M} such that $t_{\text{M}}\, \Delta\omega \gtrsim 1$. Associating this wave with a particle through $\omega = \mathscr{E}/\hbar$, with $\Delta t = t_{\text{M}}$ representing an uncertainty in time at which the particle is observed, yields the energy–time form of the uncertainty relation $\Delta\mathscr{E}\, \Delta t \gtrsim \hbar$. It is left as exercises for the reader to carry out the indicated spatial and temporal Fourier integral developments required to complete these latter approaches to obtain the two forms of the uncertainty relation.

Yet another approach to the uncertainty relation can be developed from the experimental viewpoint of a position measurement of the location of a point mass by means of a microscope using electromagnetic waves of wavelength λ_{phot} and frequency $\nu_{\text{phot}} = c/\lambda_{\text{phot}}$. It is a fact that the resolution of a microscope is limited by the wavelength of the light used for the measurement such that the uncertainty Δx_{mass} of the position measurement will be of the order of (or greater than) the wavelength, $\Delta x_{\text{mass}} \gtrsim \lambda_{\text{phot}}$. However, photons have momentum $p = h/\lambda$, and the conservation of momentum when the measuring photon scatters off of the point mass will cause an uncertainty in the final momentum of the point mass of this order, namely $\Delta p_{\text{mass}} \approx h/\lambda_{\text{phot}}$. Therefore we must conclude that, following the measurement,

$$\Delta x_{\text{mass}}\ \Delta p_{\text{mass}} \gtrsim (\lambda_{\text{phot}})(h/\lambda_{\text{phot}}) = h,$$

which once again is certainly consistent with the position–momentum form (1.173) of the Heisenberg uncertainty relation.

A rigorous development of the uncertainty relation shows that the *minimum* value of $\Delta x\ \Delta p$ or of $\Delta \mathscr{E}\ \Delta t$ is of the order of $\frac{1}{2}\hbar$, while there is no particular limit to the maximum value. One reason for this is that Δx (and thus Δt) usually increases with time as the packet undergoes dispersion. (In rigorous treatments of the Gaussian wave packet, the value of δx is evaluated explicitly as a function of time. If our knowledge of p is not considered to improve with time, then the product $\Delta x\ \Delta p$ will certainly increase with time.) Because $\frac{1}{2}\hbar$ therefore represents a *minimum* value for the uncertainty product, the uncertainty relation must always be considered to be an inequality. The factor of $\frac{1}{2}$ occurring in the minimum uncertainty value $\frac{1}{2}\hbar$ allows Eqs. (1.173) and (1.206) to be restated in the more rigorous forms given in Eq. (1.207),

$$\Delta x\ \Delta p \geqslant \tfrac{1}{2}\hbar \qquad \text{and} \qquad \Delta \mathscr{E}\ \Delta t \geqslant \tfrac{1}{2}\hbar$$

(rigorous forms of the Heisenberg uncertainty relations). (1.207)

PROJECT 1.9 Uncertainty Relations

Carry out spatial and temporal Fourier integral developments of the position–momentum and the energy–time uncertainty relations.

7.9 Wave–Packet Physics and Philosophy

The wave packet is one way of representing the wave function $\psi(x, t)$ for a given physical system. The interpretation of the wave function ψ is given by *Born's postulate*: *The probability of finding the particle described by a wave function* ψ *in a small (differential) region surrounding the position x is proportional to* $|\psi|^2 \equiv \psi^*\psi$, *the square of the magnitude of the wave function, evaluated at that position x*. This postulate relating to particles is analogous to the situation in electromagnetic wave propagation wherein the probability of finding a photon in the neighborhood of a point in space is proportional to the square of the electromagnetic wave intensity at that point. (In addition, in classical electromagnetic theory the square of the electromagnetic wave intensity is interpreted as a measure of the electromagnetic energy density at the point in question.)

Although the time dependence of the wave packet seems to indicate that it is a somewhat dubious theoretical tool for describing the location of a particle, it has the merit of furnishing us with a way to introduce at least some degree of localization into the infinitely extended plane-wave solutions of the Schrödinger equation. The spreading of the packet with time merely reflects that there is some uncertainty in initial momentum of the particle corresponding to the spread in k values, which, with progressing time, results in greater uncertainty in the position of the particle. The particle itself should not be considered to disperse with time; it is only our *knowledge* of its location which decreases with time. This is the intellectually frustrating aspect of quantum mechanics, namely, that our knowledge is restricted to that which is contained within the statistical function $\psi(\mathbf{r}, t)$. On the other hand, the intellectually satisfying aspect of quantum mechanics is that the predictions derived from the statistical function $\psi(\mathbf{r}, t)$ are in accord with experiment. In this all-important respect, quantum mechanics is superior to classical mechanics.

8 Expectation Values for Quantum-Mechanical Operators

8.1 Significance and Use of Wave Functions for Computing Statistical Averages

One of the important tenets used in our development of the Schrödinger equation in §6 was the assumption that the probability density for finding a particle at position $\mathbf{r}$ and at time t is given by $|\psi(\mathbf{r}, t)|^2 = \psi^*(\mathbf{r}, t)\psi(\mathbf{r}, t)$. The Fourier transform of $\psi(\mathbf{r}, 0)$ given by Eq. (1.169) gives the corresponding $\mathbf{k}$-vector distribution for the particle, and therefore the *momentum probability amplitude* distribution, since $\mathbf{p} = \hbar\mathbf{k}$. In essence, the physical state of a particle at time t is described as completely as is possible within the formalism of quantum mechanics by the wave function $\psi(\mathbf{r}, t)$! Moreover, the time development of this wave function is determined by the time-dependent Schrödinger equation (1.124); this equation can therefore be considered to be the fundamental *equation of motion* for the quantum-mechanical system [Mandl (1957)].

Let us presume that $\psi(\mathbf{r}, t)$ has been scaled by a constant factor of the proper magnitude for normalization. This means simply that the probability of finding the particle somewhere in space at any given time must be unity,

$$\int_{\Omega_r} \psi^*(\mathbf{r}, t)\psi(\mathbf{r}, t)\, d\mathbf{r} = 1 \qquad \textbf{(normalization condition)}. \tag{1.208}$$

If the integral in Eq. (1.208) has a value M that is not unity, then ψ must generally be replaced by the corresponding normalized wave function given by

$$\psi_{\mathrm{N}} = (M)^{-1/2}\, e^{i\delta}\psi(\mathbf{r}, t), \tag{1.209}$$

where δ can be any arbitrary real number called the *phase factor*. Occasionally the wave function cannot be normalized over the infinity of all space. Such is the case for a completely nonlocalized wave function such as the plane wave $\exp[i(\mathbf{k} \cdot \mathbf{r} - \omega t)]$. In this situation, the wave function can still be normalized over a finite region of space, such as a bounded universe or a 1-cm^3 block of metal. Henceforth we will consider that the wave function ψ has been appropriately normalized.

Since $\psi^*\psi$ denotes the *normalized particle probability density*, the *statistical average value* $\langle f(\mathbf{r})\rangle$ of a function $f(\mathbf{r})$ dependent upon the position $\mathbf{r}$ of the particle is given by

$$\langle f(\mathbf{r})\rangle = \int_{\Omega_r} \psi^*(\mathbf{r}, t)\psi(\mathbf{r}, t) f(\mathbf{r})\, d\mathbf{r} = \int_{\Omega_r} \psi^*(\mathbf{r}, t) f(\mathbf{r})\psi(\mathbf{r}, t)\, d\mathbf{r}. \tag{1.210}$$

This quantity $\langle f(\mathbf{r})\rangle$ is called in quantum mechanics the *expectation value* of f. A particular application would be the choice $f(\mathbf{r}) = \mathbf{r}$, in which case we would obtain the "expected" or "average" value for the position of the particle.

Suppose that we attempt to construct a quantum-mechanical *operator* f^{op} which is analogous to the classical function $f(\mathbf{r})$. We do this by the usual prescription of replacing $\mathbf{r}$ in $f(\mathbf{r})$ by $\mathbf{r}^{\text{op}}$. According to Eq. (1.167), however, $\mathbf{r}^{\text{op}} = \mathbf{r}$, so $f^{\text{op}} = f(\mathbf{r})$. The above equation (1.210) for the expectation value of $f(\mathbf{r})$ can thus be written in terms of the *expectation value for a quantum-mechanical operator*,

$$\langle f^{\text{op}}\rangle = \int_{\Omega_r} \psi^*(\mathbf{r}, t) f^{\text{op}}\psi(\mathbf{r}, t)\, d\mathbf{r}, \tag{1.211}$$

where f^{op} can be considered to *operate* on $\psi(\mathbf{r}, t)$.

It would be very useful if this procedure could be extended to all operators, instead of being restricted to multiplicative operators involving position only. It is quite amazing that this particular form (1.211) has indeed been found to be generally valid for other quantum-mechanical operators $\mathscr{Q}^{\text{op}}$, namely,

$$\langle \mathscr{Q}^{\text{op}}\rangle = \int_{\Omega_r} \psi^*(\mathbf{r}, t)\mathscr{Q}^{\text{op}}\psi(\mathbf{r}, t)\, d\mathbf{r}. \tag{1.212}$$

Let us adopt the approach that Eq. (1.212) *defines* what we shall refer to as the **quantum-mechanical expectation value of an operator**, and then show that for various specific cases it leads to results which can be interpreted as average values.

EXERCISE Substitute ψ_N into the normalization integral (1.208) to show the correctness of the normalization factor.

8.2 Special Case Where the System Is in an Eigenstate of the Hamiltonian $\mathscr{H}$ and the Operator $\mathscr{Q}^{\text{op}}$

First, consider the simple case in which $\psi(\mathbf{r}, t)$ happens to be an eigenfunction $\psi_i(\mathbf{r}, t)$ of a time-independent operator $\mathscr{Q}^{\text{op}}$ corresponding to the eigenvalue q_i, as well as being an eigenfunction of $\mathscr{H}$ corresponding to the eigenvalue $\mathscr{E}_i$. Then, according to the discussion in §6, it follows that

$$\mathscr{Q}^{\text{op}}\psi_i(\mathbf{r}, t) = q_i\psi_i(\mathbf{r}, t), \tag{1.213}$$

where q_i is a constant. Substitution into the above equation (1.212) gives for this case

$$\langle \mathscr{Q}^{op} \rangle = \int_{\Omega_r} \psi_i^*(\mathbf{r}, t) q_i \psi_i(\mathbf{r}, t) \, d\mathbf{r} = q_i \int_{\Omega_r} \psi_i^*(\mathbf{r}, t) \psi_i(\mathbf{r}, t) \, d\mathbf{r} = q_i, \quad (1.214)$$

where we have specifically employed the normalization condition (1.208). On intuitive grounds, this is the expected result. That is, if the system is in an eigenstate of $\mathscr{Q}^{op}$ corresponding to the eigenvalue q_i, then the average value of the physical observable represented by $\mathscr{Q}^{op}$ should be equal to the eigenvalue q_i. In fact, it is possible to show that in such a case there is no statistical uncertainty in the value of $\mathscr{Q}^{op}$. A standard measure of the deviation from the mean value in statistics is the "mean square deviation," which for a continuous function f of position $\mathbf{r}$ takes the form $\overline{\Delta f^2} = \int_{\Omega_r} (f - \bar{f})^2 \, d\mathbf{r}$, where $\bar{f}$ is the mean value of f. In the present case this leads us to compute

$$\langle (\mathscr{Q}^{op} - \langle \mathscr{Q}^{op} \rangle)^2 \rangle = \int_{\Omega_r} \psi_i^* \{ \mathscr{Q}^{op} \cdot \mathscr{Q}^{op} - 2\mathscr{Q}^{op} \langle \mathscr{Q}^{op} \rangle + \langle \mathscr{Q}^{op} \rangle^2 \} \psi_i \, d\mathbf{r}$$

$$= \int_{\Omega_r} \psi_i^* \{ q_i^2 - 2q_i^2 + q_i^2 \} \psi_i \, d\mathbf{r} = 0.$$

Thus we say that there is no deviation from the value q_i for the physical observable represented by $\mathscr{Q}^{op}$ when the system is in the eigenstate ψ_i.

PROJECT 1.10 Mean Square Deviation of Quantum Operators Leading to the Uncertainty Relation

Use the standard measure of the deviation from the mean in statistics, taking the position x as an operator, to deduce the mean square deviation for the wave packet described in §7.2. Repeat for the Gaussian wave packet described in §7.7. [*Hint*: Your results should bear some resemblance to the Heisenberg uncertainty relations (1.207).]

8.3 Case of a Superposition State with $\mathscr{H}$ and $\mathscr{Q}^{op}$ Having a Complete Set of Simultaneous Eigenfunctions

We next consider the more general case in which $\psi(\mathbf{r}, t)$ is not itself an eigenfunction of $\mathscr{Q}^{op}$. The procedure is then to attempt to expand $\psi(\mathbf{r}, t)$ in a *complete set* of eigenfunctions of the Hamiltonian $\mathscr{H}$ which are simultaneous eigenfunctions of $\mathscr{Q}^{op}$; an example specifically applicable for the momentum operator $\mathbf{p}^{op}$ would be to express $\psi(\mathbf{r}, t)$ as the superposition of a group of plane waves, each individual plane wave being an eigenfunction of $\mathbf{p}^{op}$ as shown in §6. This particular expansion would be the Fourier integral wave packet treated previously in some detail.

It might be asked at this point whether or not an arbitrary operator could be expected to possess a complete set of eigenfunctions. There is a basic tenet of quantum mechanics, called the *fundamental expansion postulate*, which states that *every physical quantity can be represented by a Hermitian operator, which is an operator with real eigenvalues* so that it has real expectation values, and *this Hermitian operator possesses normalized eigenfunctions* g_i $(i = 1, 2, \ldots)$ *of sufficient number to represent an arbitrary state* ϕ *by a linear superposition* $\sum_i b_i g_i$. Such a set of normalized functions g_i is therefore said to be *complete*. However the complete set g_i may or may not be the same as the complete set of functions ϕ_i which are eigenfunctions of the Hamiltonian operator $\mathscr{H}$. If we consider $\psi_i(\mathbf{r}, t) = \phi_i(r) \exp[-(i/\hbar)\mathscr{E}_i t]$ to represent the set of time-dependent eigenfunctions of the Hamiltonian $\mathscr{H}$, then $\psi(\mathbf{r}, t)$ can be expressed as a linear combination of such functions with weighting factors or probability amplitudes a_i,

$$\psi(\mathbf{r}, t) = \sum_i a_i \psi_i(\mathbf{r}, t). \tag{1.215}$$

Sometimes this wavefunction $\psi(\mathbf{r}, t)$ is said to represent a *superposition state*, since the system is not in a specific eigenstate $\psi_i(\mathbf{r}, t)$ of the Hamiltonian $\mathscr{H}$. [The summation sign in Eq. (1.215) is designated to *include* an integration over any portion of the energy eigenvalue spectrum which is continuous (cf. §6.3) rather than discrete.] Operating on this equation with $\mathscr{Q}^{\mathrm{op}}$, where we assume the a_i to be time independent, gives

$$\mathscr{Q}^{\mathrm{op}}\psi(\mathbf{r}, t) = \mathscr{Q}^{\mathrm{op}} \sum_i a_i \psi_i = \sum_i a_i \mathscr{Q}^{\mathrm{op}} \psi_i. \tag{1.216}$$

If we then assume that we have the very special case for which the complete set of eigenfunctions g_i of $\mathscr{Q}^{\mathrm{op}}$ are identical with the complete set of eigenfunctions ϕ_i of $\mathscr{H}$, then Eq. (1.216) becomes

$$\mathscr{Q}^{\mathrm{op}}\psi(\mathbf{r}, t) = \sum_i a_i q_i \psi_i. \tag{1.217}$$

Substituting this result and Eq. (1.215) into Eq. (1.212) gives

$$\langle \mathscr{Q}^{\mathrm{op}} \rangle = \int_{\Omega_r} \left(\sum_j a_j \psi_j \right)^* \left(\sum_i a_i q_i \psi_i \right) d\mathbf{r} = \sum_i \sum_j a_i a_j^* q_i \int_{\Omega_r} \psi_j^* \psi_i \, d\mathbf{r}. \tag{1.218}$$

Let us assume for the moment the theorem (which is proved in §8.6 of this chapter) that eigenfunctions belonging to different eigenvalues are orthogonal, namely,

$$\int_{\Omega_r} \psi_j^* \psi_i \, d\mathbf{r} = \delta_{ij}, \tag{1.219}$$

where δ_{ij} is the Kronecker delta function which is unity for $i = j$ but zero whenever $i \neq j$. Then we obtain from Eq. (1.218),

$$\langle \mathscr{Q}^{\mathrm{op}} \rangle = \sum_i \sum_j a_i a_j^* q_i \delta_{ij} = \sum_i a_i a_i^* q_i = \sum_i |a_i|^2 q_i. \tag{1.220}$$

The coefficients a_i are the weighting factors for the individual normalized eigenfunctions ψ_i constituting $\psi(\mathbf{r}, t)$, and $|a_i|^2$ gives the weighting factor for the contribution of any particular eigenstate ψ_i to the total probability. It is perhaps helpful to explain in greater detail the above statement that $|a_i|^2$ represents, in some sense, a weighting factor for the contribution of the state ψ_i to the probability density $\psi^*(\mathbf{r}, t)\psi(\mathbf{r}, t)$. Substituting the linear combination (1.215) for ψ into the normalization condition (1.208) yields

$$1 = \int_{\Omega_r} \psi^*(\mathbf{r}, t)\psi(\mathbf{r}, t)\, d\mathbf{r} = \int_{\Omega_r} \left(\sum_j a_j \psi_j\right)^* \left(\sum_i a_i \psi_i\right) d\mathbf{r}$$

$$= \sum_j \sum_i a_j^* a_i \int_{\Omega_r} \psi_j^* \psi_i \, d\mathbf{r} = \sum_j \sum_i a_j^* a_i \delta_{ij} = \sum_i a_i^* a_i = \sum_i |a_i|^2. \qquad (1.221)$$

We can therefore interpret this result in the following manner: *each coefficient a_i leads to a contribution of amount $|a_i|^2$ for state ψ_i to the total probability*. Using the fact that $|a_i|^2$ gives the weighting factor for the contribution of the state ψ_i to the total probability, together with the knowledge that q_i is the value of $\mathscr{Q}^{\text{op}}$ corresponding to this state, we can now interpret Eq. (1.220) above for $\langle \mathscr{Q}^{\text{op}} \rangle$ as the weighted average of q_i over all states ψ_i. The weighting factors $|a_i|^2$ are determined by the relative admixture of the states in the total wave function $\psi(\mathbf{r}, t)$. We thus conclude that the prescription given by Eq. (1.212) for evaluating the quantity defined as the "expectation value of the operator $\mathscr{Q}^{\text{op}}$" can be used, at least in the special cases examined above, to obtain an average value of the physical observable represented by the operator $\mathscr{Q}^{\text{op}}$.

PROJECT 1.11 Hilbert Space

Write a paper on the properties of the mathematical construct known as *Hilbert space*. Include a discussion of the relevance of Hilbert space to our formulation of quantum mechanics.

PROJECT 1.12 Matrix Theory

Prove that the eigenvalues of a Hermitian matrix are real, and that eigenvectors corresponding to different eigenvalues are orthogonal.

8.4 More General Case for the Superposition State

The expression (1.220) developed above and interpreted as the expectation value of an operator $\mathscr{Q}^{\text{op}}$ is time independent because $\mathscr{Q}^{\text{op}}$ had eigenvalues which were independent of time and we assumed that there exists a complete set of functions which are simultaneous eigenfunctions of $\mathscr{H}$ and $\mathscr{Q}^{\text{op}}$. Neither of these criteria need be met for an arbitrary Hermitian operator. It is therefore quite worthwhile to consider the somewhat more general case where the $\psi_i(\mathbf{r}, t)$ are eigenfunctions of $\mathscr{H}$ but not necessarily eigenfunctions of $\mathscr{Q}^{\text{op}}$. Using the definition (1.212) for the quantum-mechanical expectation value of an operator once again, we obtain for the present case

$$
\begin{aligned}
\langle \mathcal{Q}^{\mathrm{op}} \rangle &= \int_{\Omega_r} \psi^*(\mathbf{r}, t)\mathcal{Q}^{\mathrm{op}}\psi(\mathbf{r}, t)\, d\mathbf{r} \\
&= \int_{\Omega_r} \left\{ \sum_j a_j \psi_j(\mathbf{r}, t) \right\}^* \mathcal{Q}^{\mathrm{op}} \left\{ \sum_l a_l \psi_l(\mathbf{r}, t) \right\} d\mathbf{r} \\
&= \sum_j \sum_l a_j^* a_l \int_{\Omega_r} \psi_j^*(\mathbf{r}, t)\mathcal{Q}^{\mathrm{op}}\psi_l(\mathbf{r}, t)\, d\mathbf{r}. \qquad (1.222)
\end{aligned}
$$

Employing the stationary-state eigenfunctions $\phi_i(\mathbf{r})$, we can write $\psi_i(\mathbf{r}, t) = \phi_i(\mathbf{r}) \exp[-(i/\hbar)\mathscr{E}_i t]$. Expression (1.222) then becomes

$$
\langle \mathcal{Q}^{\mathrm{op}} \rangle = \sum_j \sum_l a_j^* a_l \exp[(i/\hbar)(\mathscr{E}_j - \mathscr{E}_l)t] \int_{\Omega_r} \phi_j^*(\mathbf{r})\mathcal{Q}^{\mathrm{op}}\phi_l(\mathbf{r})\, d\mathbf{r}, \qquad (1.223)
$$

provided $\mathcal{Q}^{\mathrm{op}}$ does not operate on the time factors. This is easily assured by requiring $\mathcal{Q}^{\mathrm{op}}$ to be time independent. The integrals

$$
\mathcal{Q}_{jl} = \int_{\Omega_r} \phi_j^*(\mathbf{r})\mathcal{Q}^{\mathrm{op}}\phi_l(\mathbf{r})\, d\mathbf{r} \qquad (1.224)
$$

are known as *matrix elements* of the operator $\mathcal{Q}^{\mathrm{op}}$, and the quantities

$$
\omega_{jl} = (\mathscr{E}_j - \mathscr{E}_l)/\hbar \qquad (1.225)
$$

represent angular frequencies. We note that the matrix elements are time independent if $\mathcal{Q}^{\mathrm{op}}$ is time independent. With these abbreviations the above expression (1.223) becomes

$$
\langle \mathcal{Q}^{\mathrm{op}} \rangle = \sum_j \sum_l a_j^* a_l \mathcal{Q}_{jl} e^{i\omega_{jl}t}. \qquad (1.226)
$$

The expectation value for the operator is therefore generally *time dependent*. The terms for which $j = l$ have $\omega_{jl} = 0$, so they are time independent if $\mathcal{Q}^{\mathrm{op}}$ is time independent; these are called the *diagonal* terms. For $j \neq l$, the terms are time dependent even if $\mathcal{Q}^{\mathrm{op}}$ is time independent; these are called the *off-diagonal* terms.

PROJECT 1.13 Parseval's Formula, Bessel's Inequality, and Schwarz's Inequality

State and prove the following: 1. Parseval's formula, 2. Bessel's inequality, 3. Schwarz's inequality. [*Hint*: See Ikenberry (1962) and Merzbacher (1970).]

8.5 Operators for Physical Observables Which Are Constants of Motion for the System

Whenever $\mathcal{Q}^{\mathrm{op}}$ is itself time independent, the additional criterion which must be met in Eq. (1.226) if $\langle \mathcal{Q}^{\mathrm{op}} \rangle$ is to be time independent is that *the off-diagonal matrix elements must be zero.* A sufficient condition for this to be true is the criterion mentioned previously, namely, that the eigenfunctions $\phi_i(\mathbf{r})$ of the Hamiltonian $\mathscr{H}$ be simultaneous eigenfunctions of the operator $\mathscr{Q}^{\text{op}}$. Then

$$\mathscr{Q}_{jl} = \int_{\Omega_r} \phi_j^*(\mathbf{r})\mathscr{Q}^{\text{op}}\phi_l(\mathbf{r})\, d\mathbf{r} = \int_{\Omega_r} \phi_j^*(\mathbf{r})q_l\phi_l(\mathbf{r})\, d\mathbf{r}$$

$$= q_l \int_{\Omega_r} \phi_j^*(\mathbf{r})\phi_l(\mathbf{r})\, d\mathbf{r} = q_l\delta_{jl}, \qquad (1.227)$$

where again we have assumed orthonormality of the eigenfunctions according to Eq. (1.219). The off-diagonal terms which give rise to the time dependence are indeed zero for this situation, and $\mathscr{Q}^{\text{op}}$ has a time-independent expectation value.

The importance of Hermitian operators which have time-independent expectation values for a given physical system is that they can represent physical observables which are constants of motion for the system. Constants of motion are just as important in characterizing quantum-mechanical systems as they are in classical mechanics [see Goldstein (1956)].

8.6 Orthogonality Proof for Eigenfunctions Having Different Eigenvalues

Let us now prove the important theorem which we used several times already that *all eigenfunctions of a Hermitian operator corresponding to different eigenvalues are orthogonal.* We do this for an arbitrary Hermitian operator $\mathscr{Q}^{\text{op}}$, so the results will be immediately applicable to the special case of the Hamiltonian operator $\mathscr{H}$ with eigenfunctions ϕ_i. The proof of the theorem is easily carried out by using the characteristic eigenvalue equation for two arbitrary eigenstates i and j,

$$\mathscr{Q}^{\text{op}}g_i = q_i g_i, \qquad \mathscr{Q}^{\text{op}}g_j = q_j g_j. \qquad (1.228)$$

We multiply the first of these equations by g_j^* and the second by g_i^* and integrate over all space to give

$$\int_{\Omega_r} g_j^* \mathscr{Q}^{\text{op}} g_i\, d\mathbf{r} = \int_{\Omega_r} g_j^* q_i g_i\, d\mathbf{r} = q_i \int_{\Omega_r} g_j^* g_i\, d\mathbf{r}, \qquad (1.229)$$

$$\int_{\Omega_r} g_i^* \mathscr{Q}^{\text{op}} g_j\, d\mathbf{r} = \int_{\Omega_r} g_i^* q_j g_j\, d\mathbf{r} = q_j \int_{\Omega_r} g_i^* g_j\, d\mathbf{r}. \qquad (1.230)$$

If $\mathscr{Q}^{\text{op}}$ is to represent a physical observable, its eigenvalues q_i must be real, which condition is satisfied if $\mathscr{Q}^{\text{op}}$ is chosen to be *Hermitian.*

One *definition of a Hermitian operator*, which is the one to be used in our proof, is that

$$\int_{\Omega_r} g_i^*(\mathscr{Q}^{\text{op}}g_j)\, d\mathbf{r} = \int_{\Omega_r} (\mathscr{Q}^{\text{op}}g_i)^* g_j\, d\mathbf{r} \qquad \textbf{(definition of Hermitian operator)} \qquad (1.231)$$

for arbitrary functions g_i and g_j. This definition (1.231) follows from the more common definition, that $\mathcal{Q}^{\mathrm{op}}$ is Hermitian if

$$\int_{\Omega_r} (\mathcal{Q}^{\mathrm{op}} f)^* f \, d\mathbf{r} = \int_{\Omega_r} f^*(\mathcal{Q}^{\mathrm{op}} f) \, d\mathbf{r} \tag{1.232}$$

for arbitrary f, by choosing f to be $g_i + \gamma g_j$, where γ is an arbitrary complex parameter. It is more clear from the latter definition (1.232) that there is a correlation between physical observables and the Hermitian property of operators. That is, whenever f_s is an eigenfunction of $\mathcal{Q}^{\mathrm{op}}$ corresponding to a *real* eigenvalue λ (so that $\lambda_s^* = \lambda_s$), as must be the case when $\mathcal{Q}^{\mathrm{op}}$ represents a physical observable, then the relation (1.232) reduces to

$$\int_{\Omega_r} (\lambda_s f_s)^* f_s \, d\mathbf{r} = \int_{\Omega_r} f_s^*(\lambda_s f_s) \, d\mathbf{r} \tag{1.233}$$

which is obviously satisfied.

EXERCISE Prove that the definition (1.231) for a Hermitian operator follows from the seemingly more restricted definition (1.232) of such. (*Hint*: Consider f to be a linear combination of g_i and g_j with complex coefficients.)

To return to our proof, using the relation (1.231), Eq. (1.230) can be written

$$\int_{\Omega_r} (\mathcal{Q}^{\mathrm{op}} g_i)^* g_j \, d\mathbf{r} = q_j \int_{\Omega_r} g_i^* g_j \, d\mathbf{r}. \tag{1.234}$$

Taking the complex conjugate of this relation gives

$$\int_{\Omega_r} (\mathcal{Q}^{\mathrm{op}} g_i) g_j^* \, d\mathbf{r} = q_j^* \int_{\Omega_r} g_i g_j^* \, d\mathbf{r}. \tag{1.235}$$

The order of the scalar factors in the products in the integrand is not important. As mentioned before, q_j is real. This equation can therefore be written

$$\int_{\Omega_r} g_j^* \mathcal{Q}^{\mathrm{op}} g_i \, d\mathbf{r} = q_j \int_{\Omega_r} g_j^* g_i \, d\mathbf{r}. \tag{1.236}$$

Subtracting from Eq. (1.229) gives immediately

$$0 = (q_i - q_j) \int_{\Omega_r} g_j^* g_i \, d\mathbf{r}. \tag{1.237}$$

If g_i and g_j correspond to different eigenvalues of $\mathcal{Q}^{\mathrm{op}}$, namely, $q_i \neq q_j$, then this relation tells us that

$$\int_{\Omega_r} g_j^* g_i \, d\mathbf{r} = 0 \qquad (i \neq j), \tag{1.238}$$

which proves the theorem and thereby justifies Eq. (1.219). If g_i is a different function from g_j but $q_i = q_j$, which is the degenerate case, Eq. (1.237) then gives no information on the value of $\int_{\Omega_r} g_j^* g_i \, d\mathbf{r}$. This integral can still be zero if the functions g_i and g_j are chosen properly. Relating to this question, there is a technique for constructing orthonormal functions, called the Gram–Schmidt process, so that we can in principle always assume that our complete set of eigenfunctions is orthogonal. Details are given in Ikenberry (1962).

8.7 Alternative Treatment of the Time Dependence of Quantum-Mechanical Operators

It is worthwhile at this point to examine the time dependence of quantum-mechanical operators from a standpoint somewhat different from Eq. (1.226). We do this by differentiating the expectation value of the arbitrary operator $\mathscr{Q}^{\text{op}}$ with respect to time, using the Leibniz rule for differentiation of an integral,

$$\begin{aligned}\frac{d}{dt}\langle\mathscr{Q}^{\text{op}}\rangle &= \frac{d}{dt}\int_{\Omega_r}\psi^*(\mathbf{r},t)\mathscr{Q}^{\text{op}}\psi(\mathbf{r},t)\,d\mathbf{r}\\ &= \int_{\Omega_r}\frac{\partial\psi^*}{\partial t}\mathscr{Q}^{\text{op}}\psi\,d\mathbf{r} + \int_{\Omega_r}\psi^*\mathscr{Q}^{\text{op}}\frac{\partial\psi}{\partial t}\,d\mathbf{r} + \int_{\Omega_r}\psi^*\frac{\partial\mathscr{Q}^{\text{op}}}{\partial t}\psi\,d\mathbf{r}. \qquad (1.239)\end{aligned}$$

We use the Schrödinger equation (1.163) and its complex conjugate to replace $\partial\psi^*/\partial t$ and $\partial\psi/\partial t$ by $(i/\hbar)\mathscr{H}\psi^*$ and $-(i/\hbar)\mathscr{H}\psi$, respectively, and note that the third integral in Eq. (1.239) is the expectation value of $\partial\mathscr{Q}^{\text{op}}/\partial t$. Of course, $\langle\partial\mathscr{Q}^{\text{op}}/\partial t\rangle$ will be zero whenever $\mathscr{Q}^{\text{op}}$ is not an explicit function of time. (We have assumed throughout that the Hamiltonian $\mathscr{H}$ is not explicitly time dependent; otherwise we would not have an energy-conserving system, since a classical Hamiltonian is time-dependent only when the total energy of the system is time dependent. Use of the stationary-state eigenfunctions $\phi_i(\mathbf{r},t) = \phi_i(\mathbf{r})\cdot\exp[(-i/\hbar)\mathscr{E}_i t]$ always implies a time-independent Hamiltonian, because our separation-of-variables technique used in §6 to deduce the time-independent Schrödinger equation is based on this assumption.) Thus we have

$$\frac{d}{dt}\langle\mathscr{Q}^{\text{op}}\rangle = \frac{i}{\hbar}\int_{\Omega_r}(\mathscr{H}\psi^*)(\mathscr{Q}^{\text{op}}\psi)\,d\mathbf{r} - \frac{i}{\hbar}\int_{\Omega_r}\psi^*\mathscr{Q}^{\text{op}}\mathscr{H}\psi\,d\mathbf{r} + \left\langle\frac{\partial\mathscr{Q}^{\text{op}}}{\partial t}\right\rangle. \qquad (1.240)$$

Since $\mathscr{H}$ is Hermitian, the first integral can be written $\int_{\Omega_r}\psi^*(\mathscr{H}\mathscr{Q}^{\text{op}})\psi\,d\mathbf{r}$, so

$$\begin{aligned}\frac{d}{dt}\langle\mathscr{Q}^{\text{op}}\rangle &= -\frac{i}{\hbar}\int_{\Omega_r}\psi^*\{\mathscr{Q}^{\text{op}}\mathscr{H} - \mathscr{H}\mathscr{Q}^{\text{op}}\}\psi\,d\mathbf{r} + \left\langle\frac{\partial\mathscr{Q}^{\text{op}}}{\partial t}\right\rangle\\ &= -\left(\frac{i}{\hbar}\right)\langle\mathscr{Q}^{\text{op}}\mathscr{H} - \mathscr{H}\mathscr{Q}^{\text{op}}\rangle + \left\langle\frac{\partial\mathscr{Q}^{\text{op}}}{\partial t}\right\rangle\\ &= -\left(\frac{i}{\hbar}\right)\langle[\mathscr{Q}^{\text{op}},\mathscr{H}]\rangle + \left\langle\frac{\partial\mathscr{Q}^{\text{op}}}{\partial t}\right\rangle. \qquad (1.241)\end{aligned}$$

The symbol

$$[\mathscr{Q}^{\text{op}},\mathscr{H}] = \mathscr{Q}^{\text{op}}\mathscr{H} - \mathscr{H}\mathscr{Q}^{\text{op}} \qquad (1.242)$$

is called the **commutator** of $\mathcal{Q}^{op}$ and $\mathcal{H}$. The commutator is an operator, and it operates on the functions $\psi_i(\mathbf{r}, t)$. Equation (1.241) is a very important result. We examine in the next section the conditions for which the time derivative of $\langle \mathcal{Q}^{op} \rangle$ is zero.

EXERCISE Show that $[p_x^{op}, x^{op}] = -i\hbar\mathcal{I}^{op}$, where $\mathcal{I}^{op}$ is the identity operator. (*Hint*: Operate on an arbitrary function of position $f(x)$ with the commutator.)

PROJECT 1.14 Commutator Algebra

Prove the following identities for the arbitrary quantum-mechanical operators $\mathcal{Q}_1^{op}, \mathcal{Q}_2^{op}$, and $\mathcal{Q}_3^{op}$:
1. $[\mathcal{Q}_1^{op}, \mathcal{Q}_2^{op} + \mathcal{Q}_3^{op}] = [\mathcal{Q}_1^{op}, \mathcal{Q}_2^{op}] + [\mathcal{Q}_1^{op}, \mathcal{Q}_3^{op}]$.
2. $[\mathcal{Q}_1^{op}, (\mathcal{Q}_2^{op}\mathcal{Q}_3^{op})] = [\mathcal{Q}_1^{op}, \mathcal{Q}_2^{op}]\mathcal{Q}_3^{op} + \mathcal{Q}_2^{op}[\mathcal{Q}_1^{op}, \mathcal{Q}_3^{op}]$.

PROJECT 1.15 Commutation Relations between Position and Linear Momentum Operators

1. Prove that $[x^{op}, p_x^{op}] = i\hbar$, $[x^{op}, p_y^{op}] = 0, \ldots,$ or in general, $[x_i^{op}, p_j^{op}] = i\hbar\delta_{ij}$ $(i = 1, 2, 3;$ $j = 1, 2, 3)$, where x_1, x_2, x_3 represent the $\hat{\mathbf{x}}$, $\hat{\mathbf{y}}$, and $\hat{\mathbf{z}}$ components of the position vector $\mathbf{r}$, and p_1, p_2, and p_3 denote the $\hat{\mathbf{x}}$, $\hat{\mathbf{y}}$, and $\hat{\mathbf{z}}$ components of the linear momentum vector $\mathbf{p}$. The relations stated above are to be interpreted in the usual operator fashion, namely, the right-hand side is the scalar multiplicative factor obtained when the differential operator on the left-hand side acts on some arbitrary function $f(\mathbf{r})$ of position $\mathbf{r}$. The square brackets identify the commutator of two operators (such as x^{op} and p_x^{op}) in accordance with Eq. (1.242), namely, for operators $\mathcal{Q}_1^{op}$ and $\mathcal{Q}_2^{op}$, $[\mathcal{Q}_1^{op}, \mathcal{Q}_2^{op}] \equiv \mathcal{Q}_1^{op}\mathcal{Q}_2^{op} - \mathcal{Q}_2^{op}\mathcal{Q}_1^{op}$.
2. Prove the commutation relations $[x_i^{op}, x_j^{op}] = 0$ $(i = 1, 2, 3; j = 1, 2, 3)$ for all combinations of x_i^{op} and x_j^{op}.
3. Prove the commutation relations $[p_i^{op}, p_j^{op}] = 0$ for all combinations of p_i^{op} and p_j^{op}.

PROJECT 1.16 Angular Momentum Commutation Relations

1. Work out the commutation relation $[\mathcal{L}_x^{op}, \mathcal{L}_y^{op}] = i\hbar\mathcal{L}_z^{op}$. (*Hint*: You may wish to refer to Project 1.6 on angular momentum operators.)
2. Evaluate all possible commutators $[\mathcal{L}_i^{op}, \mathcal{L}_j^{op}]$, where $i = 1, 2$, or 3 and $j = 1, 2$, or 3, with $\mathcal{L}_1^{op}$, $\mathcal{L}_2^{op}$, and $\mathcal{L}_3^{op}$ denoting the $\hat{\mathbf{x}}$, $\hat{\mathbf{y}}$, and $\hat{\mathbf{z}}$ components of the angular momentum operator.
3. Show that $[\mathcal{L}_x^{op}, y^{op}] = i\hbar z^{op}$.
4. Evaluate all combinations $[\mathcal{L}_i^{op}, x_j^{op}]$, where $i = 1, 2$, or 3 and $j = 1, 2$, or 3, with x_1, x_2, and x_3 denoting the $\hat{\mathbf{x}}$, $\hat{\mathbf{y}}$, and $\hat{\mathbf{z}}$ components of the position vector $\mathbf{r}$.
5. Show that $[\mathcal{L}_x^{op}, p_z^{op}] = -i\hbar p_y^{op}$.
6. Evaluate all combinations $[\mathcal{L}_i^{op}, p_j^{op}]$, where $i = 1, 2$, or 3 and $j = 1, 2$, or 3, with p_1, p_2, and p_3 denoting the $\hat{\mathbf{x}}$, $\hat{\mathbf{y}}$, and $\hat{\mathbf{z}}$ components of the linear momentum $\mathbf{p}$.
7. Organize the above results 1–6 into a meaningful table, and summarize your conclusions.
8. Evaluate $[\mathcal{L}_i^{op}, (\mathcal{L}^{op})^2]$, $[\mathcal{L}_i^{op}, (\mathbf{r}^{op})^2]$, and $[\mathcal{L}_i^{op}, (\mathbf{p}^{op})^2]$ for $i = 1, 2$, and 3.

8.8 Commuting Operators, Simultaneous Eigenfunctions, and Constants of Motion

If

$$[\mathcal{Q}^{op}, \mathcal{H}]\psi_i = 0, \tag{1.243}$$

we say that $\mathcal{Q}^{op}$ and $\mathcal{H}$ *commute*. Let us now prove two important theorems relating the concepts of commuting operators and simultaneous eigenfunctions.

The first theorem is that *commuting operators have a complete set of simultaneous eigenfunctions*. The proof proceeds as follows. If ψ_i is one of the eigenfunctions of $\mathscr{H}$, then

$$\mathscr{H}(\mathscr{Q}^{\mathrm{op}}\psi_i) = \mathscr{Q}^{\mathrm{op}}\mathscr{H}\psi_i = \mathscr{Q}^{\mathrm{op}}\mathscr{E}_i\psi_i = \mathscr{E}_i(\mathscr{Q}^{\mathrm{op}}\psi_i). \tag{1.244}$$

In this case, we see that $\mathscr{Q}^{\mathrm{op}}\psi_i$ is likewise an eigenfunction of $\mathscr{H}$ corresponding to the same eigenvalue $\mathscr{E}_i$. If the functions ψ_i represent a complete set which is nondegenerate, then ψ_i is the only eigenfunction of $\mathscr{H}$ with the eigenvalue $\mathscr{E}_i$, and we are forced to conclude that

$$\mathscr{Q}^{\mathrm{op}}\psi_i \propto \psi_i. \tag{1.245}$$

If we denote the proportionality factor by q_i, then we have

$$\mathscr{Q}^{\mathrm{op}}\psi_i = q_i\psi_i, \tag{1.246}$$

which tells us that ψ_i is also an eigenfunction of $\mathscr{Q}^{\mathrm{op}}$. Because ψ_i is an arbitrary eigenfunction of $\mathscr{H}$, we can apply the results to the complete set of eigenfunctions of $\mathscr{H}$. [A bit more detail is required (see Chap. 5, §2 on diagonalization) for the situation where some of the eigenfunctions of $\mathscr{H}$ are degenerate; the theorem can still be shown to be valid, however.] To summarize, we can say that *commuting operators have simultaneous eigenfunctions.*

Conversely, we can prove the theorem that *operators which have a complete set of simultaneous eigenfunctions are commuting operators.* The proof is easily developed. If ψ_i is a simultaneous eigenfunction of $\mathscr{Q}^{\mathrm{op}}$ and $\mathscr{H}$ corresponding to the eigenvalues q_i and $\mathscr{E}_i$, respectively, then

$$\mathscr{Q}^{\mathrm{op}}\mathscr{H}\psi_i = \mathscr{Q}^{\mathrm{op}}\mathscr{E}_i\psi_i = \mathscr{E}_i\mathscr{Q}^{\mathrm{op}}\psi_i = \mathscr{E}_i q_i\psi_i, \tag{1.247}$$

$$\mathscr{H}\mathscr{Q}^{\mathrm{op}}\psi_i = \mathscr{H}q_i\psi_i = q_i\mathscr{H}\psi_i = q_i\mathscr{E}_i\psi_i. \tag{1.248}$$

Subtracting these two relations gives

$$(\mathscr{Q}^{\mathrm{op}}\mathscr{H} - \mathscr{H}\mathscr{Q}^{\mathrm{op}})\psi_i = [\mathscr{Q}^{\mathrm{op}}, \mathscr{H}]\psi_i = 0 \tag{1.249}$$

or symbolically, $\mathscr{Q}^{\mathrm{op}}\mathscr{H} = \mathscr{H}\mathscr{Q}^{\mathrm{op}}$. By hypothesis, this holds for any one of the complete set of eigenfunctions of $\mathscr{H}$. Thus, the commutator gives zero when it operates on any one of the complete set of simultaneous eigenfunctions. If the commutator acts on any linear combination of the complete set of simultaneous eigenfunctions, it again will give a zero result. Thus the operators commute when acting on *any* arbitrary function f, provided only that the function f can be expanded as a linear combination of the complete set of eigenfunctions in question. The theorem is therefore proven.

Commuting operators can be used to represent physical observables which can be measured simultaneously; thus they are very important in quantum systems. In the expression (1.241) for the time derivative of $\langle\mathscr{Q}^{\mathrm{op}}\rangle$, the expectation value of $[\mathscr{Q}^{\mathrm{op}}, \mathscr{H}]$ involves

$$[\mathscr{Q}^{\mathrm{op}}, \mathscr{H}]\psi(\mathbf{r}, t) = [\mathscr{Q}^{\mathrm{op}}, \mathscr{H}] \sum_l a_l\psi_l(\mathbf{r}, t), \tag{1.250}$$

so that the expectation value of the commutator will be zero if each function in the complete set $\psi_i(\mathbf{r}, t)$ is also an eigenfunction of $\mathscr{Q}^{\text{op}}$. Therefore, we conclude from Eq. (1.241) that *the sufficient conditions for the time dependence of the expectation value, viz,* $(d/dt)\langle\mathscr{Q}^{\text{op}}\rangle$, *to be zero, and hence for* $\mathscr{Q}^{\text{op}}$ *to represent a constant of motion for the system, is that* $\mathscr{Q}^{\text{op}}$ *be explicitly time independent and that* $\mathscr{Q}^{\text{op}}$ *and* $\mathscr{H}$ *possess a complete set of simultaneous eigenfunctions*. Comparing this with the previous criteria we developed for $\langle\mathscr{Q}^{\text{op}}\rangle$ to be a constant of motion, we see that we have replaced the condition that the off-diagonal matrix elements be zero by the condition that the complete set ψ_i be simultaneous eigenfunctions of $\mathscr{Q}^{\text{op}}$ and $\mathscr{H}$. It is readily seen in Eq. (1.227) that whenever the complete set of functions ψ_i are also simultaneous eigenfunctions of $\mathscr{Q}^{\text{op}}$, every off-diagonal matrix element is indeed zero, so the two conditions are consistent.

Practical applications of commuting operators and constants of motion are prevalent throughout quantum mechanics, one particularly important example being that of angular momentum. However, operators are also important for the linear momentum, which is a more important quantity for understanding electron transport in solids since it is related directly to the electric current. Before treating electron transport, let us extend our understanding of matrix mechanics a bit, and formally introduce the Dirac notation.

8.9 Matrix Formulation

The term "matrix element" arises from the matrix formulation of quantum mechanics [see Heisenberg (1930)] where operators are represented by matrices. For example, if the complete set of functions ϕ_i satisfies the Schrödinger equation, $\mathscr{H}\phi_i = \mathscr{E}_i\phi_i$, then these functions can be used as the basis states for the function space in question, and the matrix elements $\mathscr{Q}_{ij}$ of the arbitrary operator $\mathscr{Q}^{\text{op}}$ with respect to this basis are defined by $\mathscr{Q}_{ij} \equiv \int_{\Omega_{\mathbf{r}}} \phi_i^*(\mathbf{r})\mathscr{Q}^{\text{op}}\phi_j(\mathbf{r})\,d\mathbf{r}$. Because the set $\{\phi_i\}$ is complete, the operator $\mathscr{Q}^{\text{op}}$ is represented totally by the matrix made up of all of the matrix elements, namely,

$$\mathscr{Q}^{\text{op}} = \begin{pmatrix} \mathscr{Q}_{11} & \mathscr{Q}_{12} & \mathscr{Q}_{13} & \cdots \\ \mathscr{Q}_{21} & \mathscr{Q}_{22} & \mathscr{Q}_{23} & \\ \mathscr{Q}_{31} & \mathscr{Q}_{32} & \mathscr{Q}_{33} & \\ \vdots & & & \ddots \end{pmatrix}.$$

Without going into detail, it can be said that the diagonal elements are related to expectation values of physical observables and the off-diagonal elements are related to transition probabilities. Thus the matrix elements prove to be very useful. Dirac introduced a shorthand notation wherein the above integral for $\mathscr{Q}_{ij}$ is written simply $\mathscr{Q}_{ij} = \langle i|\mathscr{Q}^{\text{op}}|j\rangle$. Dirac's notation lends itself to ready interpretation in the language of linear algebra. The set of $\phi_j(\mathbf{r})$ can be said to constitute a set of functions spanning a *linear vector space*. These functions are denoted by the symbol $|j\rangle$ and are called *ket vectors*. The complex conjugate set $\phi_i(\mathbf{r})^*$ are, in general, different functions which bear a one-to-one correspondence to the $\phi_i(\mathbf{r})$, so they can be said to be the basis vectors for a *dual space*. They are denoted by the symbol $\langle i|$ and are called *bra vectors*. The inner product of $\phi_i(\mathbf{r})$ and $\phi_j(\mathbf{r})$, analogous to the *scalar product* in ordinary vector space, is denoted by the bra–ket symbol $\langle i|j\rangle$ and is defined by the integral $\langle i|j\rangle \equiv \int_{\Omega_\mathbf{r}} \phi_i^*(\mathbf{r})\phi_j(\mathbf{r})\,d\mathbf{r}$. The operation of $\mathcal{Q}^{\text{op}}$ on $\phi_j(\mathbf{r})$ produces by a linear *mapping* of the vector $|j\rangle$ some other vector $\mathcal{Q}^{\text{op}}\phi_j(\mathbf{r}) \equiv \mathcal{Q}^{\text{op}}|j\rangle$ in the linear vector space. The matrix element $\mathcal{Q}_{ij}$ can then be interpreted as the *inner product* $\langle i|\mathcal{Q}^{\text{op}}|j\rangle$ of the vector $\mathcal{Q}^{\text{op}}|j\rangle$ (which is produced by this mapping) with the basis vector $\langle i|$,

$$\langle i|\mathcal{Q}^{\text{op}}|j\rangle \equiv \int_{\Omega_\mathbf{r}} \phi_i^*(\mathbf{r})\mathcal{Q}^{\text{op}}\phi_j(\mathbf{r})\,d\mathbf{r}.$$

It can be noted that the *diagonal elements* $\langle i|\mathcal{Q}^{\text{op}}|i\rangle$ represent the *expectation values* of $\mathcal{Q}^{\text{op}}$ whenever the system is in a specific eigenstate $\phi_i(\mathbf{r})$ of the Hamiltonian.

Whenever the operator $\mathcal{Q}^{\text{op}}$ has a simultaneous (viz, common) set of eigenfunctions with the Hamiltonian $\mathcal{H}$, such that $\mathcal{Q}^{\text{op}}\phi_j = Q_j\phi_j$ (all j), then the matrix elements of $\mathcal{Q}^{\text{op}}$ take the form $\langle i|\mathcal{Q}^{\text{op}}|j\rangle = \langle i|Q_j|j\rangle = Q_j\langle i|j\rangle = Q_j\delta_{ij}$, where we have presumed to have an orthonormal set ϕ_i. The operator $\mathcal{Q}^{\text{op}}$ is then said to constitute a *diagonal* operator, since its matrix representation is a *diagonal* matrix with respect to the basis functions in question,

$$\mathcal{Q}^{\text{op}} = \begin{pmatrix} \mathcal{Q}_{11} & 0 & 0 & \cdots \\ 0 & \mathcal{Q}_{22} & 0 & \\ 0 & 0 & \mathcal{Q}_{33} & \\ \vdots & & & \ddots \end{pmatrix}.$$

When the operator $\mathcal{Q}^{\text{op}}$ and the Hamiltonian have a simultaneous set of orthonormal eigenfunctions, then the expectation value of $\mathcal{Q}^{\text{op}}$ can be written

$$\begin{aligned}\langle \mathcal{Q}^{\text{op}}\rangle &= \left\langle \sum_i a_i^*\phi_i^* \middle| \mathcal{Q}^{\text{op}} \middle| \sum_j a_j\phi_j \right\rangle = \sum_i\sum_j a_i^*a_j\langle i|\mathcal{Q}^{\text{op}}|j\rangle = \sum_i\sum_j a_i^*a_j\langle i|Q_j|j\rangle \\ &= \sum_i\sum_j a_i^*a_jQ_j\langle i|j\rangle = \sum_i\sum_j a_i^*a_jQ_j\delta_{ij} = \sum_i a_i^*a_iQ_i = \sum_i |a_i|^2Q_i.\end{aligned}$$

This result can be interpreted by means of a postulate on quantum measurements due to John von Neumann: *Any ideal measurement of the value of an observable for a physical system yields one of the eigenvalues of the operator representing the observable.* Therefore the above result constitutes a weighted average of the result of many measurements on identical systems, with the weighting factors $|a_i|^2$ representing the probabilities for the system to be in the various eigenstates ϕ_i.

If the operator $\mathcal{Q}^{\text{op}}$ is not diagonal with respect to the *representation* (i.e., *the basis set*) defined by the eigenfunctions ϕ_i of the Hamiltonian, then to obtain the eigenvalues of $\mathcal{Q}^{\text{op}}$ we can proceed as follows. First of all, we know from the properties of Hermitian operators that a complete set of eigenfunctions exists for an arbitrary Hermitian operator $\mathcal{Q}^{\text{op}}$, so that we can confidently write $\mathcal{Q}^{\text{op}}g_i = Q_ig_i$ $(i = 1, 2, 3, \ldots)$. Here the g_i represent the complete set of functions and the Q_i are constants. By the fundamental expansion postulate, we next write a specific (though arbitrary) one of the functions g_i as a linear combination of the set of eigenfunctions ϕ_i of the Hamiltonian, $g_i = \sum_j a_{ij}\phi_j$ (i fixed), where the a_{ij} are the constant coefficients in the linear expansion. Substituting this expansion into the above eigenvalue equation for $\mathcal{Q}^{\text{op}}$ gives $\mathcal{Q}^{\text{op}} \sum_j a_{ij}\phi_j = Q_i \sum_j a_{ij}\phi_j$ (i fixed). To evaluate the coefficients a_{ij}, take the *inner product* (i.e., the *scalar product*) of each side of the above equation with an arbitrary basis state ϕ_k. This gives $\langle \phi_k | \mathcal{Q}^{\text{op}} \sum_j a_{ij}\phi_j \rangle = \langle \phi_k | Q_i \sum_j a_{ij}\phi_j \rangle$, or equivalently, $\sum_j a_{ij}[\langle k | \mathcal{Q}^{\text{op}} | j \rangle] = \sum_j a_{ij} Q_i \langle k | j \rangle$. Assuming orthonormality of the functions ϕ_i leads to $\langle k | j \rangle = \delta_{kj}$, which when substituted into this equation gives $\sum_j a_{ij}(\mathcal{Q}_{kj} - Q_i\delta_{kj}) = 0$ (i fixed; $k = 1, 2, 3, \ldots$). This constitutes the standard matrix form for the characteristic *eigenvalue problem*. Recognizing this as a set of simultaneous linear homogeneous algebraic equations for the unknowns $a_{i1}, a_{i2}, a_{i3}, \ldots$, we can (by employing Cramer's rule to attempt a solution) see that there exists no set of nonzero a_{ij} which satisfy this equation unless the determinant of the coefficients is zero, namely, $\det(\mathcal{Q}_{kj} - Q_i\delta_{kj}) = 0$ (i fixed), or more explicitly,

$$\begin{vmatrix} (\mathcal{Q}_{11} - Q_i) & \mathcal{Q}_{12} & \mathcal{Q}_{13} & \cdots \\ \mathcal{Q}_{21} & (\mathcal{Q}_{22} - Q_i) & \mathcal{Q}_{23} & \\ \mathcal{Q}_{31} & \mathcal{Q}_{32} & (\mathcal{Q}_{33} - Q_i) & \\ \vdots & & & \ddots \end{vmatrix} = 0.$$

This square array is referred to as the *secular determinant*, and expansion by minors leads in the usual way to an algebraic equation in the unknowns Q_i. This algebraic equation, commonly referred to as the *secular equation*, is of a high degree. Specifically, its degree is the same as the number N of orthonormal basis states ϕ_i. The solution of this algebraic equation then leads to N values of Q_i, and each such value can be used to solve the set of simultaneous algebraic equations given above for the coefficients a_{ij}. These in principle give all of the eigenfunctions g_i, since $g_i = \sum_j a_{ij}\phi_j$, and the set of Q_i constitute the set of eigenvalues of the operator $\mathcal{Q}^{\text{op}}$ which can be interpreted in light of the *von Neumann postulate* stated above. The set of g_i so obtained are referred to in matrix terminology as the *eigenvectors* of $\mathcal{Q}^{\text{op}}$. Symbolically, the set of simultaneous linear equations given above can be written in matrix form

$$\begin{pmatrix} (\mathcal{Q}_{11} - Q_i) & \mathcal{Q}_{12} & \mathcal{Q}_{13} & \cdots \\ \mathcal{Q}_{21} & (\mathcal{Q}_{22} - Q_i) & \mathcal{Q}_{23} & \\ \mathcal{Q}_{31} & \mathcal{Q}_{32} & (\mathcal{Q}_{33} - Q_i) & \\ \vdots & & & \ddots \end{pmatrix} \begin{pmatrix} a_{i1} \\ a_{i2} \\ a_{i3} \\ \vdots \end{pmatrix} = 0$$

or equivalently,

$$\begin{pmatrix} \mathcal{Q}_{11} & \mathcal{Q}_{12} & \mathcal{Q}_{13} & \cdots \\ \mathcal{Q}_{21} & \mathcal{Q}_{22} & \mathcal{Q}_{23} & \\ \mathcal{Q}_{31} & \mathcal{Q}_{32} & \mathcal{Q}_{33} & \\ \vdots & & & \ddots \end{pmatrix} \begin{pmatrix} a_{i1} \\ a_{i2} \\ a_{i3} \\ \vdots \end{pmatrix} = Q_i \begin{pmatrix} a_{i1} \\ a_{i2} \\ a_{i3} \\ \vdots \end{pmatrix},$$

or symbolically, $\mathcal{Q}^{\text{op}}\mathbf{a}^{(i)} = Q_i\mathbf{a}^{(i)}$, where $\mathbf{a}^{(i)}$ is the ith *eigenvector* of $\mathcal{Q}^{\text{op}}$,

$$\mathbf{a}^{(i)} \equiv \begin{pmatrix} a_{i1} \\ a_{i2} \\ a_{i3} \\ \vdots \end{pmatrix}.$$

It can be noted that the eigenvector is a column matrix made up of the coefficients a_{ij} appearing in the linear combination $g_i = \sum_j a_{ij}\phi_j$ of eigenstates of the Hamiltonian required for g_i to be an eigenfunction of the operator $\mathcal{Q}^{\text{op}}$ corresponding to the eigenvalue Q_i. This procedure is sometimes referred to as *diagonalizing the operator* $\mathcal{Q}^{\text{op}}$ *in the manifold spanned by the eigenfunctions of the Hamiltonian.*

8.10 Dirac Notation Summary

The use of angular brackets in quantum mechanics can be attributed to Dirac (1962). An abstract vector in function space (*Hilbert space*) is denoted by a *ket vector* symbol $|\,\rangle$. Labels, such as eigenvalues, which distinguish this vector from similar ket vectors are written in as arguments of the ket vector. For example, $|j\rangle$ could denote the eigenfunction $\psi_j(\mathbf{r}, t)$ belonging to the energy eigenvalue $\mathscr{E}_j$.

The complex conjugate of a ket vector is denoted by a *bra vector* symbol $\langle\,|$, with the arguments being unchanged from the ket. Thus $\langle j|$ could denote $\psi_j^*(\mathbf{r}, t)$.

The scalar (or *inner*) product of a function with itself is denoted by the bra-(c)-ket,

$$\langle j|j\rangle \equiv \int_{\Omega_\mathbf{r}} \psi_j^*(\mathbf{r}, t)\psi_j(\mathbf{r}, t)\, d\mathbf{r},$$

which is sometimes referred to as the square of the *norm* (absolute value) of the vector.

The scalar product of a function $\psi_i(\mathbf{r}, t)$ with a function $\psi_j(\mathbf{r}, t)$ is also denoted by a bra-(c)-ket, $\langle i|j\rangle \equiv \int_{\Omega_\mathbf{r}} \psi_i^*(\mathbf{r}, t)\psi_j(\mathbf{r}, t)\, d\mathbf{r}$. If $\psi_i(\mathbf{r}, t)$ and $\psi_j(\mathbf{r}, t)$ are orthogonal and normalized, then $\langle i|j\rangle = \delta_{ij}$, where δ_{ij} is the Kronecker delta function which equals unity when $i = j$ but is zero otherwise.

The expectation value of an operator $\mathcal{Q}^{\text{op}}$ with respect to a state $\psi_j(\mathbf{r}, t)$ is denoted in bra-(c)-ket notation by $\langle j|\mathcal{Q}^{\text{op}}|j\rangle \equiv \int_{\Omega_\mathbf{r}} \psi_j^*(\mathbf{r}, t)\mathcal{Q}^{\text{op}}\psi_j(\mathbf{r}, t)\, d\mathbf{r}$. The matrix representation of the arbitrary operator $\mathcal{Q}^{\text{op}}$ with respect to a complete set of basis functions $\psi_1, \psi_2, \psi_3, \ldots$ involves a square matrix of numbers, each number being one of the "matrix elements"

$$\mathcal{Q}_{ij} = \langle i|\mathcal{Q}^{\text{op}}|j\rangle \equiv \int_{\Omega_\mathbf{r}} \psi_i^*(\mathbf{r}, t)\mathcal{Q}^{\text{op}}\psi_j(\mathbf{r}, t)\, d\mathbf{r}.$$

The bra and ket vectors correspond, respectively, to the row and column vectors of matrix algebra.

The principal merit of Dirac notation insofar as our work is concerned lies in the simplification of writing equations. There are few who would argue the point that $\langle j|j\rangle$ is far easier to write than the integral form it represents. Moreover, writing quantum mechanics equations in Dirac notation is fun!

PROJECT 1.17 Born's Proof of the General Uncertainty Relation

1. Show that the Heisenberg uncertainty relations (1.207) are special cases of the basic theorem regarding all general quantum mechanical uncertainty relations: $(\Delta\mathcal{Q}_1^{\text{op}})^2(\Delta\mathcal{Q}_2^{\text{op}})^2 \geqslant \frac{1}{4}\langle i[\mathcal{Q}_1^{\text{op}}, \mathcal{Q}_2^{\text{op}}]\rangle^2$, where $\mathcal{Q}_1^{\text{op}}$ and $\mathcal{Q}_2^{\text{op}}$ are Hermitian operators, $[\mathcal{Q}_1^{\text{op}}, \mathcal{Q}_2^{\text{op}}]$ denotes the commutator $\mathcal{Q}_1^{\text{op}}\mathcal{Q}_2^{\text{op}} - \mathcal{Q}_2^{\text{op}}\mathcal{Q}_1^{\text{op}}$ in accordance with Eq. (1.242), the symbol $\langle\ \rangle$ denotes expectation value in accordance with Eq. (1.212), and $(\Delta\mathcal{Q}^{\text{op}})^2 \equiv \langle(\mathcal{Q}^{\text{op}} - \langle\mathcal{Q}^{\text{op}}\rangle)^2\rangle$ denotes the mean square deviation.
2. Prove the general uncertainty relation stated above (part 1). [*Hint*: You may find it helpful to refer to Born (1957, p. 387) or to some standard reference such as Ikenberry (1962, pp. 74–75).]

PROJECT 1.18 Fundamental Postulates of Quantum Mechanics

Formulate a set of postulates upon which the discipline of quantum mechanics can be erected. [*Hint*: See Rojansky (1938).]

9 Probability Current Density

9.1 Equation of Continuity

The probability density $\rho = |\psi(\mathbf{r}, t)|^2$ represents a convenient starting point for the consideration of particle and charge currents as computed quantum mechanically. If we consider this as a statistical quantity which is a continuous function of position, then the time derivative gives the rate of change of the particle density with time,

$$\frac{\partial\rho}{\partial t} = \frac{\partial}{\partial t}|\psi(\mathbf{r}, t)|^2 = \frac{\partial}{\partial t}[\psi^*(\mathbf{r}, t)\psi(\mathbf{r}, t)] = \psi^*\frac{\partial\psi}{\partial t} + \frac{\partial\psi^*}{\partial t}\psi. \tag{1.251}$$

However, a time rate of change of the probability density at any given point in space requires a difference between the particle currents flowing into and out of the differential volume surrounding the point in question. The mathematical statement of this fact is the well-known microscopic equation of continuity,

$$\partial\rho/\partial t = -\nabla\cdot\mathbf{J}, \tag{1.252}$$

where $\nabla\cdot\mathbf{J}$ is the divergence of the particle current density $\mathbf{J}$ at the point in question.

9.2 Development of a General Expression for the Particle Current Density

Equating the two expressions (1.251) and (1.252) for $\partial\rho/\partial t$ gives the relation

$$\nabla\cdot\mathbf{J} = -\left(\psi^*\frac{\partial\psi}{\partial t} + \frac{\partial\psi^*}{\partial t}\psi\right) \tag{1.253}$$

which must be obeyed by the quantum-mechanical analog of the particle current density **J**. For further development of this expression, we employ the time-dependent Schrödinger equation (1.124) and its complex conjugate,

$$i\hbar \frac{\partial \psi}{\partial t} = -\frac{\hbar^2}{2m} \nabla^2 \psi + \mathscr{V}(\mathbf{r})\psi, \tag{1.254}$$

$$-i\hbar \frac{\partial \psi^*}{\partial t} = -\frac{\hbar^2}{2m} \nabla^2 \psi^* + \mathscr{V}(\mathbf{r})\psi^*. \tag{1.255}$$

In taking the complex conjugate, we have used the fact that $\mathbf{r}^* = \mathbf{r}$, $t^* = t$, and $\mathscr{V}(\mathbf{r})^* = \mathscr{V}(\mathbf{r})$ due to the fact that we are concerned with real positions, real times, and real potentials. We multiply the Schrödinger equation by ψ^* and its complex conjugate by ψ to give

$$i\hbar\psi^* \frac{\partial \psi}{\partial t} = -\frac{\hbar^2}{2m} \psi^* \nabla^2 \psi + \psi^* \mathscr{V}(\mathbf{r})\psi, \tag{1.256}$$

$$-i\hbar\psi \frac{\partial \psi^*}{\partial t} = -\frac{\hbar^2}{2m} \psi \nabla^2 \psi^* + \psi \mathscr{V}(\mathbf{r})\psi^*. \tag{1.257}$$

Subtracting the second equation from the first gives

$$i\hbar \left(\psi^* \frac{\partial \psi}{\partial t} + \psi \frac{\partial \psi^*}{\partial t} \right) = \frac{\hbar^2}{2m} (\psi \nabla^2 \psi^* - \psi^* \nabla^2 \psi), \tag{1.258}$$

where in equating $\psi^* \mathscr{V}(\mathbf{r})\psi$ with $\psi \mathscr{V}(\mathbf{r})\psi^*$ we have used the fact that $\mathscr{V}(\mathbf{r})$ is merely a multiplicative operator so that the factors in the product $\psi^* \mathscr{V}(\mathbf{r})\psi$ commute. The right-hand side involves the Laplacian operator, so that it is reasonable to expect that it can be expressed as the divergence of some vector quantity. If we take the divergence of $\psi^* \nabla \psi$, we obtain

$$\nabla \cdot (\psi^* \nabla \psi) = \nabla \psi^* \cdot \nabla \psi + \psi^* \nabla^2 \psi, \tag{1.259}$$

whereas if we take the divergence of $\psi \nabla \psi^*$, we obtain

$$\nabla \cdot (\psi \nabla \psi^*) = \nabla \psi \cdot \nabla \psi^* + \psi \nabla^2 \psi^*. \tag{1.260}$$

Recognizing that the dot product of two vectors such as $\nabla \psi \cdot \nabla \psi^*$ is commutative, so that $\nabla \psi \cdot \nabla \psi^* = \nabla \psi^* \cdot \nabla \psi$, we can subtract the above two quantities (1.259) and (1.260) to give the relation

$$\nabla \cdot (\psi^* \nabla \psi - \psi \nabla \psi^*) = \psi^* \nabla^2 \psi - \psi \nabla^2 \psi^*. \tag{1.261}$$

The right-hand side can be identified with the factor in the right-hand side of (1.258); we thereby obtain

$$i\hbar(\psi^* \partial\psi/\partial t + \psi\, \partial\psi^*/\partial t) = -(\hbar^2/2m)\, \nabla \cdot (\psi^* \nabla \psi - \psi \nabla \psi^*). \tag{1.262}$$

Substituting into Eq. (1.253) for $\nabla \cdot \mathbf{J}$ gives

$$\nabla \cdot \mathbf{J} = (\hbar/2mi)\, \nabla \cdot (\psi^* \nabla \psi - \psi \nabla \psi^*). \tag{1.263}$$

Within an arbitrary constant, then,

$$\mathbf{J} = (\hbar/2mi)(\psi^* \nabla\psi - \psi \nabla\psi^*). \tag{1.264}$$

The arbitrary constant is zero if $\mathbf{J} = 0$ whenever $\psi = 0$, as one would expect. Knowledge of the wave function ψ therefore allows us to calculate the particle current density $\mathbf{J}$ quantum mechanically. For the case of electrons, the charge per particle is $-e$, so that the charge current density $\mathscr{J}$ follows immediately from

$$\mathscr{J} = -e\mathbf{J}. \tag{1.265}$$

9.3 Specific Application to Free Particles

It is illuminating to apply the above relation (1.264) to the specific case of plane waves, which have been shown to be eigenfunctions of the momentum operator $\mathbf{p}^{\text{op}} = -i\hbar\nabla$. Thus $\nabla\psi = (i/\hbar)\mathbf{p}\psi$ and $\nabla\psi^* = (-i/\hbar)\mathbf{p}\psi^*$, so that

$$\mathbf{J} = (\hbar/2mi)[\psi^*(i/\hbar)\mathbf{p}\psi - \psi(-i/\hbar)\mathbf{p}\psi^*] = (\mathbf{p}/2m)[\psi^*\psi + \psi^*\psi] = \psi^*\psi\mathbf{v}. \tag{1.266}$$

This is simply the product of the particle probability density $\psi^*\psi$ and the particle velocity $\mathbf{v}$, which is readily interpreted as the particle current density on the basis of physical considerations.

10 Energy Levels and Density of States

10.1 Bound States of a Particle

The plane-wave solutions of the Schrödinger equation for the motion of particles in free space were shown in §6 to be eigenfunctions of both the Hamiltonian operator and the momentum operator. The energy eigenvalue $\mathscr{E} = \hbar\omega$ and the momentum eigenvalue $\mathbf{p} = \hbar\mathbf{k}$ corresponding to a given plane wave $A_{\mathbf{k}} \exp[i(\mathbf{k} \cdot \mathbf{r} - \omega t)] = A_{\mathbf{k}} \exp[(i/\hbar)(\mathbf{p} \cdot \mathbf{r} - \mathscr{E} t)]$ are related by $\mathscr{E} = p^2/2m$, and there is no restriction on the values of $\mathscr{E}$ or p. Therefore we say that this represents a region of the energy spectrum which is a continuum. For negative energy eigenvalues, however, representing particles trapped in a potential $\mathscr{V}(\mathbf{r})$, the boundary conditions effectively restrict the eigenvalues to discrete values. The allowed values of the energy in this discrete region of the energy spectrum are of interest in themselves. Moreover, a great many physical properties of the system depend on the number of such energy levels per unit energy range and the variation of this number with energy, which is termed *the density of states as a function of energy*. That is, *the elemental number dN of states in an energy interval $d\mathscr{E}$ at the energy $\mathscr{E}$ is given by the product of the density of states $\tilde{g}(\mathscr{E})$ and the energy interval $d\mathscr{E}$*,

$$dN = \tilde{g}(\mathscr{E})\, d\mathscr{E}, \tag{1.267}$$

or equivalently,

$$\tilde{g}(\mathscr{E}) = dN/d\mathscr{E} \qquad \textbf{(density of states as a function of energy)}. \tag{1.268}$$

We can define a *density of states as a function of momentum* $\mathscr{G}(p)$ in an analogous fashion, namely,

$$dN = \mathscr{G}(p)\, dp. \tag{1.269}$$

The purpose of the present section is to work out the allowed energy levels and the density of states for a simple one-dimensional problem in order to illustrate the general approach.

10.2 Particle Trapped in a One-Dimensional Box

Consider a particle trapped in a potential well extending from $-\frac{1}{2}L \leqslant x \leqslant \frac{1}{2}L$. We can assure this mathematically by choosing the potential $\mathscr{V}(x)$ to be zero inside and infinite outside this region. The total energy $\mathscr{E}$ of the particle will be entirely kinetic inside the potential well; for finite values of the kinetic energy inside the well, the particle would never be able to penetrate into the infinitely high potential barriers in the regions $x < -\frac{1}{2}L$ and $x > \frac{1}{2}L$. Therefore we must choose $\psi(x, t)$ to be zero in these forbidden regions, which gives us the boundary conditions $\psi(-\frac{1}{2}L, t) = 0$ and $\psi(\frac{1}{2}L, t) = 0$. We must consider the allowable solutions of the Schrödinger equation inside the well. Let us consider stationary-state solutions $\psi_l(x, t) = \phi_l(x) \exp[(-i/\hbar)\mathscr{E}_l t]$. Since $\mathscr{V}(x) = 0$, the time-independent Schrödinger equation (1.164) is simply

$$-\frac{\hbar^2}{2m}\frac{d^2\phi_l}{dx^2} = \mathscr{E}_l \phi_l. \tag{1.270}$$

Exponential plane-wave solutions satisfy this equation but do not individually satisfy the boundary conditions stated above. Linear combinations of plane waves can be used to generate sine and cosine functions which can satisfy the boundary conditions. If we attempt the trial solution

$$\phi_l(x) = A_l \sin k_l x + B_l \cos k_l x, \tag{1.271}$$

with A_l and B_l being constants and k_l representing allowed values of the wave vector, we find by direct substitution into the Schrödinger equation (1.270) that the condition for the validity of such a solution is

$$\hbar^2 k_l^2/2m = \mathscr{E}_l. \tag{1.272}$$

The boundary conditions require

$$\phi_l(-\tfrac{1}{2}L) = A_l \sin(-\tfrac{1}{2}k_l L) + B_l \cos(-\tfrac{1}{2}k_l L) = 0, \tag{1.273}$$

$$\phi_l(\tfrac{1}{2}L) = A_l \sin(\tfrac{1}{2}k_l L) + B_l \cos(\tfrac{1}{2}k_l L) = 0. \tag{1.274}$$

Since the sine function is zero whenever its argument is an even multiple of $\frac{1}{2}\pi$ and the cosine function is zero whenever its argument is an odd multiple of $\frac{1}{2}\pi$, we can satisfy the above conditions in two different ways. First, we can choose

$$B_l = 0, \tag{1.275}$$

$$\tfrac{1}{2}k_l L = 2l(\tfrac{1}{2}\pi) = l\pi \qquad (l = 1, 2, 3, \ldots) \tag{1.276}$$

which leads to the *odd-parity solutions*

$$\phi_l(x) = A_l \sin(2l\pi x/L) \tag{1.277}$$

with

$$\mathscr{E}_l = \hbar^2 k_l^2/2m = (\hbar^2/2m)(2l\pi/L)^2 \qquad (l = 1, 2, 3, \ldots). \tag{1.278}$$

Second, we can choose

$$A_l = 0, \tag{1.279}$$

$$\tfrac{1}{2}k_l L = (2l + 1)(\tfrac{1}{2}\pi) = (l + \tfrac{1}{2})\pi \qquad (l = 1, 2, 3, \ldots), \tag{1.280}$$

which leads to the *even-parity solutions*

$$\phi_l(x) = B_l \cos[(2l + 1)\pi x/L] \tag{1.281}$$

with

$$\mathscr{E}_l = \hbar^2 k_l^2/2m = (\hbar^2/2m)[(2l + 1)\pi/L]^2 \qquad (l = 1, 2, 3, \ldots). \tag{1.282}$$

We can combine the even and odd parity results by introducing the single integer n, where n represents $2l$ when even and $(2l + 1)$ when odd. The integer n is called the quantum number. Then

$$k_n = n\pi/L, \tag{1.283}$$

so that

$$\phi_n(x) = \begin{cases} A_n \sin(n\pi x/L) & (n \text{ even}) \\ B_n \cos(n\pi x/L) & (n \text{ odd}), \end{cases} \tag{1.284}$$

$$\mathscr{E}_n = (\hbar^2/2m)(n\pi/L)^2 \qquad (n = 1, 2, 3, \ldots). \tag{1.285}$$

The corresponding time-dependent wave functions $\psi_n(x, t)$ are given by

$$\psi_n(x, t) = \phi_n(x)\, e^{-(i/\hbar)\mathscr{E}_n t}, \tag{1.286}$$

so the probability densities are given by

$$|\psi_n(x, t)|^2 = \psi_n^*(x, t)\psi_n(x, t) = \phi_n^*(x)\phi_n(x) = |\phi_n(x)|^2. \tag{1.287}$$

Normalization of the wave functions for the present case requires

$$\int_{-L/2}^{L/2} |\psi_n(x, t)|^2\, dx = 1. \tag{1.288}$$

This ensures that there is unit probability for finding the particle somewhere in the potential well whenever the state is occupied. Substitution of $|A_n|^2 \sin^2(n\pi x/L)$ and $|B_n|^2 \cos^2(n\pi x/L)$ into this relation determines $|A_n|$ and $|B_n|$. The integration gives

$$|A_n| = |B_n| = (2/L)^{1/2} \tag{1.289}$$

for all n. In addition, A_n aad B_n can be chosen to be real and positive with no loss in generality, in which case $A_n = |A_n|$ and $B_n = |B_n|$.

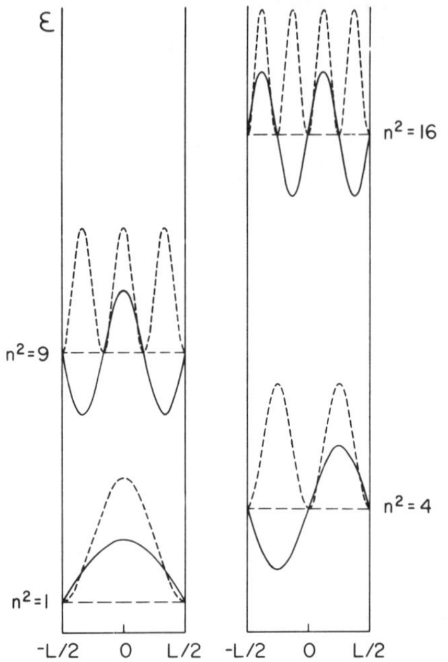

Fig. 1.30 Energy levels, stationary-state wave functions (solid curves) versus position, and probability densities (dashed curves) versus position for a particle trapped in a one-dimensional square-well potential extending from $x = -L/2$ to $x = L/2$. Even-parity solutions are represented on the left-hand side, and odd-parity solutions are represented on the right-hand side; higher energy solutions than the ones illustrated occur for larger integer values of the quantum number n.

The quantum-mechanical solutions obtained above for the problem of a particle in a one-dimensional box are sketched in Fig. 1.30. The relative location of the two lowest even-parity solutions on an energy scale are shown in the left-hand side of the figure, and the relative location of the two lowest odd-parity solutions on an energy scale are shown in the right-hand side of the figure. Also plotted are the stationary-state wave functions $\phi_n(x)$ as a function of position x for each discrete energy eigenvalue; these are the solid curves. It can be seen that the even-parity solutions have an odd number of half-wavelengths within the potential well, and the odd-parity solutions have an even number of half-wavelengths. This follows directly from the relation $k_n = 2\pi/\lambda_n$ given by Eq. (1.12) and the allowed values $k_n = n\pi/L$ given above in Eq. (1.283), namely,

$$k_n = 2\pi/\lambda_n = n\pi/L, \tag{1.290}$$

giving

$$n(\tfrac{1}{2}\lambda_n) = L, \tag{1.291}$$

where n is an odd integer for even-parity solutions and an even integer for odd-parity solutions. The dashed curves in Fig. 1.30 are the corresponding probability densities $|\psi(x, t)|^2 = |\phi(x)|^2$. It can be seen that the even- and odd-parity solutions alternate with increasing energy, and the higher energy solutions have a greater number of nodes in the wave function. From Eq. (1.285) it is seen that the energies increase as n^2, and are inversely proportional to the particle mass m and the square of the dimension L of the potential well.

PROJECT 1.19 Particle in a Box

Consider a one-dimensional potential well 4 Å on an edge with infinitely high potential barriers bounding the potential well. Compute the following quantities:
1. The ground-state energy of an electron in the well.
2. The ground-state energy of a proton in the well.
3. The first three excited states of an electron in the well.
4. The first three excited states of a proton in the well.
5. The wavelengths and frequencies of the electromagnetic radiation given off or absorbed in all electronic transitions between the ground state and the first three excited states, and among the first three excited states.
6. The wavelengths and frequencies of the electromagnetic radiation given off or absorbed in all proton transitions between the ground state and the first three excited states, and among the first three excited states.
7. Would you expect there to be any restrictions on the transitions considered above? If so, what would be the nature of such restrictions?
8. Repeat the above calculations for particles in a three-dimensional potential well 4 Å on an edge with infinitely high potential barriers bounding the well. What additional factors do you find relative to the one-dimensional case?
9. Qualitatively explain what changes might occur in the energy eigenfunctions and eigenvalues as the heights of the potential walls at the boundaries are decreased from infinity toward lower and lower values.

PROJECT 1.20 Matrix Representation of Operators

1. Deduce the matrix representation of the position coordinate x in the representation given by the energy eigenfunctions for a particle of mass m in an infinite one-dimensional square-well potential. (Consider both periodic and fixed boundary condition eigenfunctions.)
2. Deduce the matrix representation of the momentum operator in the energy representation for a particle in the infinite one-dimensional square-well potential. Attempt to extend your treatment to the three-dimensional infinite square-well potential.
3. Prove that $p_{jk} = (i/\hbar)m(\mathscr{E}_j - \mathscr{E}_k)x_{jk}$, where p_{jk} is the jk matrix element of the x momentum operator, x_{jk} is the jk element of the position operator, and m is the mass of the particle. The quantities $\mathscr{E}_j$ and $\mathscr{E}_k$ represent the energy eigenvalues for the states j and k, respectively.

PROJECT 1.21 Particle in a Superposition State: Expansion of Wave Function in Terms of a Complete Set of Eigenstates and the Expectation Value of the Energy

For the one-dimensional square-well potential with infinitely high barriers at the edges, consider a non-stationary-state wave function $\Phi(x) = Ax(L - x)$, where A is the normalization factor and L is the width of the potential well.
1. Determine A.
2. Solve for the coefficients a_n in the expansion $\psi(x, t) = \sum_n a_n\phi_n \exp[(-i/\hbar)\mathscr{E}_n t]$ of the wave function $\psi(x, t)$ in terms of a complete set of the eigenfunctions $\phi_n(x)$ of the Hamiltonian.
3. Find the average (i.e., the expectation value) of the energy.
4. Find the dispersion [i.e., the mean square deviation (as defined in §8.2)] of the energy. [*Hint*: See ter Haar (1975)].

PROJECT 1.22 Momentum Probability Distribution

For a particle in a one-dimensional square-well potential of width L with infinitely high potential-energy barriers at the edges, determine the momentum probability density $|\chi(k)|^2$ for the ground state and the first four excited states.

10.3 Density of States

To compute a density of states $\tilde{g}(\mathscr{E})$ as a function of energy $\mathscr{E}$, we consider averages over regions of energy containing many discrete levels $\mathscr{E}_n$ so that $\tilde{g}(\mathscr{E})$ can be considered to be a quasi-continuous function of $\mathscr{E}$. This is a valid approach for those macroscopic systems for which the concept of a density of states happens to be useful (cf. Chap. 3). Thus, if we consider $\mathscr{E}_n$ as a continuous function of n, we can differentiate $\mathscr{E}_n$ with respect to n. Since for the present case $\mathscr{E}_n = (\hbar^2/2m)(n\pi/L)^2$, according to Eq. (1.285), we obtain

$$d\mathscr{E}_n/dn = 2(\hbar^2/2m)\,n\pi^2/L^2. \tag{1.292}$$

We take the square root of both sides of Eq. (1.285) and solve for n,

$$n = (L/\pi)(2m/\hbar^2)^{1/2}\mathscr{E}_n^{1/2}. \tag{1.293}$$

Substituting into Eq. (1.292) for $d\mathscr{E}_n/dn$ gives

$$d\mathscr{E}_n/dn = (2\pi/L)(\hbar^2/2m)^{1/2}\mathscr{E}_n^{1/2}. \tag{1.294}$$

Since we are considering $\mathscr{E}_n$ to be a continuous function of n, we can identify it with the continuous variable $\mathscr{E}$. We can thus replace $\mathscr{E}_n$ by $\mathscr{E}$ in this last expression if we wish, and due to the fact that $\mathscr{E}_n$ is a function of the single variable n, we can also take the reciprocal of each side of this expression. We thus obtain

$$dn/d\mathscr{E} = (2m/\hbar^2)^{1/2}(L/2\pi)\mathscr{E}^{-1/2}. \tag{1.295}$$

For each different value of the integer quantum number n, we have a different quantum state, so the value of n gives the total number N of quantum states with energy equal to or less than the energy of state n. Therefore $dn/d\mathscr{E}$ gives the change in the total number of energy states per unit increase in energy at any given energy $\mathscr{E}$. Since $dn/d\mathscr{E} \simeq \Delta n/\Delta\mathscr{E}$ can be interpreted as the number of states per unit energy range at any given energy, we can identify $dn/d\mathscr{E}$ as the density of states $\tilde{g}(\mathscr{E})$,

$$\tilde{g}(\mathscr{E}) = dn/d\mathscr{E} = (2m/\hbar^2)^{1/2}(L/2\pi)\mathscr{E}^{-1/2}. \tag{1.296}$$

We thus have derived the density of states for the problem of a particle trapped in a one-dimensional square-well potential. The usefulness of the density of states concept becomes quite evident in the free electron model for metals developed in Chap. 3.

It is illustrative to derive this result using a slightly different approach. We note that the allowed k values for this problem are $k_n = n\pi/L$ with the corresponding allowed values of momentum $p_n = \hbar k_n = n\pi\hbar/L$. Again the value of n tells us how many allowed states have k values less than or equal to k_n and corresponding momentum values less than or equal to $n\pi\hbar/L$. Therefore the density of states $\tilde{w}(k) = dn/dk$ as a function of allowed k value is given by

$$\tilde{w}(k) = dn/dk_n = (d/dk_n)(k_n L/\pi) = L/\pi. \tag{1.297}$$

This represents a uniform distribution of allowed k values since it is independent of k. Because momentum is related linearly to k, we likewise have a uniform distribution of allowed momentum values for this problem. If the density of states as a function of momentum is denoted by $\mathscr{G}(p)$, then conservation of the number of allowed states in any momentum interval dp at any arbitrary value of the momentum p requires

$$\mathscr{G}(p)\, dp = \tilde{w}(k)\, dk. \tag{1.298}$$

That is, the number of states in a given momentum interval must be equal to the number of states in the corresponding k interval. Since $p = \hbar k$ and $dp = \hbar\, dk$, we obtain

$$\mathscr{G}(p) = \tilde{w}(k)\, dk/dp = \hbar^{-1}\tilde{w}(k) = L/\pi\hbar \tag{1.299}$$

for this one-dimensional problem. To obtain the corresponding density of states as a function of energy, we again employ a relation requiring that the number of states in a given energy interval $d\mathscr{E}$ be equal to the number of states in the corresponding momentum interval dp,

$$\tilde{g}(\mathscr{E})\, d\mathscr{E} = \mathscr{G}(p)\, dp, \tag{1.300}$$

where $\mathscr{E} = p^2/2m$ and $d\mathscr{E} = (p/m)\, dp$. [This equation can also be obtained by equating dN in Eq. (1.267) with dN in Eq. (1.269).] Thus we obtain

$$\tilde{g}(\mathscr{E}) = \mathscr{G}(p)\, dp/d\mathscr{E} = (L/\pi\hbar)(m/p) = (mL/\pi\hbar)(2m\mathscr{E})^{-1/2}, \tag{1.301}$$

which agrees with our previous result (1.296).

The corresponding three-dimensional problem can be solved in an analogous fashion; the results are highly useful in understanding the electronic properties of metals within the framework of the quantum-mechanical free-electron model. This model will be formulated and developed in Chap. 3. It is first necessary to understand how the energy levels of a system are statistically populated at various temperatures, since for electrons in condensed media, classical Boltzmann statistics can be used only in certain limiting cases. We will therefore develop the essentials of quantum statistics in Chap. 2.

In the following section we shall develop another classical quantum-mechanical problem which is quite different in character from the square-well potential problem developed in the present section, namely, we shall work out the mathematics which gives us the physical behavior of quantum particles in the neighborhood of potential-energy barriers of finite height. The importance of this problem can hardly be overestimated in view of the array of quantum tunnel current devices that have been recently developed and are currently under development (cf. Chap. 4, §§7.2–7.5).

EXERCISE Evaluate the one-dimensional density of states given by Eq. (1.296) for an electron system extending over a 1 cm length at an energy $\mathscr{E} = 1$ eV.

EXERCISE Work out $\tilde{w}(\mathbf{k})$, $\mathscr{G}(\mathbf{p})$, and $\tilde{g}(\mathscr{E})$ for the two-dimensional free-electron model.

11 Reflection and Transmission Coefficients for a Particle Beam at a Potential-Energy Step Discontinuity and at a Rectangular Barrier

11.1 Introduction

Sudden changes in the potential energy can produce wavelike reflections of quantum particles which are quite unlike any phenomena which would be expected classically for point particles. These effects are not only intrinsically interesting as textbook quantum phenomena, but in addition the treatment leads us directly to the consideration of quantum mechanical tunneling which is currently of the greatest interest for solid state devices.

11.2 The Step Potential

Consider the so-called *step potential* illustrated in Fig. 1.31, for which the potential energy $U(x)$ is zero for $x < 0$ (region I) and has the constant value U_0 for $x > 0$ (region II). To be specific, we choose an incident beam of particles

$$\psi_{\text{inc}} = Ae^{i(kx-\omega t)} = Ae^{(i/\hbar)[px-\mathscr{E}t]} \tag{1.302}$$

traveling to the right in region I toward the step at $x = 0$. The incident beam consists of individual particles having a well-defined momentum $p^{(\text{I})} = \hbar k$ in the x direction and a well defined kinetic energy $\mathscr{E}_{\text{K}}^{(\text{I})} = [p^{(\text{I})}]^2/2m$ equal to the total energy $\mathscr{E}$, since the potential energy is zero in region I. The particle density in the incident beam is given by

$$\psi_{\text{inc}}^*\psi_{\text{inc}} = A^*A. \tag{1.303}$$

The incident beam intensity I_{inc} will be given by the product of the particle density $\psi_{\text{inc}}^*\psi_{\text{inc}}$ and the particle velocity $v^{(\text{I})} = p^{(\text{I})}/m$,

$$I_{\text{inc}} = \psi_{\text{inc}}^*\psi_{\text{inc}}v^{(\text{I})} = A^*A(\hbar k/m), \tag{1.304}$$

where

$$\hbar k = p^{(\text{I})} = (2m\mathscr{E})^{1/2}. \tag{1.305}$$

The positive sign is to be chosen for the square root.

From the standpoint of classical mechanics, the kinetic energy $\mathscr{E}_{\text{K}}^{(\text{II})} = \mathscr{E} - U_0$ is positive for $\mathscr{E} > U_0$ so the particle would always traverse the step; however, the kinetic energy would be negative for $\mathscr{E} < U_0$, which represents an impossible situation from the standpoint of classical physics, since it would require an imaginary value for the momentum $p^{(\text{II})} = [2m(\mathscr{E} - U_0)]^{1/2}$. Thus the particle would reverse its direction (i.e., it would always be *reflected*) when incident on a

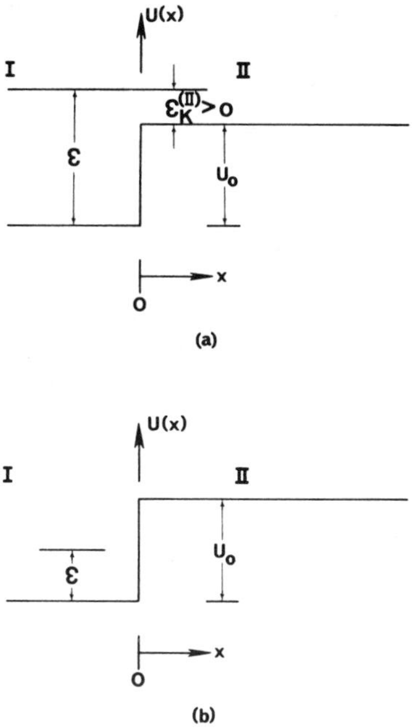

Fig. 1.31 Step potential defined by $U(x) = 0$ for $x < 0$ and $U(x) = U_0$ for $x \geqslant 0$, with the total energy per particle of the incident beam denoted by $\mathcal{E}$. (a) $\mathcal{E} > U_0$, in which case the kinetic energy remains positive as a particle from the incident beam in region I enters region II. (b) $\mathcal{E} < U_0$, in which case the kinetic energy becomes negative when a particle from the incident beam in region I enters region II.

step potential for which $\mathcal{E} < U_0$. It is interesting to compare these classical predictions with the quantum results deduced below.

Even though we must eventually consider two situations, namely, the case in which $\mathcal{E} < U_0$ and the case in which $\mathcal{E} > U_0$, for the moment let us not make a specific choice but continue insofar as we are able with a treatment applicable to both situations. In either case, let us consider the possibility that a portion of the beam will be reflected. The wave function for the reflected component can be written

$$\psi_{\text{ref}} = Be^{i(-kx-\omega t)}, \tag{1.306}$$

since the reflected component will be traveling in the negative x direction with the same total energy $\mathcal{E} = \hbar\omega$ and the same x-momentum value, with the exception of a reversal in the sign of the momentum. Thus the reflected particle intensity is given by

$$\psi_{\text{ref}}^*\psi_{\text{ref}} = B^*B, \tag{1.307}$$

and the reflected beam intensity is given by

$$I_{\text{ref}} = \psi_{\text{ref}}^*\psi_{\text{ref}}(-\hbar k/m) = -B^*B(\hbar k/m), \tag{1.308}$$

where k is defined as previously stated above.

The wave functions ψ_{inc} and ψ_{ref} were written simply from our knowledge of traveling waves (see §6.1). It is easy to show, however, that both satisfy the time-dependent Schrödinger equation for region I where $U(x) = 0$, namely,

$$-(\hbar^2/2m)\,\partial^2\psi/\partial x^2 = i\hbar\,\partial\psi/\partial t. \tag{1.309}$$

Substituting ψ_{inc} into this equation gives

$$-(\hbar^2/2m)(ik)^2\psi_{\text{inc}} = i\hbar(-i\omega)\psi_{\text{inc}} \tag{1.310}$$

which is satisfied provided $\hbar^2k^2/2m = \hbar\omega$. This is equivalent to the free particle dispersion relation $\mathscr{E} = p^2/2m$ which is indeed appropriate for region I. Alternately, the direct substitution of $\hbar k = p = (2m\mathscr{E})^{1/2}$ and $\omega = \mathscr{E}/\hbar$ into this equation yields $\mathscr{E} = \mathscr{E}$, thus verifying the applicability of the condition. Similarly, substituting ψ_{ref} into the above time-dependent Schrödinger equation for region I gives

$$-(\hbar^2/2m)(-ik)^2\psi_{\text{ref}} = i\hbar(-i\omega)\psi_{\text{ref}}, \tag{1.311}$$

which is satisfied since once again $\mathscr{E} = \hbar\omega = \hbar^2k^2/2m$. The superposition wave function

$$\psi_{\text{I}} = \psi_{\text{inc}} + \psi_{\text{ref}} = Ae^{i(kx-\omega t)} + Be^{i(-kx-\omega t)} = (Ae^{ikx} + Be^{-ikx})e^{-i\omega t} \tag{1.312}$$

likewise satisfies the above time-dependent Schrödinger equation for region I, as can be argued from the linearity of the Schrödinger equation and therefore the applicability of the principle of superposition. (The principle of superposition means that any linear combination of individual solutions will likewise be a solution.) Alternately, the direct substitution of ψ_{I} into the above Schrödinger equation for region I shows that it satisfies the equation.

Let us now consider the Schrödinger equation appropriate for region II, and examine the form of possible solutions. The total energy $\mathscr{E}$ of the particle is the same in region II as in region I because there are no external forces acting on the particle. The step potential does instantaneously accelerate the particle in the negative direction, which leads to an interchange of kinetic and potential energies for the conservative system. Since the potential energy is U_0 in region II, the time-dependent Schrödinger equation which must be satisfied is

$$-(\hbar^2/2m)\,d^2\psi/dx^2 + U_0\psi = i\hbar\,\partial\psi/\partial t. \tag{1.313}$$

It is now convenient to treat individually the cases $\mathscr{E} > U_0$ and $\mathscr{E} < U_0$.

11.2.1 Case of $\mathscr{E} > U_0$. If $\mathscr{E} > U_0$, the kinetic energy $\mathscr{E}_{\text{K}}^{(\text{II})} = \mathscr{E} - U_0$ will be positive in region II, so there should be a continuation of the beam into this region. It is reasonable to expect the momentum $p^{(\text{II})}$ for this region to be obtained in the usual manner,

$$p^{(\text{II})} = [2m\{\mathscr{E} - U_0\}]^{1/2}, \tag{1.314}$$

where the *positive* square root is taken in order to give propagation in the positive x direction (see Fig. 1.31a). A propagation vector κ can be defined in the usual way from the momentum,

$$\hbar\kappa = p^{(\mathrm{II})} \tag{1.315}$$

so that κ has the positive value

$$\kappa = \hbar^{-1}[2m\{\mathscr{E} - U_0\}]^{1/2}. \tag{1.316}$$

The wave function for this transmitted beam is therefore expected to be of the form

$$\psi_{\mathrm{trans}} = Ce^{i(\kappa x - \omega t)}, \tag{1.317}$$

where again $\omega = \mathscr{E}/\hbar$, the same as in region I. Direct substitution of this wave into the Schrödinger equation for region II gives

$$-(\hbar^2/2m)(i\kappa)^2\psi_{\mathrm{trans}} + U_0\psi_{\mathrm{trans}} = i\hbar(-i\omega)\psi_{\mathrm{trans}}, \tag{1.318}$$

which leads to

$$-(\hbar^2/2m)(-2m/\hbar^2)[\mathscr{E} - U_0] + U_0 = \mathscr{E}, \tag{1.319}$$

which indeed is true. Therefore ψ_{trans} does satisfy the Schrödinger equation. The transmitted beam intensity I_{trans} follows immediately,

$$I_{\mathrm{trans}} = \psi^*_{\mathrm{trans}}\psi_{\mathrm{trans}}v_{\mathrm{trans}} = C^*C[p^{(\mathrm{II})}/m] = C^*C(\hbar\kappa/m). \tag{1.320}$$

The alternate form of a reverse-traveling wave in region II might also be expected to satisfy the Schrödinger equation, but this wave is not considered here because physically for our present problem there is no source of particles in region II and there are no additional barriers in region II to give reflection of the forward-propagating beam. Thus for the present problem,

$$\psi_{\mathrm{II}} = \psi_{\mathrm{trans}}. \tag{1.321}$$

The application of the two boundary conditions,

$$\psi_{\mathrm{I}}|_{x=0} = \psi_{\mathrm{II}}|_{x=0}, \tag{1.322}$$

$$(d\psi_{\mathrm{I}}/dx)|_{x=0} = (d\psi_{\mathrm{II}}/dx)|_{x=0}, \tag{1.323}$$

provides two equations which determine the two constants B and C for the reflected and transmitted wave amplitudes in terms of the incident wave amplitude A. These two conditions are sufficient to ensure that the particle density and the probability current density [Eq. (1.264) of §9] are conserved across the step. More explicitly, at a position immediately inside the step the particle density is considered to be the same as immediately outside the step, and the sum of the transmitted and reflected particle beam intensities at the step are considered to be equal to the incident particle beam intensity. These are not unreasonable requirements for the physical problem at hand. Substituting ψ_{I} and ψ_{II} into the above-listed boundary conditions at the location of the step ($x = 0$) gives the following two relations,

$$Ae^{-i\omega t} + Be^{-i\omega t} = Ce^{-i\omega t}, \tag{1.324}$$

$$ikAe^{-i\omega t} - ikBe^{-i\omega t} = i\kappa Ce^{-i\omega t}, \tag{1.325}$$

or equivalently,

$$A + B = C, \tag{1.326}$$

$$A - B = (\kappa/k)C. \tag{1.327}$$

Adding the two equations gives

$$2A = [1 + (\kappa/k)]C, \tag{1.328}$$

or equivalently,

$$C = [2k/(k + \kappa)]A. \tag{1.329}$$

On the other hand, subtraction gives

$$2B = C[1 - (\kappa/k)], \tag{1.330}$$

or equivalently,

$$C = [2k/(k - \kappa)]B. \tag{1.331}$$

Equating the above two results for C gives

$$[2k/(k + \kappa)]A = [2k/(k - \kappa)]B, \tag{1.332}$$

or equivalently,

$$B = [(k - \kappa)/(k + \kappa)]A. \tag{1.333}$$

Thus the transmitted and reflected wave amplitudes have been determined in terms of the incident wave amplitude and the physical parameters of the system such as particle energy $\mathscr{E}$ and potential barrier height U_0.

The reflection coefficient $\mathscr{R}$ is given by the magnitude of the ratio of the reflected beam intensity to the incident beam intensity,

$$\mathscr{R} = |I_{\text{ref}}/I_{\text{inc}}| = |-B^*B(\hbar k/m)/A^*A(\hbar k/m)| = |B^*B/A^*A|. \tag{1.334}$$

Substituting from above gives

$$\begin{aligned}\mathscr{R} &= \left(\frac{k-\kappa}{k+\kappa}\right)^*\left(\frac{k-\kappa}{k+\kappa}\right) = \left(\frac{k-\kappa}{k+\kappa}\right)^2\\ &= \left[\frac{[(2m/\hbar^2)\mathscr{E}]^{1/2} - [(2m/\hbar^2)\{\mathscr{E} - U_0\}]^{1/2}}{[(2m/\hbar^2)\mathscr{E}]^{1/2} + [(2m/\hbar^2)\{\mathscr{E} - U_0\}]^{1/2}}\right]^2 = \left[\frac{\mathscr{E}^{1/2} - (\mathscr{E} - U_0)^{1/2}}{\mathscr{E}^{1/2} + (\mathscr{E} - U_0)^{1/2}}\right]^2.\end{aligned} \tag{1.335}$$

The transmission coefficient $\mathscr{T}$ is given by the magnitude of the ratio of the transmitted beam intensity to the incident beam intensity,

$$\mathscr{T} = |I_{\text{trans}}/I_{\text{inc}}| = |C^*C(\hbar\kappa/m)/A^*A(\hbar k/m)|. \tag{1.336}$$

Substituting from above gives

$$\mathscr{T} = \frac{\kappa}{k}\left(\frac{2k}{k+\kappa}\right)^*\left(\frac{2k}{k+\kappa}\right) = \frac{4k\kappa}{(k+\kappa)^2}$$

$$= \frac{4[(2m/\hbar^2)\mathscr{E}]^{1/2}[(2m/\hbar^2)\{\mathscr{E}-U_0\}]^{1/2}}{\{[(2m/\hbar^2)\mathscr{E}]^{1/2}+[(2m/\hbar^2)\{\mathscr{E}-U_0\}]^{1/2}\}^2} = \frac{4\mathscr{E}^{1/2}(\mathscr{E}-U_0)^{1/2}}{[\mathscr{E}^{1/2}+(\mathscr{E}-U_0)^{1/2}]^2}. \tag{1.337}$$

Therefore in the case of energies above the barrier height, some particles are transmitted and some are reflected. This conclusion reached on the basis of quantum mechanics is quite different from that which one would expect from the viewpoint of classical mechanics, since there would be no reason to expect to have reflection whenever $\mathscr{E} > U_0$ on the basis of classical mechanics.

The above results give

$$\mathscr{R} + \mathscr{T} = [(k-\kappa)/(k+\kappa)]^2 + 4k\kappa/(k+\kappa)^2 \tag{1.338}$$

or equivalently,

$$\mathscr{R} + \mathscr{T} = (k^2 - 2k\kappa + \kappa^2 + 4k\kappa)/(k+\kappa)^2 = 1. \tag{1.339}$$

It is generally true that $\mathscr{R} + \mathscr{T} = 1$, since this is equivalent to a statement of the conservation of beam intensity.

It is very interesting to examine these results for certain limiting cases. If $U_0 = 0$, we obtain immediately $\mathscr{R} = 0$ and $\mathscr{T} = 1$, which is certainly reasonable. The same results are approximately true for $\mathscr{E} \gg U_0$. On the other hand, whenever $\mathscr{E} \to U_0$, then $\mathscr{R} \to 1$ and $\mathscr{T} \to 0$, which is again very reasonable. In this latter case where the energy is not much greater than the step height, κ becomes relatively small, and the spatial wavelength of the transmitted beam is very long, corresponding to a particle with a small momentum and kinetic energy.

EXERCISE Compute the reflection and transmission coefficients $\mathscr{R}$ and $\mathscr{T}$ as a function of $\mathscr{E}/U_0$, using Eqs. (1.335), (1.337), and (1.356) for the step potential. Use any reasonable value (such as 0.1–10 eV) for the step height U_0, and employ a programmable calculator to carry out the computations for a range of values of particle energy $\mathscr{E}$ below and above U_0.

EXERCISE Plot the probability density $\psi^*\psi$ as a function of position for $x < 0$ and $x > 0$ for the step potential chosen in the above exercise.

11.2.2 Case of $\mathscr{E} < U_0$. If $\mathscr{E} < U_0$, the kinetic energy $\mathscr{E}_K^{(\mathrm{II})} = \mathscr{E} - U_0$ would be negative in region II, as can be noted from Fig. 1.32, which would preclude any

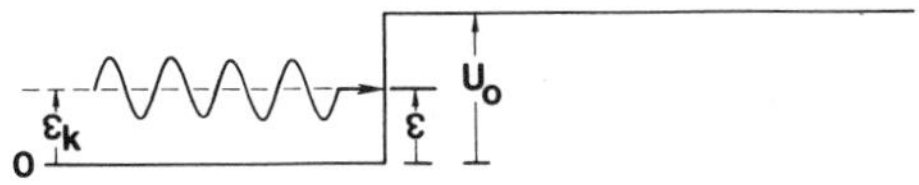

Fig. 1.32 Particle beam incident on a potential energy barrier greater than the total energy per particle in the beam.

particle penetration from a classical-mechanics viewpoint. The situation is not quite this simple in quantum mechanics, since solutions to the Schrödinger equation for region II can still be found.

Because κ as previously defined becomes imaginary for $\mathscr{E} < U_0$, let us define the new positive real quantity γ as follows,

$$\gamma = \hbar^{-1}[2m\{U_0 - \mathscr{E}\}]^{1/2}, \tag{1.340}$$

where the positive sign is to be chosen for the square root. Direct substitution of the wave function

$$\psi_{\text{II}} = De^{-\gamma x}e^{-i\omega t} \tag{1.341}$$

into the Schrödinger equation for region II gives

$$(-\hbar^2/2m)(-\gamma)^2\psi_{\text{II}} + U_0\psi_{\text{II}} = i\hbar(-i\omega)\psi_{\text{II}}, \tag{1.342}$$

or equivalently,

$$-(U_0 - \mathscr{E}) + U_0 = \hbar\omega, \tag{1.343}$$

which is satisfied when $\omega = \mathscr{E}/\hbar$. The probability density $\psi^*\psi$ given by this wave function approaches zero as $x \to \infty$. The alternate function $e^{\gamma x}$ would likewise be expected to satisfy the equation, but this function would lead to a probability density which increases with x and diverges as $x \to \infty$, and thus it must be discarded on physical grounds for the present problem. The region I wave function is the same for $\mathscr{E} < U_0$ as it is for $\mathscr{E} > U_0$. Imposing the boundary conditions

$$\psi_{\text{I}}|_{x=0} = \psi_{\text{II}}|_{x=0}, \tag{1.344}$$

$$(d\psi_{\text{I}}/dx)|_{x=0} = (d\psi_{\text{II}}/dx)|_{x=0} \tag{1.345}$$

leads in the present case to

$$Ae^{-i\omega t} + Be^{-i\omega t} = De^{-i\omega t}, \tag{1.346}$$

$$ikAe^{-i\omega t} - ikBe^{-i\omega t} = -\gamma De^{-i\omega t}, \tag{1.347}$$

or equivalently,

$$A + B = D, \tag{1.348}$$

$$A - B = -(\gamma/ik)D. \tag{1.349}$$

Adding the two equations gives

$$2A = [1 - (\gamma/ik)]D, \tag{1.350}$$

or equivalently,

$$D = \frac{2A}{1 - (\gamma/ik)} = \frac{2[1 + (\gamma/ik)]A}{1 + (\gamma^2/k^2)} = \frac{2[1 - i(\gamma/k)]A}{1 + (\gamma/k)^2}. \tag{1.351}$$

Subtraction, on the other hand, gives

$$2B = [1 + (\gamma/ik)]D, \tag{1.352}$$

or equivalently,

$$D = \frac{2B}{1 + (\gamma/ik)} = \frac{2[1 - (\gamma/ik)]B}{1 + (\gamma^2/k^2)} = \frac{2[1 + i(\gamma/k)]B}{1 + (\gamma/k)^2}. \tag{1.353}$$

Equating the two expressions for D thus obtained leads to

$$\frac{2[1 + i(\gamma/k)]B}{1 + (\gamma/k)^2} = \frac{2[1 - i(\gamma/k)]A}{1 + (\gamma/k)^2}, \tag{1.354}$$

from which it follows that

$$B = \left[\frac{1 - i(\gamma/k)}{1 + i(\gamma/k)}\right] A. \tag{1.355}$$

The reflection coefficient is given by

$$\begin{aligned}\mathscr{R} &= \left|\frac{I_{\text{ref}}}{I_{\text{inc}}}\right| = \left|\frac{-B^*B(\hbar k/m)}{A^*A(\hbar k/m)}\right| = \left|\frac{B^*B}{A^*A}\right| \\ &= \left[\frac{1 - i(\gamma/k)}{1 + i(\gamma/k)}\right]^* \left[\frac{1 - i(\gamma/k)}{1 + i(\gamma/k)}\right] = \left[\frac{1 + i(\gamma/k)}{1 - i(\gamma/k)}\right]\left[\frac{1 - i(\gamma/k)}{1 + i(\gamma/k)}\right] = 1.\end{aligned} \tag{1.356}$$

Therefore all particles are reflected, so the transmission coefficient is zero. For the case of the step barrier, then, the reflection of the particle beam having an energy below the barrier height is the same as one would expect from the viewpoint of classical mechanics.

It is now quite interesting to examine the quantum mechanical prediction that the particle density is nonzero in region II. This follows from

$$\begin{aligned}\psi_{\text{II}}^*(x)\psi_{\text{II}}(x) = D^*De^{-2\gamma x} &= 4A^*A\left[\frac{[1 + i(\gamma/k)]}{[1 + (\gamma^2/k^2)]}\right]\left[\frac{[1 - i(\gamma/k)]}{[1 + (\gamma^2/k^2)]}\right] e^{-2\gamma x} \\ &= 4(\psi_{\text{inc}}^*\psi_{\text{inc}})\frac{e^{-2\gamma x}}{1 + (\gamma^2/k^2)}.\end{aligned} \tag{1.357}$$

Therefore the particle density decays exponentially in region II with a fall-off length l given by

$$l = 1/2\gamma = \hbar/\{2[2m(U_0 - \mathscr{E})]^{1/2}\}. \tag{1.358}$$

As $U_0 \to \infty, l \to 0$, so the particle density immediately decays to zero in region II. On the other hand, as $\mathscr{E} \to U_0$, then $l \to \infty$, and so in this limit the decay is very slow in region II. Likewise in this limit, $\gamma \to 0$, and thus $\psi_{\text{II}}^*\psi_{\text{II}} \to 4\psi_{\text{inc}}\psi_{\text{inc}} = 4A^*A$. The reason for this can be found in the behavior of B, which can be noted to approach A for $\gamma \ll k$. Then $\psi_{\text{I}}(0) \to 2A$ and so $\psi_{\text{I}}(0)^*\psi_{\text{I}}(0) \to 4A^*A$, and the probability density is conserved across the step as expected from the first boundary condition.

Another way of considering the probability density in region II is to observe that the total probability density in region I at the step is

$$\begin{aligned}\psi_{\mathrm{I}}^*(0)\psi_{\mathrm{I}}(0) &= (A + B)^*(A + B)\\ &= \left\{A\left[1 + \left(\frac{1 - i(\gamma/k)}{1 + i(\gamma/k)}\right)\right]\right\}^*\left\{A\left[1 + \left(\frac{1 - i(\gamma/k)}{1 + i(\gamma/k)}\right)\right]\right\}\\ &= A^*A\left[\frac{1 + i(\gamma/k) + 1 - i(\gamma/k)}{1 + i(\gamma/k)}\right]^*\left[\frac{1 + i(\gamma/k) + 1 - i(\gamma/k)}{1 + i(\gamma/k)}\right]\\ &= A^*A\,\frac{4}{[1 - i(\gamma/k)][1 + i(\gamma/k)]} = \frac{4A^*A}{1 + (\gamma^2/k^2)} = \frac{4\psi_{\mathrm{inc}}^*\psi_{\mathrm{inc}}}{1 + (\gamma^2/k^2)}.\end{aligned} \tag{1.359}$$

Therefore we obtain for region II the result

$$\psi_{\mathrm{II}}^*(x)\psi_{\mathrm{II}}(x) = [\psi_{\mathrm{I}}^*(0)\psi_{\mathrm{I}}(0)]e^{-2\gamma x}. \tag{1.360}$$

This means that whatever probability density exists in region I, immediately adjacent to the step, also exists in region II immediately adjacent to the step, and this probability density falls off exponentially in region II with the characteristic decay length l. Thus a measurement of the particle density in region II would yield a nonzero value, in contrast to what one might expect on the basis of purely

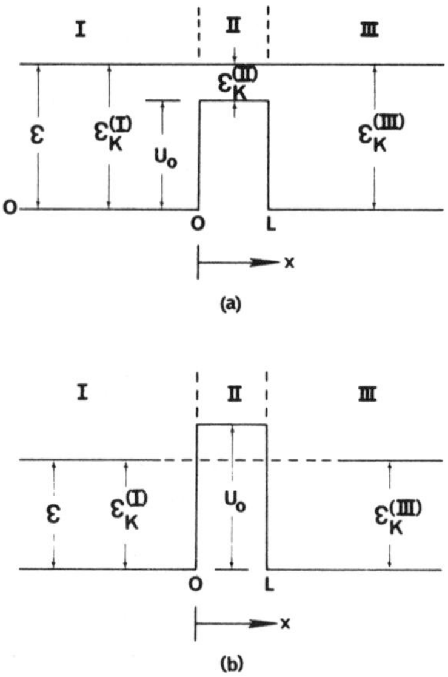

Fig. 1.33 Rectangular potential energy barrier in which the potential energy $U(x)$ is zero for $x < 0$ and for $x > L$, but has the constant value U_0 over the region $0 \leqslant x \leqslant L$. The total energy $\mathscr{E}$ per electron is the same for all three regions; it is entirely kinetic in regions I and III. (a) $\mathscr{E} > U_0$, in which case the kinetic energy $\mathscr{E}_{\mathrm{K}}^{(\mathrm{II})}$ in region II is positive. (b) $\mathscr{E} < U_0$, in which case the kinetic energy $\mathscr{E}_{\mathrm{K}}^{(\mathrm{II})}$ in region II is negative.

classical mechanics. This conclusion has great physical consequences when the step potential U_0 extends over only a finite distance L $(0 \leqslant x \leqslant L)$ (cf. Fig. 1.33) instead of extending an infinite distance $(0 \leqslant x \leqslant \infty)$, as now considered. The situation just described, for which $U(x) = U_0$ $(0 \leqslant x \leqslant L)$ and $U(x) = 0$ otherwise, is termed the *rectangular barrier* problem; it is treated in detail in the following section. The quantum solution to the rectangular-barrier problem leads to the remarkable conclusion that some particles can be transmitted across the barrier even when the particle energy $\mathscr{E}$ is less than the barrier height. This is the phenomenon of *electron tunneling*. The predictions of the theory are amply supported by the experimental results. This represents another area in which quantum mechanics provides a theoretical framework for describing phenomena that cannot be described satisfactorily within the framework of classical mechanics.

EXERCISE (a) For $\mathscr{E} < U_0$ in the step potential problem, show that the particle current density is zero. [*Hint*: Recall that $\mathbf{J} = (\hbar/2mi)(\psi^* \nabla\psi - \psi \nabla\psi^*)$.]
(b) Next, evaluate $\mathbf{J}$ for $\mathscr{E} > U_0$.

PROJECT 1.23 Reflection of Particles by a Step Potential

1. Choose the step height U_0 in the problem of a semi-infinite step potential (cf. Fig. 1.32) to be 1, 2, 3, 4, 5, or 10 eV. Plot the transmission coefficient $\mathscr{T}$ and the reflection coefficient $\mathscr{R}$ as a function of incident neutron energy $\mathscr{E}$, scanning the range $(0 < \mathscr{E} < 2U_0)$, assuming an incident beam of 10^{12} particles/cm^2 sec.
2. Plot the particle density $\psi^*\psi$ at position $x = 1$ Å and at position $x = 5$ Å as a function of incident neutron energy $\mathscr{E}$, again scanning the range $(0 < \mathscr{E} < 2U_0)$, and assuming an incident beam of 10^{12} neutrons/cm^2 sec.

11.3 The Rectangular Potential Energy Barrier

Figure 1.34 illustrates the three regions defined by the potential energy function

$$U(x) = \begin{cases} 0 & (x < 0) \\ U_0 & (0 \leqslant x \leqslant L) \\ W & (x > L). \end{cases} \tag{1.361}$$

This functional form of the potential energy is termed a *rectangular* (or, at times, a *square*) barrier in the limit $W \to 0$. For our purposes, it is better to permit W to

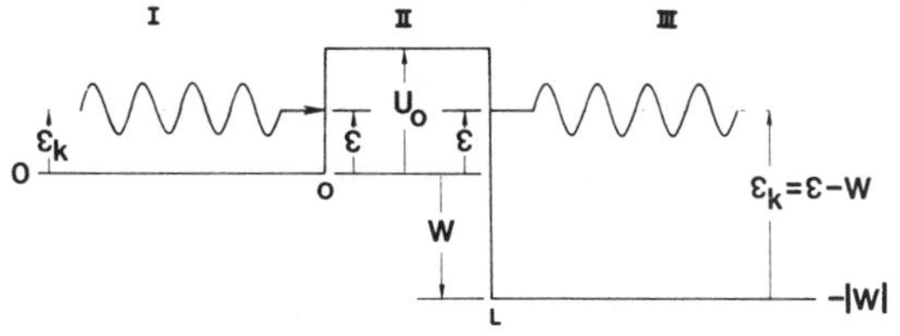

Fig. 1.34 Modified rectangular potential barrier for which the potential energy $U(x)$ is zero for $x < 0$, $U(x) = U_0$ for $0 \leqslant x \leqslant L$, and $U(x) = W$ for $x > L$. The total particle energy $\mathscr{E}$ is the same for all three regions; it is entirely kinetic in region I.

be arbitrary, so that it can take on negative as well as positive values. This yields results that are immediately applicable to the tunneling of electrons between metals separated by a thin insulator. We shall consider individually the several possible cases, and finally we shall consider the classical analog in order to highlight the different predictions of the two theories.

11.3.1 Case of $\mathscr{E} > U_0$ and $\mathscr{E} > W$. For $\mathscr{E} > U_0$, we expect the wave functions to be of the plane-wave type, because the momentum is real and thus propagation of the particle is possible even in the classical sense. As in the case of the step barrier, we let the incident wave be given by

$$\psi_{\text{inc}} = Ae^{i(kx-\omega t)}, \tag{1.362}$$

where

$$k = \hbar^{-1}(2m\mathscr{E})^{1/2}, \tag{1.363}$$

$$\omega = \mathscr{E}/\hbar. \tag{1.364}$$

The reflected wave will likewise have the same form as for the step potential, namely,

$$\psi_{\text{ref}} = Be^{i(-kx-\omega t)}. \tag{1.365}$$

Then for region I,

$$\psi_{\text{I}} = \psi_{\text{inc}} + \psi_{\text{ref}} = (Ae^{ikx} + Be^{-ikx})e^{-i\omega t}. \tag{1.366}$$

The transmitted wave for the present situation is the propagating wave in region III. Let us, in analogy with the situation for the step potential, denote the transmitted wave by

$$\psi_{\text{trans}} = Ce^{i(\kappa x-\omega t)}, \tag{1.367}$$

where

$$\kappa = \hbar^{-1}[2m(\mathscr{E} - W)]^{1/2}. \tag{1.368}$$

In the absence of sources and other variations in region III which could lead to a reflected wave in that region, we can then write

$$\psi_{\text{III}} = \psi_{\text{trans}} = Ce^{i\kappa x}e^{-i\omega t}. \tag{1.369}$$

Region II is the additional factor in the currently considered rectangular barrier problem which was absent in the preceding problem of the step potential. Due to the finite thickness of the region ($0 \leqslant x \leqslant L$) and the discontinuity at $x = L$, it is possible to have a reverse traveling (reflected) wave in this region in addition to a forward propagating wave. Denoting the forward wave by $Fe^{i(\beta x-\omega t)}$ and the reverse wave by $Ge^{i(-\beta x-\omega t)}$, where

$$\beta \equiv \hbar^{-1}[2m(\mathscr{E} - U_0)]^{1/2}, \tag{1.370}$$

we can write

$$\psi_{\text{II}} = (Fe^{i\beta x} + Ge^{-i\beta x})e^{-i\omega t}. \tag{1.371}$$

The boundary conditions at $x = 0$ of wave function continuity

$$\psi_{\text{I}}(0) = \psi_{\text{II}}(0) \tag{1.372}$$

and continuity of the first derivative of the wave function

$$(d\psi_{\text{I}}/dx)|_{x=0} = (d\psi_{\text{II}}/dx)|_{x=0} \tag{1.373}$$

lead directly to the relations

$$A + B = F + G, \tag{1.374}$$

$$ikA - ikB = i\beta F - i\beta G. \tag{1.375}$$

Rewriting this pair of equations in the form

$$A + B = F + G, \tag{1.376}$$

$$A - B = (\beta/k)\{F - G\} \tag{1.377}$$

makes it easy to obtain expressions for A and B in terms of F and G. That is, by adding the two equations, we obtain

$$A = \tfrac{1}{2}(F[1 + (\beta/k)] + G[1 - (\beta/k)]), \tag{1.378}$$

and by subtracting the two equations, we obtain

$$B = \tfrac{1}{2}(F[1 - (\beta/k)] + G[1 + (\beta/k)]). \tag{1.379}$$

Let us next apply boundary conditions at the barrier discontinuity at $x = L$. Continuity of the wave function $\psi_{\text{II}}(L) = \psi_{\text{III}}(L)$ and continuity of the first derivative of the wave function $(d\psi_{\text{II}}/dx)|_{x=L} = (d\psi_{\text{III}}/dx)_{x=L}$ lead directly to the two additional relations

$$Fe^{i\beta L} + Ge^{-i\beta L} = Ce^{i\kappa L}, \tag{1.380}$$

$$i\beta Fe^{i\beta L} - i\beta Ge^{-i\beta L} = i\kappa Ce^{i\kappa L}. \tag{1.381}$$

Rewriting this pair of equations in the form

$$Fe^{i\beta L} + Ge^{-i\beta L} = Ce^{i\kappa L}, \tag{1.382}$$

$$Fe^{i\beta L} - Ge^{-i\beta L} = (\kappa/\beta)Ce^{i\kappa L} \tag{1.383}$$

leads to expressions for F and G in terms of C. That is, by adding the two equations, we obtain

$$F = \tfrac{1}{2}e^{-i\beta L}[1 + (\kappa/\beta)]Ce^{i\kappa L} = \tfrac{1}{2}[1 + (\kappa/\beta)]Ce^{i(\kappa - \beta)L}, \tag{1.384}$$

and by subtracting the two equations, we obtain

$$G = \tfrac{1}{2}e^{i\beta L}[1 - (\kappa/\beta)]Ce^{i\kappa L} = \tfrac{1}{2}[1 - (\kappa/\beta)]Ce^{i(\kappa + \beta)L}. \tag{1.385}$$

Next, we can substitute the two expressions just obtained for F and G into expressions (1.378) and (1.379) previously obtained for A and B. This gives A and B in terms of C. With these results the transmission and reflection coefficients can be evaluated. The transmission coefficient $\mathscr{T}$ follows from the ratio of the transmitted to incident beam intensities,

$$\mathscr{T} = \frac{I_{\text{trans}}}{I_{\text{inc}}} = \frac{\psi^*_{\text{trans}}\psi_{\text{trans}}(\hbar\kappa/m)}{\psi^*_{\text{inc}}\psi_{\text{inc}}(\hbar k/m)} = \frac{C^*C}{A^*A}(\kappa/k) = \frac{\kappa/k}{[(A/C)^*(A/C)]}. \tag{1.386}$$

The reflection coefficient can be obtained from the ratio of the reflected to the incident beam intensities,

$$\mathscr{R} = \left|\frac{I_{\text{ref}}}{I_{\text{inc}}}\right| = \left|\frac{B^*B(-\hbar k/m)}{A^*A(\hbar k/m)}\right| = \left(\frac{B}{A}\right)^*\left(\frac{B}{A}\right). \tag{1.387}$$

EXERCISE Show that the wave function (1.367) and the value given for κ are consistent with the time-dependent Schrödinger equation for region III.

EXERCISE Complete the calculations in Eqs. (1.386) and (1.387) for $\mathscr{R}$ and $\mathscr{T}$. Show that $\mathscr{R} + \mathscr{T} = 1$.

EXERCISE Show that your results in the preceding exercise reduce to the usual textbook expressions in the limit $W \to 0$ [Schiff (1968)]. In particular, show that

$$\mathscr{T} = \frac{8}{[6 + (\beta/k)^2 + (k/\beta)^2] + [2 - (\beta/k)^2 - (k/\beta)^2]\cos 2\beta L}.$$

EXERCISE Take the limit $W \to U_0$ and show that the previously deduced step potential results for $\mathscr{E} > U_0$ are thereby obtained.

EXERCISE Consider the situation in which $U_0 < 0$, and follow through the above derivation to see what differences (if any) are obtained in the mathematical results and the physical predictions. The various limits for W are also of interest in this situation, as for example, $W \to U_0$ in the negative potential energy domain. [*Hint*: See Bohm (1951).]

11.3.2 Case of $W < \mathscr{E} < U_0$. For $\mathscr{E} < U_0$, the wave function in region II cannot be of the plane-wave type, because the kinetic energy would be negative and the momentum would consequently be imaginary. It is easy to show that the wave function $\psi_{\text{II}} = (De^{-\alpha x} + Ee^{\alpha x})e^{-i\omega t}$ satisfies the Schrödinger equation

$$-(\hbar^2/2m)\,d^2\psi/dx^2 + U_0\psi = i\hbar\,\partial\psi/\partial t$$

appropriate for this region, where α is obtained by substituting ψ_{II} into the equation. This leads to

$$(-\hbar^2/2m)(\alpha^2\psi_{\text{II}}) + U_0\psi_{\text{II}} = i\hbar(-i\omega)\psi_{\text{II}},$$

which gives $(-\hbar^2/2m)\alpha^2 = \hbar\omega - U_0$, or since $\hbar\omega = \mathscr{E}$, $\alpha = [(2m/\hbar^2)\cdot\{U_0 - \mathscr{E}\}]^{1/2}$. The positive sign is conventionally chosen for α; the choice of a negative sign would simply interchange the coefficients D and E.

Since $\mathscr{E} > W$, the solution in region III is again of the propagating type,

$$\psi_{\text{III}} = \psi_{\text{trans}} = Ce^{i(\kappa x - \omega t)} = Ce^{i\kappa x}e^{-i\omega t}$$

and the wave function ψ_{I} for region I is the same as before,

$$\psi_{\text{I}} = \psi_{\text{inc}} + \psi_{\text{ref}} = (Ae^{ikx} + Be^{-ikx})e^{-i\omega t},$$

where the constants κ and k have their previously defined values, $\kappa = \hbar^{-1}[2m(\mathscr{E} - W)]^{1/2}$, $k = \hbar^{-1}(2m\mathscr{E})^{1/2}$.

The boundary conditions at $x = 0$ of continuity of the wave function $\psi_{\mathrm{I}}(0) = \psi_{\mathrm{II}}(0)$ and continuity of the first derivative of the wave function $(d\psi_{\mathrm{I}}/dx)|_{x=0} = (d\psi_{\mathrm{II}}/dx)|_{x=0}$ lead directly to the two relations

$$A + B = D + E, \qquad ikA - ikB = -\alpha D + \alpha E.$$

Rewriting this pair of equations in the form

$$A + B = D + E, \qquad A - B = (-\alpha/ik)[D - E]$$

makes it easy to obtain expressions for A and B in terms of D and E. That is, by adding the two equations we obtain

$$A = \tfrac{1}{2}(D[1 - (\alpha/ik)] + E[1 + (\alpha/ik)]),$$

and by subtracting the two equations we obtain

$$B = \tfrac{1}{2}(D[1 + (\alpha/ik)] + E[1 - (\alpha/ik)]).$$

Next, we apply boundary conditions at the barrier discontinuity at $x = L$. The condition of continuity of the wave function $\psi_{\mathrm{II}}(L) = \psi_{\mathrm{III}}(L)$ and continuity of the first derivative of the wave function $(d\psi_{\mathrm{II}}/dx)|_{x=L} = (d\psi_{\mathrm{III}}/dx)|_{x=L}$ lead directly to the two additional relations

$$De^{-\alpha L} + Ee^{\alpha L} = Ce^{i\kappa L}, \qquad -\alpha De^{-\alpha L} + \alpha Ee^{\alpha L} = i\kappa Ce^{i\kappa L}.$$

Rewriting this pair of equations in the form

$$De^{-\alpha L} + Ee^{\alpha L} = Ce^{i\kappa L}, \qquad De^{-\alpha L} - Ee^{\alpha L} = (-i\kappa/\alpha)Ce^{i\kappa L}$$

leads to expressions for D and E in terms of C. That is, by adding the two equations we obtain

$$D = \tfrac{1}{2}e^{\alpha L}[1 + (-i\kappa/\alpha)]Ce^{i\kappa L} = \tfrac{1}{2}[1 - (i\kappa/\alpha)]Ce^{(i\kappa + \alpha)L},$$

and by subtracting the two equations we obtain

$$E = \tfrac{1}{2}e^{-\alpha L}[1 + (i\kappa/\alpha)]Ce^{i\kappa L} = \tfrac{1}{2}[1 + (i\kappa/\alpha)]Ce^{(i\kappa - \alpha)L}.$$

Next we can substitute the two expressions which we have just obtained for D and E into the expressions obtained above for A and B. This gives A and B in terms of C, so that the transmission and reflection coefficients can be evaluated for this case. The reflection coefficient $\mathscr{R}$ can be obtained from the ratio of the reflected to the incident beam intensities,

$$\mathscr{R} = \left|\frac{I_{\mathrm{ref}}}{I_{\mathrm{inc}}}\right| = \left|\frac{B^*B(-\hbar k/m)}{A^*A(\hbar k/m)}\right| = \left(\frac{B}{A}\right)^*\left(\frac{B}{A}\right)$$
$$= \frac{[1 - (\kappa/k)]^2 + [1 + (\kappa/\alpha)^2][1 + (\alpha/k)^2]\sinh^2 \alpha L}{[1 + (\kappa/k)]^2 + [1 + (\kappa/\alpha)^2][1 + (\alpha/k)^2]\sinh^2 \alpha L}.$$

The transmission coefficient $\mathscr{T}$ is readily evaluated from this expression for the reflection coefficient,

$$\mathscr{T} = 1 - \mathscr{R} = \frac{4(\kappa/k)}{[1 + (\kappa/k)]^2 + [1 + (\kappa/\alpha)^2][1 + (\alpha/k)^2]\sinh^2 \alpha L}.$$

In the limit $W \to 0$, then $\kappa \to k$ and the transmission coefficient reduces to

$$\mathscr{T} = \frac{4}{4 + [1 + (k/\alpha)^2 + (\alpha/k)^2 + 1] \sinh^2 \alpha L}$$

$$= \frac{1}{1 + \frac{1}{4}[(k/\alpha) + (\alpha/k)]^2 \sinh^2 \alpha L} \qquad \text{(rectangular barrier).}$$

This is the transmission coefficient for the situation illustrated in Fig. 1.33b.

In the further limit in which $\alpha L \gg 1$, then $\sinh \alpha L \simeq \frac{1}{2}e^{\alpha L}$, and we obtain the approximate form

$$\mathscr{T} \simeq \frac{1}{1 + \frac{1}{4}[(k/\alpha) + (\alpha/k)]^2 \frac{1}{4} e^{2\alpha L}} \simeq \frac{16 e^{-2\alpha L}}{[(k/\alpha) + (\alpha/k)]^2} = \left[\frac{4(\alpha/k)}{1 + (\alpha/k)^2}\right]^2 e^{-2\alpha L}.$$

Thus we have derived the remarkable quantum-mechanical result that particles can penetrate potential energy barriers which are even higher than the particle energy. This has important applicability in quantum electronic devices. It likewise explains the decay of radioactive nuclei by α-particle emission.

A sample calculation spanning the domains for $\mathscr{E} < U_0$ and $\mathscr{E} > U_0$, with $W = 0$ for both cases, has been carried out. The value of the electronic mass is used, and the rectangular barrier is chosen to have a thickness of 10 Å and a height of 10 eV. Figure 1.35 illustrates the variation of the transmission

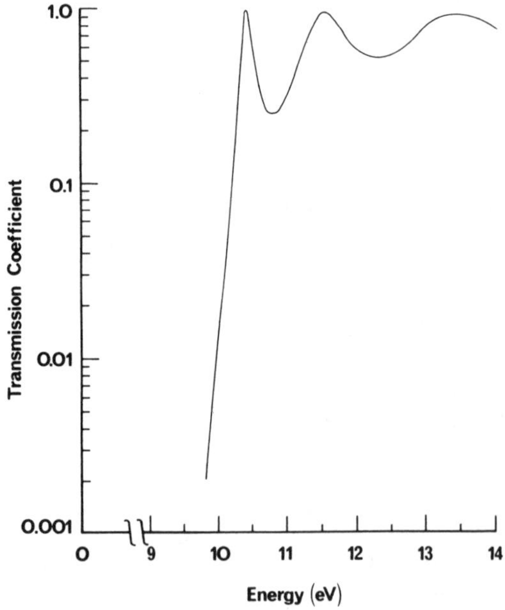

Fig. 1.35 Transmission coefficient versus energy of a particle incident on a rectangular potential energy barrier 10 eV in height and 10 Å in width.

coefficient with incident electron energy. The remarkable oscillatory behavior is due to the wavelike nature of the particle; the peaks coincide with certain relationships between the de Broglie wavelength and the barrier thickness (see corresponding exercise).

EXERCISE Complete in detail the algebraic steps leading from the above evaluation of the coefficients D and E to the evaluation of the coefficients A and B in terms of C. Carry out the indicated substitutions to obtain the above expressions for the transmission and reflection coefficients.

EXERCISE Compare the two cases $W < 0$ and $W > 0$, assuming that $W < \mathscr{E} < U_0$.

EXERCISE Take the limit $W \to U_0$, where $\mathscr{E} < U_0$, and show that the step potential results for $\mathscr{E} < U_0$ are thereby obtained.

EXERCISE Consider the situation in which $U_0 < 0$ and $0 < \mathscr{E} < W$.

EXERCISE Consider the bound state situation in which $U_0 < 0$, with $U_0 < \mathscr{E} < 0$. Various limits for W can be chosen, such as $W = 0$, $W \to \infty$, $\mathscr{E} < W < 0$, and $W > 0$.

EXERCISE Deduce the specific wavelengths (in terms of barrier thickness L) at which the transmission coefficient has a value of unity, as illustrated in Fig. 1.35.

11.3.3 Case of $\mathscr{E} < U_0$ and $\mathscr{E} < W$. As a final consideration, let us examine the situation in which the particle energy $\mathscr{E}$ is less than the potential energy in regions II and III. Since region III extends to $x = \infty$, we expect a probability density in region III which approaches zero as $x \to \infty$. The wave function for region III thus may be chosen to be

$$\psi_{\text{III}} = He^{-\gamma x}e^{-i\omega t}, \qquad \text{where} \quad \gamma = \hbar^{-1}[2m(W - \mathscr{E}_0)]^{1/2}.$$

Matching the wave functions and the first derivations at $x = 0$ yields the same results given in §11.3.2 for the relationships between A, B, D, E, since the wave functions in regions I and II are the same as for the case $W < \mathscr{E} < U_0$. However, the matching of the wave functions and their first derivations at $x = L$ is different since ψ_{III} is now different. The result obtained from this matching is

$$De^{-\alpha L} + Ee^{\alpha L} = He^{-\gamma L}, \qquad -\alpha De^{-\alpha L} + \alpha Ee^{\alpha L} = -\gamma He^{-\gamma L}.$$

Writing this pair of equations in the form

$$De^{-\alpha L} + Ee^{\alpha L} = He^{-\gamma L}, \qquad De^{-\alpha L} - Ee^{\alpha L} = (\gamma/\alpha)He^{-\gamma L}$$

facilitates the algebra. Adding the two equations leads to an evaluation of D, and subtracting the two equations leads to an evaluation of E. Substituting these two expressions for D and E into the previously derived expressions for A and B leads to an evaluation of the reflection coefficient for this case, $\mathscr{R} = (B/A)^*(B/A)$. Substituting gives $\mathscr{R} = 1$. This is the result to be expected on physical grounds for this situation.

EXERCISE Carry out the algebraic details in the above derivation.

PROJECT 1.24 Tunneling of Particles through a Rectangular Barrier

1. Choose the barrier height U_0 in the problem of a rectangular barrier (cf. Fig. 1.33) to be 1, 2, 3, 4, 5, or 10 eV, and choose the barrier thickness L to be 1 Å. Plot the transmission coefficient $\mathscr{T}$ and the

reflection coefficient $\mathcal{R}$ as a function of incident electron energy $\mathcal{E}$, scanning the range $(0 < \mathcal{E} < 2U_0)$ and assuming an incident beam of 10^{15} electrons/cm^2 sec.
2. Plot the charge density $-e\psi^*\psi$ at positions $x_1 = -\frac{1}{2}$ Å, $x_2 = \frac{1}{2}$ Å, $x_3 = 1$ Å, $x_4 = \frac{3}{2}$ Å, and $x_5 = 2$ Å as a function of incident electron energy $\mathcal{E}$, again scanning the range $(0 < \mathcal{E} < 2U_0)$ and assuming an incident beam of 10^{15} electrons/cm^2 sec.

PROJECT 1.25 Resonance Tunneling

Consider two identical rectangular barriers of height V_0 and width $2a$ separated by a distance b. (See also Problem 146, and especially the sketch of the two identical rectangular potential energy barriers.)
1. Derive a general expression for the transmission coefficient for particles with energies less than V_0.
2. For $V_0 = 25$ eV, $b = 2a = 3.07$ Å, and incident electrons with energies $\mathcal{E} = 0.5, 1, 2, 2.5, 3, 4, 5, 6, 6.5, 7, 8, 9$, and 10 eV, compute the transmission coefficient $\mathcal{T}$ and the reflection coefficient $\mathcal{R}$. Plot the results.
3. Compare your results with those derived for a single barrier. Qualitatively explain the difference.
4. Repeat Part 1 for energies greater than V_0. What are the wavelength conditions for maximum and for minimum transmission?
5. Repeat Part 4 for inverted barriers (i.e., two wells of depth V_0 replacing the barriers).

11.4 Classical-Mechanics Predictions

Let us now focus our attention on the predictions of classical physics for this example. This sharpens our understanding of the problem and highlights the differences in the predictions of quantum mechanics and classical mechanics.

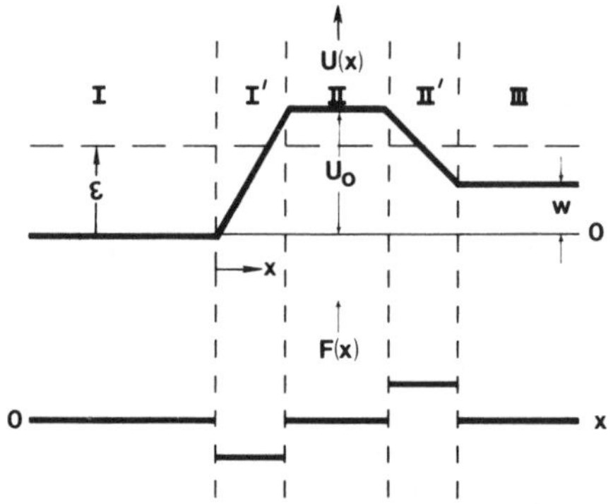

Fig. 1.36 Barrier $U(x)$ with gently sloping sides (regions I′ and II′), as opposed to the discontinuous changes in potential energy manifested in the rectangular barrier. The force $F(x)$ indicated in the lower part of the figure is nonzero only in regions I′ and II′; it is discontinuous, being given by the negative derivative of the potential energy $U(x)$ in accordance with Eq. (1.389).

Instead of an infinite slope at the barrier edges, however, let us consider a finite (but arbitrarily steep) slope, as indicated schematically in Fig. 1.36. Since the force $\mathbf{F}$ on a particle is given by the negative gradient of the potential energy,

$$\mathbf{F} = -\nabla U, \tag{1.388}$$

which is a quite generally valid physical relation, then for one dimension x the force is given by

$$F = -\,dU(x)/dx. \tag{1.389}$$

Thus the force is constant and is negatively directed in the region labeled I′ in the figure; the force is likewise constant but is positively directed in the region labeled II′. The slope being zero in regions I, II, and III, the force is likewise zero in these regions.

Consider now a charged particle of mass m incident on the barrier from the left. The force is zero and the kinetic energy is constant and equal to the total particle energy throughout region I. (This presumes that the constant energy beam has been experimentally produced in the region to the left of I, as, for example, by a hot filament and a series of accelerating grids, as illustrated in Fig. 1.37.) As the charged particle enters region I′ it experiences a reverse force which slows it down. If the potential energy difference across region I′ is greater than the kinetic energy of the particle when it enters region I, the particle velocity will be reduced to zero before the particle has traversed region I′. At that point, the force on the particle is still negatively directed and nonzero (see Fig. 1.37 diagram), so the particle will be accelerated in the reverse direction. As the particle re-enters region I in this manner, it will have regained its initial kinetic energy: the particle momentum will be the same in magnitude but reversed in sign. This process is called "reflection," and insofar as classical physics is concerned, the reflection is *total* in the sense that it is predicted to occur for *all* particles having an initial kinetic energy less than the height of the potential energy barrier encountered.

On the other hand, for a particle entering region I′ with a kinetic energy *greater* than the potential energy barrier, the particle will still have a nonzero and positively directed velocity when it reaches region II. The probability would therefore be 100% for the particle to enter region II, if classical physics could be believed. The particle would then travel through region II with its constant (though reduced) kinetic energy $\mathscr{E}_{\mathrm{K}}^{(\mathrm{II})} = \mathscr{E} - U_0$ until it reached region II′, which it would then enter, and undergo a continuous acceleration until it entered region III with the appropriate kinetic energy determined from $\mathscr{E} - W$. The only difference between the currently considered barriers and those previously considered in the quantum derivation are the slopes, and there is no reason to refrain from taking the mathematical limit in which the slopes become infinite, even though this would correspond to an impossible experimental arrangement: The accelerating grids indicated in the figure would then be separated only infinitesimally while the accelerating voltages would remain fixed, thus corresponding to an immeasurably large electric field which would cause arcing and dielectric breakdown.

We therefore find that the classical and quantum predictions are somewhat different. The classical-mechanics result is more straightforward in a sense, because the reflection or transmission, as the case may be, is total. The quantum mechanics result, on the other hand, is a bit more mysterious, there being cases of

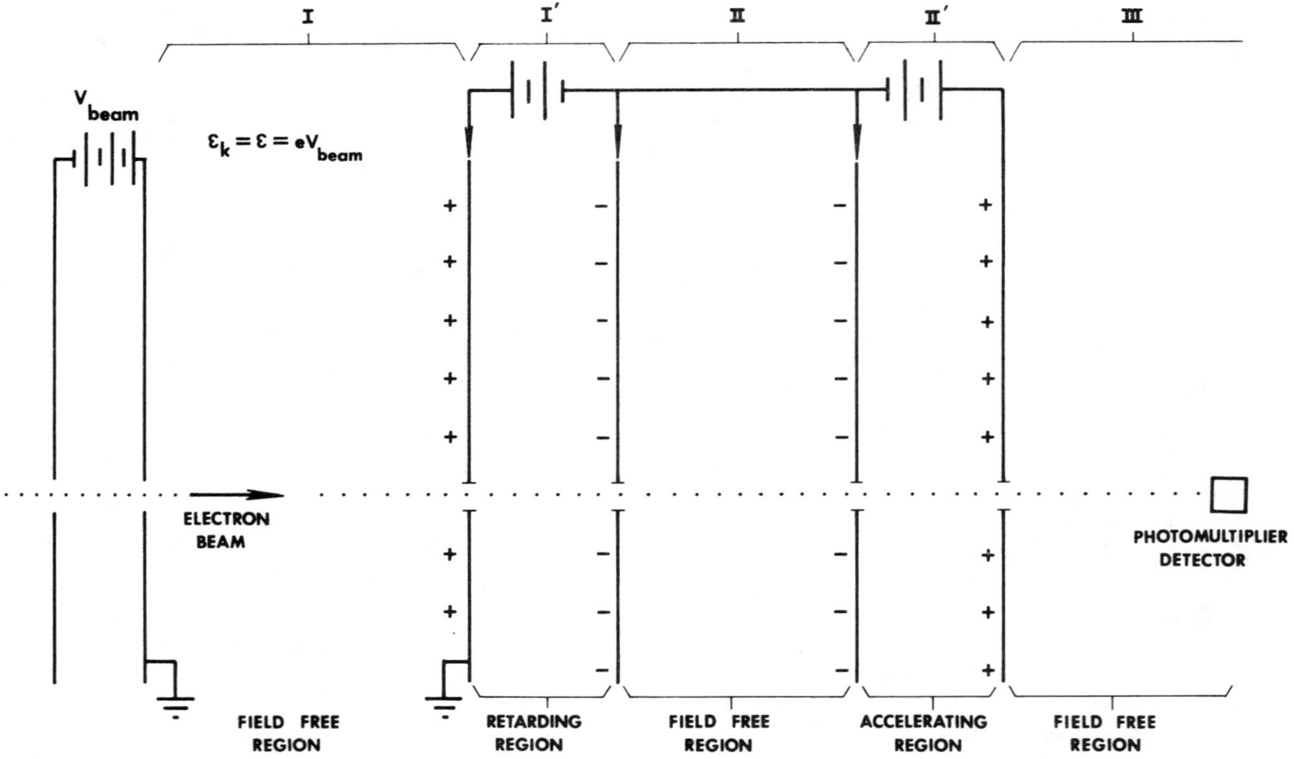

Fig. 1.37 Series of biased grids to duplicate experimentally the forces and potential energies illustrated in Fig. 1.36.

partial transmission and partial reflection of a particle beam. It is the quantum-mechanics prediction, however, that is found to agree with the experimental results.

12 Bound-State Problems

12.1 Introduction

A particle may be confined to a certain region of space by potential energy barriers which surround the particle. In one dimension, for example, the sketch in Fig. 1.38 illustrates that at positions x_1 and x_2 the potential energy $U(x)$ is equal to the total energy $\mathscr{E}$ of the particle. At these points, the kinetic energy $\mathscr{E}_K$ given by $[\mathscr{E} - U(x)]$ is necessarily zero. These points are called "classical turning points," since classically the particle would simply be reflected from the barrier at these points, with a reversal of the perpendicular momentum component similar to the elastic rebound of a rubber ball from a concrete wall. (See the discussion on the classical mechanics limit in the preceding section.) The motion of the particle would thus continue in the currently considered "potential well" delineated by the surrounding energy barriers. The total energy $\mathscr{E}$ is conserved, with a continuous interchange of kinetic and potential energies. The time dependence of the motion therefore depends upon the exact functional form of

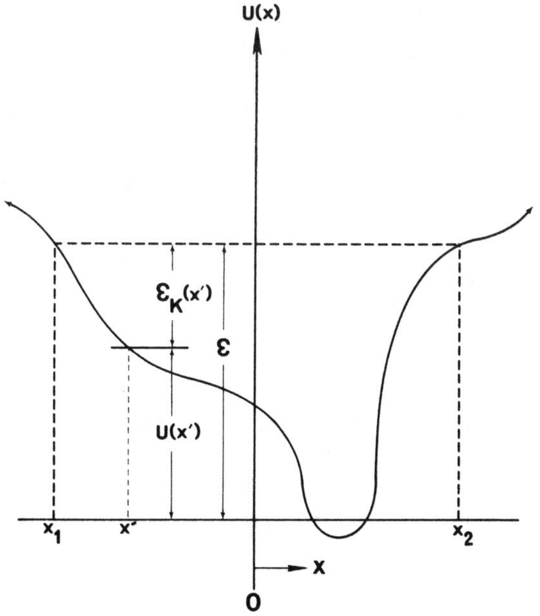

Fig. 1.38 Arbitrary potential energy well wherein a particle with total energy $\mathscr{E}$ is trapped, oscillating back and forth between the classical turning points x_1 and x_2 where the kinetic energy is zero. (At arbitrary position x', the total energy can be noted to be divided into potential energy $U(x')$ and kinetic energy $\mathscr{E}_K(x')$ portions.)

the potential energy $U(x)$ with position x. The characteristic feature of bound-state problems in quantum mechanics is that the application of realistic boundary conditions forces a restriction on the energy values, so that the eigenvalues for the total energy are confined to a discrete set. This characteristic feature is independent of the particular functional form of the potential energy, although the exact details of the energy level spectrum are dependent upon the form of $U(x)$.

12.2 The Square-Well Potential

The most readily available example of a bound-state problem is that of a "particle in a box" which we considered in §10.2. The "box" is the region of space wherein the potential energy $U(x)$ is zero, with the "walls" of the box being the infinitely high potential-energy barriers which reflect the particle and thus cause it to remain within the box. Discrete momentum and energy values,

$$p_n = \hbar k_n = n\pi\hbar/L, \tag{1.390}$$

$$\mathscr{E}_n = \hbar^2 k_n^2/2m = n^2\pi^2\hbar^2/2mL^2 \qquad (n = 1, 2, 3, \ldots, \infty), \tag{1.391}$$

were found to be required in order to meet the boundary conditions of zero probability density $\psi^*\psi$ at the "walls" where the potential energy rises to infinity. The integer n is the *quantum number* for the problem. Such quantized values for the total energy are in marked contrast to the classical viewpoint where there is no condition on the values of the momentum and energy which a particle can assume within the box. Energy *absorption* by the particle in the box occurs when the particle is promoted from a quantum level n to a higher energy level n', with the characteristic energy $\Delta\mathscr{E}(n \to n')$ required for the absorption process being

$$\Delta\mathscr{E}(n \to n') = \mathscr{E}_{n'} - \mathscr{E}_n = (n'^2 - n^2)\pi^2\hbar^2/2mL^2. \tag{1.392}$$

Similarly, energy *emission* occurs when the particle in the box drops from a quantum level n' to a lower energy level n, with the characteristic energy of emission being

$$\Delta\mathscr{E}(n' \to n) = \mathscr{E}_{n'} - \mathscr{E}_n = (n'^2 - n^2)\pi^2\hbar^2/2mL^2. \tag{1.393}$$

It is interesting to note that the lowest energy state (i.e. the *ground state*) is given by $n = 1$. (For $n = 0$, $\psi = 0$ so there is no particle state.) *The ground-state energy is therefore nonzero.* The potential energy of the particle is zero within the box, so the total energy represents kinetic energy. Thus the particle is in motion within the box even when it is in its lowest energy state. The lowest speed is given by

$$\mathscr{E}_1 = \tfrac{1}{2}mv_{\text{min}}^2, \tag{1.394}$$

so that

$$v_{\text{min}} = \pi\hbar/mL, \tag{1.395}$$

where m is the mass of the particle and L is the length of the box.

The three-dimensional square-well potential is similar to, but more involved than, the one-dimensional problem thus far treated. The spectrum of energy levels involves three quantum numbers, as shown in Chap. 3. This represents perhaps the most important problem in developing the quantum mechanics for the free electron model of metals, and also it is quite important for our development of energy band theory in Chap. 7.

PROJECT 1.26 Eigenfunctions for Particle in a Two-Dimensional Square-Well Potential

1. Solve for the energy eigenfunctions for the Schrödinger equation for a particle trapped in a two-dimensional square-well potential.
2. Construct two-dimensional sketches (or plots) illustrating contours of equal probability density $\psi^*\psi$ for the five lowest energy eigenstates.
3. Interpret these sketches in terms of location of the particle in the two-dimensional box.
4. Considering the time dependence $\theta_l(t)$ of these eigenfunctions, attempt to arrive at some classically meaningful interpretation for the actual motion of the particle inside the box.

PROJECT 1.27 Finite Square-Well Potential

Deduce the quantum energy levels and wave functions for the finite-depth square-well potential problem.

PROJECT 1.28 Ammonia Clock

Consider the following symmetric one-dimensional square-well model of an ammonia molecule, with position representing the perpendicular distance of the nitrogen atom from the plane defined by the three hydrogen atoms:

$$V(x) = V_0 \qquad (-a \leqslant x \leqslant a),$$

$$V(x) = 0 \qquad (a < x < a + b) \quad \text{and} \quad (-a - b < x < -a),$$

$$V(x) = \infty \qquad (x \geqslant a + b) \quad \text{and} \quad (x \leqslant -a - b).$$

(See also Problem 146, and especially the sketch of the symmetric double-well potential energy diagram.)
1. Deduce the eigenvalues and eigenfunctions.
2. Derive an expression for the energy eigenvalues.
3. Qualitatively describe the dependence on the parameters V_0, a, and b.
4. Compute the time dependence of the wave function.
5. Take the limit in which $2aV_0$ becomes a Dirac delta function.

PROJECT 1.29 The Asymmetric Potential Well

Consider a particle of mass m trapped in the asymmetric potential defined by

$$U(x) = U_1 \quad (x < 0), \qquad U(x) = 0 \quad (0 \leqslant x \leqslant L), \qquad U(x) = U_2 \quad (x > L).$$

1. Solve the Schrödinger equation to obtain the energy eigenfunctions.
2. Determine the energy eigenvalues.
3. Take the limit $U_2 \to U_1$ and show that your results reduce appropriately to those for the finite square-well potential.
4. Take the limit $U_1 \to \infty$, $U_2 \to \infty$ and show that your results reduce appropriately to those for the infinite square-well potential.
5. Evaluate numerically the eigenvalues in Part 2 for reasonable parameter values, such as $U_1 =$ 1-5 eV and $U_2 =$ 5–100 eV.

12.3 The Harmonic Oscillator Potential

Another of the most important potential energy functions is the *harmonic oscillator potential,*

$$U(x) = \tfrac{1}{2}Kx^2,$$

which is illustrated in Fig. 1.39. This potential is found applicable to a variety of problems in classical mechanics, such as that of a mass M attached to a spring having a stiffness constant K and displaced a distance x from its equilibrium position on an essentially frictionless horizontal plane (see Fig. 1.39). The force obtained by taking the negative gradient of $U(x)$ is $\mathbf{F} = -\nabla U(x) = -\hat{\mathbf{x}}\, dU(x)/dx = -(Kx)\hat{\mathbf{x}}$.

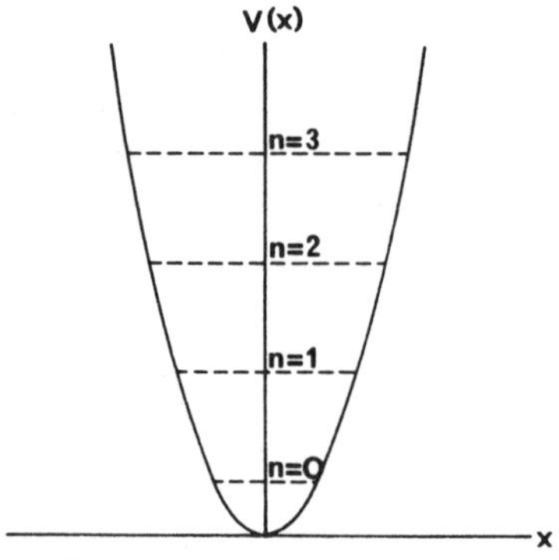

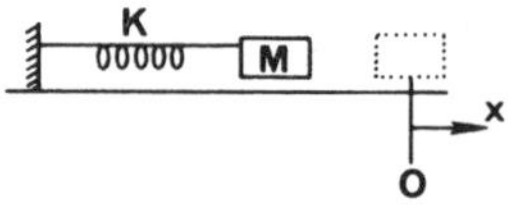

Fig. 1.39 Harmonic oscillator potential energy (left-hand side) and a typical harmonic oscillator (right-hand side).

EXERCISE Apply Newton's second law and Hooke's law for the force exerted by a spring in tension or compression to deduce the classical harmonic oscillations of the back and forth motion of a mass affixed to one end of a spring and sliding on a frictionless horizontal plane.

The Hamiltonian for the harmonic oscillator problem is

$$\mathscr{H} = -(\hbar^2/2m)\, d^2/dx^2 + \tfrac{1}{2}Kx^2,$$

and the time-independent Schrödinger equation for this problem is

$$-(\hbar^2/2m)\, d^2\phi_n(x)/dx^2 + \tfrac{1}{2}Kx^2\phi_n(x) = \mathscr{E}_n\phi_n(x).$$

This is a troublesome differential equation to solve for the uninitiated, but the solutions have been worked out and studied in great detail [see, for example, Pauling and Wilson (1935) and Bohm (1951)]. The solutions involve *Hermite polynomials.* The complete time-dependent solutions $\psi_n(x, t)$ are then given as usual by taking the product of the spatial functions with the corresponding time-dependent function $\theta_n(t) = \exp[(-i/\hbar)\mathscr{E}_n t]$,

$$\psi_n(x, t) = \phi_n(x) \exp[-(i/\hbar)\mathscr{E}_n t].$$

The application of boundary conditions to the Hermite polynomials then yields the following discrete spectrum of energy eigenvalues:

$$\mathscr{E}_n = (n + \tfrac{1}{2})\hbar\omega = (n + \tfrac{1}{2})h\nu_{\text{osc}} \qquad (n = 0, 1, 2, \ldots, \infty)$$

(harmonic oscillator energy eigenvalues),

where ω is an angular frequency, which for the well-known problem of the horizontal motion of a mass M attached to a spring having a force constant K is given by $(K/M)^{1/2}$. The integer n is the *quantum number*. These energy levels are indicated as dashed lines in Fig. 1.39. An interesting feature of the quantum solution is the fact that the ground-state energy ($n = 0$) is nonzero. Lattice vibrations in a solid can be treated in terms of the harmonic oscillator potential. That the ground-state energy is nonzero means that even at 0°K there will be some vibrational motion of the lattice.

Another interesting feature of the quantum solution is the dependence of the energy upon the *frequency* of oscillation. The increase in amplitude (see Fig. 1.39) which accompanies larger values of the energy $\mathscr{E}_n$ seems almost incidental to the quantum solution, whereas in the classical solution the dependence of energy upon amplitude appears to be a central feature. In fact, the energy $\Delta\mathscr{E}(n \to n')$ required to excite an oscillator of frequency ω from a state of quantum number n to the state with quantum number n' is

$$\Delta\mathscr{E}(n \to n') = (n' + \tfrac{1}{2})\hbar\omega - (n + \tfrac{1}{2})\hbar\omega = (n' - n)\hbar\omega.$$

Thus *energy absorption* occurs in integer multiples of a basic unit of energy $\hbar\omega$ which is characteristic of the oscillator frequency. Analogously, energy emission due to the deexcitation of an oscillator of frequency ω from a state of quantum number n' to the state with quantum number n is given by

$$\Delta\mathscr{E}(n' \to n) = (n' + \tfrac{1}{2})\hbar\omega - (n + \tfrac{1}{2})\hbar\omega = (n' - n)\hbar\omega = (n' - n)h\nu.$$

If this energy is emitted in the form of a photon of energy $\mathscr{E}_{\text{phot}} = h\nu_{\text{phot}}$, then the frequency of the photon will be $\nu_{\text{phot}} = (n' - n)\omega/2\pi = (n' - n)\nu_{\text{osc}}$, and the wavelength of the emitted photon will be $\lambda_{\text{phot}} = c/\nu_{\text{phot}} = c/[(n' - n)\nu_{\text{osc}}]$. The allowed transitions are governed by selection rules involving off-diagonal matrix elements (see §12.5).

EXERCISE Show that $\psi = A\exp[-(m\omega/2\hbar)x^2]$ is a solution to the Schrödinger equation for a particle in a harmonic oscillator potential.

EXERCISE Numerically evaluate the Schrödinger equation for the harmonic oscillator using a programmable calculator. [*Hint*: See Eisberg (1976).]

PROJECT 1.30 Solution of the Schrödinger Equation for the Harmonic Oscillator

Solve the Schrödinger equation analytically, obtaining the energy eigenvalues listed above and the energy eigenfunctions. [*Hint*: Good references for this problem are Bohm (1951) and Pauling and Wilson (1935).]

PROJECT 1.31 Eigenvalue Equations Satisfied by Harmonic Oscillator Eigenfunctions

1. By direct substitution, show that the harmonic oscillator eigenfunctions satisfy the energy eigenvalue equation with the energy eigenvalues given above.
2. Show whether or not the harmonic oscillator eigenfunctions satisfy the eigenvalue equation $p^{\text{op}}\psi_i = p_i\psi_i$ for the linear momentum, and explain physically the result you obtain.

PROJECT 1.32 Matrix Elements of Position Operator for the Harmonic Oscillator

Deduce the matrix representation of the position coordinate x in the representation given by the harmonic oscillator wave functions. (*Hint*: These wave functions can be found tabulated in many standard reference texts in quantum mechanics.)

12.4 The Coulomb Potential

12.4.1 Energy Eigenvalues and Spectroscopic Lines. The Coulomb attraction between two point charges of opposite sign gives rise to a potential energy of the form (see §5.3), $U(\mathbf{r}) = q_1q_2/4\pi\varepsilon_0 r$, where $r \equiv |\mathbf{r}|$ is the separation distance between the two charges. This potential is sketched in Fig. 1.40. The classical solution to this problem, assuming q_1 to be a proton and q_2 to be an electron, has

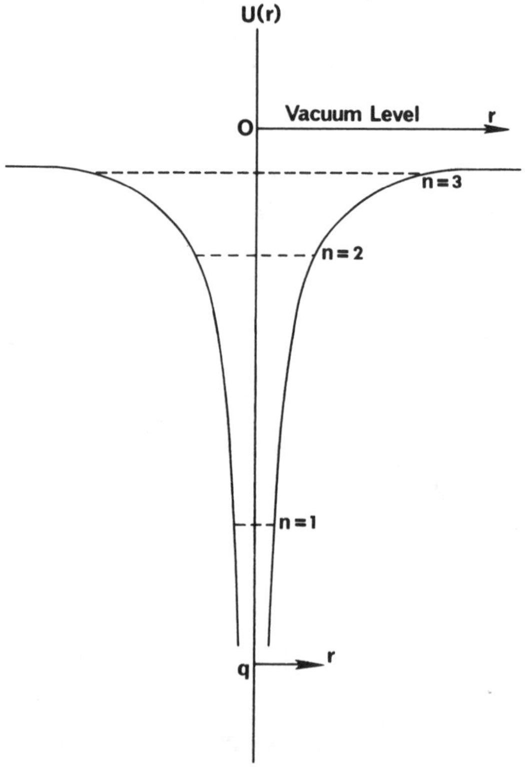

Fig. 1.40 Coulomb potential energy between two oppositely charged point charges as a function of separation r.

been worked out in detail in §5.3. The additional consideration of the de Broglie relation $\lambda = h/p$ for the electron then leads to the Bohr theory of the hydrogen atom, which is also presented in that section. The Bohr theory, though semi-classical in character, does give rise to the correct system of discrete energy levels. However, a more appropriate way to develop this problem is to use the Hamiltonian

$$\mathscr{H} = -(\hbar^2/2m)\,\nabla^2 + q_1 q_2/4\pi\varepsilon_0 r$$

to set up the appropriate time-independent Schrödinger equation $\mathscr{H}\phi_n(\mathbf{r}) = \mathscr{E}_n\phi_n(\mathbf{r})$ for this problem. Due to the spherically symmetric potential-energy term, $U(\mathbf{r}) = U(r)$, the use of spherical polar coordinates (r, θ, ϕ) enables a separation-of-variables technique to be used. This gives rise to three separated equations for the factors $R(r)$, $\Theta(\theta)$, and $\Phi(\phi)$ appearing in the product form of the spatial portion of the wave function $\phi(\mathbf{r}) = R(r)\Theta(\theta)\Phi(\phi)$. The solution of the three equations ranges from the almost trivial [for the $\Phi(\phi)$] to the almost horrendous [for the $R(r)$], but the extensive development of the detailed solutions in the literature [see Pauling and Wilson (1935) and Bohm (1951)] saves the reader from an otherwise monumental mathematical task. Very nice illustrations of these functions are given in Leighton (1959). The application of appropriate boundary conditions of single valuedness and boundedness on the wave function leads to three quantum numbers, n, l, and m, which are referred to respectively as the *principal quantum number*, the *orbital* (or *azimuthal*) *quantum number*, and the *magnetic quantum number*. The origin of these quantum numbers is the topic of §12.4.2. In the absence of a magnetic field the quantized energy levels depend only upon the principal quantum number n,

$$\mathscr{E}_n = -mZ^2e^4/32n^2\pi^2\hbar^2\varepsilon_0^2 \qquad (n = 1, 2, 3, \ldots, \infty),$$

where the charges are $q_1 = Ze$ and $q_2 = -e$ for the one-electron atom, with $Z = 1$ for the hydrogen atom. These levels agree with those deduced by means of the semiclassical Bohr theory presented in §5.3, as already mentioned. The spectrum of energy levels is indicated as dashed lines in Fig. 1.41.

Energy absorption by the atom is required to promote an electron from a given quantum state n to a higher energy quantum state n'. The amount of energy $\Delta\mathscr{E}(n \to n')$ required for this excitation is

$$\Delta\mathscr{E}(n \to n') = \mathscr{E}_{n'} - \mathscr{E}_n = -Z^2hcR_\infty(1/n'^2 - 1/n^2),$$

where R_∞ is known as the *Rydberg constant*. It has the value

$$R_\infty = me^4/64\pi^3\varepsilon_0^2\hbar^3 c = 1.0967758 \times 10^7 \quad \text{m}^{-1}.$$

Similarly, the energy emission accompanying the transition of an electron from a quantum state n' to a lower energy quantum state n is

$$\Delta\mathscr{E}(n' \to n) = \mathscr{E}_{n'} - \mathscr{E}_n = -Z^2hcR_\infty(1/n'^2 - 1/n^2).$$

If this energy is given off in the form of a photon of frequency ν_{phot} and energy

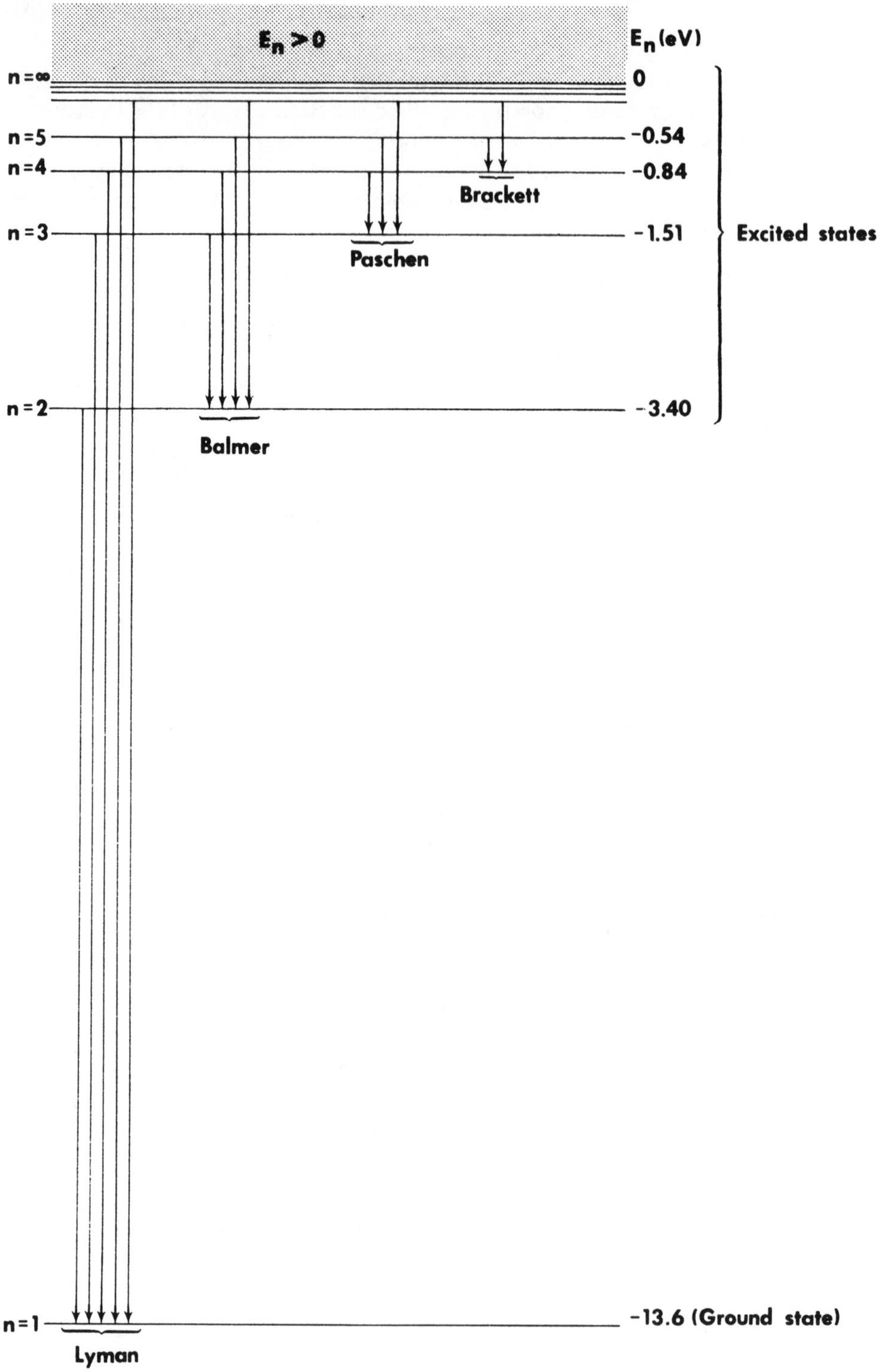

Fig. 1.41 Quantized energy levels and the associated series of spectral lines for the hydrogen atom.

$\mathscr{E}_{\text{phot}} = h\nu_{\text{phot}}$, the photon wavelength is $\lambda_{\text{phot}} \equiv c/\nu_{\text{phot}} = hc/\mathscr{E}_{\text{phot}}$, and the reciprocal wavelength is given by

$$1/\lambda_{\text{phot}} = Z^2 R_\infty (1/n^2 - 1/n'^2).$$

The spectroscopic lines (as observed photographically) involve energy absorption and emission processes. The wavelengths of the observed lines for hydrogen agree well with the above expression, once a small correction factor is introduced to take into account the finite nuclear mass. The choice of $n = 2$ with $n' = 3, 4, 5, \ldots$ gives rise to the sequence of observed lines known as the *Balmer series*, the choice $n = 3$ with $n' = 4, 5, 6, \ldots$ gives rise to the sequence of observed lines known as the *Paschen series*, *etc*. These various series are listed in Table 1.1 and are illustrated in Fig. 1.41.

Table 1.1

Hydrogen Atom Radiation Series[a]

n	Series Designation	Spectral Region
1	Lyman	Ultraviolet
2	Balmer	Near ultraviolet and visible
3	Paschen	Infrared
4	Brackett	Infrared
5	Pfund	Infrared

[a] $\nu = c/\lambda = cR_{\text{H}}[n^{-2} - n'^{-2}]$ $(n' = n+1, n+2, n+3, \ldots)$, $R_{\text{H}} = 1.0967758 \times 10^7/\text{m}$, $c = 2.997925 \times 10^8$ m/sec.

The series of energy levels is modified in the presence of a magnetic field because of two factors, namely, the electron orbital motion around the nucleus and the intrinsic spin of the electron. A charged particle in orbit constitutes a circulating current, which in turn produces a magnetic moment. This so called *orbital magnetic moment* $\boldsymbol{\mu}_{\text{orb}}$ interacts with an applied magnetic field $\mathbf{B}$ to give an additional energy term

$$U_{\text{orb}} = -\boldsymbol{\mu}_{\text{orb}} \cdot \mathbf{B}$$

which must be added to the Hamiltonian. Likewise the intrinsic spin angular momentum of the electron with respect to an axis through its center of mass gives rise to a circulating current and thus to a *spin magnetic moment* $\boldsymbol{\mu}_{\text{s}}$, due to the fact that the electronic charge has some spatial extent. Thus there is an energy term

$$U_{\text{spin}} = -\boldsymbol{\mu}_{\text{s}} \cdot \mathbf{B}$$

to consider also in the Hamiltonian. Other energy terms can also arise, such as the energy of interaction between two magnetic moments $\boldsymbol{\mu}_{\text{orb}}$ and $\boldsymbol{\mu}_{\text{s}}$, called the *spin–orbit* interaction energy. For further details and the expected modifications in the energy level spectrum, the reader is referred to the excellent treatise by Leighton (1959).

PROJECT 1.33 Solution of the Schrödinger Equation for the Hydrogen Atom

1. Substitute the separated wave function into the Schrödinger equation with the potential energy being that of the Coulomb potential with corresponding Hamiltonian. Use spherical polar coordinates for the Laplacian. Carry out the standard separation of variables process, obtaining individual differential equations for $R(r)$, $\Theta(\theta)$, and $\Phi(\phi)$.
2. Solve the differential equation for $\Phi(\phi)$, obtaining the ϕ dependence of $\psi(r, \theta, \phi, t)$.
3. Solve the differential equation for $\Theta(\theta)$, obtaining the θ dependence of $\psi(r, \theta, \phi, t)$.
4. Solve the differential equation for $R(r)$, obtaining the r dependence of $\psi(r, \theta, \phi, t)$ and the energy eigenvalues given. [*Hint*: Good references for this problem are Pauling and Wilson (1935) and Bohm (1951).]

12.4.2 Source of the Quantum Numbers in the Hydrogen Atom Solutions. It is interesting to examine the way in which the three quantum numbers arise in the hydrogen atom problem, and to understand how the electrons in the multielectron atom can be characterized by the corresponding electronic states of the one-electron atom. These states, together with the fourth quantum number of electron spin and the statistics of energy level occupation implied by the Pauli exclusion principle, enable us to understand in a rudimentary way the entire periodic table for the chemical elements.

The key to the separation of variables in the hydrogen atom problem is the recognition that the Coulomb potential energy of interaction between the electron in question and the nucleus depends only upon separation $r \equiv |\mathbf{r}|$ and is independent of the spatial orientation of the line of centers between electron and nucleus. Therefore in spherical polar coordinates (r, θ, ϕ) the Schrödinger equation can be separated (in the usual way of variables separation in partial differential equations) into three equations, each involving a function of one of these three variables, namely,

$$\psi(\mathbf{r}, t) = R(r)\Theta(\theta)\Phi(\phi) \exp[-(i/\hbar)\mathscr{E}t], \tag{1.396}$$

where only the equation for $R(r)$ contains the Coulomb potential energy of interaction between electron and nucleus. The separation constant in the equation for $\Phi(\phi)$ is found to lead to a wave function $\psi(\mathbf{r}, t) \propto \exp(im\phi)$ which is single valued whenever the angle ϕ is increased by multiples of 2π (thus reproducing $\mathbf{r}$) if and only if m^2 is equal to the square of an integer. (The single valuedness is necessary because the probability density $\psi^*\psi$ is a physical quantity which must have only one value at a given point in space.) Denoting the integer in question by the symbol m, we thereby obtain one quantum number characterizing the electronic state. It is thus conceivable that m has allowable values of $0, \pm 1, \pm 2, \ldots$, although an upper bound on $|m|$ is dictated by another consideration to be discussed shortly. The resulting wave-function factor $\Phi(\phi)$ is found to be an eigenfunction of the Hermitian operator $\mathscr{L}_z$ which represents the $\hat{\mathbf{z}}$ component of the electron orbital angular momentum, with the eigenvalue being $m\hbar$,

$$\mathscr{L}_z\Phi(\phi) = m\hbar\Phi(\phi) \qquad (m = 0, \pm 1, \pm 2, \ldots), \tag{1.397}$$

which in turn leads to

$$\mathscr{L}_z\psi(\mathbf{r}, t) = m\hbar\psi(\mathbf{r}, t) \qquad (m = 0, \pm 1, \pm 2, \ldots). \tag{1.398}$$

These eigenvalues are illustrated in Fig. 1.42. The energy $-\boldsymbol{\mu} \cdot \mathbf{B}$ of interaction of the corresponding magnetic moment $\boldsymbol{\mu} = -(e/2m)\mathscr{L}$ (produced by the circulating electrical current due to electron orbital motion) with a magnetic field $\mathbf{B} = B_z\hat{\mathbf{z}}$ oriented along the $\hat{\mathbf{z}}$ axis is $m\hbar B_z$. For this reason, the integer m is called the *magnetic quantum number*.

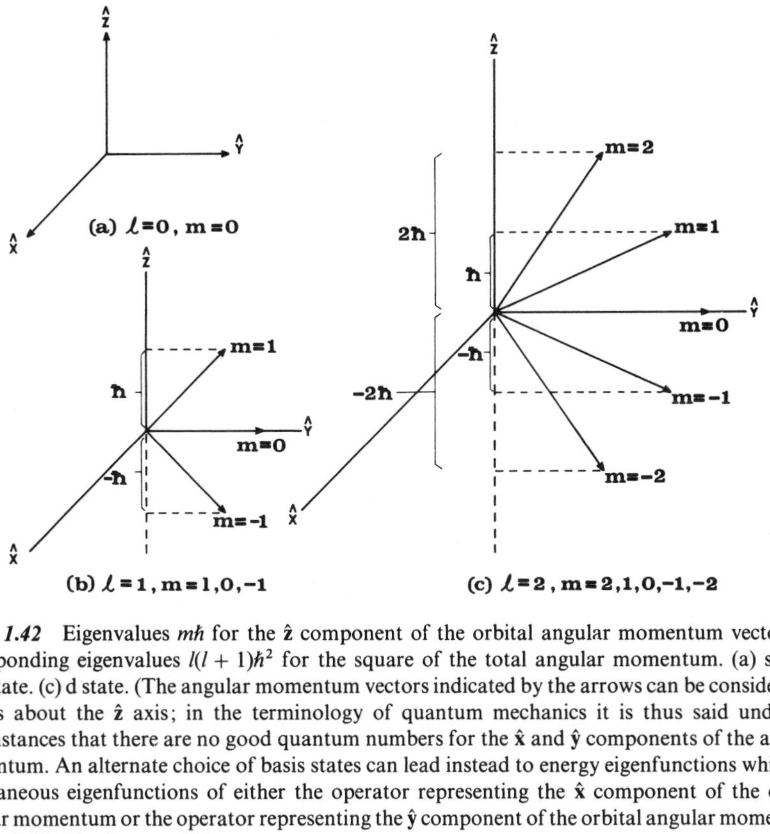

Fig. 1.42 Eigenvalues $m\hbar$ for the $\hat{\mathbf{z}}$ component of the orbital angular momentum vector and corresponding eigenvalues $l(l+1)\hbar^2$ for the square of the total angular momentum. (a) s state. (b) p state. (c) d state. (The angular momentum vectors indicated by the arrows can be considered to precess about the $\hat{\mathbf{z}}$ axis; in the terminology of quantum mechanics it is thus said under the circumstances that there are no good quantum numbers for the $\hat{\mathbf{x}}$ and $\hat{\mathbf{y}}$ components of the angular momentum. An alternate choice of basis states can lead instead to energy eigenfunctions which are simultaneous eigenfunctions of either the operator representing the $\hat{\mathbf{x}}$ component of the orbital angular momentum or the operator representing the $\hat{\mathbf{y}}$ component of the orbital angular momentum, but not both simultaneously.)

The differential equation for $\Theta(\theta)$ representing the θ component of the wave function $\psi(\mathbf{r}, t)$ contains m^2 (discussed above) as well as a second separation constant λ. Whenever $m = 0$, the equation can be cast in a form known as *Legendre's differential equation*. The solutions diverge unless $\lambda = l(l+1)$, where l represents a nonnegative integer. Therefore physically meaningful probability densities are obtained only for this choice for the second separation constant. The solutions $P_l(\cos\theta)$ obtained are known as the *Legendre polynomials*. These solutions turn out to be eigenfunctions of the Hermitian operator $\mathscr{L}^2$ representing the square of the orbital angular momentum of the electron, with eigenvalues $l(l+1)\hbar^2$,

$$\mathscr{L}^2\Theta(\theta) = l(l+1)\hbar^2\Theta(\theta) \qquad (l = 0, 1, 2, 3, \ldots) \tag{1.399}$$

which in turn leads to

$$\mathscr{L}^2\psi(\mathbf{r}, t) = l(l+1)\hbar^2\psi(\mathbf{r}, t) \qquad (l = 0, 1, 2, 3, \ldots). \tag{1.400}$$

Therefore l is called the *orbital angular momentum quantum number*. The orbital angular momentum eigenvalues are illustrated in Fig. 1.42.

Whenever $m \neq 0$, the corresponding solutions to the $\Theta(\theta)$ equation are the *associated Legendre functions* $P_l^m(\cos\theta)$, where $m^2 \leqslant l^2$. The restriction on m is thus a mathematical one, although it ties in very well with the corresponding physics since the square of the $\hat{\mathbf{z}}$ component of the orbital angular momentum, namely, $m^2\hbar^2$, cannot exceed the square of the total orbital angular momentum, namely, $l(l+1)\hbar^2$, which in turn requires $|m| \leqslant l$.

The differential equation for $R(r)$ representing the r component of the wave function $\psi(\mathbf{r}, t)$ contains the quantum number l (discussed above) as well as a third separation constant. The solutions can be expressed in terms of the *associated Laguerre polynomials*. The requirement that the solutions not diverge in order to have a physically meaningful probability density $\psi^*\psi$ places once again severe restrictions on the separation constant. This in turn requires an integer quantum number n, known as the *principal quantum number*, together with the condition $n > l$. This leads immediately to the quantized energy eigenvalues

$$\mathscr{E}_n = -\tfrac{1}{2}Z^2e^2/4\pi\varepsilon_0 a_0 n^2 \qquad (n = 1, 2, 3, \ldots), \tag{1.401}$$

where a_0 is the parameter known as the *Bohr radius*,

$$a_0 \equiv 4\pi\varepsilon_0\hbar^2/me^2; \tag{1.402}$$

that is,

$$\mathscr{H}\psi(\mathbf{r}, t) = \mathscr{E}_n\psi(\mathbf{r}, t) \qquad (n = 1, 2, 3, \ldots), \tag{1.403}$$

where the $\psi(\mathbf{r}, t)$ depends upon the n value in question.

Because $\psi(\mathbf{r}, t)$ is characteristic of the three quantum numbers (n, l, m), it is appropriate to indicate this explicitly by writing

$$\psi(\mathbf{r}, t) = \psi_{nlm}(\mathbf{r}, t). \tag{1.404}$$

To summarize, three integer quantum numbers (n, l, m) are required to characterize a given energy eigenfunction for the hydrogen atom, with the allowed values being

$$n = 1, 2, 3\ldots, \tag{1.405}$$

$$0 \leqslant l \leqslant n - 1, \tag{1.406}$$

$$-l \leqslant m \leqslant l. \tag{1.407}$$

It can be seen from these results that for a given n value, there are n allowable l values, and for a given l value, there are $(2l+1)$ possible m values. Since the energy eigenvalues $\mathscr{E}_n$ given above depend only on the value of n and are independent of the values of l and m, it can be seen that there are many eigenfunctions for a given energy eigenvalue. The solutions for the hydrogen atom are therefore highly degenerate except for the ground state ($n = 1$, $l = 0$, $m = 0$). When the fourth quantum number m_s, representing electron spin, is taken into account ($m_s = \pm\frac{1}{2}$), corresponding to spin angular momentum values of $\pm m_s\hbar$, it is found that two electrons can be accommodated in the ground state without violating the Pauli exclusion principle, which requires only that no two electrons have the same set of quantum numbers. Table 1.2 illustrates some allowed sets of quantum numbers.

Table 1.2

Allowed Combinations of Quantum Numbers

n	l	m_l	m_s	
1	0	0	$\pm\frac{1}{2}$	1s electrons
2	0	0	$\pm\frac{1}{2}$	2s electrons
2	1	0	$\pm\frac{1}{2}$	2p electrons
2	1	1	$\pm\frac{1}{2}$	
2	1	−1	$\pm\frac{1}{2}$	
3	0	0	$\pm\frac{1}{2}$	3s electrons
3	1	0	$\pm\frac{1}{2}$	3p electrons
3	1	1	$\pm\frac{1}{2}$	
3	1	−1	$\pm\frac{1}{2}$	
3	2	0	$\pm\frac{1}{2}$	3d electrons
3	2	1	$\pm\frac{1}{2}$	
3	2	2	$\pm\frac{1}{2}$	
3	2	−1	$\pm\frac{1}{2}$	
3	2	−2	$\pm\frac{1}{2}$	

Figures 1.43 and 1.44 illustrate the results of the quantum treatment of the hydrogen atom for the specific cases of $l = 0$, $l = 1$, and $l = 3$. The case $l = 0$ is

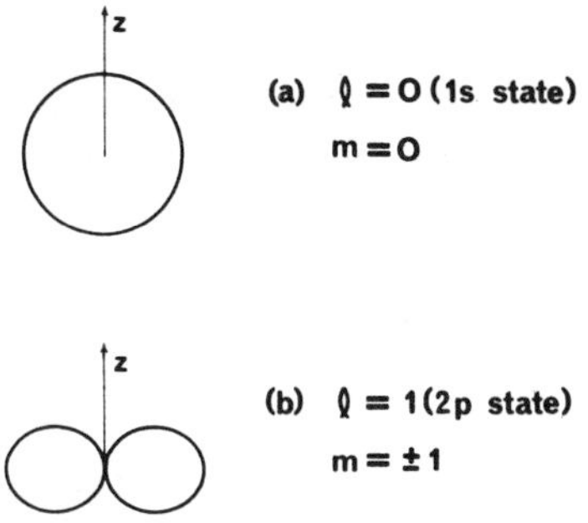

Fig. 1.43 Angular probability density distributions for the 1s and the 2p states of the electron in the hydrogen atom.

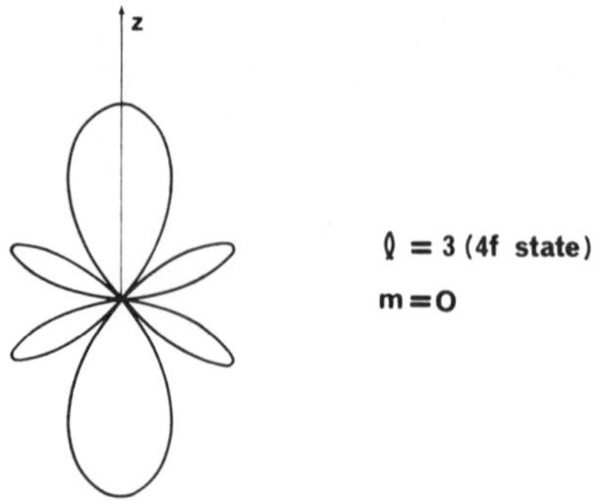

Fig. 1.44 Angular probability density distribution for the 4f state of the electron in the hydrogen atom.

interesting in that it constitutes a state of zero angular momentum. Classically this could happen only if the electron were oscillating along the line of centers between the electron and the nucleus. Examination of the eigenfunctions for this case [McGervey (1971)] shows that they are spherically symmetric and decay more or less exponentially with increasing distance from the origin. Sketches illustrating both the angular and the radial probability density distributions for the $n = 1, l = 0$ and the $n = 2, l = 0$ eigenfunctions (known respectively as the 1s and 2s *states*) are shown in Fig. 1.4.

PROJECT 1.34 Eigenvalue Equations Satisfied by Hydrogen Atom Wave Functions

1. By direct substitution, show that the hydrogenic wave functions ψ_{nlm} [see Bohm (1951); Pauling and Wilson (1935)] satisfy the energy eigenvalue equation (1.403) with energy eigenvalues given by Eq. (1.401).
2. Show also that the ψ_{nlm} satisfy the eigenvalue equation (1.397) for the $\hat{\mathbf{z}}$ component of the orbital angular momentum.
3. Show that the ψ_{nlm} likewise satisfy the eigenvalue equation (1.400) for the square of the total orbital angular momentum.
4. Show that the functions ψ_{nlm} do or do not satisfy the eigenvalue equation $\mathscr{L}^{\text{op}} g_i = \mathbf{L}_i g_i$ for the total orbital angular momentum vector, and explain the physical consequences of your finding.
5. What are the constants of motion in the hydrogen atom problem?

12.4.3 Multielectron Atoms. First, it is important to recognize that the potential energy in a multielectron atom depends upon the electron–electron Coulomb energies as well as on the Coulomb energy of interaction of the electron in question with the nucleus. Although the electron–electron interaction can be viewed as a perturbation for purposes of very crude estimates (see Project 5.2 for the helium atom), it is in fact much too large to be treated within the framework of perturbation theory. If only a few electrons are involved, as is the case for the lighter atoms, a *variational treatment* can be used to obtain approximate solutions to the many-electron Schrödinger equation [see Schiff (1968)]. However, for the heavier atoms where a larger number of electrons are involved (cf. Fig. 1.45), the starting point for most calculations is the approximation that the total potential energy of interaction of a given electron

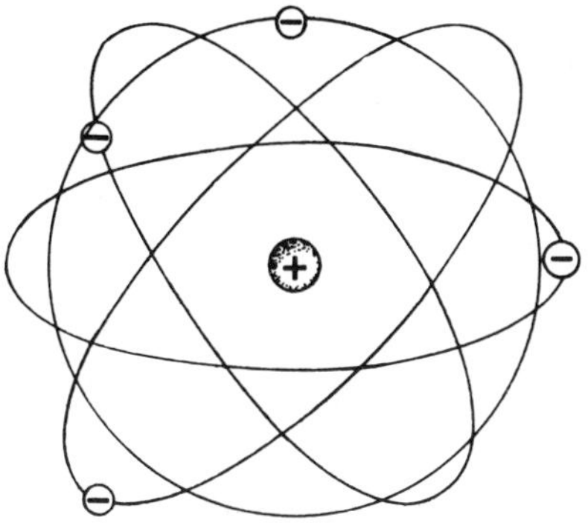

Fig. 1.45 Classical picture of the multielectron atom in which the electrons interact among themselves and with the central oppositely charged nucleus.

with the nucleus and the other electrons can be represented by a spherically symmetric potential $V(r)$. This is referred to as the *central-field approximation*. In practice, then, one of the most difficult parts of the problem is to estimate or calculate the potential. The details are quite beyond the scope of our present treatment; however, the important results of such an approach are of interest to us, since they justify using one-electron eigenvalues and eigenfunctions as a semantic framework for describing the multielectron atom. Including spin, there is once again a set of four quantum numbers (n, l, m_l, and m_s) required to specify an electronic state. The wave function specified by a given set of quantum numbers is called an *orbital*, or more specifically, *an atomic orbital*, in analogy with the older Bohr theory in which electrons were considered to travel in planetary orbits in accordance with classical mechanics. The Pauli exclusion principle again requires that no two electrons have the same set of quantum numbers [viz., the *principal quantum number n* characteristic of the total energy of the electron, the *angular momentum (azimuthal) quantum number l* characteristic of the total orbital angular momentum of the electron, the *magnetic quantum number* m_l characteristic of the orientation of the magnetic moment with respect to the $\hat{z}$ axis, and the *spin quantum number* m_s characteristic of the orientation of the electron spin magnetic moment]. The *orbital angular momentum* and *magnetic* quantum numbers l and m_l are the same as the quantum numbers l and m in the hydrogen atom since the separation of variables with the more general central potential $V(r)$ proceeds in exactly the same way as for the single electron (or hydrogen atom) where $U(r) = -Ze^2/4\pi\varepsilon_0 r$, thus yielding the same equations for the $\Theta(\theta)$ and $\Phi(\phi)$ factors in the product wave function $\psi(\mathbf{r}, t)$,

$$\psi(\mathbf{r}, t) = R(r)\Theta(\theta)\Phi(\phi) \exp[-(i/\hbar)\mathscr{E}t]. \qquad (1.408)$$

The electron spin quantum number $m_s = \pm\frac{1}{2}$ is likewise the same as in the hydrogen atom. The radial equation, containing as it does the generalized central potential $V(r)$ instead of simply the electron–nucleus Coulomb potential, requires a generalized *total* quantum number n which is quite analogous to the *principal* quantum number n for the hydrogen atom. One very important difference between the results for the general central potential problem in which the potential no longer varies as $1/r$, and the hydrogen atom problem in which the potential varies strictly as $1/r$, is the fact that electronic states characterized by different values of the orbital angular momentum quantum number l and the same total quantum number n generally correspond to different energy

Table 1.3

Electron Configurations of the Elements[a] (*in Gaseous Phase*)

Shell		*K*	*L*		*M*			*N*			
Number of Electrons	Element	1s	2s	2p	3s	3p	3d	4s	4p	4d	4f
1	H	1									
2	He	2									
3	Li	2	1								
4	Be	2	2								
5	B	2	2	1							
6	C	2	2	2							
7	N	2	2	3							
8	O	2	2	4							
9	F	2	2	5							
10	Ne	2	2	6							
11	Na	2	2	6	1						
12	Mg	2	2	6	2						
13	Al	2	2	6	2	1					
14	Si	2	2	6	2	2					
15	P	2	2	6	2	3					
16	S	2	2	6	2	4					
17	Cl	2	2	6	2	5					
18	Ar	2	2	6	2	6					
19	K	2	2	6	2	6		1			
20	Ca	2	2	6	2	6		2			
21	Sc	2	2	6	2	6	1	2			
22	Ti	2	2	6	2	6	2	2			
23	V	2	2	6	2	6	3	2			
24	Cr	2	2	6	2	6	5	1			
25	Mn	2	2	6	2	6	5	2			
26	Fe	2	2	6	2	6	6	2			
27	Co	2	2	6	2	6	7	2			
28	Ni	2	2	6	2	6	8	2			
29	Cu	2	2	6	2	6	10	1			
30	Zn	2	2	6	2	6	10	2			
31	Ga	2	2	6	2	6	10	2	1		
32	Ge	2	2	6	2	6	10	2	2		
33	As	2	2	6	2	6	10	2	3		
34	Se	2	2	6	2	6	10	2	4		
35	Br	2	2	6	2	6	10	2	5		
36	Kr	2	2	6	2	6	10	2	6		

(*continued*)

Table 1.3 (continued)

Shell		K	L	M	N				O					P			Q
Number of Electrons	Element				4s	4p	4d	4f	5s	5p	5d	5f	5g	6s	6p	6d	7s
37	Rb	2	8	18	2	6			1								
38	Sr	2	8	18	2	6			2								
39	Y	2	8	18	2	6	1		2								
40	Zr	2	8	18	2	6	2		2								
41	Nb	2	8	18	2	6	4		1								
42	Mo	2	8	18	2	6	5		1								
43	Tc	2	8	18	2	6	(5)		(2)								
44	Ru	2	8	18	2	6	7		1								
45	Rh	2	8	18	2	6	8		1								
46	Pd	2	8	18	2	6	10										
47	Ag	2	8	18	2	6	10		1								
48	Cd	2	8	18	2	6	10		2								
49	In	2	8	18	2	6	10		2	1							
50	Sn	2	8	18	2	6	10		2	2							
51	Sb	2	8	18	2	6	10		2	3							
52	Te	2	8	18	2	6	10		2	4							
53	I	2	8	18	2	6	10		2	5							
54	Xe	2	8	18	2	6	10		2	6							
55	Cs	2	8	18	2	6	10		2	6				1			
56	Ba	2	8	18	2	6	10		2	6				2			
57	La	2	8	18	2	6	10		2	6	1			2			
58	Ce	2	8	18	2	6	10	2	2	6				2			
59	Pr	2	8	18	2	6	10	3	2	6				2			
60	Nd	2	8	18	2	6	10	4	2	6				2			
61	Pm	2	8	18	2	6	10	5	2	6				2			
62	Sm	2	8	18	2	6	10	6	2	6				2			
63	Eu	2	8	18	2	6	10	7	2	6				2			
64	Gd	2	8	18	2	6	10	7	2	6	1			2			
65	Tb	2	8	18	2	6	10	8	2	6	1			2			
66	Dy	2	8	18	2	6	10	9	2	6	1			2			
67	Ho	2	8	18	2	6	10	10	2	6	1			2			
68	Er	2	8	18	2	6	10	11	2	6	1			2			
69	Tm	2	8	18	2	6	10	13	2	6				2			
70	Yb	2	8	18	2	6	10	14	2	6				2			
71	Lu	2	8	18	2	6	10	14	2	6	1			2			
72	Hf	2	8	18	2	6	10	14	2	6	2			2			
73	Ta	2	8	18	2	6	10	14	2	6	3			2			
74	W	2	8	18	2	6	10	14	2	6	4			2			
75	Re	2	8	18	2	6	10	14	2	6	5			2			
76	Os	2	8	18	2	6	10	14	2	6	6			2			
77	Ir	2	8	18	2	6	10	14	2	6	7			2			
78	Pt	2	8	18	2	6	10	14	2	6	9			1			
79	Au	2	8	18	2	6	10	14	2	6	10			1			
80	Hg	2	8	18	2	6	10	14	2	6	10			2			
81	Tl	2	8	18	2	6	10	14	2	6	10			2	1		
82	Pb	2	8	18	2	6	10	14	2	6	10			2	2		

(*continued*)

Table 1.3 (continued)

Shell		K	L	M	N				O					P			Q
Number of Electrons	Element				4s	4p	4d	4f	5s	5p	5d	5f	5g	6s	6p	6d	7s
83	Bi	2	8	18	2	6	10	14	2	6	10			2	3		
84	Po	2	8	18	2	6	10	14	2	6	10			2	4		
85	At	2	8	18	2	6	10	14	2	6	10			2	5		
86	Rn	2	8	18	2	6	10	14	2	6	10			2	6		
87	Fr	2	8	18	2	6	10	14	2	6	10			2	6		1
88	Ra	2	8	18	2	6	10	14	2	6	10			2	6		2
89	Ac	2	8	18	2	6	10	14	2	6	10			2	6	1	2
90	Th	2	8	18	2	6	10	14	2	6	10			2	6	2	2
91	Pa	2	8	18	2	6	10	14	2	6	10	2		2	6	1	2
92	U	2	8	18	2	6	10	14	2	6	10	3		2	6	1	2
93	Np	2	8	18	2	6	10	14	2	6	10	5		2	6		2
94	Pu	2	8	18	2	6	10	14	2	6	10	6		2	6		2
95	Am	2	8	18	2	6	10	14	2	6	10	7		2	6		2
96	Cm	2	8	18	2	6	10	14	2	6	10	7		2	6	1	2
97	Bk	2	8	18	2	6	10	14	2	6	10	8		2	6	1	2
98	Cf	2	8	18	2	6	10	14	2	6	10	9		2	6	1	2
99	Es	2	8	18	2	6	10	14	2	6	10	10		(2	6	1	2)
100	Fm	2	8	18	2	6	10	14	2	6	10	11		(2	6	1	2)
101	Md	2	8	18	2	6	10	14	2	6	10	12		(2	6	1	2)
102	No	2	8	18	2	6	10	14	2	6	10	13		(2	6	1	2)
103	Lw	2	8	18	2	6	10	14	2	6	10	14		(2	6	1	2)

[a] We note here the permeating influence of the Pauli exclusion principle in determining the ground state configurations of atoms in nature's beautiful array of chemical elements. For this reason, the exclusion principle is sometimes referred to as Pauli's *aufbau* (i.e., building up) principle for atomic structure.

eigenvalues. In the hydrogen atom problem the energy eigenfunctions for a given n but different l values are degenerate. In the multielectron atom, states of lower l value consistent with a fixed n value lie at a lower energy. The combined values of l and n for a given eigenfunction again determine its radial nodes, these being $n - l - 1$ in number. As in the hydrogen atom, n must be a positive integer, and the magnitude of the integer l cannot exceed $n - 1$. An atomic *shell* is specified by a given value for n, and an atomic *subshell* is specified by a given set of values for both n and l. Taking into account the two possible spin quantum numbers $m_s = \pm\frac{1}{2}$ and the $2l + 1$ values for m_l ($m_l = -l, -l + 1, \ldots, 0, 1, \ldots, l$), one deduces the result that a given subshell contains $2(2l + 1)$ degenerate electronic states. The series of shells are denoted (cf. Table 1.3) by K, L, M, N, etc., in standard spectroscopic notation.

The ground state of a many-electron atom is the one in which a sufficient number of electrons populate the orbitals of lowest energy consistent with the Pauli exclusion principle to give a neutral entity. The *ground-state configuration* of the electrons in an atom is specified by the number of electrons in each shell

Table 1.4

Elements in the Periodic Table with the Corresponding Number of Electrons per Atom

H^{1}																	He^{2}
Li^{3}	Be^{4}											B^{5}	C^{6}	N^{7}	O^{8}	F^{9}	Ne^{10}
Na^{11}	Mg^{12}											Al^{13}	Si^{14}	P^{15}	S^{16}	Cl^{17}	Ar^{18}
K^{19}	Ca^{20}	Sc^{21}	Ti^{22}	V^{23}	Cr^{24}	Mn^{25}	Fe^{26}	Co^{27}	Ni^{28}	Cu^{29}	Zn^{30}	Ga^{31}	Ge^{32}	As^{33}	Se^{34}	Br^{35}	Kr^{36}
Rb^{37}	Sr^{38}	Y^{39}	Zr^{40}	Nb^{41}	Mo^{42}	Tc^{43}	Ru^{44}	Rh^{45}	Pd^{46}	Ag^{47}	Cd^{48}	In^{49}	Sn^{50}	Sb^{51}	Te^{52}	I^{53}	Xe^{54}
Cs^{55}	Ba^{56}	La^{57}	Hf^{72}	Ta^{73}	W^{74}	Re^{75}	Os^{76}	Ir^{77}	Pt^{78}	Au^{79}	Hg^{80}	Tl^{81}	Pb^{82}	Bi^{83}	Po^{84}	At^{85}	Rn^{86}
Fr^{87}	Ra^{88}	Ac^{89}															
Lanthanide series				Ce^{58}	Pr^{59}	Nd^{60}	Pm^{61}	Sm^{62}	Eu^{63}	Gd^{64}	Tb^{65}	Dy^{66}	Ho^{67}	Er^{68}	Tm^{69}	Yb^{70}	Lu^{71}
Actinide series				Th^{90}	Pa^{91}	U^{92}	Np^{93}	Pu^{94}	Am^{95}	Cm^{96}	Bk^{97}	Cf^{98}	Es^{99}	Fm^{100}	Md^{101}	No^{102}	Lw^{103}

(Table 1.3). The chemical properties of the different atoms (or *elements*) are determined principally by the uppermost filled energy levels, since these higher energy electrons, being less tightly bound to the atomic core, most easily share themselves with adjacent atomic cores for the formation of chemical bonds in molecules and in solids. If the uppermost occupied shell is full, there is generally an appreciable difference in energy between the occupied and next higher unoccupied state, and the atom tends to be chemically inert. It is standard in spectroscopic notation to give the n value of a shell as a number and the l value as a lower case letter, with $l = 0, 1, 2, 3, 4, \ldots$ being denoted respectively by the letters s, p, d, f, g, ... The periodic filling of successive shells as Z increases explains the use of a *periodic table* (Table 1.4) for listing the chemical elements. The number of electrons in a given shell is generally denoted by a superscript. Table 1.3 shows that sodium has 2 electrons in the 1s subshell, 2 electrons in the 2s subshell, 6 electrons in the 2p subshell, and one electron in the 3s subshell. This ground state configuration for sodium ($Z = 11$) would be denoted by Na: $1s^2 2s^2 2p^6 3s$. The rule that the maximum number of electrons in a shell be $2(2l + 1)$ with $l \leqslant n - 1$ can be consulted in conjunction with this configuration to illustrate that atomic sodium consists of two filled shells (or three filled subshells) containing the *core* electrons, and an outermost partly filled shell containing the single valence electron.

This ends our discussion of typical bound-state problems. The next question which naturally arises is how the various bound state energy levels will be populated in a thermal equilibrium situation, since the system at temperatures above 0°K need not be in its ground state configuration. To answer this question requires a knowledge of quantum statistics, the subject of the following chapter.

EXERCISE Substitute the product solution into the time-dependent Schrödinger equation using the Coulomb potential and carry out the separation of variables, employing the spherical polar coordinate system.

PROJECT 1.35 The Periodic Table

Using Table 1.4, locate the following series of elements:

1. Those elements having a single s electron in the outer shell.
2. Those elements having an incomplete d shell.
3. Those elements having incomplete p shells.
4. The magnetic metals iron, cobalt, and nickel.
5. The alkali metals lithium, sodium, potassium, rubidium, and cesium.
6. Hydrogen. (Is this an alkali metal? Give a sound reason for your answer.)
7. Beryllium, magnesium, and calcium. (What do these elements have in common?)
8. Chromium, molybdenum, and tungsten. (What do these three elements have in common?)
9. Rhodium, palladium, and silver. (What do these elements have in common?)
10. The noble metals copper, silver, and gold. (Why are these called noble? Do they have similar chemical properties? Do they have similar electronic structures?)
11. The rare-earth metals. (How are the electronic structures of these elements similar?)
12. The rare gases. (How are the electronic structures of these elements different?)
13. Silicon and germanium. (Are these useful materials? Why?)
14. Zinc, cadmium, and mercury. (How are the physical properties of these elements dissimilar?)
15. The lanthanide series. (Why is it seemingly out of place?)

16. Uranium and plutonium. (What physical properties do these elements have in common? Why is the actinide series, of which they are members, seemingly displaced in the periodic table?) (*Hint*: A dictionary or chemistry textbook will be of aid. Also, Table 1.3 will prove helpful.)

12.5 Selection Rules for Quantum Transitions

Atomic spectral lines are due to radiative electronic transitions in which an electron in an atom in an excited state undergoes a transition to a lower energy state with the attendant emission of a photon of electromagnetic radiation having an energy equal to the energy difference between the two atomic states. Quantum electrodynamics gives us the result that the matrix element for a radiative transition in which a single spinless particle of charge e changes from a state ϕ_i to a state ϕ_f, emitting or absorbing a quantum of radiation of wavelength λ and momentum $\hbar k$ in the z direction with polarization vector in the x direction, is

$$\langle f|ep_x^{\text{op}}|i\rangle \equiv \int_{\Omega_r} \phi_f^*(\mathbf{r})ep_x^{\text{op}}e^{\pm ikz}\phi_i(\mathbf{r})\,d\mathbf{r},$$

where $p_x^{\text{op}} = -i\hbar(\partial/\partial x)$ is the x-momentum operator and the plus and minus signs denote, respectively, processes in which a photon is absorbed or created. Whenever the wavelength of the radiation is large relative to the size of the atom, it can be shown that the lowest-order contribution is proportional to the matrix element

$$\langle f|ex|i\rangle \equiv \int_{\Omega_r} \phi_f^*(\mathbf{r})ex\phi_i(\mathbf{r})\,d\mathbf{r}.$$

This latter matrix element is said to be the *electric dipole* interaction matrix element, and the transition probability is in this case proportional to the square of the magnitude of this quantity. This provides us with *selection rules* for quantum transitions. (See also Chap. 5, §7.2.)

PROBLEMS

1. List two underlying causes of discrepancy between the experimental facts concerning microscopic phenomena and the predictions of classical mechanics.
2. Summarize the experimental evidence (a) for wave–particle duality in both matter and radiation, (b) against wave–particle duality in both matter and radiation.
3. What is the energy (in electron volts) of a photon of wavelength 5461 Å? (Numerical values of the physical constants are given in the Appendix.) A. 0.374, B. 1.02, C. 2.27, D. 7.93, E. 894.
4. Compute the approximate energy (or range of energies) of photons of red, orange, green, yellow, and blue light. Tabulate your results with the respective frequencies (cycles per second and radians per second), wavelengths, and free space velocities.
5. Find solutions to the classical wave equation $(\partial^2\psi/\partial x^2) = (1/c^2)\,\partial^2\psi/\partial t^2$, where ψ is the wave amplitude, c the wave velocity, x position, and t time.
6. Write an expression for a transverse wave having a frequency of 200 Hz and an amplitude of 20 (in appropriate units) propagating at a phase velocity of 500 m/sec.

7. A piano wire 1 m in length has a mass of 1 g. Compute the numerical values of the allowed vibrational frequencies (in cycles per second) of this wire when it is stretched to a tension of 1 N with both ends anchored tightly. (*Hint*: Dimensional analysis will give you a needed equation.)
8. Given the two waves $y_1 = 4 \sin \omega t$ and $y_2 = 3 \sin(\omega t + 60^\circ)$, what is the amplitude and phase of the superposition wave $y_1 + y_2$?
9. Show that energy flow depends upon ψ^2 in classical wave motion, where ψ is the amplitude. (This to a certain extent justifies the same assumption for the motion of quantum particles.)
10. Prove that $g(x, t) = \lambda \coth[\ln(\cos\{e^{i\gamma x^2} e^{i\gamma\beta^2 t^2}(-i \sin 2\gamma\beta xt + \cos 2\gamma\beta xt)\})]$ is a valid solution of the classical wave equation. (The parameters λ, γ, and β may be considered to be constants.)
11. (a) Compute the energy of a photon of wavelength 6328 Å corresponding to the red line of the helium neon laser. (b) Explain how a light source consisting of a stream of such photons could be used to separate a group of metals into two categories which could be classified as relatively high and relatively low work function metals.
12. What is the maximum kinetic energy which can be observed for ejected electrons when incident electromagnetic radiation of wavelength 2460 Å is incident on a nickel surface having a work function of 5 eV?
13. If 4500 Å photons strike a metal having a work function of 3 eV, determine whether one can reasonably expect electron ejection, and if so, compute the maximum velocity of the ejected electrons.
14. Look up values for the work function of five different metals in a handbook, and compute the wavelengths and frequencies of the electromagnetic radiation required to eject photoelectrons from these metals.
15. Compute the number of photons emitted in a laser pulse having a power of 10^6 W which lasts for 10^{-4} sec, assuming a wavelength of 6328 Å.
16. A total of 38,000 J of energy is reflected perpendicularly from a mirror. Determine the momentum transferred to the mirror if the energy reflected is in each of the following forms: (a) photons, (b) electrons, (c) neutrons, (d) helium atoms.
17. Repeat the above calculations of Problem 16, assuming an angle of incidence of 45°.
18. Reconsider the above calculations in Problems 16 and 17 if the energy is absorbed by the mirror instead of being reflected from the mirror.
19. Is it possible to observe an atomic nucleus having a diameter of 2×10^{-15} m with a light microscope? Justify your answer quantitatively.
20. Research de Broglie waves in the library. Write a paper on your findings.
21. Determine the wavelength (in meters) of a 1-g sphere of iron traveling with a speed of 270 m/sec. A. 2.98×10^{-68}, B. 8.56×10^{-51}, C. 2.45×10^{-33}, D. 7.04×10^{-16}, E. 202.
22. (a) What is the de Broglie wavelength and the frequency of an electron which has been accelerated through a potential difference of 3.1 V? (b) Repeat the calculation for a proton.
23. (a) Compute the de Broglie wavelengths for an electron and for a neutron which are traveling at 10^6 m/sec. (b) Repeat the calculation if the two particles each have energies of 1 eV.
24. Estimate the wavelength of the de Broglie wave associated with a 1-cm-radius lead sphere which has fallen from rest from a 10 story building.
25. (a) What is the de Broglie wavelength of an electron traveling at a speed of 1 km/sec? (b) What range of the electromagnetic spectrum (e.g., x-ray, radio wave, light, microwave) has wavelengths of this order? (c) What frequency is associated with the above electron? (d) What does this correspond to in the electromagnetic spectrum? (e) Is the frequency correspondence the same as the above wavelength correspondence?
26. (a) Compute the wavelength of a 1-eV photon. (b) Compute the wavelength of a 1-eV electron. (c) Compute the wavelength of a 1-eV neutron.
27. (a) Compute the wavelength and frequency of a photon having the same energy as a 4-eV electron. (b) Compare the value of the momentum of the 4-eV electron with the 4-eV photon.
28. If one wishes to probe the lattice spacings in crystals by means of diffraction techniques, it is generally necessary to use waves having wavelengths which are comparable to the dimensions of an atom (1–4Å). Compute (or estimate) the particle energies (in electron volts) which yield such wavelengths for the following types of particle: (a) neutrons, (b) electrons, (c) x rays, (d) baseballs.
29. Express the wavelength in terms of the total energy and potential energy of a particle.

30. For what wavelength (in angstroms) of incident light will photoelectrons ejected from silver (with a work function $\phi = 4.8$ eV) have a maximum velocity of 10^6 m/sec? A. 988, B. 1622, C. 5460, D. 9420, E. 17,310.
31. Consider a metal surface having a workfunction $\phi = 5$ eV. (a) Compute the wavelength and frequency of the electromagnetic radiation at which photoemission just begins. (b) Could photoelectrons be produced from this surface by radiation of 2536 Å emitted in electronic transitions from the first excited state to the ground state in vaporized mercury atoms? (Explain your conclusion.) (c) Calculate the wavelength and frequency of electromagnetic radiation required if photoelectrons emitted from this surface are to have maximum energies of 4 eV.
32. Carefully describe the meaning of Bohr's complementarity principle.
33. What is the correspondence principle?
34. Work out the equations for describing the Compton effect.
35. The limit of the resolving power (viz, the smallest distance separating two points in space which can be distinguished under optimum conditions) of a microscope is the wavelength of the light (in a light microscope) or the wavelength of the electrons (in an electron microscope). (a) Compute the minimum uncertainty in the position of a particle observed with 4-eV photons. (b) Compute the minimum uncertainty in the position of a particle observed with 4-eV electrons. (c) Following the position measurements in these two cases, what are the corresponding uncertainties in momentum?
36. Derive the real and the complex Fourier series coefficients for the functions sketched in Fig. 1.46, considering $l = 1$ cm, $d = 1$ cm, $h = 1$ cm, and $f_0 = 1$ cm.

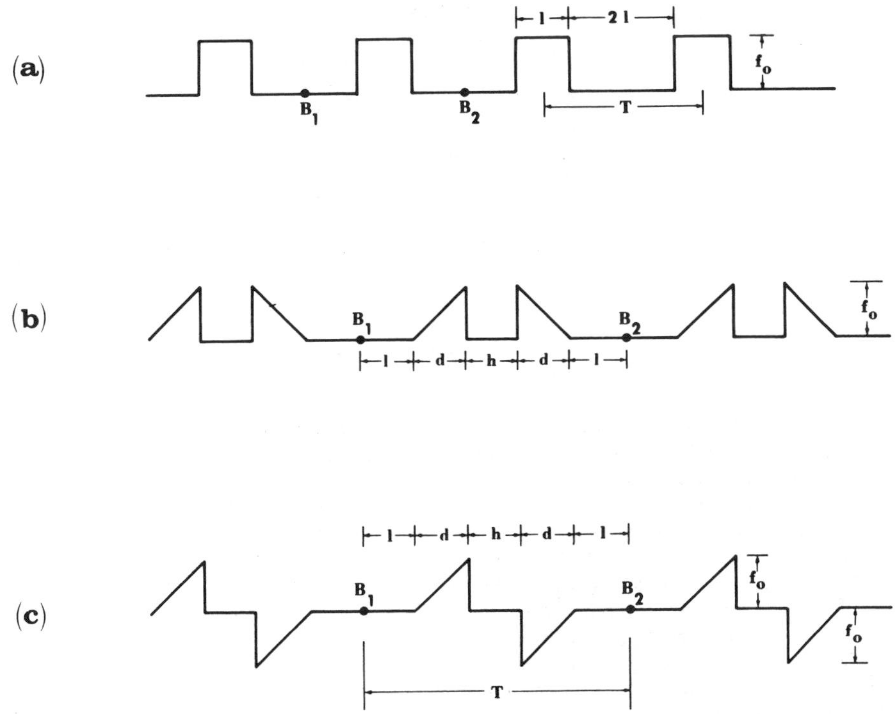

Fig. 1.46 Periodic functions for Fourier series expansion (see Problem 36).

37. Derive the Fourier integral expressions for the portions of the functions sketched in the preceding problem (Problem 36) between the points B_1 and B_2 in the limit in which the functions become aperiodic.
38. Write the Fourier series and evaluate the Fourier coefficients for the following: (a) periodic

time-dependent voltage $V(t)$ with a period of 3 sec,

$$V(t) = \begin{cases} 1 \text{ volt} & (0 < t < 1 \quad \text{sec}) \\ 0 \text{ volt} & (1 < t < 3 \quad \text{sec}); \end{cases}$$

(b) spatial displacement $g(x)$ having period of 4 m and a maximum value of $g_{\text{max}} = 1$ m,

$$g(x) = \begin{cases} Ax & (0 < x < 2 \quad \text{m}) \\ 0 & (2 < x < 4 \quad \text{m}) \end{cases} \qquad \text{where } A \text{ is a constant.}$$

39. A piano wire 1 m in length is deformed into a semicircle. Set up the Fourier integral for the position of the wire as a function of position along the axis across the diameter of the semicircle.

40. (a) Consider a function $M(\mathbf{r})$ which in three-dimensional space is constant over the region $a \leqslant |\mathbf{r}| \leqslant b$ (see Fig. 1.47) and is zero otherwise, where a and b are constant distances which may be chosen to be 1 and 2 m, respectively. The nonzero value of the function can be chosen as unity and it can have any units desired. Derive the Fourier integral expression for this function $M(\mathbf{r})$. (b) Consider $M(\mathbf{r})$ to be an electric field magnitude in free space. Considering the time-dependent wave equation derived from Maxwell's equations, can we say anything regarding the time-dependence of $M(\mathbf{r})$? (c) Derive the Fourier integral expression for subsequent propagation of a burst of light occurring at the origin of a coordinate system at $t = 0$.

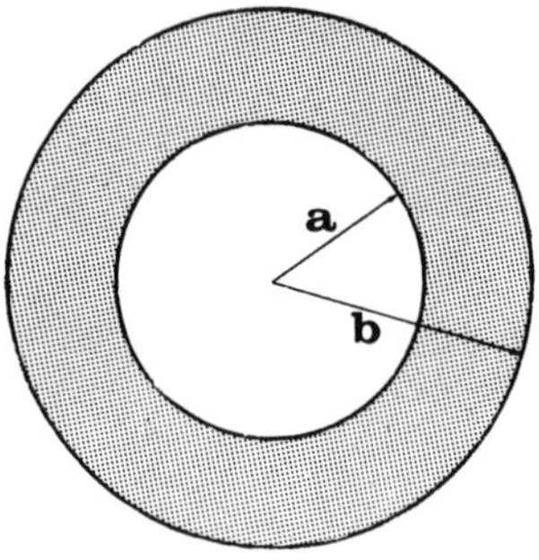

Fig. 1.47 Aperiodic function in three dimensions for expansion in a Fourier integral (see Problem 40).

41. Let the square-integrable functions $f(x)$ and $F(u)$ be Fourier transforms of each other: (a) What is the transform of $df(x)/dx$? (b) What is the transform of $xf(x)$?

42. Wave interference and diffraction are easily understood from algebraic approaches. [See Halliday and Resnick (1974), for example.] Assuming equal overall exposure, give formulas for the intensity $I(y)$ as a function of position y on a photographic plate separated a distance L from a double slit array with the slits symmetrically located at $y = \pm \frac{1}{2}d$ for the following physical situations: (a) both slits open for a given time T; (b) upper slit open and lower slit closed for a time T; (c) upper slit closed and lower slit open for a time T; (d) exposure (b) followed by exposure (c). Can you extend these considerations to multiple equally spaced slits?

43. Choose one or more sets of values in Problem 42 for a, d, λ, and L, such as $a = 0.010$, 0.020, or 0.050 mm, $d = 0.10$, 0.20, 0.50, or 1.00 mm, $\lambda = 4800$ or 5460 Å, $L = 50$, 100, or 200 cm. Numerically evaluate (by means of a calculator or computer) the patterns, and carefully plot them on graph paper (or use a computer-driven mechanical plotter).

44. If a beam of particles having a momentum per particle of $p = 2.3 \times 10^{-20}$ kg m/sec is incident perpendicularly to a slit, what is the angle (relative to the beam axis) of the first and second diffraction maxima produced by the particles which pass through the slit?

45. An x-ray beam having an energy of 3 keV incident on a crystal lattice at an angle of 45° is observed to undergo second-order diffraction. What is the spacing between the lattice planes?

46. When electrons having a kinetic energy of 6 eV are incident at an angle θ on a set of crystal lattice planes having spacing of 5 Å, first-order electron diffraction is observed to occur. Compute the angle θ (in degrees). A. 13.2, B. 30.0, C. 41.7, D. 68.4, E. 72.5.

47. Solve for the wavelength of the neutron beam which exhibits first-order diffraction when incident at an angle of 40° with respect to a set of crystal lattice planes having a 2.85 Å spacing. What is the kinetic energy per neutron?

48. (a) Compute the incident angle between a beam of 1-eV electrons and a set of crystal lattice planes having an interplanar spacing of 3.2 Å, assuming that first-order ($n = 1$) diffraction is being observed. (b) Repeat the calculations if the diffraction is second order ($n = 2$). (c) Repeat the above calculations for a 1-eV beam of neutrons.

49. A beam of electrons is diffracted by a single crystal of nickel when incident at an angle of 60° relative to a set of lattice planes having a spacing of 2.1 Å. Compute the value (or set of values) of the momentum per particle exchanged with the nickel crystal in the diffraction process.

50. (a) Compute the kinetic energy of x-ray photons which give first-order Bragg diffraction from NaCl lattice planes having a spacing of 2.82 Å which are oriented at an angle of 30° with respect to the incident x-ray beam. (b) Compute the kinetic energy of neutrons in a beam which likewise give first-order Bragg diffraction from NaCl lattice planes having a spacing of 2.82 Å which are oriented at an angle of 30° with respect to the incident neutron beam.

51. (a) Consider the problem of Bragg diffraction by a set of parallel crystal planes. If the spacing between the two successive planes is 2.0 Å, what is the minimum photon energy (in electron volts) in an x ray required to produce a diffraction line? (b) What are the minimum energies (also in electron volts) required to do so for a neutron and for an electron?

52. Apply the semiclassical quantization rules to obtain the allowed orbits for a charged particle undergoing circular motion in a uniform magnetic field.

53. In the hydrogen atom there is a photon emitted when an electron in the first excited state falls into the ground state. Compute the frequency (in cycles per second) of this photon. A. 3.29×10^{15}, B. 2.47×10^{15}, C. 1.85×10^{15}, D. 1.39×10^{15}, E. 1.04×10^{15}.

54. Find the wavelength (in meters) of the photon emitted when the electron in a hydrogen atom makes a transition from the $n = 2$ state to the $n = 1$ state. A. 1.22×10^{-7}, B. 3.84×10^{-7}, C. 5.46×10^{-7}, D. 7.13×10^{-7}, E. 9.46×10^{-7}.

55. From the standpoint of the correspondence principle and classical physics, an electron in an atomic state characterized by very large quantum numbers should emit electromagnetic radiation due to its centripetal acceleration. Is it possible to deduce a relation between electron orbital frequencies and the frequency of electromagnetic radiation emitted in a transition from one orbit to another, considering two adjacent Bohr orbits in the limit of large quantum number? What insight does this give into the correspondence principle?

56. A free particle has the wave function

$$\psi(\mathbf{r}, t) = Ae^{i(\mathbf{k}\cdot\mathbf{r} - \omega t)}$$

with $\mathbf{k} = (5\hat{\mathbf{x}} + 7\hat{\mathbf{y}} + 10\hat{\mathbf{z}})$ Å^{-1}, $\omega = 4.59 \times 10^{13}$ rad/sec. What is the momentum of the free particle (in units of kilogram meters per second)? A. 1.39×10^{-23}, B. 5.44×10^{-20}, C. 1.88×10^{-17}, D. 6.74×10^{-14}, E. Cannot be determined.

57. In Problem 56, what is the mass (in kilograms) of the free particle? A. 9.10×10^{-31}, B. 1.67×10^{-27}, C. 1.99×10^{-26}, D. 1.602×10^{-19}, E. Cannot be determined.

58. A free particle has the wave function

$$\psi(\mathbf{r}, t) = Ae^{i(\mathbf{k}\cdot\mathbf{r} - \omega t)}$$

with $\mathbf{k} = (5\hat{\mathbf{x}} + 17\hat{\mathbf{y}} + 10\hat{\mathbf{z}})$ Å^{-1}, $\omega = 1.36 \times 10^{7}$ rad/sec. What is the momentum of the free particle (in units of kilogram meters per second)? A. 1.36×10^{-27}, B. 2.15×10^{-23}, C. 1.88×10^{-19}, D. 8.65×10^{-15}, E. 3.22×10^{-11}.

59. In Problem 58, what is the mass (in kilograms) of the free particle? A. 9.109×10^{-31}, B. 1.673×10^{-27}, C. 1.990×10^{-26}, D. 1.602×10^{-19}, E. 7.362×10^{-12}.

60. (a) Given a wave function $\psi = 15 \exp\{i[27x - 13t]\}$, where the units are standard for the meter–kilogram–second (mks) system, compute the wavelength and the frequency of particles described by this wave function. (b) Determine also the direction of propagation and the phase velocity.

61. Write the wave function for a beam of 4-eV electrons traveling with a velocity of 10^5 m/sec in a direction making angles of 30° with respect to the $\hat{\mathbf{x}}$ axis and 40° with respect to the $\hat{\mathbf{z}}$ axis.

62. Obtain solutions to the time-dependent Schrödinger wave equation for free particles.

63. For the following cases, attempt to write the time-dependent and time-independent Schrödinger equations: (a) One-dimensional potential step. (b) One-dimensional potential barrier. (c) One-dimensional square well with rigid walls, i.e., the potential goes to infinity at the boundaries. (d) Repeat (c) with finite walls. (e) Repeat (c) for three dimensions. (f) Repeat (d) in three dimensions. (g) Hydrogen atom. (h) Helium atom. (i) Oxygen atom. (j) Water molecule. (k) 1 mole of helium gas. (l) 1 cm^3 of water. (m) 1 g of metallic silver.

64. Suggest appropriate boundary conditions for each case in Problem 63.

65. Construct electromagnetic wave analogs to (a)–(f) in Problem 63.

66. (a) Apply the technique of separation of variables to the time-dependent Schrödinger equation for any arbitrary potential $\mathscr{V}$ which depends only on position **r**. Obtain thereby separate equations for the position dependence and the time dependence of $\psi(\mathbf{r}, t)$. (b) Solve the equation deduced in (a) for the time-dependent factor. (c) Can the separation in (a) be carried out when $\mathscr{V} = \mathscr{V}(\mathbf{r}, \mathbf{p})$?

67. In the Schrödinger equation for free particles, attempt trial solutions having the following traveling-wave forms: A. $\psi = B_1 \sin(kx - \omega t)$, B. $\psi = B_2 \cos(kx - \omega t)$, C. $\psi = C \cos(kx - \omega t + \delta)$, D. $\psi = A_1 \cos(kx - \omega t) + A_2 \sin(kx - \omega t)$. (a) Which of these trial solutions are good? (b) What requirements must be imposed upon the constants $B_1, B_2, C, \delta, A_1, A_2$ for the valid solutions? (c) What can you conclude about the functional form of traveling wave solutions to the Schrödinger equation? [*Hint*: See Eisberg (1967).]

68. Are all plane-wave solutions of the Schrödinger equation simultaneous eigenfunctions of the momentum and energy operators?

69. Consider the functions $A \sin kx$, $B \cos kx$, $C \tan kx$, where A, B, C, and k are constants. Which of these (if any) are eigenfunctions of the following operators: (a) linear momentum, (b) kinetic energy.

70. (a) Solve the Schrödinger equation in a region of space where the potential energy $V(x)$ has a constant (though nonzero) value, and interpret your results by comparing with those obtained for $V(x) = 0$. (b) Repeat for the three-dimensional case, namely, $V(\mathbf{r}) =$ const.

71. Write the Schrödinger equation for the following physical situations: (a) particle of mass m in a gravitational field in which the acceleration g is uniform, (b) particle of mass m under the force of gravity of a far-distant and much larger mass M, (c) electron subject to the Coulomb force of a proton.

72. Consider a change of reference frame for the Schrödinger equation, using a Galilean transformation. Is the equation invariant under such transformation? (*Hint*: Consider physically the individual variation of $\mathbf{p}, \mathscr{E}, p^2/2m, h\nu, \lambda = h/p$, etc., under such transformation.)

73. (a) Develop the Schrödinger equation using the total energy as obtained from the relativistic expression $\mathscr{E} = mc^2$ in the nonrelativistic limit, namely, $\mathscr{E} \simeq m_0c^2 + V(\mathbf{r}) + p^2/2m_0$, where m_0 is the rest mass and c the velocity of light. (b) What is the interpretation of the wave frequency ν for this situation? (Specifically, does it depend upon the rest mass energy m_0c^2?)

74. Write down 10 functions of position x [such as $A \tan(\beta x + \theta)$, with A, β, and θ being constants], and test to see which of these are eigenfunctions of the linear momentum operator.

75. (a) Show that the superposition of any two eigenfunctions of the linear momentum operator corresponding to different wavelength particle beams does not constitute a momentum eigenfunction. (b) Can the superposition of wave functions for two particle beams having different kinetic energy per particle be a momentum eigenfunction? (c) Can the superposition described in (b) be an energy eigenfunction?

76. Consider the superposition of two waves ψ_1 and ψ_2, where

$$\psi_i = A_i \cos(k_i x - \omega_i t) \qquad (i = 1, 2)$$

with $A_1 = 30$, $A_2 = 40$, $k_1 = 4$, $k_2 = 5$, $\omega_1 = 12$, and $\omega_2 = 15$. Deduce all information which you possibly can about the packet $\psi = \psi_1 + \psi_2$ made up by linear superposition of the two waves ψ_1 and ψ_2 in question.

77. In the example illustrated in Figs. 1.27–1.29, show mathematically that $\psi(x, t)$ is normalized whenever $\chi(k)$ is normalized.
78. Given the particular spatial wave function $\psi(x) = \psi_0$ ($\psi_0 =$ const) from $x = x_1$ to $x = x_2$, and $\psi(x) = 0$ otherwise, deduce the corresponding momentum function $\chi(k)$.
79. Given the wave function stated in Problem 78, deduce the expectation values for the position, the linear momentum, and the kinetic energy. (Choose the specific values $x_1 = 10$ Å and $x_2 = 20$ Å if you wish.)
80. If an oxygen molecule is trapped somewhere within a region of space 1 cm in length, what is the minimum uncertainty in the value of its speed in units of meters per second? The mass of the oxygen molecule is approximately 32 amu, where 1 amu $= 1.6604 \times 10^{-27}$ kg. A. 9.51×10^{-4}, B. 1.98×10^{-7}, C. 4.14×10^{-11}, D. 8.65×10^{-15}, E. 1.80×10^{-18}.
81. If a molecule of mass 4×10^{-26} kg is ascertained to be somewhere within a region 0.25 cm in length, what is the minimum uncertainty in the velocity of the molecule?
82. A 200-kg meteorite enters the earth's atmosphere with a velocity (known to an accuracy of 0.01%) of 2000 m/sec. What is the limit of precision with which the position of the meteorite can be located? How does this compare with a reasonable estimate of the size of the meteorite?
83. (a) The kinetic energies of a traveling jet plane and a traveling electron are each measured over time intervals of 10^{-3} sec. Compute the minimum uncertainties in energy for each. (b) In each case, compute the ratio of the energy uncertainty to a reasonable estimate of the total energy.
84. (a) An electron is known to have a speed of 300 m/sec to an accuracy of 1%. What is its minimum uncertainty in position? (b) What is the minimum uncertainty in the time it would take this electron to travel 1 km?
85. (a) Compute the minimum uncertainty in the speed of a 1-eV electron which has its position determined to within an uncertainty of 1 Å. (b) Compare this uncertainty in speed with the speed itself. (c) Compute the minimum uncertainty in the speed of a 1-eV neutron assuming its position known to within an uncertainty of 1 Å, and compare this uncertainty with the speed itself.
86. If a neutron is confined to a region of the order of 10^{-15} m in extent, such as the nucleus of an atom, what is its minimum uncertainty in momentum?
87. (a) Suppose one wished to set up some experiment to "follow" an electron in its Bohr "orbit" about the proton in a hydrogen atom, assuming the hydrogen atom to be in its ground state. A satisfactory mapping of the electron trajectory might require, for example, 100 or more position measurements of the electron on the orbit. Analyze this problem in detail, keeping in mind such uncertainties as momentum exchange of the photons (or particles) used for the position measurement with the electron whose position is being measured. (b) Extend your considerations to the various excited states of the hydrogen atom.
88. The Fourier transform pair relating momentum and position probability distributions differs from the standard form of the Fourier series involving the wave vector and position only slightly, since $\mathbf{p} = \hbar\mathbf{k}$. Write the Fourier transform pair relating frequency and time, and deduce the appropriate Fourier transform pair relating energy and time, using $\mathscr{E} = \hbar\omega$. Tell how the various forms of the uncertainty relation have the same origin.
89. Consider an electronic state having a lifetime of 10^{-15} sec. (a) Estimate the minimum uncertainty in the energy of an electron populating such a state. (b) Can you draw some conclusions regarding the finite width of spectral lines?
90. The energy emitted when electronic transitions between stationary states occur in atoms is not quite monochromatic in frequency. Estimate the wavelength spread (in angstroms) for light given off at a central wavelength of 5461 Å whenever the lifetime of the excited state is of the order of 10^{-10} sec. A. 1.58×10^{-2}, B. 0.995, C. 631, D. 7140, E. 8.91×10^{10}.
91. Assuming that π^+ mesons have a lifetime of 2.54×10^{-8} sec, estimate the maximum precision possible for a measurement of the energy of a π^+ meson.
92. Obtain a measure of the momentum uncertainty $\Delta p = \hbar\,\Delta k$ for a particle restrained to a length L by assuming a sinusoidal wave function over the length and expanding this wave function in a Fourier integral. Deduce in this way the position–momentum form of the uncertainty relation.
93. (a) Obtain a measure of the energy uncertainty $\Delta\mathscr{E} = \hbar\,\Delta\omega$ for a particle observed sometime over the time interval $(0, t_0)$ with uniform probability over this interval. Use the procedure of expanding a uniform sinusoidal or plane-wave function over this time interval in a Fourier integral having

frequency $\omega = \mathscr{E}/\hbar$ as the dummy variable. (b) Deduce the width of $g(\omega)$ for values of t_0 of $1, 2, 10, 100$, and 1000 sec. (You should be able to deduce the time–energy form of the uncertainty relation in this manner.)

94. A quantized particle is made to pass through an opening of diameter d equipped with a shutter which opens for a time τ. Show that the particle necessarily exchanges with this device (diaphragm plus shutter) a momentum of the order of $\hbar/d$ and an energy of the order of $\hbar/\tau$.

95. Consider as a packet the plane wave $\psi(\mathbf{r}, t) = A_{\mathbf{p}} \exp\{(i/\hbar)[\mathbf{p}\cdot\mathbf{r} - \mathscr{E}t]\}$. (a) Solve for $\chi(\mathbf{p})$. (b) Use the $\chi(\mathbf{p})$ derived in part (a) to compute $\psi(\mathbf{r}, 0)$. Add the time dependence to regain $\psi(\mathbf{r}, t)$ stated above, thus showing self-consistency in your mathematics.

96. Consider an initial wave packet, $\psi(x, 0) = C\exp[-\{x^2/4(\Delta x)^2\} + i\bar{k}_x x]$, where $\bar{k}_x$ is a mean wave number. (a) Calculate the corresponding k_x distribution and $|\psi(x, t)|^2$. (b) Does it violate the Heisenberg uncertainty relations? (c) Does the wave packet advance according to classical laws? (d) Does the packet spread in time? (If so, at what rate?) (e) Apply the results to calculate effects in some typical microscopic and macroscopic experiments.

97. Extend Problem 96 to three dimensions.

98. (a) Compute the variance in position and the variance in momentum for the Gaussian wave packet, and take the product of the two. Relate this result to the Heisenberg uncertainty principle. (b) Evaluate the dispersion of the packet as a function of time.

99. (a) Extend the general wave packet development to three dimensions. (b) Deduce the appropriate vector expression for the group velocity.

100. Give a good definition of completeness as it relates to a linear vector space.

101. Show that all eigenfunctions of Hermitean operators correspond to real eigenvalues. (*Hint*: Use the general definition of a Hermitian operator.)

102. Prove that the expectation value $\mathscr{Q}^{\text{op}}$ of any Hermitian operator $\mathscr{Q}^{\text{op}}$ is real. (*Hint*: If $\mathscr{Q}^{\text{op}}$ is real, then $\langle\mathscr{Q}^{\text{op}}\rangle^* = \langle\mathscr{Q}^{\text{op}}\rangle$.)

103. Suppose that an arbitrary function ψ is represented by a linear combination of a complete orthonormal set of basis vectors $\phi_n^{(r)}$ which are eigenfunctions of the Hermitian operator A corresponding to a discrete spectrum of eigenvalues. (a) What is $\langle A\rangle$? (b) What is the probability that the corresponding physical observable will have the particular eigenvalue a_j when measured? (c) If $f(A)$ is an arbitrary function of A, what is $\langle f(A)\rangle$? (d) How would $\langle f(A)\rangle$ be modified if the eigenvalue spectrum had in addition a continuous portion?

104. Prove the theorem: "The physical quantity associated with the Hermitian operator A possesses with certainty a well-defined value if and only if the dynamical state of the physical system is represented by an eigenfunction ψ_a of A, and the value assumed by this quantity is the eigenvalue a associated with that function."

105. (a) How would one go about determining experimentally the *maximum information* about a quantum system? (In other words, how does one experimentally determine the complete dynamical state of the system?) (b) How would you define complementary and compatible variables for such a system? (c) Prove that a necessary condition for two observables to be simultaneously measurable is that the corresponding operators commute.

106. A beam of free electrons is described by the wave function

$$\psi = 109e^{i(\mathbf{k}\cdot\mathbf{r} - \omega t)} \quad (\text{electrons/m}^3)^{1/2}.$$

Compute the charge current density $\mathscr{J}$ in A/m^2, given that $\mathbf{k} = -[27\hat{\mathbf{x}} + 19\hat{\mathbf{y}} + 24\hat{\mathbf{z}}]\ \text{m}^{-1}$, and $\omega = 9.64 \times 10^{-2}\ \text{sec}^{-1}$. A. 9×10^{-18}, B. 7×10^{-15}, C. 5×10^{-12}, D. 3×10^{-9}, E. 1×10^{-6}.

107. Consider two beams of electrons with wave functions $\psi_1 = Ae^{(i/\hbar)[p_1 x - \mathscr{E}_1 t]}$ and $\psi_2 = Be^{(i/\hbar)[p_2 x - \mathscr{E}_2 t]}$. (a) Compute the net current $J_1 + J_2$. (b) Compute the total probability density from $\psi_1 + \psi_2$. (c) Estimate the total probability (i.e., the integrated probability density) over an interval of space L, assuming $L \gg \hbar/p_1$, $L \gg \hbar/p_2$.

108. Evaluate the quantum mechanical electric current density $\mathscr{J}$ for the Bloch functions, defined as $\psi_{\mathbf{k}}(\mathbf{r}, t) = u_{\mathbf{k}}(\mathbf{r})e^{i(\mathbf{k}\cdot\mathbf{r} - \omega t)}$, where $\mathbf{k} = \mathbf{p}/\hbar$ and $\omega = \mathscr{E}/\hbar$. The time and position are represented by t and $\mathbf{r}$, respectively. The function $u_{\mathbf{k}}(\mathbf{r})$ is time independent.

109. The Bloch functions defined in Problem 108 are appropriate for conduction electrons in a crystalline substance such as a metal (cf. Chap. 7). The function $u_{\mathbf{k}}(\mathbf{r})$ is a periodic function having the lattice periodicity, and the allowed values of **k** are determined by applying periodic boundary conditions with respect to the dimensions of the metal itself. Determine the allowed values of **k**, assuming a rectangular parallelepiped block of metal having dimensions L_x, L_y, and L_z in the $\hat{\mathbf{x}}, \hat{\mathbf{y}}$, and $\hat{\mathbf{z}}$ directions, respectively.

110. Substitute the Bloch function given in Problem 108 into the Schrödinger equation for a periodic potential to obtain an appropriate equation for obtaining the functions $u_{\mathbf{k}}(\mathbf{r})$. A periodic potential can be represented by $V(\mathbf{r}) = V(\mathbf{r} + \mathbf{R_n})$, where $\mathbf{R_n}$ is defined by $\hat{\mathbf{x}}n_x d_x + \hat{\mathbf{y}}n_y d_y + \hat{\mathbf{z}}n_z d_z$, with n_x, n_y, n_z representing arbitrary integers and d_x, d_y, d_z representing the atomic lattice periodicities in the $\hat{\mathbf{x}}, \hat{\mathbf{y}}, \hat{\mathbf{z}}$ directions, respectively.

111. For an infinite potential well of width L extending from $x = -\frac{1}{2}L$ to $x = \frac{1}{2}L$, what is the probability that an electron in its lowest energy state will be in the center half (namely, $-\frac{1}{4}L < x < \frac{1}{4}L$) of the well? A. 0.818, B. 0.655, C. 0.491, D. 0.327, E. 0.164.

112. Consider an electron trapped in a one-dimensional square-well potential with the length of the potential well being 5.6 Å and the heights of the potential barriers bounding the potential well being infinitely high. Solve for the ground state energy (in electron volts) of the electron, assuming that the potential energy is chosen to be zero within the potential well. (Assume fixed boundary conditions, for which the electron wave function is zero at the edges of the potential well.) A. 0.000, B. 0.487, C. 1.20, D. 36.8, E. 56.0.

113. What is the speed (in meters per second) of an electron in the *first excited state* within a one-dimensional square-well potential of length 3.8 Å, assuming zero potential energy within the potential well, infinitely high potential energy barriers at the boundaries, and fixed boundary conditions? (Note that this well is a different length from that in the preceding problem.) A. 51.3, B. 713, C. 9.91×10^3, D. 1.38×10^5, E. 1.91×10^6.

114. An electron is trapped in a one-dimensional square-well potential. It is in its ground state, and the ground-state energy is 15 eV. With what frequency (in cycles per second) does the electron oscillate back and forth in the box? (One complete cycle requires a return of the electron to its starting position.) A. 1.88×10^{17}, B. 1.50×10^{16}, C. 1.11×10^{16}, D. 7.25×10^{15}, E. 3.39×10^{15}.

115. Repeat Problem 114 for a ground state energy of 7 eV. A. 3.39×10^{15}, B. 7.25×10^{15}, C. 1.11×10^{16}, D. 1.50×10^{16}, E. 1.88×10^{17}.

116. An organic molecule weighing 10^{-6} g is placed in a one-dimensional box 1 mm on edge. What is the minimum energy (in joules) of this molecule? A. 7.96×10^{-67}, B. 5.49×10^{-53}, C. 3.78×10^{-39}, D. 2.61×10^{-25}, E. 1.80×10^{-11}.

117. What is the smallest speed (in meters per second) for the organic molecule in the box in Problem 116? (*Suggestion*: Use the quantized momentum for this calculation to avoid propagating any errors you might have made in the last problem.) A. 3.99×10^{-29}, B. 3.31×10^{-22}, C. 2.75×10^{-15}, D. 2.28×10^{-8}, E. 0.190.

118. What would be the approximate quantum number n if the organic molecule in the box in Problem 116 has a velocity of approximately 30 m/sec? A. 3, B. 17, C. 1700, D. 6.0×10^{10}, E. 9.06×10^{22}.

119. What would be the approximate *increase* in velocity (in units of meters per second) if the organic molecule in Problem 118 were promoted from level n to the next higher level $(n + 1)$ by absorption of energy? A. 1.15×10^{-48}, B. 1.95×10^{-35}, C. 3.31×10^{-22}, D. 5.63×10^{-9}, E. 95,600.

120. (a) What is the quantum number of a 500-eV electron in a one-dimensional square-well potential having length of 1 cm? (b) What is the quantum number of a 1-mg particle with an energy of 500 eV in this same potential well?

121. (a) What is the separation in energy between the ground state and the first excited state for an electron confined to a linear molecule 48 Å in length, assuming that the linear molecule can be represented by a one-dimensional square-well potential? (b) Compute the wavelength of the electromagnetic radiation emitted when this electron undergoes a transition from the second excited state to the ground state.

122. A proton is confined to a one-dimensional box 10,000 Å in length. (a) Compute the momentum for the ground state, assuming periodic boundary conditions. (b) Repeat the calculation for fixed boundary conditions. (c) Explain what physically is happening to the motion of the proton with time.

123. Light having a wavelength of 5000 Å promotes a particle in a one-dimensional box 20,000 Å in length from the ground state to the second excited state. What is the mass of the particle?
124. (a) If a 1-kg mass is confined to a one-dimensional square-well potential 1 m in length, compute the velocity of the mass in the ground state. (b) Is this result in accordance with the requirements of the Heisenberg uncertainty principle? (c) Repeat the above calculation if an electron replaces the 1-kg mass.
125. A neutron having a momentum of 2×10^{-10} kg m/sec is said to be trapped in a one-dimensional square-well potential of length 1.6 cm. Is this possible? If so, what is the approximate quantum number?
126. Compute the minimum value of $\Delta p \, \Delta x$ for the ground state and the first three excited states for a neutron in a one-dimensional square-well potential of length L. How do these results compare with the predictions of the Heisenberg uncertainty relations?
127. Show that the density of states functions $\tilde{g}(\mathscr{E})$ and $\mathscr{G}(\mathbf{p})$ derived for the one-dimensional potential well problem using periodic boundary conditions (which are appropriate for traveling-wave solutions to the Schrödinger equation) are the same as those functions derived utilizing fixed boundary conditions (which are appropriate for the standing-wave solutions).
128. Derive the density of states functions $\tilde{g}(\mathscr{E})$ and $\mathscr{G}(\mathbf{p})$ for a two-dimensional quantum system, assuming a square-well potential having lengths L_x and L_y and having walls which are infinitely high.
129. Consider a one-dimensional square-well potential $V(x) = 0$ over the domain $-(L - \delta) < x < \delta$ and $V(x) = \infty$ otherwise. (a) Deduce the allowed wave vectors k_n and the corresponding values of A_n/B_n for the stationary-state eigenfunctions $\phi_n = A_n \cos k_n x + B_n \sin k_n x$. (b) Choosing $\delta/L = 0.1$, compute the values of A_n and B_n for the five lowest normalized eigenfunctions. (c) Repeat (b) with $\delta/L = 0.2$. (d) Show numerically that the boundary conditions are met in (b) and (c). (e) Assuming that (a) was carried out using an analytical technique, show how you could deduce the same results graphically. (f) Quantitatively compare your results in (a) through (c) for an electron trapped in a box of length $L = 1$ cm with the results for an electron trapped in a box of length $L = 10$ Å.
130. (a) Fourier-analyze the wave function for the lowest allowable state for an electron in a one-dimensional box of length 4 Å (with infinitely high walls) into a sequence of plane waves. Plot $\chi(k)$ versus k. (b) Repeat for the second allowable state. (c) Compute the product of the variances in position and momentum. (d) What happens if you double the length of the box? (e) Relate your findings to the Heisenberg uncertainty principle.
131. A simple model used to describe the interaction between the neutron and the proton that make up a deuteron is a square-well potential with $d = 2.0 \times 10^{-13}$ cm. What is the minimum depth of the potential well? In experimental studies of the spectrum of the deuteron, no excited states have been found. What does this imply?
132. An electron beam is incident from the left on a semi-infinite step potential exactly 3 eV in height (cf. Fig. 1.31a). The electrons have zero potential energy and individual kinetic energies of 3.01 eV to the left of the step (i.e., before reaching the step potential). What is the relative probability that any given electron will be reflected by the step? A. 0.000, B. 0.287, C. 0.528, D. 0.794, E. 1.000.
133. If the incident electrons in Problem 132 had energies of only 2.99 eV (cf. Fig. 1.32), compute the electron density 15 Å past the edge of the 3-eV step, assuming unit electron density in the region immediately to the left of the step. A. 0.000, B. 0.215, C. 0.500, D. 0.785, E. 1.000.
134. An electron beam is incident from the left on a semi-infinite step potential exactly 3 eV in height. Suppose the electrons have individual kinetic energies of 3.05 eV to the left of the step (i.e., before reaching the step potential). What is the relative probability that any given electron will be reflected by the step? A. 0.012, B. 0.267, C. 0.598, D. 0.724, E. 0.998.
135. If the incident electrons in Problem 134 had energies of only 2.95 eV, compute the electron density 5 Å past the edge of the 3-eV step, assuming unit electron density in the region immediately to the left of the step. A. 0.164, B. 0.318, C. 0.522, D. 0.792, E. 0.947.
136. (a) Compute the transmission and reflection coefficients for a semi-infinite step potential 5.00 eV in height, assuming incident electrons with kinetic energies of 4.95 eV. (b) Compute the relative probability density distribution as a function of position within the barrier. (c) Repeat the calculation for incident electrons with kinetic energies of 5.05 eV.

137. A horsefly weighing 0.025 g has a velocity of 120 m/sec directed perpendicular to a thin transparent cellophane sheet. Assume that the cellophane barrier can be modeled by a one-dimensional square barrier of height U_0 joules with a width of 2 Å. What is the maximum value of U_0 which the fly could cross from the standpoint of classical mechanics? A. 0.180, B. 0.720, C. 2.88, D. 11.5, E. 46.0.

138. In Problem 137, what is the quantum-mechanical probability that the fly would cross a barrier 2×10^{18} eV in height with a width of 10^{-30} m? A. 5.16×10^{-6}, B. 5.33×10^{-10}, C. 5.51×10^{-14}, D. 5.69×10^{-18}, E. 5.89×10^{-22}.

139. (a) Compute the transmission and reflection coefficients for a rectangular barrier 5.00 eV in height and 5 Å in thickness, assuming incident electrons with kinetic energies of 4.95 eV. (b) Repeat the calculation for incident electrons with kinetic energies of 5.05 eV. (c) Replace the electrons by neutrons and carry out the corresponding calculations.

140. (a) A 10-eV electron tunnels through a 10-Å-thick square-barrier potential with a transmission coefficient of 10^{-10}. What is the height of the potential barrier? (b) Repeat the calculation for a neutron. (c) Repeat the calculation for an argon atom.

141. An elementary-particle physicist postulates that a strange particle exists which has a rest mass which is $\frac{1}{3}$ that of the electron and a charge which is $\frac{2}{3}$ that of the electron. If such a particle were incident perpendicularly on a rectangular potential energy barrier 5 eV in height and 6.25 Å in width while traveling at a speed of 1.50×10^6 m/sec, what would be the probability that this particle would penetrate the barrier? A. 1.64×10^{-9}, B. 3.18×10^{-7}, C. 5.85×10^{-5}, D. 7.42×10^{-3}, E. 0.921.

142. A beam of electrons having speeds of 10^6 m/sec in free space is incident perpendicularly upon a step barrier 3 eV in height which begins at $x = 0$ and continues to $x = \infty$. The electron probability density $\psi^*\psi$ is given as 100 electrons/cm^3 in the incident beam. Compute the electron probability density (in electrons per cubic centimeter) at $x = 10$ Å within the barrier. A. 6.51, B. 1.73, C. 4.60×10^{-3}, D. 1.22×10^{-5}, E. 3.24×10^{-8}.

143. In Problem 142 above, compute the total probability of finding an electron in the region $100 \text{ Å} \leqslant x \leqslant \infty$. A. 4.21×10^{-31}, B. 2.09×10^{-25}, C. 1.04×10^{-17}, D. 5.18×10^{-12}, E. 2.58×10^{-6}.

144. Repeat Problem 142 for $\psi^*\psi = 10{,}000$ electrons/cm^3 in the incident beam. A. 3.24×10^{-8}, B. 1.22×10^{-5}, C. 4.60×10^{-3}, D. 1.73, E. 651.

145. Repeat Problem 143 for $\psi^*\psi = 10{,}000$ electrons/cm^3 in the incident beam and consider the interval $100 \text{ Å} \leqslant x \leqslant 200$ Å. A. 2.58×10^{-6}, B. 5.18×10^{-12}, C. 1.04×10^{-17}, D. 2.09×10^{-23}, E. 4.21×10^{-29}.

146. Deduce the momentum conditions for minimum and maximum transmission of a particle across one-dimensional potential wells and barriers. In particular, consider the barriers illustrated in Figs. 1.48 and 1.49 which are appropriate for Projects 1.25 and 1.28, respectively.

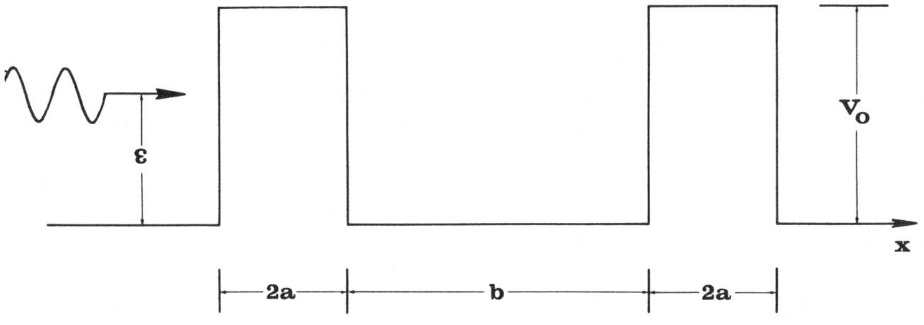

Fig. 1.48 Particle incident on two identical rectangular potential energy barriers. (See Problem 146 and Project 1.25.)

147. Set up and solve the harmonic oscillator problem classically and compare results with the quantum-mechanical solution.

148. A mass of 10 g is attached to a spring having a force constant $K = 3$ J/m^2, and the mass is set into simple harmonic motion on the surface of a frictionless laboratory bench. Compute the ground

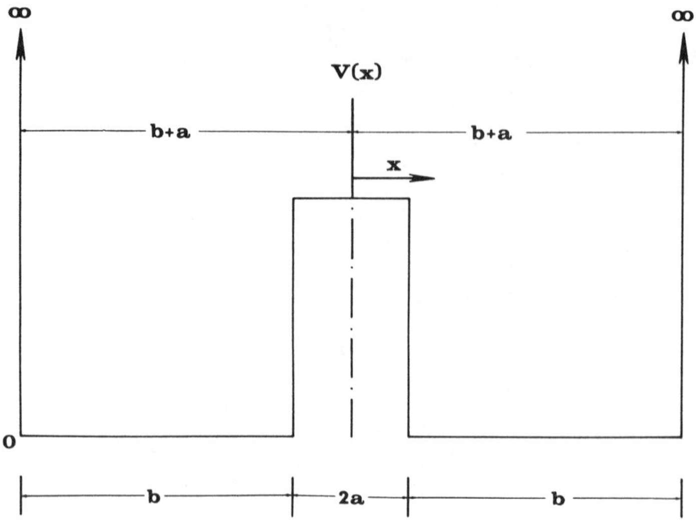

Fig. 1.49 Symmetric double well potential energy diagram. (See Problem 146 and Project 1.28.)

state energy of the system in joules. A. 0.939, B. 9.13×10^{-34}, C. 3.42×10^{-29}, D. 1.31×10^{-9}, E. 15.1.

149. In Problem 148, an incident photon is absorbed by the system, and as a consequence the system is promoted into the first excited state. What is the frequency (in cycles per second) of this photon? A. 4.12×10^{13}, B. 4.14×10^{-6}, C. 2.76, D. 242, E. 6.37×10^{15}.

150. Set up and solve as far as you can the quantum-mechanical problem of the hydrogen atom, obtaining the energy levels, eigenfunctions, etc.

151. (a) What are the quantized energy levels for the singly ionized helium atom (viz., He^{+})? (b) What are the quantized energy levels for the doubly ionized lithium atom (viz., Li^{2+})?

152. Find the frequency (in cycles per second) of the electromagnetic radiation which will promote an electron in the hydrogen atom from the ground state to the second excited state (i.e., the state for which the quantum number $n = 3$). A. 6.22×10^{13}, B. 2.48×10^{14}, C. 2.92×10^{15}, D. 7.29×10^{16}, E. 8.42×10^{17}.

153. Find the energy in joules given off when an electron with zero kinetic energy falls from the vacuum level (i.e., $n = \infty$) to the first excited state ($n = 2$) in the hydrogen atom. A. 1.60×10^{-19}, B. 5.45×10^{-19}, C. 2.18×10^{-18}, D. 7.61×10^{-18}, E. 6.27×10^{-17}

154. (a) Compute the wavelength of the photon which would be required to excite an electron from the ground state of the hydrogen atom to the vacuum level. (b) What is the frequency of this photon?

155. Calculate the wavelength of the electromagnetic radiation emitted when a hydrogen atom undergoes a transition from the first excited state to the ground state.

156. Compute the value of the static dielectric constant ε_0 in F/m, given that the quantized energy levels $\mathscr{E}_n$ of the hydrogen atom found by solving the Schrödinger equation are given by $\mathscr{E}_n = -me^4/8\varepsilon_0^2h^2n^2$, with all symbols having their usual meaning. A. 0.204, B. 4.15×10^{-2}, C. 1.73×10^{-3}, D. 2.98×10^{-6}, E. 8.85×10^{-12}.

157. Compare the energy difference between the ground state and the first excited state of the electron in the hydrogen atom with the energy difference between the ground state and the first excited state of an electron trapped in a one-dimensional square-well potential with its dimension equal to twice the Bohr radius.

158. A helium atom ($Z = 2$) is stripped of one electron, leaving it as a one-electron quantum system. Compute the values for the five lowest quantized energy levels.

159. (a) What is the separation in energy between the first and second excited states of the hydrogen atom? (b) What is the separation in energy between the first and second excited states of the singly ionized helium atom?

160. (a) An experimental measurement of the Rydberg constant R is carried out by measuring the frequency of light emitted when electrons in ionized helium atoms drop from the second excited state to the ground state. What value is obtained if the frequency is measured to be 2.9258×10^{15} Hz? (b) Calculate the theoretical value of the Rydberg constant and compare it with the value obtained from the measurements. (c) What comment(s) can you make?

161. Construct operators for the angular momentum components.

162. What is your idea of a "good" quantum number?

163. If the components of the angular momentum of a particle are given by L_x, L_y, L_z, what is $[L_x, L_y]$?

164. As a very crude model of an atom, consider the electrons to be trapped in a one-dimensional square-well potential 5 Å in width. What would be the frequency ν (in cycles per second) of the light wave which would be strongly absorbed in promoting an electron from its lowest energy state (i.e., the ground state) to the next higher energy state (i.e., to the first excited state)? A. 5.75×10^{11}, B. 1.09×10^{15}, C. 2.07×10^{18}, D. 3.93×10^{21}, E. 7.46×10^{24}.

165. (a) Write the Schrödinger equation for the helium atom. (b) What simplification results whenever the Coulomb potential energy of interaction between the two electrons in the helium atom can be neglected? (c) Use the ground-state wave function of the hydrogen atom to deduce the ground-state wave function of the helium atom in the limit in which the electron–electron interaction energy is neglected in the helium atom.

166. Compare the results for the problem of a harmonic oscillator, of electron motion in the hydrogen atom, and of the motion of a particle in a box. Give the relevant Schrödinger equations, and point out similarities and differences. From the similarities, and with the aid of insight into expected properties of quantum-mechanical solutions gained from other problems, deduce as many general properties of quantum-mechanical bound-state systems as you can.

ANSWERS TO MULTIPLE CHOICE PROBLEMS

3. C, **21.** C, **30.** B, **46.** B, **53.** B, **54.** A, **56.** A, **57.** C, **58.** B, **59.** D, **80.** B, **90.** A, **106.** A, **111.** A, **112.** C, **113.** E, **114.** D, **115.** A, **116.** B, **117.** B, **118.** E, **119.** C, **132.** D, **133.** B, **134.** C, **135.** B, **137.** A, **138.** E, **141.** D, **142.** A, **143.** B, **144.** E, **145.** D, **148.** B, **149.** C, **152.** C, **153.** B, **156.** E, **164.** B.

PART II

Quantum Statistics of Many-Particle Systems; Formulation of the Free-Electron Model for Metals

CHAPTER 2

MANY-PARTICLE SYSTEMS AND QUANTUM STATISTICS

A satisfactory theory ought, of course, to count two observationally indistinguishable states as the same state and to deny that any transition does occur when two similar particles exchange places. P. A. M. Dirac (1930)

1 Wave Functions for a Many-Particle System

Consider a system consisting of a large number of identical quantum-mechanical constituents that interact very weakly or not at all with one another. Assume that energy is being slowly transferred between these constituents in such a way that an equilibrium condition exists, so the number of elements having a given energy is not changing systematically with time. For purposes of classification, let us consider three model systems:

(1) systems composed of identical but distinguishable particles (or other elements),
(2) systems composed of identical indistinguishable particles of half-integral spin,
(3) systems composed of identical indistinguishable particles of integral spin.

The treatment of *distinguishable particles* for the first model system represents the classical limit which leads to *Maxwell–Boltzmann statistics*, whereas the treatments of the two types of *indistinguishable particles* in the second and third model systems yield respectively *the Fermi–Dirac and Bose–Einstein distribution functions*. [In the category of distinguishable particles we can include other elements, such as degrees of freedom of a system, which are not actually particles and which may even be identical in their physical behavior, but yet are distinguishable by means of their spatial location or orientation. Examples would be the normal modes of atom vibration (phonons) in a crystalline solid and spin waves in a magnetically ordered material.]

Let us first consider the quantum mechanics of a system composed of a large number N of identical particles that do not interact with one another. Assume that all of the particles are moving under the influence of the same potential function $\mathscr{V}(\mathbf{r})$, so that when the particles are treated quantum mechanically on an individual basis, as prescribed in §6 of Chap. 1, there are deduced a number of stationary states ϕ_i of energy $\mathscr{E}_i$ $(i = 0, 1, 2, \ldots)$ which are available for occupation by the particles. For example, if the particles are all subjected to forces due only to the nucleus of a given atom, the wave functions and energy levels would be those of a one-electron atom. Similarly, if the particles are all confined as a gas inside a given container, the wave functions and energy levels are those of a particle in a potential well.

The actual quantum state of an N-particle system will be somewhat different in general from that constructed from a superposition of N single-particle states, due to the potential energy of interaction between the particles. For example, weak gravitational forces between the particles exist, and there is frequently present the stronger electron–electron Coulomb interaction. We neglect these fluctuating perturbing effects on the potential in the present section for reasons of simplicity, and as a consequence, we derive the statistics that yield the occupation probability for energy levels of a system of N *noninteracting* particles in the presence of some average potential $\mathscr{V}(\mathbf{r})$. Furthermore we assume, again for simplicity, that the single-particle energy levels are *nondegenerate*. By definition, nondegeneracy implies that each different single-particle wave function corresponds to a different energy of the particle, so that there is a unique correspondence between energy levels and wave functions $[\mathscr{E}_i \leftrightarrow \phi_i$ $(i = 0, 1, 2, \ldots)]$. Nondegeneracy of wave functions means, for example, that we cannot have $\mathscr{E}_2 = \mathscr{E}_3$, in which case ϕ_2 and ϕ_3 would both correspond to the same energy $\mathscr{E}_2$. Nondegenerate single-particle energy levels can therefore be listed in the order of increasing energy. Cases involving degenerate levels can be approximated by regarding such levels to be very closely spaced, but not absolutely identical, in energy. Alternatively, degenerate levels may be included by assigning to these levels a statistical weighting factor determined by the degree of degeneracy.

1.1 Systems of Distinguishable Particles

Let us now consider the functional form of the total wave function for the N-particle system of distinguishable particles. One possible wave function for the total system of N particles is the product of the single-particle eigenfunctions:

$$\Phi_{i_1 i_2 \cdots i_N} = \phi_{i_1}(1)\phi_{i_2}(2) \cdots \phi_{i_N}(N). \tag{2.1}$$

This function satisfies the Schrödinger equation (1.164) appropriate to each individual particle, as can be seen by direct substitution. The subscript i_1 is the index for the first particle, which by hypothesis can be distinguished from all other particles. The index i_1 can have the values $1, 2, \ldots, r, \ldots$, corresponding respectively to the possible single-particle wave functions $\phi_1, \phi_2, \ldots, \phi_r, \ldots$ and energy levels $\mathscr{E}_1, \mathscr{E}_2, \ldots, \mathscr{E}_r, \ldots$. The index i_2 is the index for the second particle, i_3 is the index for the third particle, etc. Thus ϕ_{i_j} would indicate that the jth particle is in the eigenstate represented by ϕ_k, where k is the specific value of the index i_j. Since the single-particle stationary-state wave functions are functions of the spatial position $\mathbf{r}$, this functional dependence should be denoted in some way, as for example $\phi(\mathbf{r})$. The Hamiltonian operator in quantum mechanics is obtained from the classical Hamiltonian representing the total energy of the system, as discussed in Chap. 1, §6, and this classical Hamiltonian contains the vectors $\mathbf{r}_1, \mathbf{r}_2, \ldots, \mathbf{r}_n$ which specify the classical location of each of the N particles at time t relative to a common origin; in addition the classical Hamiltonian contains the time derivatives of these position vectors. Thus particle 1, which is located at $\mathbf{r}_1$ and has velocity $\dot{\mathbf{r}}_1 \equiv d\mathbf{r}_1/dt$ at time t in the classical limit, has $\mathbf{r}_1$ as the argument in its quantum-mechanical wave function. Likewise, the jth particle has $\mathbf{r}_j$ as the argument of its wave function, so the symbol ϕ_{i_j} has argument $\mathbf{r}_j$. Thus $\phi_{i_j} = \phi_{i_j}(\mathbf{r}_j)$, which indicates that the jth particle is in the eigenstate represented by $\phi_k(\mathbf{r}_j)$, where k is the value of the index i_j. For simplicity, we denote the argument $\mathbf{r}_j$ by j, so $\phi_{i_j}(\mathbf{r}_j)$ is written simply $\phi_{i_j}(j)$. Likewise, the spin coordinate can be considered to be included in the symbol j.

To generalize to an N-particle wave function, $\Phi_{111111\cdots 1}$ would correspond to the quantum state $\phi_1(1)\phi_1(2)\cdots\phi_1(N)$ of the N-particle system in which each of the individual particles is in the single-particle state ϕ_1 with energy $\mathscr{E}_1$. The total energy of the N-particle system for this state is $N\mathscr{E}_1$. Similarly, $\Phi_{211111\cdots 1}$ corresponds to the state $\phi_2(1)\phi_1(2)\phi_1(3)\cdots\phi_1(N)$ in which the first particle is in the state ϕ_2 and all of the other $N-1$ particles are in the state ϕ_1. The total energy corresponding to this wave function is $\mathscr{E}_2+(N-1)\mathscr{E}_1$. Likewise, $\Phi_{12111\cdots 1}$ corresponds to the state in which the first particle is in the state ϕ_1, the second particle is in the state ϕ_2, and all the remaining particles are in the state ϕ_1. The total energy is again $\mathscr{E}_2 + (N-1)\mathscr{E}_1$. Thus the quantum states of the system correspond to the various particles occupying certain of the eigenstates available to a single particle. Although the N-particle wave functions $\Phi_{2111\cdots 1}$ and $\Phi_{1211\cdots 1}$ are degenerate in energy, they do represent entirely different physical states of the system. This is true since, by the hypothesis underlying the development in this section, we can distinguish particle 1 from particle 2, and can therefore tell whether it is particle 1 or particle 2 that we observe to be in a certain eigenstate at some given time.

Note that our N-particle wave function discussed above is a function of many variables $\mathbf{r}_1, \mathbf{r}_2, \ldots, \mathbf{r}_N$, and thus it can be considered to be a function in a *hyperspace*, which is a space having more than three dimensions. This is an interesting result which proceeds from converting the classical Hamiltonian to the quantum operator form by means of our usual prescription (see §6 of Chap. 1). The wave function given by Eq. (2.1) consists only of product factors of the form $\phi_{i_j}(\mathbf{r}_j)$, with no "off-diagonal" factors $\phi_{i_k}(\mathbf{r}_j)$ with $k \neq j$. The reason is that, by hypothesis, we can distinguish the particles one from the other, so we know that it is the jth particle that is located at $\mathbf{r}_j$ in a distinguishable-particle description, and definitely it is not the kth particle that is located at $\mathbf{r}_j$.

Thus the function $\phi_{i_k}(\mathbf{r}_j)$, meaning an evaluation of the eigenstate of the kth particle at the position coordinate of the jth particle, would not be physically meaningful for the situation in question where by hypothesis the particles are completely distinguishable. Since the particles are admittedly identical, distinguishability in practice requires *spatial separation*, and this can be achieved in a quantum-mechanical description only in the limit of completely nonoverlapping wave functions. This by definition means that ϕ_{i_j}, the eigenstate of the jth particle, must have a zero value at $\mathbf{r}_k$, which represents the location of the kth particle in the classical Hamiltonian. (Consider, for example, the case of one electron traveling in an oscilloscope tube located in Los Angeles, California, and a second electron bombarding a single-crystal metal surface in a research apparatus located in Ithaca, New York. Clearly the wave functions for these two electrons do not overlap appreciably, so that the two electrons are essentially distinguishable because of their spatial separation.)

EXERCISE Show that the wave function given by Eq. (2.1) satisfies each single-particle Schrödinger wave equation $\mathscr{H}_j\phi_s(j) = \mathscr{E}_s\phi_s(j)$. Show that it likewise satisfies, in the limit of no interaction between particles, the many-particle Schrödinger wave equation $\mathscr{H}\Phi = \mathscr{E}\Phi$, where for the presently considered case of noninteracting particles,

$$\mathscr{H} = \sum_{j=1}^{N} \mathscr{H}_j, \qquad \mathscr{E} = \sum_{j=1}^{N} \mathscr{E}_{i_j}.$$

1.2 Systems of Indistinguishable Particles

Suppose that we now look at an entirely different physical system that differs from the above one only insofar as the particles are indistinguishable instead of being distinguishable. Identical particles become indistinguishable in a quantum description whenever the overlap of the wave functions for the individual particles is nonzero. (That is, the wave functions of the jth and the kth particles are both nonzero over some common spatial domain, as illustrated in Fig. 2.1, so that a physical observation which detects a particle in this region could not distinguish whether the jth or the kth particle was observed.) Since overlap of wave functions is a property that is dependent on a probability density description for particle location, the property of indistinguishability of identical particles is characteristic of the quantum nature of particles. (In classical mechanics, the exact trajectory of individual particles is describable as a function of time so that even identical particles can be considered to be distinguishable.) Thus for the case of indistinguishable particles, whenever we observe a particle at some time t we generally cannot in any way tell for certain which particle it happens to be relative to some earlier observation. If we could indeed tell which particle it is relative to an earlier observation, then we would have a system of distinguishable particles, in contrast to a system of indistinguishable particles. Our wave function for the N-particle system must in some manner represent the fact that the particles are indistinguishable; that is, we must have a wave function that is invariant to any permutation of the particles. However, the permutation $1 \to 2 \to 3 \to 4 \to \cdots \to N \to 1$ transforms the distinguishable-particle wave function

$$\phi_{i_1}(1)\phi_{i_2}(2) \cdots \phi_{i_N}(N) \tag{2.2}$$

to the new (and generally different) function

$$\phi_{i_1}(2)\phi_{i_2}(3) \cdots \phi_{i_N}(1) \tag{2.3}$$

of the independent coordinates $\mathbf{r}_1, \mathbf{r}_2, \ldots, \mathbf{r}_N$. In a classical description, this permutation means that we place particle 1 at position $\mathbf{r}_2$, while simultaneously placing particle 2 at position $\mathbf{r}_3, \ldots,$ while simultaneously placing particle N at position $\mathbf{r}_1$.

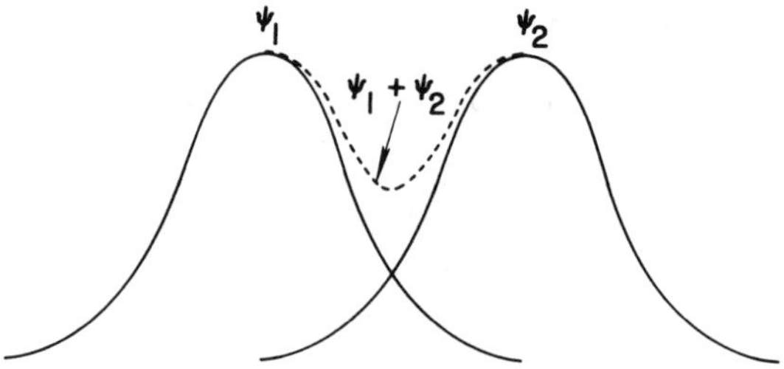

Fig. 2.1 Overlapping wave functions ψ_1 and ψ_2. (A particle found within the region of overlap may be either particle 1 or particle 2.)

But suppose we had a gigantic function of the independent variables $\mathbf{r}_1, \mathbf{r}_2, \ldots, \mathbf{r}_N$ made up of a sum of the above two functions (2.2) and (2.3) plus similar terms sufficient to cover all possible permutations of the N particles over all classical position vectors $\mathbf{r}_j$. Then any permutation would leave this colossal function invariant; this function would therefore constitute an appropriate total wave function for the system of N noninteracting indistinguishable particles. That is,

$$\begin{aligned}\Phi_{i_1 i_2 i_3 \cdots i_N} \propto\ & \phi_{i_1}(1)\phi_{i_2}(2)\phi_{i_3}(3) \cdots \phi_{i_{N-1}}(N-1)\phi_{i_N}(N) \\ & + \phi_{i_1}(2)\phi_{i_2}(1)\phi_{i_3}(3) \cdots \phi_{i_{N-1}}(N-1)\phi_{i_N}(N) + \cdots \\ & + \phi_{i_1}(2)\phi_{i_2}(3)\phi_{i_3}(4) \cdots \phi_{i_{N-1}}(N)\phi_{i_N}(1) + \cdots \\ & + (\text{sum of all other possible permutations}).\end{aligned} \tag{2.4}$$

We then choose the proportionality factor to effect normalization of $\Phi_{i_1 i_2 i_3 \cdots i_N}$. For a given set of N different eigenstates $\{i_1, i_2, \ldots, i_N\}$, there are $N!$ different permutations of the vectors $\mathbf{r}_1, \mathbf{r}_2, \ldots, \mathbf{r}_N$, so that there are a maximum of $N!$ terms in the above sum. If some of the particles are in the same eigenstate, we have fewer than $N!$ distinct terms. For example, the indistinguishable-particle wave function analogous to the distinguishable-particle wave function $\Phi_{111 \cdots 1}$

would likewise contain only the one term $\Phi_{111\cdots1}$, while that analogous to $\Phi_{2111\cdots1}$ would contain N terms

$$\{\Phi_{2111\cdots1} + \Phi_{1211\cdots1} + \Phi_{1121\cdots1} + \cdots + \Phi_{1111\cdots2}\} \tag{2.5}$$

corresponding to the permutation of the N vectors $\mathbf{r}_1, \mathbf{r}_2, \ldots, \mathbf{r}_N$ in the argument of the single-particle eigenstate $\phi_2(\mathbf{r})$.

It should be kept in mind that the order of factors in a product of single-particle eigenstates has no particular physical significance when the particles are indistinguishable. For example,

$$\Phi_{1111\cdots1} = \{\phi_1(\mathbf{r}_1)\phi_1(\mathbf{r}_2)\phi_1(\mathbf{r}_3)\cdots\phi_1(\mathbf{r}_N)\} \tag{2.6}$$

may be equally well represented by the product

$$\{\phi_1(\mathbf{r}_2)\phi_1(\mathbf{r}_1)\phi_1(\mathbf{r}_3)\phi_1(\mathbf{r}_4)\cdots\phi_1(\mathbf{r}_N)\}. \tag{2.7}$$

Likewise the term

$$\Phi_{2111\cdots1} = \{\phi_2(\mathbf{r}_1)\phi_1(\mathbf{r}_2)\phi_1(\mathbf{r}_3)\phi_1(\mathbf{r}_4)\cdots\phi_1(\mathbf{r}_N)\} \tag{2.8}$$

is the same as the term

$$\{\phi_1(\mathbf{r}_2)\phi_2(\mathbf{r}_1)\phi_1(\mathbf{r}_3)\phi_1(\mathbf{r}_4)\cdots\phi_1(\mathbf{r}_N)\}. \tag{2.9}$$

On the other hand, this is not the same as the term

$$\Phi_{1211\cdots1} = \{\phi_1(\mathbf{r}_1)\phi_2(\mathbf{r}_2)\phi_1(\mathbf{r}_3)\phi_1(\mathbf{r}_4)\cdots\phi_1(\mathbf{r}_N)\}. \tag{2.10}$$

The term $\Phi_{2111\cdots1}$ denotes a possible state with particles $2, 3, 4, \ldots, N$ in state ϕ_1 but particle 1 in state ϕ_2, whereas the term $\Phi_{1211\cdots1}$ denotes a possible state with particles $1, 3, 4, \ldots, N$ in state ϕ_1 but particle 2 in state ϕ_2. The linear combination of such terms then takes care of the indistinguishability of the particles as regards a macroscopic quantum state of the system.

The possible energy levels for the complete N-particle system will remain the same as before the symmetrizing process, namely,

$$\mathscr{E}_{i_1 i_2\cdots i_N} = \mathscr{E}_{i_1} + \mathscr{E}_{i_2} + \cdots + \mathscr{E}_{i_N}, \tag{2.11}$$

since ϕ_{i_j} has the energy $\mathscr{E}_{i_j}$ associated with it regardless of the particular position coordinate which may happen to be its argument. Therefore we now have one possibility for a wave function that describes the system of N particles. Let us study the properties required of such a wave function from a more general standpoint in the following section.

1.3 Symmetry under Particle Exchange

That the particles of the system are physically indistinguishable places certain restrictions on the mathematical form of the total wave function for the system. The restrictions are based on the invariance of the Hamiltonian operator under all possible interchanges of identical particles in the system. The classical Hamiltonian function for the system is of the form

$$H = (1/2m)(p_1^2 + p_2^2 + \cdots + p_N^2) + \mathscr{V}(\mathbf{r}_1, \mathbf{r}_2, \ldots, \mathbf{r}_N), \tag{2.12}$$

where m is the particle mass and $\mathbf{p}_i$ is the momentum of the ith particle. (In classical mechanics, we say that $\mathbf{p}_i$ is *canonically conjugate* to $\mathbf{r}_i$.) The Hamiltonian operator $\mathscr{H}$ for the system of N particles as obtained by the usual prescription given in Chap. 1 §6 for converting the classical Hamiltonian function to the corresponding quantum operator is thus

$$\mathscr{H} = -(\hbar^2/2m)(\nabla_1^2 + \nabla_2^2 + \cdots + \nabla_N^2) + \mathscr{V}(\mathbf{r}_1, \mathbf{r}_2, \ldots, \mathbf{r}_N). \tag{2.13}$$

The wave function Ψ for the system will in general depend upon all coordinates $\mathbf{r}_i$ and the time t,

$$\Psi = \Psi(\mathbf{r}_1, \mathbf{r}_2, \ldots, \mathbf{r}_N, t), \tag{2.14}$$

where the time dependence is introduced by adding the individual time factors $\exp[-(i/\hbar)\mathscr{E}_j t]$ to each of the individual stationary-state single-particle eigenfunctions ϕ_j making up the stationary-state many-particle eigenfunction $\Phi(\mathbf{r}_1, \mathbf{r}_2, \ldots, \mathbf{r}_N)$. If particles 1 and 2 are interchanged, the Hamiltonian operator will be

$$\mathscr{H}' = -(\hbar^2/2m)(\nabla_2^2 + \nabla_1^2 + \cdots + \nabla_N^2) + \mathscr{V}(\mathbf{r}_2, \mathbf{r}_1, \mathbf{r}_3, \mathbf{r}_4, \ldots, \mathbf{r}_N). \tag{2.15}$$

Clearly the kinetic energy contribution to the Hamiltonian operator is unchanged. Since the particles behave entirely identically, the potential energy must also be the same as before. Therefore,

$$\mathscr{H}' = \mathscr{H}. \tag{2.16}$$

This is called *exchange invariance*. We can define an exchange operator $\mathscr{P}_{jk}$ which interchanges the coordinates of any pair of particles j, k in any function of these coordinates. Let $\mathscr{F}(\mathbf{r}_1, \mathbf{r}_2, \ldots, \mathbf{r}_j, \ldots, \mathbf{r}_k, \ldots)$ be an arbitrary function of the independent vector variables $\mathbf{r}_1, \mathbf{r}_2, \ldots, \mathbf{r}_j, \ldots, \mathbf{r}_k, \ldots$, with

$$\mathscr{P}_{jk}\mathscr{F}(\mathbf{r}_1, \mathbf{r}_2, \ldots, \mathbf{r}_j, \ldots, \mathbf{r}_k, \ldots) = \mathscr{F}(\mathbf{r}_1, \mathbf{r}_2, \ldots, \mathbf{r}_k, \ldots, \mathbf{r}_j, \ldots). \tag{2.17}$$

For shorthand notation, let the original function be represented by the symbol F_{jk} and let F_{jk} with $\mathbf{r}_j$ and $\mathbf{r}_k$ interchanged be denoted as F_{kj}. Then the above equation becomes $\mathscr{P}_{jk}F_{jk} = F_{kj}$. It follows from the definition of $\mathscr{P}_{jk}$ that $\mathscr{P}_{jk}F_{kj} = F_{jk}$, so that

$$\mathscr{P}_{jk}(\mathscr{P}_{jk}F_{jk}) \equiv \mathscr{P}_{jk}^2 F_{jk} = F_{jk}. \tag{2.18}$$

That is, any arbitrary function $\mathscr{F}$ is an eigenfunction of the operator $\mathscr{P}_{jk}^2$, with the eigenvalue being unity.

Next, we will show that the permutation operator $\mathscr{P}_{jk}$ commutes with the Hamiltonian. Consider the quantity $\mathscr{P}_{jk}(\mathscr{H}\mathscr{F})$, where $\mathscr{H}$ is the Hamiltonian operator

$$\mathscr{H} = \mathscr{H}(\mathbf{r}_1, \mathbf{r}_2, \ldots, \mathbf{r}_j, \ldots, \mathbf{r}_k, \ldots),$$

including the dependence upon the gradient operators $\nabla_1, \nabla_2, \ldots, \nabla_j, \ldots, \nabla_k, \ldots$; $\mathscr{H}$ can be denoted in shorthand as H_{jk}. Then from the definition of $\mathscr{P}_{jk}$ it follows that

$$\mathscr{P}_{jk}(\mathscr{H}\mathscr{f}) \equiv \mathscr{P}_{jk}(H_{jk}F_{jk}) = (H_{kj}F_{kj}).$$

Alternatively, let us consider the quantity $\mathscr{H}(\mathscr{P}_{jk}\mathscr{f})$, which can be written in the form

$$\mathscr{H}(\mathscr{P}_{jk}\mathscr{f}) = H_{jk}(\mathscr{P}_{jk}F_{jk}) = H_{jk}(F_{kj}).$$

Subtracting the two equations gives

$$\mathscr{P}_{jk}(\mathscr{H}\mathscr{f}) - \mathscr{H}(\mathscr{P}_{jk}\mathscr{f}) = H_{kj}F_{kj} - H_{jk}F_{kj},$$

or equivalently,

$$(\mathscr{P}_{jk}, \mathscr{H})\mathscr{f} = (H_{kj} - H_{jk})F_{kj}.$$

Thus the condition required for the permutation operator to commute with the Hamiltonian is that $H_{kj} = H_{jk}$. This is satisfied for any system of N noninteracting particles, since in this case the Hamiltonian for the system is the sum of the Hamiltonians for the individual particles,

$$\begin{aligned}\mathscr{H} &= \mathscr{H}_1 + \mathscr{H}_2 + \cdots + \mathscr{H}_j + \cdots + \mathscr{H}_k + \cdots \\ &= \mathscr{H}_1 + \mathscr{H}_2 + \cdots + \mathscr{H}_k + \cdots + \mathscr{H}_j + \cdots .\end{aligned}$$

(The two right-hand sides are equal because the order of the terms is unimportant, thus giving $H_{jk} = H_{kj}$.)

One consequence of the fact that the Hamiltonian commutes with the permutation operator is that we can use the theorem proven in Chap. 1 §8 that commuting Hermitian operators possess a complete set of simultaneous eigenfunctions. (We will prove shortly that the permutation operator is indeed Hermitian.) Thus we can choose the eigenfunctions Φ_l for the system of N particles in such a way that the eigenvalue equation

$$\mathscr{P}_{jk}\Phi_l = \gamma_l^{(jk)}\Phi_l \tag{2.19}$$

is satisfied, where $\gamma_l^{(jk)}$ is the appropriate eigenvalue. Operating on this equation with $\mathscr{P}_{jk}$ gives

$$\begin{aligned}\mathscr{P}_{jk}^2\Phi_l &\equiv \mathscr{P}_{jk}\mathscr{P}_{jk}\Phi_l = \mathscr{P}_{jk}\gamma_l^{(jk)}\Phi_l \\ &= \gamma_l^{(jk)}\mathscr{P}_{jk}\Phi_l = \gamma_l^{(jk)}\gamma_l^{(jk)}\Phi_l = \{\gamma_l^{(jk)}\}^2\Phi_l.\end{aligned}$$

However, it has been shown above that the eigenvalue of $\mathscr{P}_{jk}^2$ is always unity, so that we conclude that $\{\gamma_l^{(jk)}\}^2 = 1$, which leads to the result $\gamma_l^{(jk)} = \pm 1$. Thus the only possible eigenvalues of the permutation operator are $+1$ and -1. For the $+1$ eigenvalue,

$$\mathscr{P}_{jk}\Phi_l = \Phi_l \qquad \text{(symmetric } \Phi_l\text{)}, \tag{2.20}$$

which requires that the wave function be *symmetric* under exchange of any two particles in the system of N particles. For the -1 eigenvalue,

$$\mathscr{P}_{jk}\Phi_l = -\Phi_l \qquad \text{(antisymmetric } \Phi_l\text{)}, \tag{2.21}$$

which requires that the wave function be *antisymmetric* under exchange of any two particles in the system of N particles.

Let us now suppose that two of the particles (namely, j and k) are in the same single-particle eigenstate, such that $\phi_{i_j} = \phi_{i_k}$. This leads to the relation $\mathscr{P}_{jk}\Phi_l = \Phi_l$, using only the definition of the permutation operator and the equality $i_j = i_k$. This relation causes no difficulty if the many-particle wave function Φ_l is symmetric. If Φ_l is antisymmetric, however, this new relation together with the above antisymmetry relation (2.21) leads to the requirement $\Phi_l = -\Phi_l$. This equation can be satisfied only for the situation in which $\Phi_l = 0$, in which case there can be no particles. The conclusion, then, is that the antisymmetric wave function is zero unless every particle in the system of N noninteracting particles is in a different single-particle eigenstate. Particles that have such many-particle wave functions are said to obey the *Pauli exclusion principle*, which postulates that *no two half-odd-integer spin particles in a system can have the same set of quantum numbers*. In terms of wave functions, the *Pauli exclusion principle* can be stated thus: *The wave function for a system of half-odd-integer spin particles must always be totally antisymmetric*. This is noted to be a rather remarkable situation when we consider that the N particles in the system are completely noninteracting. Particles such as these are known as *Fermi particles*, and the statistical distribution function which we derive for occupation of the energy levels of a system by such particles is known as the *Fermi–Dirac distribution function*.

The *symmetric* wave function has no such restriction on the occupation of the various quantum states, so there can be any number of the N particles in a given single-particle eigenstate. Particles such as these are known as *Bose particles*, and the statistical distribution function which we derive for occupation of the energy levels of a system by such particles is known as the *Bose–Einstein distribution function*.

There remains one question to clarify, namely, is the permutation operator $\mathscr{P}_{ij}$ a Hermitian operator? Let us consider two arbitrary functions $\tilde{f}$ and $\tilde{g}$ of the coordinates $\mathbf{r}_1, \mathbf{r}_2, \mathbf{r}_3, \ldots, \mathbf{r}_j, \ldots, \mathbf{r}_k, \ldots$, and denote the functional dependences on $\mathbf{r}_j$ and $\mathbf{r}_k$ explicitly by writing $\tilde{f}(\mathbf{r}_j, \mathbf{r}_k)$ and $\tilde{g}(\mathbf{r}_j, \mathbf{r}_k)$, but only carry along the functional dependences on the other coordinates implicitly. Then let us consider separately the two integrals involved in the general definition (1.231) of a Hermitian operator. (We extend the general definition to include all independent variables of the system.) Using simultaneously the definition of the permutation operator $\mathscr{P}_{ij}$ leads to

$$\int \tilde{g}^*(\mathbf{r}_j, \mathbf{r}_k)[\mathscr{P}_{jk}\tilde{f}(\mathbf{r}_j, \mathbf{r}_k)]\, d\mathbf{r}_j\, d\mathbf{r}_k = \int \tilde{g}^*(\mathbf{r}_j, \mathbf{r}_k)\tilde{f}(\mathbf{r}_k, \mathbf{r}_j)\, d\mathbf{r}_j\, d\mathbf{r}_k,$$

$$\int [\mathscr{P}_{jk}\tilde{g}(\mathbf{r}_j, \mathbf{r}_k)]^*\tilde{f}(\mathbf{r}_j, \mathbf{r}_k)\, d\mathbf{r}_j\, d\mathbf{r}_k = \int \tilde{g}^*(\mathbf{r}_k, \mathbf{r}_j)\tilde{f}(\mathbf{r}_j, \mathbf{r}_k)\, d\mathbf{r}_j\, d\mathbf{r}_k.$$

In this latter expression let us interchange the dummy variables $\mathbf{r}_k$ and $\mathbf{r}_j$ in the integral on the right-hand side, which leads to the result

$$\int [\mathscr{P}_{jk}\tilde{g}(\mathbf{r}_j, \mathbf{r}_k)]^* \tilde{f}(\mathbf{r}_j, \mathbf{r}_k)\, d\mathbf{r}_j\, d\mathbf{r}_k = \int \tilde{g}^*(\mathbf{r}_j, \mathbf{r}_k)\tilde{f}(\mathbf{r}_k, \mathbf{r}_j)\, d\mathbf{r}_k\, d\mathbf{r}_j.$$

However, the right-hand side of this result has the same value as the right-hand side of the first equation in the present paragraph, so that the corresponding left-hand sides must be equal,

$$\int \tilde{g}^*(\mathbf{r}_j, \mathbf{r}_k)\{\mathscr{P}_{jk}\tilde{f}(\mathbf{r}_j, \mathbf{r}_k)\}\, d\mathbf{r}_j\, d\mathbf{r}_k = \int [\mathscr{P}_{jk}\tilde{g}(\mathbf{r}_j, \mathbf{r}_k)]^* \tilde{f}(\mathbf{r}_j, \mathbf{r}_k)\, d\mathbf{r}_j\, d\mathbf{r}_k.$$

Since $\tilde{g}(\mathbf{r}_j, \mathbf{r}_k)$ and $\tilde{f}(\mathbf{r}_j, \mathbf{r}_k)$ are arbitrary functions of $\mathbf{r}_j$ and $\mathbf{r}_k$ as well as any other independent variables which are included as parameters in the functions, this proves, through the general definition (1.231) of a Hermitian operator (as extended to several independent variables), that the permutation operator $\mathscr{P}_{ij}$ is Hermitian.

Why does nature require the wave function to be symmetric or antisymmetric? This follows from a thought experiment in which two identical particles are interchanged surreptitiously while an observer looks away from the system. Upon looking at the system once again, the observer could in no way tell that the system had been modified. This requires that $|\Phi|^2$ remain invariant under any exchange of particles if the particles are truly indistinguishable. This in itself imposes special conditions on the functional form of the wave function. However, $|\Phi|^2$ is automatically invariant under particle exchange if we require that the entire wave function be either totally symmetric or totally antisymmetric, where a *symmetric* function corresponds to the eigenvalue $+1$ of $\mathscr{P}_{jk}$ and an *antisymmetric* function corresponds to the eigenvalue -1. (The minus sign brought about by the particle exchange in the above thought experiment causes no difficulty because it is only the probability density which is physically meaningful. Only $|\Phi|^2$ is physically meaningful; $-\Phi$ is thus physically equivalent to Φ.) That in the absence of other compensating factors the system wave function must be totally symmetric or totally antisymmetric in order to maintain the same value for $|\Phi|^2$ under exchange follows directly from the fact that any arbitrary function can be written as a sum of functions which are, respectively, symmetric and antisymmetric in exchange; an exchange then yields the transformation

$$\Phi_{\text{sym}} + \Phi_{\text{antisym}} \rightarrow \Phi_{\text{sym}} - \Phi_{\text{antisym}}, \tag{2.22}$$

so in general $|\Phi|^2$ would not be preserved. Thus we limit our choice of wave function for a system containing identical particles to functions which have the proper symmetry.

EXERCISE Prove the above statement that any arbitrary function can be written as the sum of functions which are respectively symmetric and antisymmetric in exchange. You may restrict your considerations to a function involving the coordinates of two particles only. (*Hint*: Consider an arbitrary function Φ involving the coordinates of two particles which is neither symmetric nor antisymmetric under the two-particle exchange. Let us designate as Φ' the function obtained from the particle coordinate exchange in Φ. New functions can easily be constructed which are respectively symmetric and antisymmetric under the two-particle exchange in question. Namely,

$$\Phi_{\text{sym}} = \tfrac{1}{2}(\Phi + \Phi'), \qquad \Phi_{\text{antisym}} = \tfrac{1}{2}(\Phi - \Phi').$$

Adding these new functions together then gives

$$\Phi = \Phi_{\text{sym}} + \Phi_{\text{antisym}}.$$

It remains only for you to demonstrate that Φ_{sym} as constructed above is indeed symmetric under the two-particle exchange, and that Φ_{antisym} is indeed antisymmetric under the two-particle exchange.)

It is found *experimentally* that all eigenfunctions for a given type of particle show the *same* exchange symmetry. The exchange symmetry is found to be related to the intrinsic angular momentum of the type of particle in question according to the following *exchange–symmetry rules*:

(1) A system of identical particles, each of which has an integral quantum number for the intrinsic spin, can be described only by wave functions which are *symmetric* with respect to an interchange of the space and spin coordinates of any two such identical particles:

$$\Phi(1, 2, \ldots, i, \ldots, j, \ldots, N) = \Phi(1, 2, \ldots, j, \ldots, i, \ldots, N). \qquad (2.23)$$

Such particles are called *Bose particles*, or *bosons*. Equation (2.23) imposes no restriction as to the number of bosons in a given single-particle eigenstate.

(2) A system of identical particles, each of which has a half-integral quantum number for the intrinsic spin, can be described only by wave functions which are *antisymmetric* with respect to an interchange of the space and spin coordinates of any two such identical particles:

$$\Phi(1, 2, \ldots, i, \ldots, j, \ldots, N) = -\Phi(1, 2, \ldots, j, \ldots, i, \ldots, N). \qquad (2.24)$$

Such particles are called *Fermi particles*, or *fermions*. Equation (2.24) leads to the result that the system wave function is zero if any two particles are in the same single-particle eigenstate. It can thus be said that *no two half-odd-integer spin particles (Fermi particles) in a system can have the same set of quantum numbers*, or equivalently, *the wave function for a system of half-odd-integer spin particles must be totally antisymmetric*. These constitute equivalent statements of the extremely important *Pauli exclusion principle*.

The statistics appropriate for bosons are referred to as *Bose–Einstein statistics*. Some examples of Bose particles are photons (spin 1), neutral helium atoms in the ground state (spin 0), and alpha particles (spin 0). Relative to the case of distinguishable particles, Bose particles (as we show later) exhibit a quantum-mechanical "attraction" for one another and thus tend to be found spatially near one another. Bose particles tend to occupy the same low-energy quantum states. This is not prohibited for these particles since they do *not* obey the Pauli exclusion principle.

The statistics appropriate for fermions are referred to as *Fermi–Dirac statistics*. Examples of Fermi particles are plentiful; for instance, electrons, protons, neutrons, and μ mesons all have spin $\frac{1}{2}$ and are consequently fermions. Relative to the case of distinguishable particles, Fermi particles exhibit a quantum-mechanical "repulsion" for one another, and thus do not tend to be found spatially near one another. This "repulsion" can be attributed to the fact that Fermi particles cannot occupy the same quantum state because of the Pauli exclusion principle.

From these statements we can conclude that at a given temperature, the internal energy and the pressure in a spatially confined system of Bose particles are both less than the corresponding internal energy and pressure of a system of equivalent but distinguishable particles, whereas the internal energy and the pressure in a spatially confined system of Fermi particles are both greater than the corresponding quantities in a similar distinguishable-particle system.

Before demonstrating some of the above-mentioned properties of Bose and Fermi particles for a simple two-particle system, it is worthwhile to note that the $N!$ possible terms in the appropriately antisymmetrized or symmetrized wave function Φ, as obtained by adding all terms generated by the $N!$ permutations of $\mathbf{r}_1, \mathbf{r}_2, \ldots, \mathbf{r}_N$ in the function

$$\phi_{i_1}(\mathbf{r}_1)\phi_{i_2}(\mathbf{r}_2) \cdots \phi_{i_N}(\mathbf{r}_N), \tag{2.25}$$

can be grouped in an orderly fashion by writing the following determinant,

$$\Phi \propto \begin{vmatrix} \phi_{i_1}(1) & \phi_{i_1}(2) & \phi_{i_1}(3) & \cdots & \phi_{i_1}(N) \\ \phi_{i_2}(1) & \phi_{i_2}(2) & \phi_{i_2}(3) & \cdots & \phi_{i_2}(N) \\ \vdots & & & & \\ \phi_{i_N}(1) & \phi_{i_N}(2) & \phi_{i_N}(3) & \cdots & \phi_{i_N}(N) \end{vmatrix}. \tag{2.26}$$

This is sometimes referred to as a *Slater determinant*. Each term in the expansion of the determinant represents one term of the sum. The signs of the terms as obtained by ordinary expansion of a determinant alternate, but if we change all signs to be positive (or negative), then the resulting set of $N!$ terms is a symmetric function of the $\mathbf{r}_j$, and this gives us an appropriate wave function for a system of Bose particles (bosons). Alternatively, if we do not change the sign of any of the terms in the expanded Slater determinant, we have an antisymmetric function of the $\mathbf{r}_j$ which thus represents an appropriate wave function for a system of Fermi particles (fermions).

The existence of the property of antisymmetry for the Slater determinant can be justified by noting that permutation of the coordinates of any two particles i and j in the system has the effect on the determinant of interchanging columns i and j. It is proven in the theory of determinants that such an interchange of columns (or rows) results in an overall change in the sign of the determinant.

EXERCISE Show that the Slater determinant for a two-particle system changes sign upon interchange of rows or columns.

Finally, let us ask what happens to the Slater determinant if any two particles i and j happen to be in the same eigenstate. It can be seen that this causes rows i

and j in the determinant to be identical, and it is proven in the theory of determinants that whenever any two rows (or columns) are equal, then the determinant has the value zero. This corresponds to no wave function at all. *Thus there can be no antisymmetric wave function for the system which allows any two particles to be in the same eigenstate.* This is equivalent to our former statement of the important *Pauli exclusion principle.*

Thus it can be said that in a system of Fermi particles, which is characterized by an antisymmetric wave function, the particles must occupy the allowable eigenstates in accordance with the Pauli exclusion principle. On the other hand, it can be said that in a system of Bose particles, which is characterized by a symmetric wave function, there is no restriction to the occupation of the allowable eigenstates, and the Pauli exclusion principle is therefore inapplicable to such particles.

For systems containing a great many particles, the number of terms in the symmetric or the antisymmetric wave function becomes enormous. It seems evident that such complexity will preclude making significant progress in extended calculations; thus a shorthand method of writing wave functions is desirable. The salient information is that there is a system of N particles which must be treated alike, with a resulting wave function which is symmetrized for bosons or antisymmetrized for fermions. The *occupation number representation* takes this redundant information for granted, and uses the notation

$$|1, 1, 1, 0, 0, \ldots\rangle,$$

for example, to indicate simply that the first three single-particle eigenstates $\phi_1(\mathbf{r})$, $\phi_2(\mathbf{r})$, and $\phi_3(\mathbf{r})$ are occupied by a single electron, with the remaining single-particle eigenstates $\phi_4(\mathbf{r}), \phi_5(\mathbf{r}), \ldots$ of the system being unoccupied. A wave function Φ for any number of particles can be specified in this notation,

$$\Phi \equiv |n_1, n_2, n_3, \ldots\rangle \equiv |\{n_j\}\rangle,$$

where $\{n_j\}$ is the *set* of occupation numbers $n_1, n_2, n_3, \ldots$ for the single-particle eigenstates. For fermions, the n_j have only two possible values (0 and 1) consistent with the Pauli exclusion principle. For bosons, the n_j can take on the value of any positive integer number, consistent only with conservation of the total number N of particles in the system, $N = \sum_j n_j$.

It can be proved that *distinct* Φs corresponding to the same N particles are orthogonal. Likewise Φs corresponding to different values of N are orthogonal. In general, the Φs form a complete set in the space of *any number of particles. Many particle operators* can be defined in terms of *creation* and *annihilation operators* which increase or decrease the number of particles in a given single-particle eigenstate. Further details can be found in standard references, such as Taylor (1970) and Callaway (1976).

EXERCISE Show that the Slater determinant for a two-particle system is zero when the rows or columns are identical.

PROJECT 2.1 Wave Functions for a Noninteracting Many-Particle System

Suppose that the electrons in a many-electron atom could be considered as completely noninteracting in the sense of completely negligible Coulomb interactions between electrons. (a) Write the wave function $\Phi(\mathbf{r}_1, \mathbf{r}_2)$ for the helium atom in this limit. (b) Write the wave function $\Phi(\mathbf{r}_1, \mathbf{r}_2, \mathbf{r}_3)$ for the lithium atom in this limit. (c) Write the wave function $\Phi(\mathbf{r}_1, \mathbf{r}_2, \ldots)$ for any other atom in this limit.

1.4 A Simple Two-Particle System

1.4.1 Wave Functions. Now we treat a simple model, namely, a system composed of two identical particles which are both acted upon by the same outside force, but which, in the first approximation, do not interact with one another. The Hamiltonian of particle 1 when treated separately is

$$H_1 = (p_1^2/2m) + \mathscr{V}(1), \tag{2.27}$$

and the Hamiltonian of particle 2 when treated separately is

$$H_2 = (p_2^2/2m) + \mathscr{V}(2). \tag{2.28}$$

Using the correspondence $p \to -i\hbar\nabla$, developed in §6 of Chap. 1, the total Hamiltonian operator $\mathscr{H}$ for the system is the sum of $\mathscr{H}_1$ and $\mathscr{H}_2$, with

$$\mathscr{H}_1 = -(\hbar^2/2m)\nabla_1^2 + \mathscr{V}(1), \tag{2.29}$$

$$\mathscr{H}_2 = -(\hbar^2/2m)\nabla_2^2 + \mathscr{V}(2). \tag{2.30}$$

The ∇_i^2 is the Laplacian with respect to the coordinates $\mathbf{r}_i$ of particle i ($i = 1, 2$), and

$$\mathscr{V}(i) = \mathscr{V}(\mathbf{r}_i, \sigma_{zi}) \qquad (i = 1, 2), \tag{2.31}$$

where σ_{zi} is the z component of the spin angular momentum of particle i. We allow for a functional dependence on σ_{zi} in order to include effects due to particle spin. Because the Hamiltonian operator is a sum of two parts,

$$\mathscr{H} = \mathscr{H}_1 + \mathscr{H}_2, \tag{2.32}$$

where $\mathscr{H}_1$ and $\mathscr{H}_2$ are the Hamiltonian operators for the two separate particles, the energy eigenfunctions for the entire system can be expressed as some type of product of eigenfunctions for the individual particles,

$$\Phi(1, 2) = \phi(1)\phi(2). \tag{2.33}$$

Since the particles are identical, the forces of the system must act similarly on each, so the potential functions $\mathscr{V}(1)$ and $\mathscr{V}(2)$ must have the same analytic form for each particle. Thus the possible wave functions for the individual particles are the same for the two particles. Let $\phi_n(i)$ denote one of the normalized energy eigenfunctions for the ith particle alone, where n stands for the complete set of quantum numbers needed to describe a given state. With this notation, a normalized wave function for the system may be written

$$\Phi'(1, 2) = \phi_a(1)\phi_b(2). \tag{2.34}$$

This wave function describes a state in which particle 1 is in the state a and particle 2 is in the state b. This wave function, however, is not necessarily of the correct form for identical particles, since there is no assurance that $\Phi'(1,2) = \pm \Phi'(2,1)$, as required by exchange symmetry for indistinguishable particles. This can be remedied by writing instead the Slater determinant

$$\Phi(1,2) = \frac{1}{\sqrt{2}} \begin{vmatrix} \phi_a(1) & \phi_a(2) \\ \phi_b(1) & \phi_b(2) \end{vmatrix}. \tag{2.35}$$

Upon expansion of the determinant, we change all signs to be positive to describe a Bose system, or leave the signs alone to describe a Fermi system. Thus we obtain

$$\Phi(1,2) = 2^{-1/2}[\phi_a(1)\phi_b(2) \pm \phi_a(2)\phi_b(1)], \tag{2.36}$$

which does satisfy the requirements of exchange symmetry. The plus sign is used for Bose particles, and the minus sign is used for Fermi particles. The factor $2^{-1/2}$ normalizes the total wave function to unity, assuming ϕ_a to be different from ϕ_b. The wave function (2.36) describes a state of the system in which one particle is in state a and one particle is in state b, such that either of the two particles is equally likely to be found in either state. It corresponds to an energy $\mathscr{E}_a + \mathscr{E}_b$ for the system. If both particles are in the same state, the minus sign characteristic of Fermi particles causes $\Phi(1,2)$ to be zero. [This is not the case for Bose particles; however, the normalization factor must be changed (from $2^{-1/2}$ to 2^{-1}).] Therefore we again see that two noninteracting Fermi particles cannot be in the same energy eigenstate. Equivalently, we may say that two noninteracting Fermi particles cannot both be in states described by the same set of quantum numbers, in accordance with the Pauli exclusion principle. Note from Eq. (2.36) that even if ϕ_a is different from ϕ_b, $\Phi(1,2)$ is zero for Fermi particles whenever $\mathbf{r}_1 = \mathbf{r}_2$, corresponding to both particles being simultaneously at a given position in space. This is not the case for Bose particles.

1.4.2 Electron Spin. It is worthwhile to examine in an elementary manner how spin enters into the formalism, and how it can affect the symmetry of the wave function. Recall that in Chap. 1, §6 we mentioned that spin for a particle is analogous to polarization for an electromagnetic wave. In both cases we must have some coordinates and a technique for including the physical property in the formulation. The spin function $\chi(\sigma_i)$ in terms of spin coordinates σ_i provides a convenient way of doing this for particles; there are analogous ways for adding a description of polarization to a scalar function describing an electromagnetic wave, although the more widely known method is simply to formulate electric and magnetic fields as vector quantities. We restrict our consideration of quantum spin to cases for which the total wave function for the system can be written as the product of a function of the space coordinates (e.g., a Slater determinant) and a function of the spin coordinates,

$$\Phi = G(\mathbf{r}_i)\chi(\sigma_i), \tag{2.37}$$

where σ_i represents the spin coordinates of the system. The functions $G(\mathbf{r}_i)$ and $\chi(\sigma_i)$ in Eq. (2.37) are each either symmetric or antisymmetric in order that Φ be completely symmetric or completely antisymmetric,

$$G(\mathbf{r}_i) = 2^{-1/2}[u_a(1)u_b(2) \pm u_a(2)u_b(1)], \tag{2.38}$$

$$\chi(\sigma_i) = 2^{-1/2}[s_c(1)s_d(2) \pm s_c(2)s_d(1)], \tag{2.39}$$

where u and s are the individually normalized spatial and spin wave functions. The signs in $G(\mathbf{r}_i)$ and $\chi(\sigma_i)$ must now be correlated to obtain the required exchange symmetry for the total wave function: For Bose particles the signs must be the same, but for Fermi particles they must be opposite. For convenience, we replace the factor of ± 1 in $G(\mathbf{r}_i)$ by δ_1 and replace the factor of ± 1 in $\chi(\sigma_i)$ by δ_2. Then δ_1 and δ_2 are both equal to $+1$ or else both are equal to -1 for a boson system, while δ_1 and δ_2 are equal respectively to $+1$ and -1, or else -1 and $+1$, for a fermion system. Multiplying out the factors $G(\mathbf{r}_i)$ and $\chi(\sigma_i)$ then gives

$$\begin{aligned} 2\Phi = {} & u_a(1)u_b(2)s_c(1)s_d(2) + \delta_2 u_a(1)u_b(2)s_c(2)s_d(1) \\ & + \delta_1 u_a(2)u_b(1)s_c(1)s_d(2) + \delta_1\delta_2 u_a(2)u_b(1)s_c(2)s_d(1). \end{aligned} \tag{2.40}$$

Now under the transformation $1 \to 2$ and $2 \to 1$,

$$\begin{aligned} 2\Phi^{(\text{trans})} \to {} & u_a(2)u_b(1)s_c(2)s_d(1) + \delta_2 u_a(2)u_b(1)s_c(1)s_d(2) \\ & + \delta_1 u_a(1)u_b(2)s_c(2)s_d(1) + \delta_1\delta_2 u_a(1)u_b(2)s_c(1)s_d(2). \end{aligned} \tag{2.41}$$

By comparing the first term of Φ with the fourth term of $\Phi^{(\text{trans})}$, and vice versa, and by comparing the second term of Φ with the third term of $\Phi^{(\text{trans})}$, and vice versa, it is seen immediately that Φ in this form does have even parity for boson systems and odd parity for fermion systems.

1.4.3 Quantum Exchange Forces. As an illustration of the physical consequences of the symmetry character of wave functions, we evaluate the average value of the square of the distance between the two particles. For simplicity, let us neglect the spin factor (2.39) in the wave function, and consider only the spatial factor given by Eq. (2.38), or equivalently, Eq. (2.36). Using Eq. (1.211) to compute the expectation value of the square of the distance $(\mathbf{r}_2 - \mathbf{r}_1)$ between the two particles, we obtain by integrating over the coordinates of both particles,

$$\begin{aligned} \langle(\mathbf{r}_2 - \mathbf{r}_1)^2\rangle = {} & \frac{1}{2}\int_{\Omega_{\mathbf{r}_1}}\int_{\Omega_{\mathbf{r}_2}} [\phi_a^*(1)\phi_b^*(2) \pm \phi_a^*(2)\phi_b^*(1)] \\ & \times (\mathbf{r}_2^2 + \mathbf{r}_1^2 - \mathbf{r}_2\cdot\mathbf{r}_1 - \mathbf{r}_1\cdot\mathbf{r}_2)[\phi_a(1)\phi_b(2) \pm \phi_a(2)\phi_b(1)]\, d\mathbf{r}_1\, d\mathbf{r}_2. \end{aligned} \tag{2.42}$$

This can be expanded to give

$$\langle(\mathbf{r}_2-\mathbf{r}_1)^2\rangle = \frac{1}{2}\Bigg[\int_{\Omega_{\mathbf{r}_1}} d\mathbf{r}_1\,\phi_a^*(1)\phi_a(1)\int_{\Omega_{\mathbf{r}_2}} d\mathbf{r}_2\,\phi_b^*(2)r_2^2\phi_b(2)$$

$$\pm\int_{\Omega_{\mathbf{r}_1}} d\mathbf{r}_1\,\phi_b^*(1)\phi_a(1)\int_{\Omega_{\mathbf{r}_2}} d\mathbf{r}_2\,\phi_a^*(2)r_2^2\phi_b(2)$$

$$\pm\int_{\Omega_{\mathbf{r}_1}} d\mathbf{r}_1\,\phi_a^*(1)\phi_b(1)\int_{\Omega_{\mathbf{r}_2}} d\mathbf{r}_2\,\phi_b^*(2)r_2^2\phi_a(2)$$

$$+\int_{\Omega_{\mathbf{r}_1}} d\mathbf{r}_1\,\phi_b^*(1)\phi_b(1)\int_{\Omega_{\mathbf{r}_2}} d\mathbf{r}_2\,\phi_a^*(2)r_2^2\phi_a(2)\Bigg]$$

$$+\left[\tfrac{1}{2}(\text{similar terms in } r_1^2)\right]-\left[\tfrac{1}{2}(\text{similar terms in } \mathbf{r}_2\cdot\mathbf{r}_1)\right]$$

$$-\left[\tfrac{1}{2}(\text{similar terms in } \mathbf{r}_1\cdot\mathbf{r}_2)\right]. \tag{2.43}$$

We have averaged over both sets of coordinates in order to allow for every possible value of $\mathbf{r}_1$ and every possible value of $\mathbf{r}_2$. Employing the relation (1.238) expressing that eigenfunctions belonging to different eigenvalues are orthogonal, we see that two of the integrals over $\Omega_{\mathbf{r}_1}$ explicitly given in Eq. (2.43) are zero, yielding a zero value for the second and third terms involving r_2^2. Assuming normalization of the individual wave functions, we see that the remaining two integrals over $\Omega_{\mathbf{r}_1}$ explicitly given in Eq. (2.43) are unity. Letting $\langle r^2\rangle_n$ denote $\int_{\Omega_{\mathbf{r}_i}}\phi_n^* r_i^2\phi_n\,d\mathbf{r}_i$, we see the nonzero r_2^2 terms in Eq. (2.43) are simply

$$\tfrac{1}{2}[\langle r^2\rangle_b+\langle r^2\rangle_a]. \tag{2.44}$$

By symmetry, we see we will duplicate these with the r_1^2 terms, to yield a total of

$$\langle r^2\rangle_b+\langle r^2\rangle_a. \tag{2.45}$$

(We must of course keep in mind that r_1 and r_2 are simply dummy variables in their respective integrals over $\Omega_{\mathbf{r}_1}$ and $\Omega_{\mathbf{r}_2}$.)

Now we look at a cross term $\mathbf{r}_1\cdot\mathbf{r}_2$. Writing the terms in the same order as those explicitly written out in Eq. (2.43), we obtain for the first term

$$-\frac{1}{2}\int_{\Omega_{\mathbf{r}_1}} d\mathbf{r}_1\int_{\Omega_{\mathbf{r}_2}} d\mathbf{r}_2\,\phi_a^*(1)\phi_b^*(2)\mathbf{r}_1\cdot\mathbf{r}_2\phi_a(1)\phi_b(2). \tag{2.46}$$

Because integration and the vector dot product are linear operations, this can be written in the form,

$$-\frac{1}{2}\left[\int_{\Omega_{\mathbf{r}_1}} d\mathbf{r}_1\,\phi_a^*(1)\mathbf{r}_1\phi_a(1)\right]\cdot\left[\int_{\Omega_{\mathbf{r}_2}} d\mathbf{r}_2\,\phi_b^*(2)\mathbf{r}_2\phi_b(2)\right], \tag{2.47}$$

which in turn can be written in shorthand notation

$$-\tfrac{1}{2}\langle\mathbf{r}\rangle_a\cdot\langle\mathbf{r}\rangle_b. \tag{2.48}$$

From symmetry, the $\mathbf{r}_2 \cdot \mathbf{r}_1$ terms will yield a corresponding contribution $-\frac{1}{2}\langle \mathbf{r} \rangle_b \cdot \langle \mathbf{r} \rangle_a$. Since the dot product commutes, we then obtain a total contribution

$$-\langle \mathbf{r} \rangle_a \cdot \langle \mathbf{r} \rangle_b. \tag{2.49}$$

The fourth cross term is very similar to the first cross term which we have just evaluated, since it is equivalent to the first term with an interchange of ϕ_a and ϕ_b. Since the result (2.49) is symmetrical in ϕ_a and ϕ_b, and the dot product is commutative, we conclude that the contribution of the fourth cross term is also given by Eq. (2.49). We add the two contributions together to obtain a total contribution

$$-2\langle \mathbf{r} \rangle_a \cdot \langle \mathbf{r} \rangle_b. \tag{2.50}$$

Now we look at the third cross term in $\mathbf{r}_1 \cdot \mathbf{r}_2$ (or equivalently in $\mathbf{r}_2 \cdot \mathbf{r}_1$ since $\mathbf{r}_2 \cdot \mathbf{r}_1 = \mathbf{r}_1 \cdot \mathbf{r}_2$), which can be written

$$-(\pm)\frac{1}{2}\left[\int_{\Omega_{r_2}} d\mathbf{r}_2\, \phi_b^*(2)\mathbf{r}_2\phi_a(2)\right] \cdot \left[\int_{\Omega_{r_1}} d\mathbf{r}_1\, \phi_a^*(1)\mathbf{r}_1\phi_b(1)\right]. \tag{2.51}$$

In the abbreviated *Dirac notation* (Chap. 1, §8.10) this can be written

$$\mp \tfrac{1}{2}\langle b|\mathbf{r}|a\rangle \cdot \langle a|\mathbf{r}|b\rangle, \tag{2.52}$$

which is the same as

$$\mp \tfrac{1}{2}|\langle a|\mathbf{r}|b\rangle|^2. \tag{2.53}$$

Adding the equivalent results for the $\mathbf{r}_1 \cdot \mathbf{r}_2$ and $\mathbf{r}_2 \cdot \mathbf{r}_1$ terms then gives a total contribution of

$$\mp |\langle a|\mathbf{r}|b\rangle|^2. \tag{2.54}$$

Finally the second cross term will also yield the same result; this represents the third term with $\mathbf{r}_1$ and $\mathbf{r}_2$ interchanged, which introduces no change in the result (2.54). Thus the total contribution of the second and third cross terms is

$$\mp 2|\langle a|\mathbf{r}|b\rangle|^2. \tag{2.55}$$

Adding all contributions (2.45), (2.50), and (2.55) together, we finally obtain

$$\langle (\mathbf{r}_2 - \mathbf{r}_1)^2 \rangle = \langle r^2 \rangle_a + \langle r^2 \rangle_b - 2\langle \mathbf{r} \rangle_a \cdot \langle \mathbf{r} \rangle_b \mp 2|\langle a|\mathbf{r}|b\rangle|^2. \tag{2.56}$$

It is informative to compare this with the corresponding result for distinguishable particles, for which the wave function would be simply

$$\phi_a(1)\phi_b(2). \tag{2.57}$$

In this case, the expectation value of $(\mathbf{r}_2 - \mathbf{r}_1)^2$ is given by

$$\begin{aligned}
\langle(\mathbf{r}_2 - \mathbf{r}_1)^2\rangle &= \int_{\Omega_{r_1}}\int_{\Omega_{r_2}} d\mathbf{r}_1\, d\mathbf{r}_2\, \phi_a^*(1)\phi_b^*(2)(\mathbf{r}_2 - \mathbf{r}_1)^2\phi_a(1)\phi_b(2) \\
&= \int_{\Omega_{r_1}} d\mathbf{r}_1\, \phi_a^*(1)\phi_a(1) \int_{\Omega_{r_2}} d\mathbf{r}_2\, \phi_b^*(2)r_2^2\phi_b(2) \\
&\quad + \int_{\Omega_{r_1}} d\mathbf{r}_1\, \phi_a^*(1)r_1^2\phi_a(1) \int_{\Omega_{r_2}} d\mathbf{r}_2\, \phi_b^*(2)\phi_b(2) \\
&\quad - \left[\int_{\Omega_{r_1}} d\mathbf{r}_1\, \phi_a^*(1)\mathbf{r}_1\phi_a(1)\right]\cdot\left[\int_{\Omega_{r_2}} d\mathbf{r}_2\, \phi_b^*(2)\mathbf{r}_2\phi_b(2)\right] \\
&\quad - \left[\int_{\Omega_{r_2}} d\mathbf{r}_2\, \phi_b^*(2)\mathbf{r}_2\phi_b(2)\right]\cdot\left[\int_{\Omega_{r_1}} d\mathbf{r}_1\, \phi_a^*(1)\mathbf{r}_1\phi_a(1)\right] \\
&= \langle r^2\rangle_b + \langle r^2\rangle_a - 2\langle\mathbf{r}\rangle_a\cdot\langle\mathbf{r}\rangle_b. \qquad (2.58)
\end{aligned}$$

Thus in the quantum-mechanical case of *indistinguishable* particles, we have an additional term over and above the *distinguishable* particle terms. This additional term is called the *exchange* term,

$$\mp 2|\langle a|\mathbf{r}|b\rangle|^2. \qquad (2.59)$$

This term arises specifically from the requirement that $\Phi(1, 2)$ have the symmetric or antisymmetric form, and the sign of the term is opposite for the two cases. The sign is negative for Bose particles and positive for Fermi particles, corresponding to the quantity $\langle(\mathbf{r}_2 - \mathbf{r}_1)^2\rangle$ being algebraically smaller if the wave functions are symmetric (therefore representing an "attraction"), whereas it is algebraically larger if the wave functions are antisymmetric (therefore representing a repulsion). The latter gives us some insight into the physical consequences of the Pauli exclusion principle. Although these attractions and repulsions (called *exchange forces*) are physically just as real as if they were the result of a classical force appearing in the Hamiltonian function, they are of course nonclassical effects.

1.4.4 Joint Probability Density. In treating physical systems that contain two or more indistinguishable particles, one often wishes to know how the particles are distributed in space. That more than one set of coordinates is involved in the wave function, and hence in the probability density $\rho = |\psi|^2$, can be initially quite confusing. Thus it is worthwhile to look briefly at this problem for the simple two-particle system. Although the coordinates of the particles appear distinct from one another in the Hamiltonian function, the particles themselves move in the same three-dimensional space, and one cannot distinguish which of the particles has been found in a given volume element $(dx\, dy\, dz)$ symbolized by $d\mathbf{r}$. What we wish to know is the probability that one of the particles is in a given volume element $d\mathbf{r}$ regardless of where the other particle may be. The joint probability-density distribution for the two particles is

$$\Phi^*\Phi = \tfrac{1}{2}[\phi_a^*(1)\phi_b^*(2) \pm \phi_a^*(2)\phi_b^*(1)][\phi_a(1)\phi_b(2) \pm \phi_a(2)\phi_b(1)]. \quad (2.60)$$

If we integrate $\Phi^*\Phi$ over all space $\Omega_{\mathbf{r}_1}$ we get

$$\begin{aligned}\int_{\Omega_{\mathbf{r}_1}} \Phi^*\Phi \, d\mathbf{r}_1 &= \frac{1}{2}\int_{\Omega_{\mathbf{r}_1}} d\mathbf{r}_1 \, [\phi_a^*(1)\phi_b^*(2)\phi_a(1)\phi_b(2) \pm \phi_a^*(1)\phi_b^*(2)\phi_a(2)\phi_b(1) \\ &\quad \pm \phi_a^*(2)\phi_b^*(1)\phi_a(1)\phi_b(2) + \phi_a^*(2)\phi_b^*(1)\phi_a(2)\phi_b(1)] \\ &= \tfrac{1}{2}[\phi_b^*(2)\phi_b(2) + 0 + 0 + \phi_a^*(2)\phi_a(2)] \\ &= \tfrac{1}{2}[|\phi_b(2)|^2 + |\phi_a(2)|^2]. \end{aligned} \quad (2.61)$$

A second integration, this time over $\Omega_{\mathbf{r}_2}$, gives

$$\int_{\Omega_{\mathbf{r}_1}}\int_{\Omega_{\mathbf{r}_2}} \Phi^*\Phi \, d\mathbf{r}_1 \, d\mathbf{r}_2 = \frac{1}{2}\int_{\Omega_{\mathbf{r}_2}} [\,|\phi_b(2)|^2 + |\,\phi_a(2)|^2]\, d\mathbf{r}_2 = \tfrac{1}{2}[1 + 1] = 1. \quad (2.62)$$

Thus the total wave function is normalized such that there is unit probability of finding both particles somewhere in the domain of space considered.

The first integral over $\Omega_{\mathbf{r}_1}$ given by Eq. (2.61) represents the probability density for finding particle 2 at a given point $\mathbf{r}_2$ in space independent of where particle 1 may be. Due to symmetry, the probability density for finding particle 1 in $d\mathbf{r}_2$, independent of where particle 2 may be, must also be given by the same quantity. Thus the total probability density of finding one of the two particles at a position $\mathbf{r}$ is

$$2\{\tfrac{1}{2}[|\phi_b|^2 + |\phi_a|^2]\} = |\phi_b|^2 + |\phi_a|^2. \quad (2.63)$$

This is the physical probability density, and the product of this quantity and a volume element $d\mathbf{r}$ is the probability that *either* of the particles is in the given volume element, independent of where the other particle may be. Note that the physically measurable probability distribution is therefore just the sum of the single-particle distributions for the states ϕ_a and ϕ_b. This particular result is independent of the symmetry character of the spatial wave function and is the same as for distinguishable particles.

PROJECT 2.2 Understanding Two-Particle Wave Functions

1. Construct an *antisymmetric* two-particle wave function from the two lowest single-particle eigenstates obtained in solving the one-dimensional infinite square-well potential problem.
2. Construct a *symmetric* two-particle wave function from the two lowest single-particle eigenstates obtained in solving the one-dimensional infinite square-well potential problem.
3. Compute the expectation value of the square of the separation distance between the two particles for the two wave functions given in Parts 1 and 2.
4. Compare results obtained in Part 3 with those obtained using an unsymmetrized wave function (viz., use the two-particle wave function appropriate for two identical but distinguishable particles).
5. What are your qualitative conclusions?

6. Integrate the two-particle wave functions Φ_j of Parts 1 and 2 over the coordinate of one of the particles, and plot the magnitude of the result versus the coordinate of the second particle. (If you wish, you may choose a definite width such as 4 Å for the potential well, or you may choose to work in terms of a normalized position x/L.)
7. Repeat the calculation in Part 6 by first integrating over the coordinate of the second particle and plotting the magnitude of this result versus the coordinate of the first particle.
8. Repeat Parts 6 and 7 for the distinguishable-particle wave function.
9. Repeat Parts 1–8 with the exception that the first and the third single-particle eigenstates are to be used instead of the first and the second.
10. Compute the expectation value of the Hamiltonian for the wave functions obtained in Parts 1 and 2. Repeat for the wave functions obtained in Part 9. Compare the results to corresponding results deduced using distinguishable-particle wave functions.

2 Statistics for a Many-Particle System

We now seek to obtain expressions, somewhat analogous to the Boltzmann distribution law of classical statistical mechanics, which describe the statistically probable distribution of N Fermi or Bose particles among the various states ϕ_i of a one-particle quantum system. Assume that the system is in statistical equilibrium with total energy of the N particles equal to $\mathscr{E}_T$, with a small but finite uncertainty $\delta\mathscr{E}_T$. We then study the various ways in which the particles might be distributed among the various energy levels $\mathscr{E}_i$ available to them, and in so doing, find out what is the statistically most probable such distribution. We divide the entire energy range into adjoining energy "cells" (cf. Fig. 2.2) $\Delta\mathscr{E}_1, \Delta\mathscr{E}_2, \Delta\mathscr{E}_3, \ldots, \Delta\mathscr{E}_s, \ldots$, such that each cell is very narrow in comparison with the error $\delta\mathscr{E}_T$ we are likely to make in measuring the total energy $\mathscr{E}_T$, but large enough to contain a large number of energy levels. We denote the number of energy levels in the sth cell by g_s. Let us not restrict ourselves to nondegenerate levels. For degenerate levels, we consider each to contribute to g_s, as outlined at the beginning of the chapter. In statistical equilibrium, let n_s denote the number of particles with energies in the range $\mathscr{E}_s$ to $\mathscr{E}_s + \Delta\mathscr{E}_s$ of the sth cell. The possible

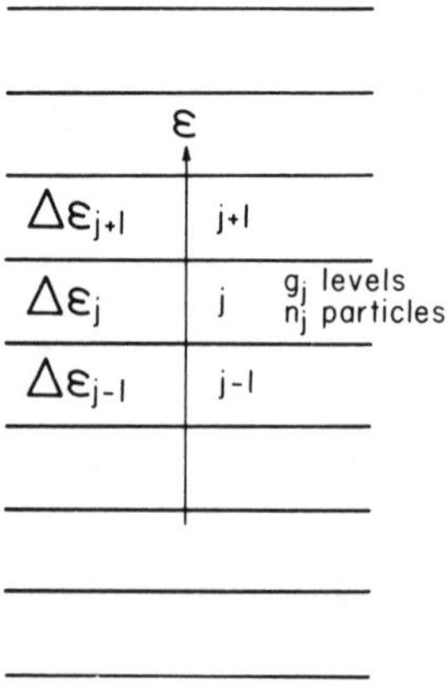

Fig. 2.2 A division of the energy range into adjoining energy cells, each containing a large number of available energy levels.

values of n_s must satisfy the conditions of conservation of the total number of particles and conservation of the total energy of the system of particles,

$$\sum_{s=1}^{\infty} n_s = N, \tag{2.64}$$

$$\sum_{s=1}^{\infty} n_s \mathscr{E}_s = \mathscr{E}_\mathrm{T}. \tag{2.65}$$

Since the n_i particles in the cell $\Delta\mathscr{E}_i$ do not all have energies exactly equal to $\mathscr{E}_i$, but instead have energies in the range $\mathscr{E}_i + \Delta\mathscr{E}_i$, the above equation (2.65) for conservation of the total energy is accurate to within $\delta\mathscr{E}_\mathrm{T}$, provided we choose the various cells $\Delta\mathscr{E}_i$ small enough.

If particles are indistinguishable and are of half-integral spin, then the Pauli exclusion principle (cf. §1) *must be considered to hold for the distribution. In this case no level can contain more than one particle*, assuming that each level is characterized by a complete set of quantum numbers.

There are many ways in which the N particles can be distributed among the cells, and many ways in which the n_s particles in a given cell can be distributed among the various energy levels of that cell. *A given distribution of cell populations* n_s ($s = 1, 2, \ldots$) is called a *macroscopic* (or *coarse-grained*) distribution, while *a given detailed distribution of the particles among the various energy levels of the cells* is called a *microscopic* (or *fine-grained*) distribution. We must deduce the probability for the occurrence of a given macroscopic distribution of particles, and find for what distribution $n_s(\mathscr{E}_s)$ this probability is a maximum. This distribution will then be our best estimate of the actual condition of the system. Insofar as the statistical properties of the system of particles are concerned, the microscopic distribution of the n_s particles in a given cell among the g_s energy levels of that cell is unimportant. Therefore we are interested in predicting statistically the coarse-grained distribution only. However, to predict the coarse-grained distribution properly, we must carefully consider all possible fine-grained distributions.

We introduce the following postulate relating to the probability of a given microscopic distribution of the N particles: *Every physically distinct microscopic distribution of the N particles among the various energy levels* $\mathscr{E}_i$ *which satisfies both the condition that the total energy be the quantity* $\mathscr{E}_\mathrm{T} \pm \delta\mathscr{E}_\mathrm{T}$ *and likewise satisfies the requirements of the exclusion principle, if it applies, is equally likely to occur*. By a *physically distinct distribution* we mean one that has a wave function which is different from that of any other distribution. Thus even degenerate wave functions are considered to represent physically distinct distributions.

Thus we may conclude that *the relative likelihood of occurrence of any given macroscopic distribution* $n_s(\mathscr{E}_s)$ *of the N particles among the various cells is proportional to the number of distinguishable ways in which such a distribution can be constructed.* Let us denote the number of ways in which a given macroscopic distribution can be constructed by $P(n_1, n_2, \ldots, n_s, \ldots)$, which we can simply call P. We must compute P for each type of particle considered. Then if we *maximize* *this quantity with respect to each n_s, subject to the auxiliary conditions of conservation of total number of particles and total energy of the system of particles*, we will have deduced the *statistically most probable distribution.*

2.1 Identical but Distinguishable Particles

In the case of *distinguishable* particles, P is equal to the product of the number of different ways of selecting the groups of particles to be put into each cell with the number of ways in which the particles can be arranged within the cells. We first consider the number of different ways of selecting the groups of particles, and only then do we turn our attention to counting the number of ways in which the particles can be arranged within each cell. We label all particles consecutively:

$$1 \quad 2 \quad 3 \quad 4 \quad 5 \quad 6 \quad 7 \quad 8 \quad 9 \quad 10 \quad 11 \quad \cdots. \tag{2.66}$$

Let us suppose that we take n_1 of these and put them in cell 1, take n_2 of the remaining ones and put them in cell 2, take n_3 of the remaining ones and put them in cell 3, etc., subject to the restrictions imposed by Eqs. (2.64) and (2.65). For the fixed set of numbers $n_1, n_2, n_3, n_4, \ldots$, we first ask ourselves how many distinguishable arrangements are possible, irrespective of the many other fine-grained distributions (obtained by arranging the particles *within* each cell) which are compatible with this coarse-grained arrangement.

From the N particles, we can select any one to put in cell 1 as the first of the n_1. There are $N - 1$ remaining particles from which to select the second particle of the n_1, etc. Hence there are

$$N(N-1)(N-2)\cdots(N-n_1+1) \tag{2.67}$$

ways of choosing the first n_1 particles. Call this P'_1. Thus

$$P'_1 = N!/(N-n_1)!. \tag{2.68}$$

In this expression we have counted as a different arrangement each separate sequence in which the first n_1 particles could have been selected. However, we need only to know which particles are in the quota n_1, but not in what sequence they appear. (For example, selecting particle 20 first and particle 24 second is equivalent to selecting particle 24 first and particle 20 second. We consider the distribution of the particles among the g_1 levels later.) We therefore divide the above number by the number of different sequences in which n_1 objects can be arranged. This is readily seen to be $n_1!$. This result follows from elementary counting: There are n_1 possible choices for the first particle, $n_1 - 1$ possible choices for the second particle once the first particle has been chosen, $n_1 - 2$ possible choices for the third particle once the first and second particles have been chosen, and so on until there remains only a single choice for the last of the n_1 particles. Considering that for *each* of the n_1 possible choices for the first particle we have the $n_1 - 1$ possible choices for the second particle, we have a total number of choices of n_1 added to itself $n_1 - 1$ times, or equivalently,

$n_1(n_1 - 1)$ independent choices for the first two particles. For each of these, we have the $n_1 - 2$ independent choices for the third particle, thus giving $n_1(n_1 - 1)$ added to itself $n_1 - 2$ times to give a total number of $n_1(n_1 - 1)(n_1 - 2)$ choices for the first three particles. Continuing this procedure down to the last of the n_1 particles yields a total of $n_1!$ independent sequences for the n_1 particles. That is, the individual arrangements in (2.67) can be placed into groups, each group containing $n_1!$ microscopic arrangements of the same set of particles. Hence the number of different ways of choosing the first n_1 particles out of the total N particles leading to different sets of n_1 particles is

$$P_1 = P_1'/n_1! = N!/[n_1!(N - n_1)!]. \tag{2.69}$$

The second quota n_2 is formed in the same manner, the only difference being that there are only $(N - n_1)$ particles left to choose from. Thus

$$P_2 = [(N - n_1)!]/[n_2!(N - n_1 - n_2)!]. \tag{2.70}$$

Similarly,

$$P_3 = [(N - n_1 - n_2)!]/[n_3!(N - n_1 - n_2 - n_3)!], \tag{2.71}$$

etc.

Remembering that all of the above numbers P_1, P_2, P_3, etc., are independent, we thus have for the number of ways of distributing the N particles among the various cells the product $P_1P_2P_3 \cdots P_s \cdots$. This product can be deduced from elementary counting in the manner now to be described.

Consider first that the number of cell 1 groups is P_1, independent of all other cells. For any given cell 2 group, we have the P_1 cell 1 groups, for a second of the cell 2 groups, we again have P_1 cell 1 groups to give $P_1 + P_1$ possibilities, for a third of the cell 2 groups we again have P_1 cell 1 groups to give $P_1 + P_1 + P_1$ possibilities, and so on until we have the sum of P_1 with itself P_2 times, namely, one time for each of the P_2 groupings. This number of possibilities is therefore P_1P_2, each requiring a different *distinguishable-particle* wave function.

Now let us consider cell 3. For any given cell 3 group, we have the P_1P_2 combined total number of cell 1 and cell 2 distinctly different groups, for a second of the cell 3 groups, we again have P_1P_2 combined cell 1 and cell 2 groups to give $P_1P_2 + P_1P_2$ possibilities, for a third of the cell 3 groups, we again have P_1P_2 cell 1 and cell 2 groups to give $P_1P_2 + P_1P_2 + P_1P_2$ possibilities, and so on until we have the sum of P_1P_2 with itself P_3 times, namely, one time for each of the P_3 groupings. This number of possibilities is therefore $P_1P_2P_3$, each requiring a different distinguishable-particle wave function.

Now let us consider cell 4. For any given cell 4 group, we have the $P_1P_2P_3$ combined total number of cell 1, cell 2, and cell 3 distinctly different groups, and thus and so for each of the P_4 different cell 4 groupings. Therefore the number of possibilities is $P_1P_2P_3$ added to itself P_4 times to give the number $P_1P_2P_3P_4$, each requiring a different distinguishable-particle wave function.

It is easy to continue the counting in this fashion, and we find $P_1P_2P_3P_4P_5$ as the number of possibilities when we include consideration of cell 5,

$P_1P_2P_3P_4P_5P_6$ as the number of possibilities when we also consider cell 6, $\prod_{s=1}^{7} P_s$ possibilities when we also consider cell 7, $\prod_{s=1}^{8} P_s$ possibilities when we also consider cell 8, and in general $\prod_{s=1}^{\infty} P_s$ possibilities when we consider all possible cells in the energy range extending from zero to infinity.

Substituting Eqs. (2.69), (2.70), (2.71),... into this product then gives

$$P_1P_2P_3\cdots P_s\cdots$$

$$= \frac{N!(N-n_1)!(N-n_1-n_2)!\cdots(N-n_1-n_2\cdots-n_s\cdots)!\cdots}{[n_1!n_2!n_3!\cdots n_s!\cdots][(N-n_1)!(N-n_1-n_2)!\cdots(N-n_1-n_2-\cdots-n_s\cdots)!\cdots]}$$

$$= \frac{N!}{n_1!n_2!n_3!\cdots n_s!\cdots} = \frac{N!}{\prod_{s=1}^{\infty} n_s!} = N!\prod \frac{1}{n_s!}\,. \tag{2.72}$$

This is in essence the number of ways in which N distinguishable objects can be put into an ordered array of boxes with prescribed numbers $n_1, n_2, \ldots, n_s \ldots$ in each box.

We must now calculate the number of ways in which the n_s particles in each cell can be arranged in the g_s energy levels included in that cell. Each such arrangement constitutes a different fine-grained arrangement. Since the exclusion principle does not apply to distinguishable particles, there is no limit on the number of particles in each energy state. Each of the n_s particles is equally likely to be in any one of the g_s states. Thus there are g_s ways in which the first particle can be put into the sth cell, and for each of these there are also g_s ways for the second particle, and so on. Remembering that we are presently considering distinguishable particles, the total number of distinct distributions of the n_s particles among the g_s levels of the sth cell is therefore just $g_s^{n_s}$. Each of these distinct distributions gives rise to a different microscopic distribution for the total system, and we have all of these possibilities for each cell. Therefore we must have the product of all of the $g_s^{n_s}$ with the previously deduced quantity $P_1P_2P_3 \cdots P_s \cdots$ if we are to have a number representing the total number of distinct microscopic distributions.

Again this result can be deduced by elementary counting. For each coarse-grained distribution, we have a different fine-grained distribution for *each* of the $g_1^{n_1}$ distinct distributions of the n_1 particles among the g_1 levels in cell 1. This gives $(P_1P_2P_3 \cdots P_s \cdots)g_1^{n_1}$ distinct arrangements, each requiring a different wave function whenever the particles are considered to be distinguishable. Fixing upon a given distribution in cell 1, then we have $g_2^{n_2}$ distinct distributions of the n_2 particles among the g_2 levels in cell 2, so the total number of arrangements (ignoring for the moment the possibility of different distributions in cells $3, 4, 5, 6, \ldots$) is $(P_1P_2P_3 \cdots P_s \cdots)g_1^{n_1}$ added to itself $g_2^{n_2}$ times, or equivalently, $(P_1P_2P_3, \ldots, P_s, \ldots)g_1^{n_1}g_2^{n_2}$ different wave functions whenever the particles are considered to be distinguishable. Fixing upon given distributions for cells 1 and 2, then we have $g_3^{n_3}$ distinct distributions of the n_3 particles among the g_3 levels in cell 3, so the total number of arrangements (ignoring for the moment the possibility of different distributions in cells $4, 5, 6, \ldots$) is $(P_1 P_2 P_3 \cdots P_s \cdots)$ $g_1^{n_1} g_2^{n_2}$ added to itself $g_3^{n_3}$ times, or equivalently, $(P_1 P_2 P_3 \cdots P_s \cdots) g_1^{n_1} g_2^{n_2} g_3^{n_3}$. It is an evident extension to reach the conclusion that the $g_4^{n_4}$ distinct distributions of the n_4 particles among the g_4 levels in cell 4 leads to a total number of arrangements (ignoring for the moment the possibility of different distributions in cells $5, 6, 7, \ldots$) equal to $(P_1 P_2 P_3 \cdots P_s \cdots) g_1^{n_1} g_2^{n_2} g_3^{n_3} g_4^{n_4}$. Continuing this process in order to include cells $5, 6, \ldots$ in sequence finally leads to

$$P = (P_1 P_2 P_3 \cdots P_s \cdots) g_1^{n_1} g_2^{n_2} g_3^{n_3} g_4^{n_4} g_5^{n_5} g_6^{n_6} \cdots .$$

Employing Eq. (2.72) then leads to the result $P = \prod_{s=1}^{\infty} (P_s g_s^{n_s}) = N! \times \prod_{s=1}^{\infty} (g_s^{n_s}/(n_s!))$, as shown in greater detail in Eq. (2.73).

We therefore have for the number of distinct microscopic distributions:

$$\begin{aligned} P &= N! \left(\prod_{s=1}^{\infty} \frac{1}{n_s!} \right) (g_1^{n_1} g_2^{n_2} \cdots g_s^{n_s} \cdots) = N! \left(\prod_{s=1}^{\infty} \frac{1}{n_s!} \right) \left(\prod_{k=1}^{\infty} g_k^{n_k} \right) \\ &= N! \left(\frac{1}{n_1!} \frac{1}{n_2!} \frac{1}{n_3!} \cdots \frac{1}{n_s!} \cdots \right) (g_1^{n_1} g_2^{n_2} \cdots g_s^{n_s} \cdots) \\ &= N! \left(\frac{g_1^{n_1}}{n_1!} \frac{g_2^{n_2}}{n_2!} \frac{g_3^{n_3}}{n_3!} \cdots \frac{g_s^{n_s}}{n_s!} \cdots \right) \\ &= N! \prod_{s=1}^{\infty} \frac{g_s^{n_s}}{n_s!} \qquad \text{(distinguishable particles).} \end{aligned} \tag{2.73}$$

2.2 Identical Indistinguishable Particles of Half-Integral Spin

For the case of indistinguishable particles, the indistinguishability prevents us from knowing which of the N particles have been placed in each cell. From a slightly different viewpoint, it can be said that the symmetrizing (or anti-symmetrizing) process places each occupied single-particle eigenstate into every subspace $\mathbf{r}_j$ of the hyperspace $\mathbf{r}_1, \mathbf{r}_2, \ldots, \mathbf{r}_N$ constituting the field of the N-particle wave function. Thus we cannot have the factor $P_1 P_2 \cdots P_s \cdots$ deduced above for the case of distinguishable particles. The only distinguishing feature of a given microscopic distribution is which of the g_s levels of a given cell are occupied by the n_s particles. The exclusion principle applicable for the present case requires that not more than one particle occupy a given energy level.

The number of distinguishable arrangements of the n_s particles among the g_s energy levels of the sth cell may be found as follows. If we first imagine the particles to be distinguishable, we see that the "first" particle can be put in any one of the g_s levels, and for each one of these choices the "second" particle can be put in any one of the $g_s - 1$ remaining levels, etc. (Clearly it is necessary that $g_s \geqslant n_s$.) Thus the number of sequences in which the n_s particles can be put into the g_s levels is

$$g_s(g_s - 1)(g_s - 2) \cdots (g_s - n_s + 1) = g_s!/(g_s - n_s)!. \tag{2.74}$$

However, the indistinguishability of the particles now requires that we shall not count as distinct distributions the various possible permutations of the particles among themselves, so that we must divide the above number by the number of possible permutations of the n_s particles, which is $n_s(n_s - 1)(n_s - 2) \cdots = n_s!$. Thus we deduce that the number of distinct distributions for each cell is

$$g_s!/[n_s!(g_s - n_s)!]. \tag{2.75}$$

Finally, we have for the total number of microscopic distributions that can lead to the given macroscopic distribution the product of these various independent arrangements,

$$P = \prod_{s=1}^{\infty} \frac{g_s!}{n_s!(g_s - n_s)!} \quad \text{(Fermi particles).} \tag{2.76}$$

2.3 Identical Indistinguishable Particles of Integral Spin

The indistinguishability of the particles again prevents us from knowing which of the N particles have been placed in each cell. The exclusion principle does not act in this case to limit the population of a given energy level. The number of distinct arrangements of the n_s particles among the g_s energy levels may be found by the following device [cf. Leighton (1959)]. Consider the cell $\Delta\mathscr{E}_s$ to consist of a linear array of black and white pegs, the black pegs representing partitions and the white pegs representing particles. There must be n_s white pegs for the n_s particles, and $g_s - 1$ black pegs for partitions which separate the g_s energy levels in $\Delta\mathscr{E}_s$. Consider now the various distinguishable permutations of the n_s white pegs and the $g_s - 1$ black pegs among the various holes, the number of holes being just sufficient to accommodate all of the pegs. (Note that the $g_s - 1$ partitions separate the entire cell into g_s intervals, and that the n_s remaining holes, which are separated into groups by these partitions, then represent the distribution of the particles among the various energy levels.) The number of distinct permutations of the black and white pegs among the holes is equal to the number of distinct arrangements of the n_s indistinguishable particles among the g_s energy levels.

From a slightly different viewpoint, consider the black pegs as partitions separating the discrete energy levels $\mathscr{E}_s$, with the number of white pegs between any two adjacent partitions giving the particle occupation number for that energy level. Considering the set of energy levels $\{\mathscr{E}_s\}$ as being ordered, there is a corresponding set of occupation numbers $\{m_s\}$, each number m_s representing the number of particles in the corresponding energy level $\mathscr{E}_s$. It is then clear that interchange of any white peg with any black peg effects some change in the set of occupation numbers $\{m_s\}$ and thus is equivalent to a distinguishable rearrangement of the n_s particles among the g_s energy levels in cell s.

Now, the number of permutations of $n_s + g_s - 1$ distinguishable objects is just $(n_s + g_s - 1)!$. This, combined with the fact that in these various sequences a permutation of the particles among themselves or a permutation of the partitions among themselves does not lead to a distinguishably different arrangement, tells us that the number of distinct arrangements for a given cell is

$$[(n_s + g_s - 1)!]/[n_s!(g_s - 1)!]. \tag{2.77}$$

Taking the product of these numbers for the various cells gives the total number of microscopic distributions consistent with the given macroscopic distribution,

$$P = \prod_{s=1}^{\infty} \frac{(n_s + g_s - 1)!}{n_s!(g_s - 1)!} \quad \text{(Bose particles).} \tag{2.78}$$

2.4 Maxwell–Boltzmann, Fermi–Dirac, and Bose–Einstein Distribution Laws

2.4.1 The Approach. With the above three expressions for the number of distinct microscopic states leading to a given macroscopic distribution, we are prepared to determine for what coarse-grained distribution the number of microscopic states is a maximum. This will be different for each of the three cases. For each case, our problem is to determine $n_s(\mathscr{E}_s)$ so that

$$P = \text{a maximum} \tag{2.79}$$

subject to the auxiliary conditions (2.64) and (2.65). Let us consider statistical systems containing large numbers of particles ($N \simeq 10^{23}$) with large numbers of energy levels in a single cell ($g_s \simeq 10^8$), so that we may regard the various expressions as being continuous functions of continuous variables. (Avogadro's number, for example, giving the number of molecules in a mole, is 6.02×10^{23}. The density of states given by Eq. (1.296) for a one-dimensional square-well potential is 8.15×10^8 levels per electron volt at an energy $\mathscr{E} = 1$ eV for a length $L = 1$ m.) For convenience, we now change the problem of obtaining the maximum of P to the equivalent one of maximizing the logarithm of P. (A maximum in P also gives rise to a maximum in $\ln P$ because the logarithm is a monotonic function of its argument.) From Eqs. (2.73), (2.76), and (2.78), we obtain

(a) Classical particles:

$$\ln P = \ln(N!) + \sum_{s=1}^{\infty} n_s \ln g_s - \sum_{s=1}^{\infty} \ln(n_s!) = \ln(N!) + \sum_{s=1}^{\infty} \{n_s \ln g_s - \ln(n_s!)\}. \tag{2.80}$$

(b) Fermi particles:

$$\ln P = \sum_{s=1}^{\infty} \{\ln(g_s!) - \ln(n_s!) - \ln[(g_s - n_s)!]\}. \tag{2.81}$$

(c) Bose particles:

$$\ln P = \sum_{s=1}^{\infty} \{\ln[(n_s + g_s - 1)!] - \ln(n_s!) - \ln[(g_s - 1)!]\}. \tag{2.82}$$

We must now maximize the value of $\ln P$ with respect to all small variations of the various cell populations, while still satisfying the auxiliary conditions (2.64) and (2.65) for N and $\mathscr{E}_T$. That is, if each n_s changes by a small amount δn_s, we must have

$$\delta(\ln P) = 0 \tag{2.83}$$

for any small values of δn_s that satisfy

$$\delta N = \sum_{s=1}^{\infty} \delta n_s = 0, \tag{2.84}$$

$$\delta \mathscr{E}_T = \sum_{s=1}^{\infty} \mathscr{E}_s \, \delta n_s = 0. \tag{2.85}$$

This problem can be solved with the use of *Lagrange's method of undetermined multipliers* [Houston (1948)], as illustrated below. We first introduce two fixed parameters, α and β, called *undetermined multipliers*, and form the following equation from Eqs. (2.83)–(2.85),

$$\delta(\ln P) - \alpha\,\delta N - \beta\,\delta\mathscr{E}_T = 0. \tag{2.86}$$

This equation certainly holds if the conditions (2.83)–(2.85) are satisfied.

We have a problem with considering variations in ln P, since differentials of factorials are frequently unknown to students who have had only elementary calculus. However, the values of n_s and g_s in our problem are very large, so that we can use Stirling's approximation to eliminate the factorials.

Stirling's formula for large n [see Burington (1956)] is

$$(2n\pi)^{1/2}(n/e)^n < n! < (2n\pi)^{1/2}(n/e)^n[1 + (12n - 1)^{-1}]. \tag{2.87}$$

Taking the logarithm gives

$$\tfrac{1}{2}\ln(2n\pi) + n \ln n - n \ln e < \ln(n!)$$
$$< \tfrac{1}{2}\ln(2n\pi) + n \ln n - n \ln e + \ln[1 + (12n - 1)^{-1}]. \tag{2.88}$$

Suppose $n \simeq 10^8$. Then $(12n - 1)^{-1} \simeq 10^{-9}$ and $\ln[1 + (12n - 1)^{-1}] \simeq \ln 1 \simeq 0$ relative to n. Also, $\ln n = 8 \ln 10 \simeq 18 \ll n$. The above approximation (2.88) becomes

$$n \ln n - n \lesssim \ln(n!) \lesssim n \ln n - n, \tag{2.89}$$

or equivalently,

$$\ln(n!) \simeq n \ln n - n \qquad \textbf{(Stirling's approximation)}. \tag{2.90}$$

Using Stirling's approximation (2.90) for $n_s!$ and $g_s!$ in Eqs. (2.80)–(2.82) gives

(a) Distinguishable particles:

$$\ln P \simeq \ln(N!) + \sum_{s=1}^{\infty} (n_s \ln g_s - n_s \ln n_s + n_s). \tag{2.91}$$

(b) Fermi particles:

$$\ln P \simeq \sum_{s=1}^{\infty} [g_s \ln g_s - g_s - n_s \ln n_s + n_s - (g_s - n_s) \ln(g_s - n_s) + (g_s - n_s)]. \tag{2.92}$$

(c) Bose particles:

$$\begin{aligned}\ln P \simeq \sum_{s=1}^{\infty} [(n_s + g_s - 1)\ln(n_s + g_s - 1) - (n_s + g_s - 1) \\ - n_s \ln n_s + n_s - (g_s - 1)\ln(g_s - 1) + (g_s - 1)].\end{aligned} \tag{2.93}$$

2.4.2 Distinguishable Particles. We now concentrate our attention on case (a) for distinguishable particles. We consider variations in $\ln P$ brought about by variations in n_s for fixed g_s. With $\ln P$ given by Eq. (2.91), we obtain

$$\delta \ln P \simeq 0 + \sum_{s=1}^{\infty} [(\delta n_s) \ln g_s - (\delta n_s) \ln n_s - n_s(\delta n_s/n_s) + \delta n_s]. \tag{2.94}$$

Thus

$$\delta \ln P \simeq \sum_{s=1}^{\infty} (\ln g_s - \ln n_s)\, \delta n_s. \tag{2.95}$$

In addition,

$$\alpha\, \delta N = \alpha\, \delta \sum_{s=1}^{\infty} n_s = \alpha \sum_{s=1}^{\infty} \delta n_s, \tag{2.96}$$

$$\beta\, \delta \mathscr{E}_{\mathrm{T}} = \beta\, \delta \sum_{s=1}^{\infty} \mathscr{E}_s n_s = \beta \sum_{s=1}^{\infty} \mathscr{E}_s\, \delta n_s. \tag{2.97}$$

Equation (2.86) thus becomes

$$0 = \delta \ln P - \alpha\, \delta N - \beta\, \delta \mathscr{E}_{\mathrm{T}} = \sum_{s=1}^{\infty} (\ln g_s - \ln n_s - \alpha - \beta \mathscr{E}_s)\, \delta n_s. \tag{2.98}$$

The two auxiliary conditions (2.64) and (2.65) may be considered to constrain only two of the parameters δn_s in any variation, so this allows all except two values of the δn_s to be chosen arbitrarily. Suppose, to be specific, that δn_1 and δn_2 are the two dependent quantities and that all the others are independent. We then see that if we assign values to α and β such that the coefficients of δn_1 and δn_2 vanish, then the coefficient of every δn_s must vanish, considering the independence of the remaining quantities δn_s. The introduction of the undetermined multipliers has the effect of replacing conditions (2.64) and (2.65) by the multipliers, thus allowing each δn_s to be treated as independent so that its coefficient can be set equal to zero. That is,

$$a_1\, \delta n_1 + a_2\, \delta n_2 + \cdots + a_r\, \delta n_r = 0 \tag{2.99}$$

implies that each $a_1, a_2, \ldots, a_r$ must vanish if the δn_s are all independent.

Thus, from Eq. (2.98) the maximum value of ln P is given by the distribution $n_s(\mathscr{E}_s)$, α, and β which satisfies the relations

$$\ln g_s - \ln n_s - \alpha - \beta\mathscr{E}_s = 0 \tag{2.100}$$

for all values of s. This gives $\ln n_s = \ln g_s - \alpha - \beta\mathscr{E}_s$, or equivalently,

$$n_s = g_s \exp(-\alpha - \beta\mathscr{E}_s) = g_s/[\exp(\alpha + \beta\mathscr{E}_s)]$$

(distinguishable-particle distribution function). (2.101)

We note from Eq. (2.101) that the number of particles in the sth cell is directly proportional to g_s, the number of energy states in that cell. In general, g_s will be a function of the system in question. However, the direct proportionality between n_s and g_s does allow us to interpret the ratio n_s/g_s as an occupation probability which is a function of the energy but which is independent of the detailed properties of the system. Furthermore we can consider

$$p(\mathscr{E}_s) = n_s/g_s \tag{2.102}$$

to be a quasi-continuous function of the energy $\mathscr{E}_s$. Employing the result (2.101), we thus obtain for distinguishable particles,

$$p(\mathscr{E}) = \exp(-\alpha - \beta\mathscr{E}). \tag{2.103}$$

The occupation probability $p(\mathscr{E})$ is simply the average number of particles per energy level at the energy $\mathscr{E}$.

Since in the classical Hamiltonian we consider particles to be spatially separated and hence distinguishable, Eq. (2.103) is most applicable for the "classical" limit and should therefore be in accord with the predictions of classical statistical mechanics. This theoretical result for the case of distinguishable particles can thus be compared with the experimentally observed variation of the classical occupation probability with energy, namely

$$p(\mathscr{E}) \propto \exp(-\mathscr{E}/k_{\mathrm{B}}T). \tag{2.104}$$

This latter expression is the Boltzmann occupation probability, which constitutes the basis of a Boltzmann distribution for occupied energy levels. A comparison of Eq. (2.104) with Eq. (2.103) serves to evaluate one of the undetermined multipliers, thus yielding

$$\beta = 1/k_{\mathrm{B}}T. \tag{2.105}$$

Instead of invoking Eq. (2.104) directly, however, there is an alternative way to evaluate β. This is to employ the auxiliary relations (2.64) and (2.65) directly for the system under consideration, with n_s given by Eq. (2.101). We can convert the sums in Eqs. (2.64) and (2.65) to integrals since n_s can be considered to be a quasi-continuous function of energy. That is, if $n(\mathscr{E})$ is the number of particles per unit energy range at energy $\mathscr{E}$, and $\tilde{g}(\mathscr{E})$ is the number of energy levels per unit energy range at energy $\mathscr{E}$, then the continuum forms of Eqs. (2.102), (2.64), and

(2.65) are respectively

$$n(\mathscr{E}) = \tilde{g}(\mathscr{E})p(\mathscr{E}), \tag{2.106}$$

$$N = \int_0^\infty n(\mathscr{E})\, d\mathscr{E} = \int_0^\infty \tilde{g}(\mathscr{E})p(\mathscr{E})\, d\mathscr{E}, \tag{2.107}$$

$$\mathscr{E}_{\mathrm{T}} = \int_0^\infty \mathscr{E} n(\mathscr{E})\, d\mathscr{E} = \int_0^\infty \mathscr{E}\tilde{g}(\mathscr{E})p(\mathscr{E})\, d\mathscr{E}. \tag{2.108}$$

Using Eq. (2.103) for $p(\mathscr{E})$ then gives for the latter two relations the expressions

$$N = \int_0^\infty \tilde{g}(\mathscr{E})e^{(-\alpha-\beta\mathscr{E})}\, d\mathscr{E} = e^{-\alpha}\int_0^\infty \tilde{g}(\mathscr{E})e^{-\beta\mathscr{E}}\, d\mathscr{E}, \tag{2.109}$$

$$\mathscr{E}_{\mathrm{T}} = \int_0^\infty \mathscr{E}\, \tilde{g}(\mathscr{E})e^{(-\alpha-\beta\mathscr{E})}\, d\mathscr{E} = e^{-\alpha}\int_0^\infty \mathscr{E}\tilde{g}(\mathscr{E})e^{-\beta\mathscr{E}}\, d\mathscr{E}. \tag{2.110}$$

Taking the ratio of the second to the first yields the average energy per particle of the system,

$$\mathscr{E}_{\mathrm{T}}/N = \int_0^\infty \mathscr{E}\tilde{g}(\mathscr{E})e^{-\beta\mathscr{E}}\, d\mathscr{E} \bigg/ \int_0^\infty \tilde{g}(\mathscr{E})e^{-\beta\mathscr{E}}\, d\mathscr{E}. \tag{2.111}$$

We see that this involves only β, and is independent of α. Equation (2.111) provides us with a general expression for evaluating β. We expect from thermodynamics that the average energy per particle of a system in thermodynamic equilibrium will be determined by the temperature of the system; hence we expect β to be related in some manner to the temperature.

Consider now the special case of a system with a uniform density of states as a function of energy. [This particular functional form for the density of states is found, for example, in a two-dimensional treatment of electrons in the free-electron model (see Project 3.2). The derivation is quite analogous to the one-dimensional treatment presented in Chap. 1, § 10.] With $\tilde{g}(\mathscr{E})$ independent of $\mathscr{E}$, we can denote this constant quantity by $\tilde{g}_0$. The above relation of $\mathscr{E}_{\mathrm{T}}/N$ then is readily evaluated,

$$\mathscr{E}_{\mathrm{T}}/N = 1/\beta. \tag{2.112}$$

We know from classical thermodynamics, however, that the average energy per particle for such a system is $k_{\mathrm{B}}T$. (This follows from the fact that a two-dimensional system has two degrees of freedom, each degree of freedom having an energy $\frac{1}{2}k_{\mathrm{B}}T$.) We thus deduce that $\beta = 1/k_{\mathrm{B}}T$, in agreement with our previous conclusion.

Using $\beta = 1/k_{\mathrm{B}}T$ in the above expression (2.109) for N yields an evaluation for α,

$$\alpha = -\ln N + \ln \int_0^\infty \tilde{g}(\mathscr{E})e^{-\mathscr{E}/k_{\mathrm{B}}T}\, d\mathscr{E}. \tag{2.113}$$

We see from this relation that α depends upon the number of particles N, the temperature T, and the properties of the system through the density of states $\tilde{g}(\mathscr{E})$. Thus it is noted that α does not have such a universal character as the multiplier β; nevertheless, it can be evaluated for any given system and hence our statistical function (2.103) is completely determined.

For the special case of a uniform density of states, $\tilde{g}(\mathscr{E}) = \tilde{g}_0$, then α is readily determined from Eq. (2.113) to have the value $\alpha = \ln(\tilde{g}_0 k_B T/N)$. In this case, we find the following energy distribution for the system of distinguishable particles,

$$n(\mathscr{E}) = (N/k_B T)\exp(-\mathscr{E}/k_B T). \tag{2.114}$$

EXERCISE Apply the above results to the special case of a one-dimensional system. [*Hint*: In Chap. 1, §10, the expression for the one-dimensional density of states is derived; from Eq. (1.296), $\tilde{g}(\mathscr{E}) = (2m/\hbar^2)^{1/2}(L/2\pi)\mathscr{E}^{-1/2}$. Substituting this expression into Eq. (2.111) enables one to evaluate $\mathscr{E}_T/N$ readily by a parts integration of either the numerator or the denominator. The result is $(2\beta)^{-1}$. The average energy per particle for a one-dimensional classical system is $\frac{1}{2}k_B T$. Equating, it is found that $\beta = (k_B T)^{-1}$. Equation (2.113) can be used to evaluate α for this hypothetical case. Using the fact that the definite integral $\int_0^\infty \eta^{-1/2}e^{-\eta}\,d\eta$ has the value $\pi^{1/2}$, one finds that $\alpha = \ln[(k_B Tm/2\pi\hbar^2)^{1/2}L/N]$, or equivalently, $\exp(-\alpha) = (N/L)(2\pi\hbar^2/k_B Tm)^{1/2}$. The factor N/L is simply the number of particles per unit length for the one-dimensional system. The point illustrated by this example is the fact that the evaluation of β for this particular system again gives $(k_B T)^{-1}$, and α is again found to depend upon the number of particles in the system, as well as upon $k_B T$ and the system parameters m and L. It is shown later in this chapter that use of the presently deduced distribution function (2.101) is valid only in the restricted domain of high temperatures and low particle densities.]

2.4.3 Fermions. Let us now derive the corresponding statistical function for the case of Fermi particles. Using Eq. (2.92) for $\ln P$, we obtain

$$\begin{aligned}\delta \ln P &\simeq \sum_{s=1}^{\infty} \{0 + 0 - \ln n_s - n_s(1/n_s) + 1 \\ &\quad - (g_s - n_s)[-1/(g_s - n_s)] - (-1)\ln(g_s - n_s) - 1\}\,\delta n_s \\ &\simeq \sum_{s=1}^{\infty} \ln[(g_s - n_s)/n_s]\,\delta n_s. \end{aligned}\tag{2.115}$$

Substituting this result together with Eqs. (2.96) and (2.97) into Eq. (2.86) gives

$$0 = \delta \ln P - \alpha\,\delta N - \beta\,\delta\mathscr{E}_T = \sum_{s=1}^{\infty} \{\ln[(g_s - n_s)/n_s] - \alpha - \beta\mathscr{E}_s\}\,\delta n_s. \tag{2.116}$$

Following the same argument relating to the undetermined multipliers used previously, we set each coefficient of δn_s equal to zero. This gives

$$\ln[(g_s - n_s)/n_s] = \alpha + \beta\mathscr{E}_s, \tag{2.117}$$

or equivalently,

$$(g_s - n_s)/n_s = \exp(\alpha + \beta\mathscr{E}_s). \tag{2.118}$$

Solving for the ratio g_s/n_s, we obtain

$$g_s/n_s = 1 + \exp(\alpha + \beta\mathscr{E}_s), \tag{2.119}$$

so that

$$p(\mathscr{E}) = n_s/g_s = [\exp(\alpha + \beta\mathscr{E}) + 1]^{-1}. \qquad (2.120)$$

Again α and β must be determined either from experiment or else from the conditions (2.64) and (2.65) requiring conservation of the number of particles and conservation of the energy of the system. Before evaluating α and β in Eq. (2.120), let us derive the analogous intermediate result for Bose particles.

2.4.4 Bosons. Using Eq. (2.93) for $\ln P$, we obtain

$$\delta \ln P = \sum_{s=1}^{\infty} \{(n_s + g_s - 1)(n_s + g_s - 1)^{-1} + \ln(n_s + g_s - 1)$$
$$- 1 - n_s(1/n_s) - \ln n_s + 1\} \, \delta n_s$$
$$= \sum_{s=1}^{\infty} \ln[(n_s + g_s - 1)/n_s] \, \delta n_s. \qquad (2.121)$$

Substituting this result together with Eqs. (2.96) and (2.97) into Eq. (2.86) gives

$$0 = \delta \ln P - \alpha \, \delta N - \beta \, \delta\mathscr{E}_{\mathrm{T}} = \sum_{s=0}^{\infty} \{\ln[(n_s + g_s - 1)/n_s] - \alpha - \beta\mathscr{E}_s\} \, \delta n_s. \qquad (2.122)$$

Setting each coefficient equal to zero in accordance with the same argument used previously, we obtain

$$\ln[(n_s + g_s - 1)/n_s] = \alpha + \beta\mathscr{E}_s, \qquad (2.123)$$

or equivalently

$$(n_s + g_s - 1)/n_s = \exp(\alpha + \beta\mathscr{E}_s). \qquad (2.124)$$

Since unity can be neglected with respect to g_s, we obtain

$$g_s/n_s \simeq \exp(\alpha + \beta\mathscr{E}_s) - 1. \qquad (2.125)$$

Thus for Bose particles

$$p(\mathscr{E}) = n_s/g_s \simeq [\exp(\alpha + \beta\mathscr{E}) - 1]^{-1}. \qquad (2.126)$$

2.4.5 Determination of the Lagrange Multipliers for the Fermion and Boson Cases. We can summarize our results for the three types of particle as follows:

$$p(\mathscr{E}) = [\exp(\alpha_X + \beta_X\mathscr{E}) + \delta_X]^{-1}, \qquad (2.127)$$

where α_X and β_X are the coefficients for the particle type in question and δ_X is a constant which is zero for distinguishable particles, $+1$ for Fermi particles and -1 for Bose particles. We thus suspect that α and β have much the same meaning for the two indistinguishable particle systems as they have for the Boltzmann system. We determined for the Boltzmann system that $\beta = 1/k_{\mathrm{B}}T$, and subsequently deduced α by the auxiliary condition equivalent to Eq. (2.64).

Indeed, β can be shown to have a universal character associated with the thermodynamic property of temperature, independent of the particular particle type being considered. To show this, consider a system which consists of a mixture of two different kinds of identical particles of the above types, either distinguishable particles plus Fermi particles or distinguishable particles plus Bose particles. In either case we divide the energy range into cells $\Delta\mathscr{E}_s$ for the first kind of particle (as distinguished by the variable index s) and into cells $\Delta\mathscr{E}_l$ for the second kind of particle (as distinguished by the variable index l), and then maximize the joint relative probability

$$P(n_s, n_l) = P(n_s)P(n_l) \tag{2.128}$$

subject to the auxiliary conditions

$$\sum_{s=1}^{\infty} n_s = N_1, \tag{2.129}$$

$$\sum_{l=1}^{\infty} n_l = N_2, \tag{2.130}$$

$$\sum_{s=1}^{\infty} n_s\mathscr{E}_s + \sum_{l=1}^{\infty} n_l\mathscr{E}_l = \mathscr{E}_T. \tag{2.131}$$

The quantities $P(n_s)$ and $P(n_l)$ are the number of distinct microscopic distributions giving rise to the macroscopic distributions $n_s(\mathscr{E}_s)$ and $n_l(\mathscr{E}_l)$ for the two kinds of particle, N_1 and N_2 give the number of each of the two kinds of particle present in the mixture, and $\mathscr{E}_T$ is the total energy of the system. An application of the method of undetermined multipliers to this new problem requires three undetermined multipliers, which can be called α_1, α_2, and β. This procedure leads to distributions for the two kinds of particle that are similar in form to those obtained previously, where each distribution involves a different α but the same parameter β. Such an analysis shows that β has a universal character independent of the nature of the particles or the system, being $1/k_B T$ for all three systems.

We proceed to carry out the derivation just outlined for the specific case of a mixture of classical particles and Fermi particles. A consideration of variations in the cell populations and cell energies subject to the constraints imposed by Eqs. (2.129), (2.130), and (2.131) leads to the following relations:

$$0 = \alpha_1 \, \delta N_1 = \sum_{s=1}^{\infty} \alpha_1 \, \delta n_s, \tag{2.132}$$

$$0 = \alpha_2 \, \delta N_2 = \sum_{l=1}^{\infty} \alpha_2 \, \delta n_l, \tag{2.133}$$

$$0 = \beta \, \delta\mathscr{E}_T = \sum_{s=1}^{\infty} \beta\mathscr{E}_s \, \delta n_s + \sum_{l=1}^{\infty} \beta\mathscr{E}_l \, \delta n_l. \tag{2.134}$$

The variation in the joint relative probability (2.128) must be zero in order that it be a maximum,

$$0 = \delta P(n_s, n_l) = P(n_s)[\delta P(n_l)] + [\delta P(n_s)]P(n_l). \tag{2.135}$$

Since $P(n_s)$ and $P(n_l)$ are independent, then each independently must have zero variation in order that Eq. (2.135) be satisfied,

$$\delta P(n_l) = 0, \tag{2.136}$$

$$\delta P(n_s) = 0. \tag{2.137}$$

If we consider s to denote the classical particles and l to denote the Fermi particles, then we have from Eqs. (2.95) and (2.115),

$$\delta \ln P(n_s) \simeq \sum_{s=1}^{\infty} (\ln g_s - \ln n_s)\, \delta n_s, \tag{2.138}$$

$$\delta \ln P(n_l) \simeq \sum_{l=1}^{\infty} \ln[(g_l - n_l)/n_l]\, \delta n_l. \tag{2.139}$$

The relations (2.132)–(2.139) can be combined to give

$$\begin{aligned} 0 &= -\beta\, \delta\mathscr{E}_\mathrm{T} + \delta \ln P(n_s) + \delta \ln P(n_l) - \alpha_1\, \delta N_1 - \alpha_2\, \delta N_2 \\ &= \sum_{s=1}^{\infty} (\ln g_s - \ln n_s - \alpha_1 - \beta\mathscr{E}_s)\, \delta n_s \\ &\quad + \sum_{l=1}^{\infty} \left[\ln\left(\frac{g_l - n_l}{n_l}\right) - \alpha_2 - \beta\mathscr{E}_l\right] \delta n_l = 0. \end{aligned} \tag{2.140}$$

Now the set $\{n_s\}$ is independent of the set $\{n_l\}$. Hence the coefficients of δn_s must be zero and the coefficients of δn_l must be zero. Setting the coefficients of δn_s equal to zero gives

$$\ln g_s - \ln n_s - \alpha_1 - \beta\mathscr{E}_s = 0. \tag{2.141}$$

This leads to

$$g_s/n_s = \exp(\alpha_1 + \beta\mathscr{E}_s), \tag{2.142}$$

or equivalently,

$$p(\mathscr{E}_s) = n_s/g_s = \exp(-\alpha_1 - \beta\mathscr{E}_s). \tag{2.143}$$

This result for the set of distinguishable particles is identical to the result (2.103) previously derived. Furthermore, $\beta = 1/k_\mathrm{B}T$ for this set of distinguishable particles according to our previous derivation.

Setting the coefficients of δn_l in Eq. (2.140) equal to zero gives

$$\ln[(g_l - n_l)/n_l] - \alpha_2 - \beta\mathscr{E}_l = 0. \tag{2.144}$$

This leads to

$$(g_l - n_l)/n_l = \exp(\alpha_2 + \beta\mathscr{E}_l), \tag{2.145}$$

or equivalently,

$$g_l/n_l = 1 + \exp(\alpha_2 + \beta \mathscr{E}_l). \tag{2.146}$$

Thus

$$p(\mathscr{E}_l) = n_l/g_l = 1/[1 + \exp(\alpha_2 + \beta \mathscr{E}_l)] = 1/[\exp(\alpha_2 + \beta \mathscr{E}_l) + 1]. \tag{2.147}$$

Since β is the same quantity in Eq. (2.147) as in Eq. (2.143), it has the value $1/k_B T$. A comparison of Eq. (2.147) with our previous result (2.126) shows them to have the identical functional form, so we are forced to conclude that β has the same value for all types of particle. Thus Eq. (2.127) can be written

$$p(\mathscr{E}) = \{\exp[\alpha_X + (\mathscr{E}/k_B T)] + \delta_X\}^{-1}, \tag{2.148}$$

where α_X depends on the system in question and δ_X has the value $0, + 1,$ and $- 1$, respectively, for classical, Fermi, and Bose particles. It is now convenient to replace the constant α_X by an equivalent constant which we call $- \mathscr{E}_X/k_B T$, thus converting Eq. (2.148) to the form

$$p(\mathscr{E}) = \{\exp[(\mathscr{E} - \mathscr{E}_X)/k_B T] + \delta_X\}^{-1}. \tag{2.149}$$

This change of variable for the constant has no physical consequences because the constant must be evaluated for any system by employing the relation (2.64), or equivalently Eq. (2.107), for the conservation of the total number of particles in the system.

EXERCISE   Considering a mixture of α particles (spin 0), neutrons (spin $\frac{1}{2}$), and classions (fictitious "classical" particles), carry out a derivation analogous to that outlined in Eqs. (2.128)–(2.147) for a system containing fermions and classions in order to show that the undetermined Lagrange multiplier β has the same universal character involving the thermodynamic temperature for all three types of particle.

2.4.6   *Summary and Correlation of Results.*   To distinguish between the distribution functions (2.149) and the appropriate constants for the three types of particle, it is convenient to introduce different notation for the occupation probability functions $p(\mathscr{E})$ in each of the three cases. We denote $p(\mathscr{E})$ for classical particles by $w(\mathscr{E})$, for Fermi particles by $\tilde{f}(\mathscr{E})$, and for Bose particles by $\mathscr{B}(\mathscr{E})$, and denote $\exp(- \alpha)$ for these cases by w_0, $\exp(- \mathscr{E}_F/k_B T)$ and $\exp(- \mathscr{E}_B/k_B T)$, respectively. We call $w(\mathscr{E})$ the Boltzmann occupation probability, $\tilde{f}(\mathscr{E})$ the Fermi–Dirac distribution function, and $\mathscr{B}(\mathscr{E})$ the Bose–Einstein distribution function, and designate the statistics denoted by each of these functions as *Maxwell–Boltzmann statistics, Fermi–Dirac statistics,* and *Bose–Einstein statistics.* Thus we have derived the following results.

(a)   Classical (distinguishable) particles occupy the energy levels of a system with a probability given by the Boltzmann occupation probability $w(\mathscr{E})$,

$$w(\mathscr{E}) = w_0 e^{-\mathscr{E}/k_B T} \quad \textbf{(Maxwell–Boltzmann statistics).} \tag{2.150}$$

A typical plot of this function is shown in Fig. 2.3.

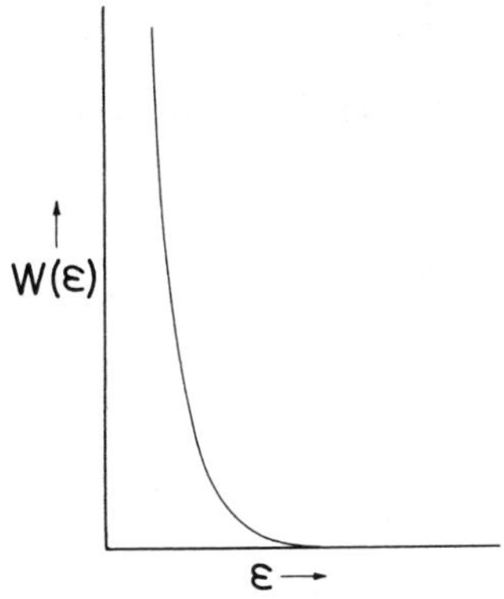

Fig. 2.3 Typical plot of the Boltzmann occupation probability $w(\mathscr{E})$.

(b) Nonclassical (indistinguishable) particles of half-integral spin, which obey the Pauli exclusion principle, occupy the energy levels of a system with a probability given by the Fermi–Dirac distribution function $\mathscr{f}(\mathscr{E})$,

$$\mathscr{f}(\mathscr{E}) = (e^{(\mathscr{E} - \mathscr{E}_F)/k_B T} + 1)^{-1} \qquad \textbf{(Fermi–Dirac statistics).} \qquad (2.151)$$

A typical plot of this function is shown in Fig. 2.4.

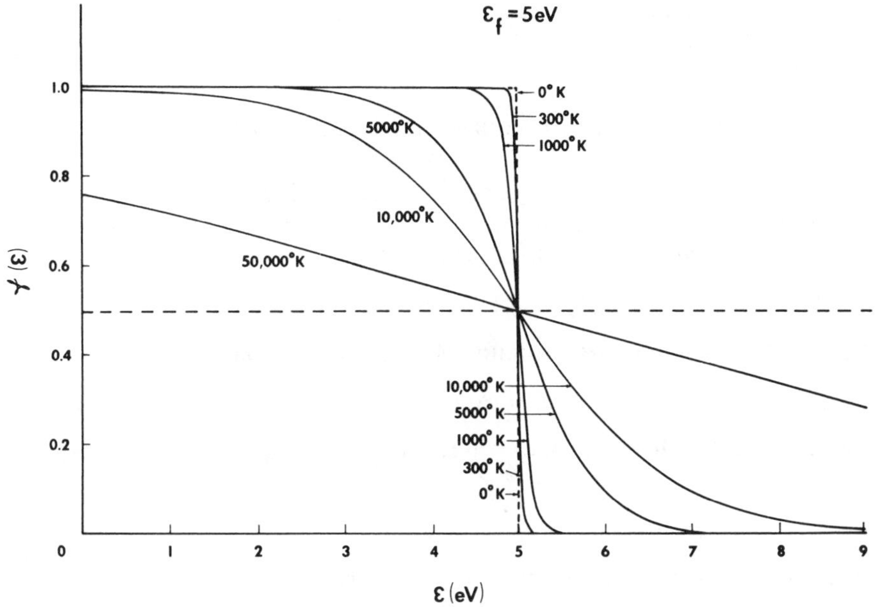

Fig. 2.4 Typical plots of the Fermi–Dirac distribution function $\mathscr{f}(\mathscr{E})$.

(c) Nonclassical (indistinguishable) particles of integral spin, which do not obey the Pauli exclusion principle, occupy the energy levels of a system with a probability given by the Bose–Einstein distribution function $\mathscr{B}(\mathscr{E})$,

$$\mathscr{B}(\mathscr{E}) = (e^{(\mathscr{E} - \mathscr{E}_B)/k_B T} - 1)^{-1} \qquad \textbf{(Bose–Einstein statistics).} \tag{2.152}$$

A typical plot of this function is shown in Fig. 2.5.

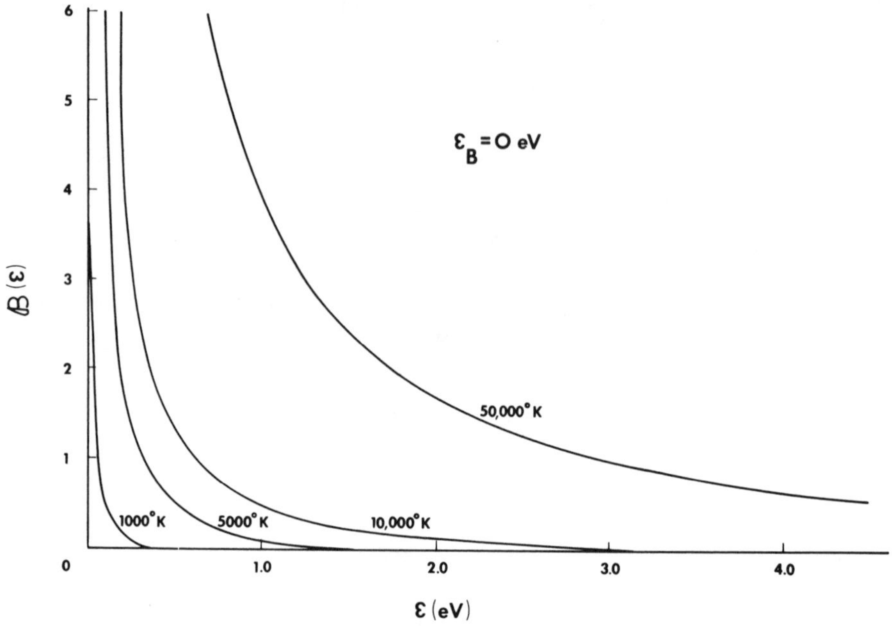

Fig. 2.5 Typical plots of the Bose–Einstein distribution function $\mathscr{B}(\mathscr{E})$.

It remains for us to show the conditions for which Fermi–Dirac and Bose–Einstein statistics can be approximated by Maxwell–Boltzmann statistics. If we compare the intermediate result (2.118) for Fermi particles

$$n_s/(g_s - n_s) = \exp(-\alpha - \beta\mathscr{E}_s) \tag{2.153}$$

with the corresponding intermediate result (2.101) for classical particles

$$n_s/g_s = \exp(-\alpha - \beta\mathscr{E}_s), \tag{2.154}$$

we see that Fermi–Dirac statistics will reduce to Maxwell–Boltzmann statistics whenever

$$n_s \ll g_s, \tag{2.155}$$

namely, whenever the energy levels at a given energy are very sparsely populated. The corresponding intermediate result (2.124) for Bose particles

$$n_s/(g_s + n_s - 1) = \exp(-\alpha - \beta\mathscr{E}_s) \tag{2.156}$$

is seen to reduce to the corresponding intermediate result (2.101) for classical particles under the same conditions. If we refer to Figs. 2.4 and 2.5 we see that the occupation probability falls to values much less than unity at higher values of the energy. In particular, for the Fermi–Dirac distribution function illustrated in Fig. 2.4, this requires energies greater than the Fermi energy $\mathscr{E}_F$ for the system in question. A glance at Eq. (2.151) shows that for

$$\exp[(\mathscr{E} - \mathscr{E}_F)/k_B T] \gg 1, \tag{2.157}$$

which is met whenever

$$(\mathscr{E} - \mathscr{E}_F)/k_B T \gg 1, \tag{2.158}$$

the Fermi–Dirac distribution function reduces to the functional form (2.150) for the Boltzmann occupation probability. Likewise, Eq. (2.152) reduces to the Boltzmann functional form (2.150) whenever

$$(\mathscr{E} - \mathscr{E}_B)/k_B T \gg 1. \tag{2.159}$$

Our conclusion, then, is that indistinguishable particles appear to obey classical statistics whenever the energy levels of the system are sparsely populated, and this depends on the energy level spectrum for the system in question as well as the temperature.

In the following chapter we examine in detail the physical consequences of quantum statistics (so painstakingly derived above!) for a system of electrons. This is done within the framework of the *three-dimensional free-electron model for metals*.

PROJECT 2.3 Maxwell–Boltzmann, Fermi–Dirac, and Bose–Einstein Statistics

Using a programmable calculator and employing reasonable values for the fixed parameters, compute and plot a series of curves (with temperature as a parametric variable) for the following statistical occupation probabilities as a function of energy:

(a) Maxwell–Boltzmann Statistics: $w(\mathscr{E})$ vs $\mathscr{E}$. [*Hint*: See Eq. (2.150).]
(b) Fermi–Dirac Statistics: $\mathcal{J}(\mathscr{E})$ vs $\mathscr{E}$. [*Hint*: See Eq. (2.151).]
(c) Bose–Einstein Statistics: $\mathscr{B}(\mathscr{E})$ vs $\mathscr{E}$. [*Hint*: See Eq. (2.152).]

PROBLEMS

1. The difference between distinguishable identical particles and indistinguishable identical particles lies in the degree of wave function overlap. Estimate the relative overlap at the midpoint between two Gaussian wave packets separated by distances of 1 Å, 10 Å, 100 Å, 1 cm, 1 m, and 1 km, choosing the width of the Gaussian packet as 5 Å. [See Eq. (1.203) in Chap. 1, §7.] Express the relative overlap in terms of probability amplitude at the midpoint relative to probability amplitude at the peak.

2. Three electrons are confined to a one-dimensional box 3 Å in length. Considering the particles to be noninteracting, compute the ground-state energy of the system.

3. For a system of three noninteracting neutrons in a one-dimensional square-well potential with infinitely high potential energy barriers, write the normalized wave function and compute the expectation value of the energy. (Carry out this problem separately for single-particle wave functions satisfying fixed and periodic boundary conditions.) Next, compute the expectation values of the momentum operator in both cases and interpret your results from the viewpoint of classical mechanics.

4. Compute the expectation values $\langle(\mathbf{r}_2 - \mathbf{r}_1)^2\rangle$, $\langle(\mathbf{r}_3 - \mathbf{r}_1)^2\rangle$, and $\langle(\mathbf{r}_3 - \mathbf{r}_2)^2\rangle$ for a three-particle quantum system for the cases of distinguishable particles, Fermi particles, and Bose particles.

5. Outline the *basic logic* underlying the development of quantum statistics. [Begin with the reasons underlying the need for consideration of wave functions for many-particle systems. Describe the criterion (or postulate) used for finding the most probable distribution.] List all major considerations in the development.

6. In the classical Hamiltonian the vectors $\mathbf{r}_1$ and $\mathbf{r}_2$ represent the specific locations of point masses with reference to some given coordinate system. These vectors are of course time-dependent: $\mathbf{r}_1 = \mathbf{r}_1(t), \mathbf{r}_2 = \mathbf{r}_2(t)$. At some given instant of time t_0 these vectors have the values $\mathbf{r}_1(t_0)$, $\mathbf{r}_2(t_0)$. Quantum mechanically, we have only a probability-density description, so usually we consider only expectation values $\langle \mathbf{r}_1(t) \rangle$, $\langle \mathbf{r}_2(t) \rangle$, which at time t_0 must be $\langle \mathbf{r}_1(t_0) \rangle$, $\langle \mathbf{r}_2(t_0) \rangle$. These quantities give the most likely positions of the point masses. It thus appears that $\langle \mathbf{r}_1(t_0) \rangle$ and $\langle \mathbf{r}_2(t_0) \rangle$ are in many respects the nearest quantum-mechanical analogs of the classical quantities $\mathbf{r}_1(t_0)$ and $\mathbf{r}_2(t_0)$. Since t_0 is arbitrary, we further can say that $\langle \mathbf{r}_1(t) \rangle$ and $\langle \mathbf{r}_2(t) \rangle$ are the quantum-mechanical analogs of the classical quantities $\mathbf{r}_1(t)$ and $\mathbf{r}_2(t)$. (a) What, then, are the *classical* analogs of the two quantum-mechanical operators $\mathbf{r}_1$ and $\mathbf{r}_2$? (b) The expectation values of $\mathbf{r}_1$ and $\mathbf{r}_2$ are obtained by permitting $\mathbf{r}_1$ and $\mathbf{r}_2$ to take on all possible values and using the probability density $\psi^*\psi$ as a weighting factor to obtain an average value. In this sense, $\mathbf{r}_1$ and $\mathbf{r}_2$ are only dummy variables in the respective integrals for the expectation values. How can this be compatible with the model used in setting up the classical Hamiltonian of two point particles located at specific positions $\mathbf{r}_1$ and $\mathbf{r}_2$? (c) In view of (a) and (b) above, can one reasonably justify the usual operation of generating a quantum-mechanical Hamiltonian operator $\mathscr{H}$ from the classical Hamiltonian H?

7. (a) Prove that a product of N single-particle eigenfunctions satisfies the Schrödinger equation for an N-particle system of noninteracting particles. (b) Is this product wave function, then, a valid wave function for all such N-particle systems? Explain the reasons for your answer. (c) Suppose in part (a) that the N particles occupy energy levels in a physical system which has a total of M energy levels, where $N < M \leqslant \infty$. How do you select the N single-particle eigenfunctions from the M possible ones? What does this tell you with regard to forming the most general wave function for a system of indistinguishable particles?

8. (a) Slater determinants are useful for describing the wave functions for systems of particles because of their antisymmetry and because they automatically satisfy the condition imposed by the Pauli exclusion principle. Prove that the permutation of the coordinates of any two particles i and j in the system gives an overall change in the sign of the Slater determinant. (Do not invoke any theorems from the theory of determinants.) (b) Prove that the Slater determinant is zero if any two particles i and j in the system are in the same single-particle eigenstate. (Again do not invoke any theorems from the theory of determinants.)

9. (a) Give a fairly detailed proof (which can be found outlined in most elementary treatments of the theory of determinants) that an interchange of any two columns or any two rows in the determinant gives an overall change in the sign of the determinant. (b) Prove that a determinant is zero if any two columns or any two rows are identical. (c) Show that the conclusion reached in Part (b) is no longer valid when the negative signs conventionally introduced in expanding the determinant are made positive. (d) State the implications of Part (a) for wave functions of many-particle systems. (e) State the implications of Parts (b) and (c) for wave functions of many-particle systems.

10. (a) Construct and plot several of the lowest energy two-particle antisymmetrical wave functions for a two-dimensional square-well potential. The dimensions of the square well can be considered to be L_x and L_y in the two directions, and the walls can be considered to be infinitely high. (b) Compute $\langle (\mathbf{r}_2 - \mathbf{r}_1)^2 \rangle$ for these wave functions.

11. The quantity $g_s^{n_s}$ gives the total number of distinct distributions of n_s distinguishable particles among g_s levels of the sth cell. Of course, this number would be too large (perhaps by $n_s!$) if the particles were indistinguishable. Attempt to construct P for bosons using this alternate counting approach. Compare your result with Eq. (2.78) and point out the reasons for any discrepancy in the two results. (Clearly the two results must agree if both counting procedures are correct; if the results do not agree, it stands to reason that one of the counting procedures must be faulty.)

12. Attempt to use the counting method for the number of microscopic distributions of Bose particles for a given energy level configuration for the case of distinguishable particles. If you run into difficulty, point out explicitly the nature of the difficulty and where one must modify the procedure. If you do not run into difficulty, do you get the same result as obtained for the distinguishable-particle method? (Need one add that if you do not obtain the same result, then obviously there is an error! Such an error could be in logic, in procedure, or in the mathematics. What can you conclude?)

13. Try to modify the counting method used for Fermi particles so that it is applicable to Bose particles.

14. Try to modify the counting method used for Bose particles so that it is applicable to Fermi particles.

15. Consider a mixture of fermions, bosons, and distinguishable particles. Carry through the evaluation of the Lagrange multiplier β_X (assuming $\beta = 1/k_B T$ for distinguishable particles) for the fermions and the bosons.

16. Electrons obey Fermi–Dirac statistics, with the occupation probability of an energy level at energy $\mathscr{E}$ having a temperature dependence given by the Fermi–Dirac distribution function $\mathscr{F}(\mathscr{E})$, $\mathscr{F}(\mathscr{E}) = [\exp(\mathscr{E} - \mathscr{E}_F)/k_B T + 1]^{-1}$. Temperature is denoted by T, Boltzmann's constant is denoted by k_B, and $\mathscr{E}_F$ is a constant which has units of energy and a value which depends upon electron density. Consider 10^{22} fermions distributed over energy levels in a system with a uniform density of states $\tilde{g}(\mathscr{E}) = \tilde{g}_0 = 4 \times 10^{23}$ energy levels/eV. Evaluate the Fermi energy $\mathscr{E}_F$. Compute a parametric set of curves for $\mathscr{F}(\mathscr{E})$ versus $\mathscr{E}$ for temperatures of 2000, 1000, 500, 300, 150, 75, 20, 4, and 0°K. Your computed points should be sufficiently close together to be able to see the functional form of the curve without drawing a line through the points, and should extend over energies from zero to $2\mathscr{E}_F$ for each curve.

17. Repeat Problem 16 if the particles are bosons instead of fermions. Employ the form $\mathscr{B}(\mathscr{E}) = [\exp(\mathscr{E} - \mathscr{E}_B)/k_B T - 1]^{-1}$ for the Bose–Einstein distribution function and evaluate $\mathscr{E}_B$.

18. Repeat Problem 16 if the particles are distinguishable particles instead of fermions or bosons.

19. Explain the fundamental importance of quantum statistics in physics, and tell how it underlies the entire realm of quantum device applications.

CHAPTER 3

FREE-ELECTRON MODEL AND THE BOLTZMANN EQUATION

> ...[*The*] *Einstein–Bohr frequency condition* (*which is valid in all cases*)... *represents such a complete departure from classical mechanics, or rather* (*using the viewpoint of wave theory*) *from the kinematics underlying this mechanics, that even for the simplest quantum–theoretical problems the validity of classical mechanics simply cannot be maintained. In this situation it seems sensible to discard all hope of observing hitherto unobservable quantities, such as the position and period of the electron,*... .
> W. Heisenberg (1925)

1 Free-Electron Gas in Three Dimensions

1.1 Conduction Electrons in a Metal

The most weakly bound electrons of the atoms constituting a metal move about freely through the entire volume of the metal. These electrons are the valence (or outer-shell) electrons in the free atoms; they become the conductors of electricity in the metal, and thus are called the *conduction* electrons. In the *free-electron model*, all calculations are performed as if the conduction electrons were free to move everywhere within the specimen. The total energy is the kinetic energy; the potential energy is neglected. The forces between the conduction electrons and the ion cores are neglected. Likewise the Coulomb forces between the conduction electrons themselves are neglected.

1.2 Electrical Forces in a Metal

The drastic approximations involved in the free-electron model relative to the true state of affairs in a metal become quite clear when order-of-magnitude estimates are made. For example, the neglect of the electron–electron interaction between conduction electrons separated by an atomic spacing of 4 Å is of the order of 3.6 eV. This is evidently a large effect when one considers that for an electron to have this amount of *kinetic* energy in free space it would have to be traveling at a speed of 1.1×10^6 m/sec, which is within a factor of 300 of the speed of light. (See Appendix for values of physical constants.) Similarly, the neglect of the Coulomb interaction energy of the conduction electrons with the positively charged ion cores is of the same order of approximation.

1.3 Atomistic Nature of a Metal

From a slightly different viewpoint, if one considers the ion cores to be of the nature of rigid spheres which exclude part of the volume of the metal from being accessible to the conduction electrons, the neglect of the ion cores is again seen to be quite a drastic approximation. Consider sodium atoms, for example, with the atomic configuration $1s^2 2s^2 2p^6 3s$ (cf. Table 1.3). The outermost shell which contains the valence electron is not closed. The atoms condense to form a metal. The atomic cores are Na^+ ions containing 10 electrons. In the simplest model these core electrons can be considered to have the configuration $1s^2 2s^2 2p^6$, essentially the same as in the free ion. The ions are immersed in a sea of conduction electrons which can be considered to be derived from the 3s valence electrons of the free atoms. In an *alkali* metal (Li, Na, K, Cs, Rb), the atomic cores typically occupy about 10% of the total volume of the crystal, as illustrated schematically in Fig. 3.1. For example, the radius of the free Na^+ ion is 0.98 Å, whereas one half of the nearest-neighbor distance in sodium metal is 1.85 Å. In a *noble* metal (Cu, Ag, Au), the atomic cores are relatively larger and may be in contact with one another. These two groups of metals are the simplest, exhibiting properties which are more nearly free-electron-like than other metals. Even in the alkali group, however, the neglect of 10% of the metal volume which

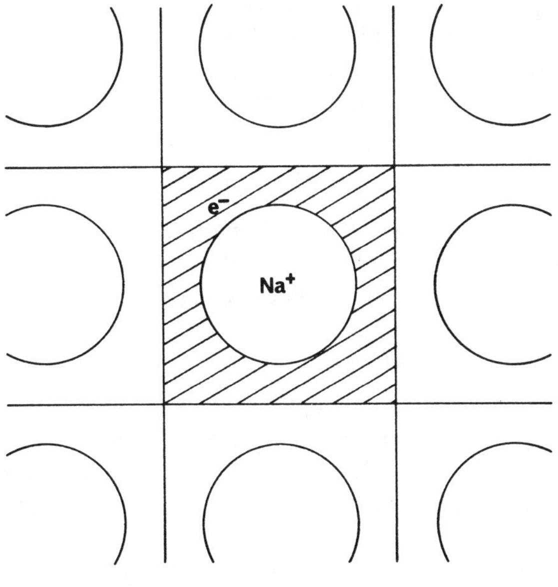

Fig. 3.1 Schematic of ion cores and intervening volume occupied by the conduction electrons in a metal such as sodium.

is inaccessible to the conduction electrons constitutes a drastic assumption. The free electron model, wherein the presence of the ion cores is entirely ignored, is therefore a very approximate model.

EXERCISE In sodium, the distance between nearest-neighbor atoms is 1.85 Å. Assuming an effective radius of 0.98 Å for the sodium ion Na^+, compute the percentage of the volume occupied by the ion cores and the percentage remaining for the conduction electrons. (*Hint*: See Chap. 6, §4.)

EXERCISE If the ion cores in metallic silver are in contact with one another but are still roughly equivalent to undeformed hard spheres, determine the fraction of the volume of the metal which can be considered as reserved for the ion cores and the fraction remaining which is available as essentially open space for motion of the conduction electrons.

1.4 Periodic Aspects of a Metal

The fact that the electrons are not actually free to travel right through the large closed-shell ions arranged in a periodic array (i.e., the lattice) in the solid causes marked departures from our deductions based on the purely free electron model outlined above. This interaction of the conduction electrons with the periodic lattice gives rise to *electron energy bands*, which we discuss first in Chap. 5 and characterize in detail in Chap. 7.

The actual spatial charge distribution of the conduction electrons in a metal does in fact reflect the strong electrostatic potential of the ion cores. The ions exert Coulomb forces on the conduction electrons, giving rise to a *periodic potential*, namely, a potential energy function which oscillates periodically in space from lattice site to lattice site. One approach which can be used to obtain some intuitive understanding of the effects to be expected is to treat the periodic potential as a perturbation on the otherwise free-electron motion. This is the approach taken in Chap. 5, §13. Sometimes this approach is referred to as the "nearly-free-electron model" [Sachs (1963)]. The purely-free-electron model, however, considers the periodic potential to have a very minimal effect on electronic motion; in fact, the choice

$$\mathscr{V}(\mathbf{r}) = \text{const} \tag{3.1}$$

is made, where the constant is for convenience taken to be zero inside the metal. The remarkable success of such a naive model is due primarily to the fact that it incorporates many of the essential quantum properties of the electrons.

1.5 Wavelike Behavior of Conduction Electrons

Because the electronic mass is small, electrons have pronounced wavelike properties, in contrast to particlelike properties normally associated with bulk masses. This follows from the de Broglie relation $\lambda = h/p = h/mv$, which shows that the de Broglie wavelength of an electron is greater than an atom spacing of 4 Å for speeds less than 1.8×10^6 m/sec. This means that a given electron cannot be considered to have a definitely known position in space; instead, it must be considered to have a *probability density* $\rho(\mathbf{r}) = \psi^*\psi$ in space determined by the electronic wave function ψ, as discussed in some detail in Chap 1, §6 [also see

Bloch (1976)]. This means that $\rho(\mathbf{r})$ gives the relative probability of finding the electron at position $\mathbf{r}$. Let us assume that the wave function ψ is appropriately normalized over the region accessible to the electron, so that

$$\int \rho(\mathbf{r})\, d\mathbf{r} = 1. \tag{3.2}$$

We thus can consider the distribution of conduction electrons to be "smeared out" in space (i.e., much as a continuous "sea" of negative charge density), instead of conceiving of the conduction electrons as a large group of point particles which zip rapidly from place to place. The quantum description of electrons is rather analogous to that appropriate for the classical problem of a swarm of insects for which we construct a function which gives the *probability* of finding an insect at a given position in the swarm. We note that the quantum description does not necessarily imply that each particle is spread out unduly (namely, over several atoms) throughout space, but merely that our knowledge is only that which is of the nature of a probability.

1.6 Successes and Failures of the Classical Mechanical Approach

Before the advent of quantum mechanics, there was formulated a classical free-electron theory. There is in this case no restriction on electron energies, which contrasts markedly with the discrete quantized energy levels which we deduce for the quantum free-electron model. In the classical free-electron model the continuum of allowable energies can be considered to represent a continuum of electronic states, and classically these states are considered to be populated or unpopulated with electrons in a statistical way in accordance with Boltzmann statistics (cf. Chap. 2). The primary successes of this classical approach were the prediction of Ohm's law and the prediction of the experimentally measured ratio of electrical to thermal conductivity [McKelvey (1966)]. Among the many failures of this classical approach were erroneous predictions for both the specific heat and the paramagnetic susceptibility contributions from the conduction electrons, and failure to explain extremely long electronic mean free paths which can be observed experimentally. As regards the latter, experiments show that conduction electrons in a metal can move freely in a straight path over many atomic distances, undeflected by collisions with other conduction electrons and undeflected by collisions with the atom cores. In a very pure specimen at low temperatures the mean free path may be as long as 10^8 or 10^9 interatomic spacings, thus exceeding a centimeter [Kittel (1971)]. This is vastly longer than we would predict if the collision probability were proportional to the relative cross-sectional area of the atomic cores in the metal. In this respect, we can say that the conduction electrons act much like a gas of noninteracting particles which are traveling through a very transparent medium. To look ahead a bit, there are actually two factors involved in the quantum-mechanical answer to the question of why solids are so transparent to conduction electrons. First, a conduction electron is not deflected by ion cores arranged in a perfectly periodic manner because matter waves can propagate freely in a periodic structure [Brillouin (1953)]. Second, a conduction electron is deflected only infrequently by other conduction electrons due to Fermi–Dirac statistics based on the Pauli exclusion principle which maintains that it is impossible for an electron to be scattered into a state which is already occupied by another electron. These two factors cause the number of possible scattering events to be reduced enormously.

We must conclude therefore that quantum effects are important for many solid state properties, and these quantum effects are responsible for the failure of the classical free-electron model. This is the reason we must be familiar with the quantum-mechanical free-electron theory. This theory consists of the properties of a *free-electron Fermi gas*, which is defined to be a gas of free and noninteracting electrons which are subject to the Pauli exclusion principle (see Chap. 2).

1.7 Three-Dimensional Potential Well Problem

Consider an electron of mass m confined to a macroscopic metal in the shape of a rectangular parallelepiped or a cube. Imagine the electrons therein to be contained due to infinite potential energy barriers at the faces of the parallelepiped. This is referred to as a "square-well potential" because the potential energy rises so sharply (namely, with infinite slope) at the boundaries of the parallelepiped. The stationary-state wave functions

$$\psi_{\mathbf{k}}(\mathbf{r}, t) = \phi_{\mathbf{k}}(\mathbf{r}) \exp[-(i/\hbar)\mathscr{E}_{\mathbf{k}}t] \tag{3.3}$$

of the electron must satisfy the time-independent Schrödinger equation

$$\mathscr{H}\phi = \mathscr{E}\phi. \tag{3.4}$$

With neglect of the potential energy, we have $\mathscr{H} = -(\hbar^2/2m)\nabla^2$, so that the free-particle Schrödinger equation in three dimensions is

$$-\frac{\hbar^2}{2m}\left(\frac{\partial^2}{\partial x^2} + \frac{\partial^2}{\partial y^2} + \frac{\partial^2}{\partial z^2}\right)\phi_{\mathbf{k}}(\mathbf{r}) = \mathscr{E}_{\mathbf{k}}\phi_{\mathbf{k}}(\mathbf{r}). \tag{3.5}$$

Considering the electrons to be confined by infinitely large potential energy barriers to a region of space delineated by a rectangular parallelepiped with edges of length L_x, L_y, and L_z in the x, y, and z directions, respectively, one form of the analog to the one-dimensional normalized standing wave function given by Eq. (1.271) in §10 of Chap. 1 is

$$\phi_{\mathbf{n}}(\mathbf{r}) = (8/V)^{1/2} \sin(n_x \pi x/L_x) \sin(n_y \pi y/L_y) \sin(n_z \pi z/L_z), \tag{3.6}$$

where n_x, n_y, and n_z are a triplet of positive integers represented by the symbol $\mathbf{n}$. Only the positive integers are chosen since the corresponding negative values yield the same wave function to within a factor of -1, which for all practical quantum-mechanical calculation purposes represents exactly the same state. (For example, the particle probability density $\psi^*\psi$ would be the same.) Linearly dependent eigenfunctions are therefore redundant, whereas linearly independent eigenfunctions (even if degenerate) are not redundant since they can represent different physical properties such as the probability density distribution. The above product of sine functions represents standing waves which by direct substitution can be shown to be perfectly good solutions to the three-dimensional Schrödinger equation, though such a solution does not represent a state having a definite momentum value. The product of sine functions given by Eq. (3.6) satisfies the fixed boundary conditions that the wave function vanishes on six faces of a rectangular parallelepiped located at $x = 0$, $y = 0$, $z = 0$, $x = L_x$, $y = L_y$, and $z = L_z$.

Suppose, however, that we choose the alternative of exponential solutions to the Schrödinger equation,

$$\phi_{\mathbf{k}}(\mathbf{r}) = (1/V)^{1/2} \exp[i\mathbf{k} \cdot \mathbf{r}]. \tag{3.7}$$

These spatial functions, when combined with the time factor $\exp[-(i/\hbar)\mathscr{E}_{\mathbf{k}}t]$, represent running waves. The **k** values can be chosen so that the functions satisfy the following boundary conditions (cf. Chap. 1, §3),

$$\begin{gathered} \phi(x + L_x, y, z) = \phi(x, y, z), \qquad \phi(x, y + L_y, z) = \phi(x, y, z), \\ \phi(x, y, z + L_z) = \phi(x, y, z), \end{gathered} \tag{3.8}$$

which are known as *periodic boundary conditions*. Direct substitution of Eq. (3.7) into Eq. (3.8) shows that these conditions are met if

$$\exp(ik_xL_x) = 1, \qquad \exp(ik_yL_y) = 1, \qquad \exp(ik_zL_z) = 1, \tag{3.9}$$

which in turn requires k_xL_x, k_yL_y, and k_zL_z to be integral multiples of 2π. Hence

$$\begin{aligned} k_x &= 2m_x\pi/L_x \qquad (m_x = \pm 1, \pm 2, \ldots), \\ k_y &= 2m_y\pi/L_y \qquad (m_y = \pm 1, \pm 2, \ldots), \\ k_z &= 2m_z\pi/L_z \qquad (m_z = \pm 1, \pm 2, \ldots). \end{aligned} \tag{3.10}$$

Both positive and negative integers are allowable, and moreover must be included, since the exponential function with a negative argument is linearly independent of the corresponding exponential function with a positive argument.

Substitution of the exponential functions given by Eq. (3.7) into the time-independent Schrödinger equation (3.5) yields

$$\mathscr{E}_{\mathbf{k}} = (\hbar^2/2m)(k_x^2 + k_y^2 + k_z^2) \equiv \hbar^2k^2/2m. \tag{3.11}$$

This represents a condition that the functions (3.7) be eigenfunctions of the Hamiltonian operator $\mathscr{H}$ with eigenvalues $\mathscr{E}_{\mathbf{k}}$. Therefore the energy eigenvalues $\mathscr{E}_{\mathbf{k}}$ are required by Eqs. (3.10) and (3.11) to be quantized,

$$\mathscr{E}_{\mathbf{m}} = (\hbar^2\pi^2/2m)[(2m_x/L_x)^2 + (2m_y/L_y)^2 + (2m_z/L_z)^2], \tag{3.12}$$

where **m** stands for the triplet of integers (m_x, m_y, m_z).

A corresponding substitution of the standing wave eigenfunctions (3.6) into the time-independent Schrödinger equation (3.5) yields a very similar result $\mathscr{E}_{\mathbf{n}}$ for the energy eigenvalues, except that the positive or negative integers $2m_x, 2m_y$, and $2m_z$ in Eq. (3.12) are replaced by the positive integers n_x, n_y, n_z. Although only half of the energy levels turn out to have the same corresponding values in the two cases, the number of levels contained in any finite energy interval which is broad relative to the spacing between levels is the same. That is, there is first of all a one-to-one correspondence between positive even integer quantum numbers in the two cases (viz., the even integers $2m_x$ and the even integer subset of the n_x). Then for each odd positive value of n_x, n_y, or n_z there is one corresponding negative even integer $2m_x$, $2m_y$, or $2m_z$ differing in magnitude by only one from the corresponding n_x value. The energy levels for the odd values of n_x, n_y, or n_z are interspersed between those for the even values of n_x, n_y, or n_z, whereas the energy levels for the negative values of $2m_x$, $2m_y$, or $2m_z$ are

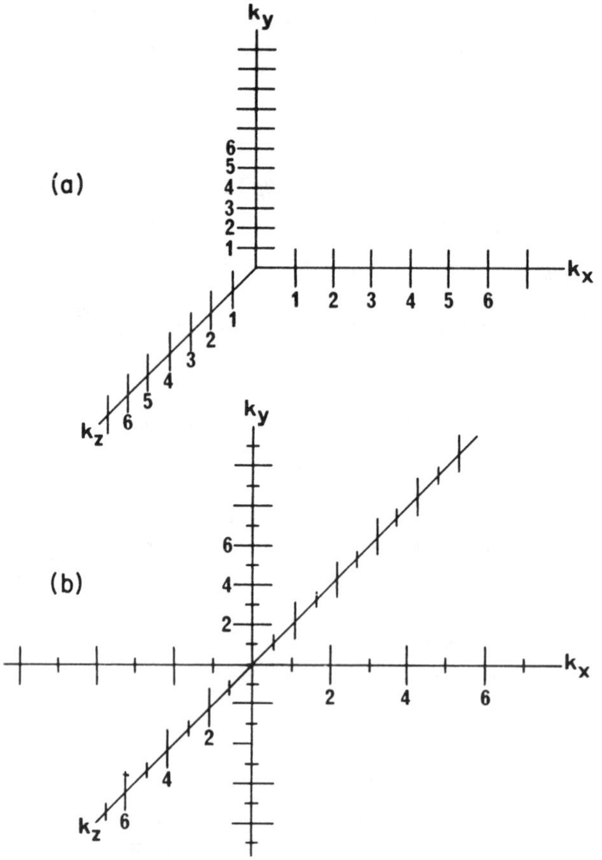

Fig. 3.2 Quantized values of the **k** vector for a particle in a box in terms of the $\hat{\mathbf{x}}$, $\hat{\mathbf{y}}$, and $\hat{\mathbf{z}}$ components k_x, k_y, and k_z. (a) Deduced by applying *fixed* boundary conditions to the energy eigenfunctions. (b) Deduced by applying *periodic* boundary conditions to the energy eigenfunctions.

degenerate with those for the corresponding positive values of $2m_x$, $2m_y$, $2m_z$. These results are illustrated in the diagram in Fig. 3.2. To summarize, we can say that the allowed values of **k** are more closely spaced for the standing wave solutions but must be restricted to the one quadrant in which k_x, k_y, k_z are all positive, whereas the allowed values of **k** are less closely spaced for the traveling wave solutions but this is compensated for precisely by the fact that they can have negative as well as positive values for k_x, k_y, and k_z.

The density of states $\tilde{g}(\mathscr{E})$ derived by using either set of energy levels will be the same. Thus one should not be too concerned about the fact that the energy levels do not occur at the same energies in the two cases. The type of boundary condition chosen, rather than the type of eigenfunction, is responsible for the slight difference in the energy values for the various levels. This becomes clear if we apply *periodic* boundary conditions to sine and cosine standing waves instead of using the more common fixed boundary conditions for these functions. Both sine and cosine functions are then allowable since neither type is eliminated by the fixed boundary condition requirement that the eigenfunctions be zero at the boundaries. The allowable quantum numbers for these two functions are found to be degenerate in energy; there is a one-to-one correspondence between sine and cosine functions having the same number of wavelengths along each of the three major directions in the solid. The negative integers are eliminated because of the requirement that the different basis states be linearly independent, so that the density of states at a given energy comes out to be the same for the two types of boundary conditions, independent of the functional form of the wave functions.

As pointed out previously, the traveling wave eigenfunctions of the Hamiltonian having the form $\exp[(i/\hbar)(\mathbf{p}\cdot\mathbf{r} - \mathscr{E}t)]$ are simultaneous eigenfunctions of the linear momentum operator $\mathbf{p}^{\text{op}} = -i\hbar\nabla$, whereas the standing-wave functions given by Eq. (3.6) are not momentum eigenfunctions, as one can easily prove using the linear momentum operator. Direct substitution of the exponential form gives

$$\begin{aligned}\mathbf{p}^{\text{op}}\psi_{\mathbf{k}}(\mathbf{r}) &= -i\hbar\nabla\ \{(1/V)^{1/2}\exp[i\mathbf{k}\cdot\mathbf{r}]\} = (-i\hbar)[\nabla\ (i\mathbf{k}\cdot\mathbf{r})]\psi_{\mathbf{k}}(\mathbf{r}) \\ &= \hbar[\nabla\ (k_x x + k_y y + k_z z)]\psi_{\mathbf{k}}(\mathbf{r}) = \hbar[\hat{\mathbf{x}}k_x + \hat{\mathbf{y}}k_y + \hat{\mathbf{z}}k_z]\psi_{\mathbf{k}}(\mathbf{r}) \\ &= \hbar\mathbf{k}\psi_{\mathbf{k}}(\mathbf{r}). \end{aligned} \tag{3.13}$$

Thus the momentum eigenvalues $\mathbf{p}_{\mathbf{k}}$ consistent with periodic boundary conditions are

$$\mathbf{p}_{\mathbf{k}} = \hbar\mathbf{k} = 2\pi\hbar[\hat{\mathbf{x}}(m_x/L_x) + \hat{\mathbf{y}}(m_y/L_y) + \hat{\mathbf{z}}(m_z/L_z)]. \tag{3.14}$$

PROJECT 3.1 Spherical Potential Well

1. Deduce the energy eigenvalues for a *spherical* three-dimensional potential well.
2. Interpret your results graphically and pictorially.
3. What well depth is necessary in order to have a bound state? Discuss this in terms of a real metal.

1.8 Density of States

The density of states $\tilde{w}(\mathbf{k})$ as a function of wave vector $\mathbf{k}$ for our three-dimensional system is now readily obtained. (The domain of the set of $\mathbf{k}$ vectors is referred to as $\mathbf{k}$ space, or *wave vector* space; it is described more fully in Chap. 7.) Since the spacings between adjacent allowed values of k_x, k_y, and k_z are $2\pi/L_x$, $2\pi/L_y$, and $2\pi/L_z$, respectively, the volume of $\mathbf{k}$ space per allowed value of the $\mathbf{k}$ vector is

$$(2\pi/L_x)(2\pi/L_y)(2\pi/L_z) = 8\pi^3/V, \tag{3.15}$$

where V is the actual volume of our metal. That is, the tip of each allowed $\mathbf{k}$ vector can be considered to constitute the corner for eight adjacent elemental rectangular parallelepipeds with dimensions $2\pi/L_x$, $2\pi/L_y$, and $2\pi/L_z$ in $\mathbf{k}$ space (cf. Fig. 3.3), and since there are eight corners required for a parallelepiped, there is one allowed wave vector per elementary parallelepiped on the average. The reciprocal of the volume $8\pi^3/V$ per parallelepiped thus gives the density of states $\tilde{w}(\mathbf{k})$ in $\mathbf{k}$ space,

$$\tilde{w}(\mathbf{k}) = V/8\pi^3 \qquad \textbf{(density of states versus wave vector for each electron spin direction)}, \tag{3.16}$$

since it represents the number of allowed $\mathbf{k}$ vectors per unit volume of $\mathbf{k}$ space.

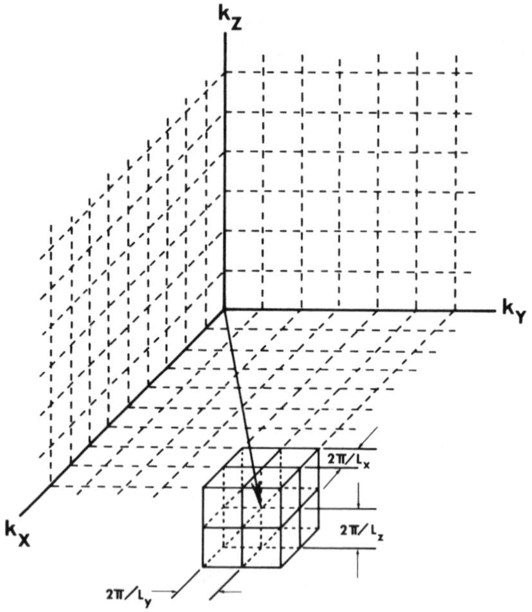

Fig. 3.3 The $\mathbf{k}$ vectors which satisfy periodic boundary conditions map out a rectangular lattice of points in $\mathbf{k}$ space which can be viewed as the corners of a contiguous stacked array of tiny parallelepipeds having sides of length $2\pi/L_x$, $2\pi/L_y$, and $2\pi/L_z$, where L_x, L_y, and L_z are the lengths of the three-dimensional box in real space representing the metal block confining the conduction electrons which is utilized for periodic boundary conditions.

Note from this result that $\tilde{w}(\mathbf{k})$ for our three-dimensional free electron model is independent of both magnitude and direction of the wave vector $\mathbf{k}$. Furthermore, since $\mathbf{p_k} = \hbar\mathbf{k}$, the volume element $d\mathbf{k} = dk_x\,dk_y\,dk_z$ in $\mathbf{k}$ space corresponds to the volume element

$$d\mathbf{p} \equiv dp_x\,dp_y\,dp_z = (\hbar\,dk_x)(\hbar\,dk_y)(\hbar\,dk_z) \equiv \hbar^3 d\mathbf{k} \tag{3.17}$$

in momentum space. Since the density of states $\mathscr{G}(\mathbf{p})$ in momentum space and the density of states $\tilde{w}(\mathbf{k})$ in $\mathbf{k}$ space are related by

$$\mathscr{G}(\mathbf{p})\,d\mathbf{p} = \tilde{w}(\mathbf{k})\,d\mathbf{k}, \tag{3.18}$$

we thus deduce that

$$\mathscr{G}(\mathbf{p}) = \hbar^{-3}\tilde{w}(\mathbf{k}) = V/(2\pi\hbar)^3 \qquad \textbf{(density of states versus momentum for each electron spin direction).} \tag{3.19}$$

For unit volume of metal, there are therefore $1/(2\pi\hbar)^3 = 1/h^3$ allowed momentum states, so that each allowed momentum state requires a volume of h^3 in momentum space for such a metal crystal.

The *constant energy surfaces* in $\mathbf{k}$ space are spheres, according to Eq. (3.11). Thus the number of allowed states in an energy interval $d\mathscr{E}$ at energy $\mathscr{E} = \hbar^2k^2/2m$ is equal to the number of allowed states in $\mathbf{k}$ space lying between the surface of the sphere with radius k and the surface of the larger sphere of radius $k + dk$. The differential dk represents a change in the magnitude of $\mathbf{k}$, namely, $dk \equiv d|\mathbf{k}|$, and the states lying in an energy interval $d\mathscr{E}$ at energy $\mathscr{E}$ are those in the corresponding differential volume $4\pi k^2\,dk$ in $\mathbf{k}$ space. Considering that each state requires a volume $8\pi^3/V$, we thus find for the number of states in the energy interval $d\mathscr{E}$ the value

$$dn = 4\pi k^2\,dk/(8\pi^3/V) = (V/2\pi^2)k^2\,dk. \tag{3.20}$$

However, $\mathscr{E} = \hbar^2k^2/2m$, so that

$$d\mathscr{E} = (\hbar^2k/m)\,dk. \tag{3.21}$$

The density of states $\tilde{g}(\mathscr{E}) = dn/d\mathscr{E}$ as a function of energy is therefore

$$\begin{aligned}\tilde{g}(\mathscr{E}) &= \frac{dn}{d\mathscr{E}} = \frac{(V/2\pi^2)k^2\,dk}{(\hbar^2k/m)\,dk} = (mV/2\pi^2\hbar^2)k \\ &= (mV/2\pi^2\hbar^2)(2m/\hbar^2)^{1/2}\mathscr{E}^{1/2} = (V/4\pi^2)(2m/\hbar^2)^{3/2}\mathscr{E}^{1/2}\end{aligned}$$

(density of states versus energy for each electron spin direction).

(3.22)

The *total* density of *electronic* states is a factor of two larger than this result, since there are two possible orientations of the electron spin, each of which corresponds to a different spin quantum number (cf. Chap. 1, §6) insofar as the Pauli exclusion principle is concerned.

It is well to remember the square root dependence of the density of states on energy, since it is frequently encountered in the theory of metals. To emphasize this result, let us derive it using a slightly different approach. For an arbitrary energy $\mathscr{E} = \hbar^2k^2/2m$, all electronic states with energies below this value lie within a sphere of radius **k** in wave vector space, and all electronic states with energies above this value lie outside the sphere. Considering that the volume of the sphere is $\frac{4}{3}\pi k^3$ and each allowed wave vector requires an elemental volume $8\pi^3/V$, we see that the number of electronic states n with energies equal to or less than $\mathscr{E}$ is given by

$$n = \tfrac{4}{3}\pi k^3/(8\pi^3/V) = (V/6\pi^2)k^3. \tag{3.23}$$

However, the density of states $\tilde{g}(\mathscr{E})$ is simply

$$\tilde{g}(\mathscr{E}) = dn/d\mathscr{E} = (dn/dk)\, dk/d\mathscr{E}. \tag{3.24}$$

It follows from $\mathscr{E} = \hbar^2k^2/2m$ that $d\mathscr{E}/dk = \hbar^2k/m$. Furthermore, $dn/dk = (V/2\pi^2)k^2$ from the above relation. Thus

$$\tilde{g}(\mathscr{E}) = [(V/2\pi^2)k^2][m/\hbar^2k] = (mV/2\pi^2\hbar^2)k = (V/4\pi^2)(2m/\hbar^2)^{3/2}\mathscr{E}^{1/2}, \tag{3.25}$$

which is the same result derived just above.

PROJECT 3.2 Density of States for a Two-Dimensional Free-Electron System

Consider a two-dimensional free electron system with momentum of the electrons given by $\mathbf{p} = \hat{\mathbf{x}}p_x + \hat{\mathbf{y}}p_y$. Assume the conduction electrons to be confined to a spatial region of length L_x in the x direction and L_y in the y direction. Solve the Schrödinger equation and apply periodic boundary conditions to obtain the energy eigenfunctions, energy eigenvalues, and momentum eigenvalues for the system. Derive the associated density of states versus momentum and the density of states versus energy for this system.

1.9 Fermi Surface and Conduction Electron Degeneracy

If we fill these quantum states deduced for a three-dimensional square-well potential with electrons, using both spin directions, then the maximum number of electrons which can be accommodated for energies less than or equal to $\mathscr{E}$ is

$$2n = (V/3\pi^2)k^3, \tag{3.26}$$

in accordance with Eq. (3.23). Thus N conduction electrons in a metal crystal filling the lowest energy states will require a sphere in **k** space of radius k_F given by

$$N = (V/3\pi^2)k_F^3. \tag{3.27}$$

The quantity k_F is designated the *Fermi wave vector*, and it can be seen to depend only on the average electron density N/V in the crystal,

$$k_F = (3\pi^2N/V)^{1/3}. \tag{3.28}$$

The sphere in **k** space with this radius (cf. Fig. 3.4) is called the *Fermi surface*. (A spherical shape for the Fermi surface is characteristic of the free-electron model

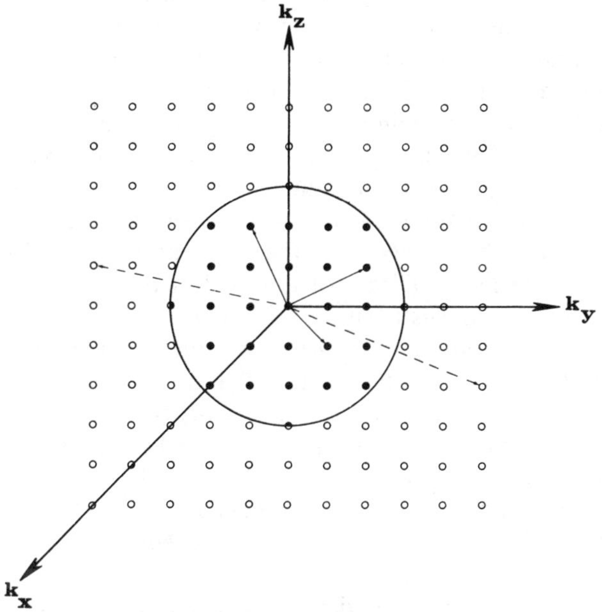

Fig. 3.4 Spherical Fermi surface which is characteristic of the free-electron model for metals. (The n lowest energy states per spin direction delineated by the allowed set of **k** vectors determined from periodic boundary conditions lie within the indicated spherical "Fermi" surface; higher energy states lie outside of the sphere.)

for metals.) The energy $\mathscr{E}_F$ corresponding to wave vector k_F is called the *Fermi energy*,

$$\mathscr{E}_F = \hbar^2 k_F^2/2m = (\hbar^2/2m)(3\pi^2 N/V)^{2/3}, \tag{3.29}$$

and the corresponding magnitude of the momentum $p_F = \hbar k_F$ is called the Fermi momentum. The magnitude of the electron velocity at the Fermi surface is given by $v_F = p_F/m = \hbar k_F/m$. For example, an electron density of $2.5 \times 10^{22}/\text{cm}^3$ as in sodium metal gives values of $\mathscr{E}_F = 3.1$ eV and $v_F = 1.0 \times 10^8$ cm/sec. The electron density in silver is more than a factor of 2 higher. The Fermi energy and the electron velocity increase accordingly, since these quantities vary as the $\frac{2}{3}$ and the $\frac{1}{3}$ power of the electron density, in accordance with Eq. (3.29) and the velocity $v_F = (2m\mathscr{E}_F)^{1/2}$ obtained from it. The values of $\mathscr{E}_F$ and v_F for silver are 5.5 eV and 1.4×10^8 cm/sec.

EXERCISE Compute values of $\mathscr{E}_F$, v_F, $\lambda_F \equiv 2\pi/k_F$, and $T_F \equiv \mathscr{E}_F/k_B$ for several metals, using a calculator.

The physical significance of the parameter $\mathscr{E}_F$ is that it delineates the point in energy separating those electronic states ($\mathscr{E} < \mathscr{E}_F$) having an occupation probability exceeding $\frac{1}{2}$ from those ($\mathscr{E} > \mathscr{E}_F$) with occupation probability less than $\frac{1}{2}$. This follows directly from the Fermi–Dirac distribution function (2.151) derived in Chap. 2. The initially surprising feature is the prediction of high velocities for some of the electrons even when the electrons are in the lowest energy states that are consistent with Fermi–Dirac statistics (cf. Chap. 2).

The exact manner in which the energy levels in the free-electron model are populated in the thermal equilibrium state is determined by the Fermi–Dirac distribution function which we derived in Chap. 2. The free-electron model neglects the interactions of the conduction electrons with the lattice, so the properties of a metal computed on the basis of this model are in fact the properties of a *dense electron plasma*. If the temperature of any electron plasma is high enough that the Fermi–Dirac distribution function is well approximated by the Maxwell–Boltzmann distribution (cf. Chap. 2), the system is said to be *nondegenerate*. This would require a temperature far in excess of the vaporization temperature for metals, however, so that the majority of the electrons in a metal must be considered to be degenerate. (Only those relatively few electrons in a metal occupying the higher sparsely populated energy levels, such as those involved in the thermal emission process to be treated in §4, can be considered to be nondegenerate.) Thus the statistical properties of a metal deduced from the free electron model can be considered to be equivalent to the statistical properties of a highly degenerate electron plasma. One of the statistical properties of the free-electron model which is very important, as well as being quite illustrative, is the low temperature specific heat. This topic is treated thoroughly in the following section.

2 Electronic Specific Heat

2.1 Classical Mechanics (Kinetic Theory) Picture

The specific heat of any macroscopic system is the increase in energy of the system per degree increase in temperature of the system under equilibrium conditions. For a dilute gas of atoms in a container, for example, classical kinetic theory coupled with the ideal gas law gives the result that each atom is in rapid translational motion with an average kinetic energy of $\mathscr{E}_{\text{atom}} = 3k_BT/2$, where k_B is Boltzmann's constant and T is the absolute temperature [Halliday and Resnick (1974)]. The energy $\mathscr{E}_T$ for N atoms is thus

$$\mathscr{E}_T = \tfrac{3}{2}Nk_BT. \tag{3.30}$$

The increase $\Delta\mathscr{E}_T$ in this energy for an increase ΔT in the temperature is $\Delta\mathscr{E}_T = \Delta(3Nk_BT/2) = (3Nk_B/2)\Delta T$. The specific heat C given by $\Delta\mathscr{E}_T/\Delta T$ is thus predicted to be $3Nk_B/2$, which can be seen to be temperature independent. More generally, the specific heat is written as the temperature derivative of the energy, so that in this example

$$C = d\mathscr{E}_T/dT = \tfrac{3}{2}Nk_B \qquad \text{(classical gas of noninteracting particles).} \tag{3.31}$$

The classical approach to a free-electron model would be to consider the conduction electrons in the metal to be just such a system of free particles; as a consequence, the specific heat would be predicted on the basis of such a classical model to be $\frac{3}{2}Nk_B$. That such a model fails catastrophically for metals is manifested by the fact that this prediction is larger than the experimentally measured electronic specific heat by a factor of the order of 10,000, and furthermore the experimental measurements show that the electronic specific heat varies linearly with the temperature instead of being temperature independent as predicted by Eq. (3.31). Thus we are strongly motivated to turn our attention to a better model, namely, the *quantum-mechanical* free-electron model.

2.2 Quantum Statistics Approach

The number of occupied states per unit energy range at any particular energy $\mathscr{E}$ for our quantum system of free electrons in thermal equilibrium at a given temperature T is

$$n(\mathscr{E}) = 2\mathscr{f}(\mathscr{E})\tilde{g}(\mathscr{E}). \tag{3.32}$$

The factor of 2 takes into account the fact that there are two allowed directions for electron spin, $\tilde{g}(\mathscr{E})$ is the density of states per spin direction as given by Eq. (3.22), and $\mathscr{f}(\mathscr{E})$ is the Fermi–Dirac distribution function given by Eq. (2.151). Figure 3.5 illustrates Eq. (3.32). The temperature dependence of the occupation probability is confined primarily to energies within several k_BT of the Fermi energy. For lower energies than this, the states remain occupied for the most part as the temperature is increased. Likewise, states lying several k_BT above $\mathscr{E}_F$ remain unoccupied for the most part as the temperature is increased. The effect of the temperature is thus to adjust the population of the energy levels in the neighborhood of $\mathscr{E}_F$, and the change in total energy of the free-electron gas with increasing temperature should be due primarily to this population adjustment. This differs markedly from the classical picture in which a temperature increase of the system results in a gain in thermal energy of each particle in the system.

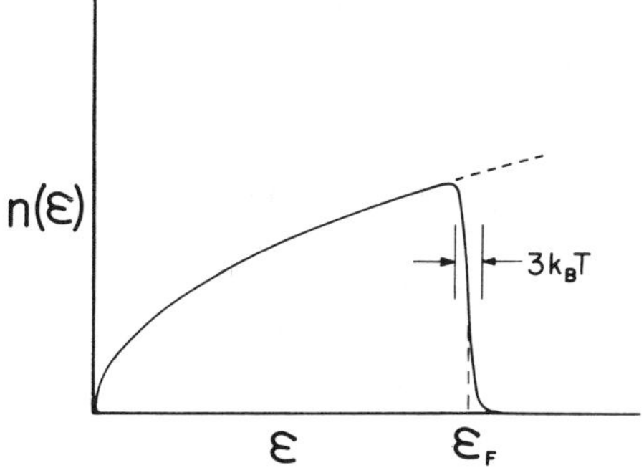

Fig. 3.5 Number $n(\mathscr{E})$ of occupied states per unit energy range versus energy $\mathscr{E}$. (This number is given by the product of the density of states $\tilde{g}(\mathscr{E})$ with the occupation probability $\mathscr{f}(\mathscr{E})$; the factor of 2 is due to the two allowed values of the electron spin quantum number.)

Since an increase in the temperature apparently does not change the energy of most of the lower energy electrons on the basis of our quantum picture, the quantum-mechanical electronic specific heat can be expected to be much less than the classical value, in accordance with experiment. The discrepancy of a factor of 10^4 or so between the classical value and the experimental value for a metal like sodium, together with the failure of the classical model to predict the experimental result that the electronic specific heat is a linear function of temperature, are both resolved by the quantum derivation. The quantum-mechanical free-electron model thus lends valuable insight into the low temperature specific heat of metals.

2.3 Formulation in Terms of Total Energy of Conduction Electrons

A quantitative derivation should be based on an integral involving Eq. (3.32) to give the total energy $\mathscr{E}_T$ of the electron system,

$$\mathscr{E}_T = \int_0^\infty 2\tilde{g}(\mathscr{E})\mathscr{f}(\mathscr{E})\mathscr{E}\, d\mathscr{E}, \tag{3.33}$$

with $\tilde{g}(\mathscr{E})$ given by Eq. (3.22). The electronic specific heat is then given by

$$C_{el} = \frac{d\mathscr{E}_T}{dT} = \int_0^\infty 2\tilde{g}(\mathscr{E})\frac{d\mathscr{f}}{dT}\mathscr{E}\, d\mathscr{E}. \tag{3.34}$$

Since the Fermi energy $\mathscr{E}_F$ is to a first approximation independent of temperature, as will subsequently be justified in §6,

$$\frac{d\mathscr{f}}{dT} = \frac{d}{dT}[e^{(\mathscr{E}-\mathscr{E}_F)/k_BT} + 1]^{-1} \simeq \frac{[(\mathscr{E}-\mathscr{E}_F)/k_BT^2]\exp[(\mathscr{E}-\mathscr{E}_F)/k_BT]}{\{\exp[(\mathscr{E}-\mathscr{E}_F)/k_BT]+1\}^2}. \tag{3.35}$$

However,

$$\frac{d\mathscr{f}}{d\mathscr{E}} = \frac{-(1/k_BT)\exp[(\mathscr{E}-\mathscr{E}_F)/k_BT]}{\{\exp[(\mathscr{E}-\mathscr{E}_F)/k_BT]+1\}^2}, \tag{3.36}$$

so that

$$d\mathscr{f}/dT = [(\mathscr{E}_F - \mathscr{E})/T]\, d\mathscr{f}/d\mathscr{E}. \tag{3.37}$$

Substituting this result into the integral in Eq. (3.34) gives

$$C_{el} = \int_0^\infty 2\tilde{g}(\mathscr{E})\mathscr{E}\left(\frac{\mathscr{E}_F - \mathscr{E}}{T}\right)\frac{d\mathscr{f}}{d\mathscr{E}}\, d\mathscr{E}. \tag{3.38}$$

A glance at the plot of $\mathscr{f}(\mathscr{E})$ versus $\mathscr{E}$ given in Fig. 2.4 is sufficient to convince oneself that $\mathscr{f}(\mathscr{E})$ has approximately zero slope everywhere except in the energy interval of width several k_BT surrounding $\mathscr{E}_F$. A plot of the derivative $d\mathscr{f}(\mathscr{E})/d\mathscr{E}$ is shown in Fig. 3.6. Therefore almost the entire contribution to the above integral for C_{el} comes from this region of integration involving energies within several multiples of $k_B T$. This is in accord with the above qualitative discussion of the change in the total energy of the system with an increase in temperature. Various approximations to the above integral can be based on this fact. First, $\tilde{g}(\mathscr{E})$ is almost the same as $\tilde{g}(\mathscr{E}_F)$ over the energy range in which $d\bar{f}/d\mathscr{E}$ is significantly different from zero. This greatly simplifies the task of evaluating some of the integrals for statistical quantities involving $\tilde{g}(\mathscr{E})$ and $\bar{f}(\mathscr{E})$. Second, series expansions which converge rapidly over this energy range could prove very useful. A general approximation technique for evaluating integrals containing the Fermi–Dirac distribution function $\bar{f}(\mathscr{E})$ is carefully developed in §5, and its usefulness is demonstrated by evaluating the temperature dependence of the Fermi energy and the electronic specific heat in §6.

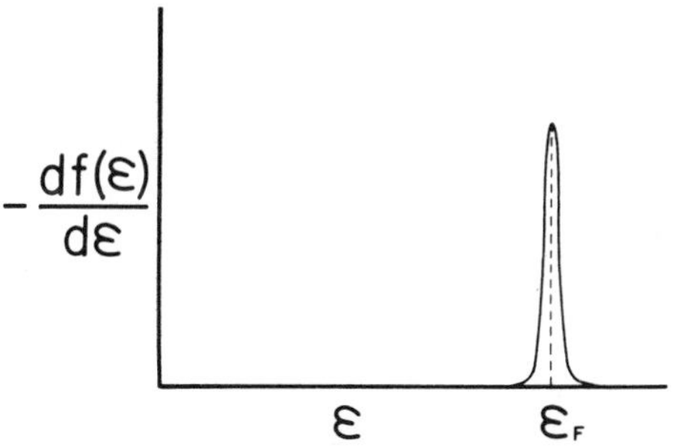

Fig. 3.6 Negative derivative of the Fermi–Dirac function $\bar{f}(\mathscr{E})$ with respect to energy $\mathscr{E}$. (The value is nearly zero except for energies within $k_B T$ or so of the Fermi energy $\mathscr{E}_F$.)

2.4 Alternate Formulation for Total Energy

Before going into the matter of general approximation techniques for integrals containing the derivative of the Fermi function, let us first use a different technique [cf. Kittel (1971)] involving physical intuition to evaluate the electronic specific heat. This technique is based on the *difference* in the total energy of the electron system at a finite temperature T and the corresponding total energy at 0°K. Let us assume that the Fermi energy $\mathscr{E}_F$ does not vary markedly with temperature, which is a valid approximation for our present purposes as we shall later show in §6. Then the change $\Delta\mathscr{E}_T$ in total energy in heating the electron system from 0°K to T can be arbitrarily but conveniently divided into two parts, the energy needed to excite an electron to an energy $\mathscr{E}_F$ plus the energy needed to excite the same electron to an energy level above $\mathscr{E}_F$. That is, consider the fact that at 0°K all levels below $\mathscr{E}_F$ are filled and all levels above $\mathscr{E}_F$ are empty; then the additional kinetic energy of the electron system at temperature T is due entirely to thermal excitation of some of the electrons below $\mathscr{E}_F$ to states above $\mathscr{E}_F$. The occupation probability is $\bar{f}(\mathscr{E})$, so the probability that a state at energy $\mathscr{E}$ below $\mathscr{E}_F$ is empty at temperature

T is $1-\tilde{f}(\mathscr{E})$. Hence the energy to excite all the electrons leaving states below $\mathscr{E}_F$ to the energy $\mathscr{E}_F$ is given by

$$\int_0^{\mathscr{E}_F} 2\tilde{g}(\mathscr{E})[1-\tilde{f}(\mathscr{E})](\mathscr{E}_F-\mathscr{E})\,d\mathscr{E}. \tag{3.39}$$

Similarly, the energy to further excite these electrons to the states above $\mathscr{E}_F$ is given by

$$\int_{\mathscr{E}_F}^{\infty} 2\tilde{g}(\mathscr{E})\tilde{f}(\mathscr{E})(\mathscr{E}-\mathscr{E}_F)\,d\mathscr{E}. \tag{3.40}$$

The sum of the two contributions (3.39) and (3.40) is the total energy change $\Delta\mathscr{E}_T$ resulting from an increase in system temperature from 0°K to T,

$$\begin{aligned}\Delta\mathscr{E}_T &= \int_0^{\mathscr{E}_F} 2\tilde{g}(\mathscr{E})[1-\tilde{f}(\mathscr{E})](\mathscr{E}_F-\mathscr{E})\,d\mathscr{E} + \int_{\mathscr{E}_F}^{\infty} 2\tilde{g}(\mathscr{E})\tilde{f}(\mathscr{E})(\mathscr{E}-\mathscr{E}_F)\,d\mathscr{E}\\ &= \int_0^{\mathscr{E}_F} 2\tilde{g}(\mathscr{E})(\mathscr{E}_F-\mathscr{E})\,d\mathscr{E} + \int_0^{\infty} 2\tilde{g}(\mathscr{E})\tilde{f}(\mathscr{E})(\mathscr{E}-\mathscr{E}_F)\,d\mathscr{E}.\end{aligned} \tag{3.41}$$

On the other hand, the total energy at 0°K is

$$\mathscr{E}_T^{(0)} = \int_0^{\mathscr{E}_F} 2\tilde{g}(\mathscr{E})\mathscr{E}\,d\mathscr{E}. \tag{3.42}$$

Thus the total energy $\mathscr{E}_T$ at temperature T is the sum of Eqs. (3.41) and (3.42),

$$\mathscr{E}_T = \mathscr{E}_T^{(0)} + \Delta\mathscr{E}_T = \int_0^{\mathscr{E}_F} 2\tilde{g}(\mathscr{E})\mathscr{E}_F\,d\mathscr{E} + \int_0^{\infty} 2\tilde{g}(\mathscr{E})\tilde{f}(\mathscr{E})(\mathscr{E}-\mathscr{E}_F)\,d\mathscr{E}. \tag{3.43}$$

A comparison of this result with our previous expression (3.33) for the total energy leads to the disturbing conclusion that they appear to be different. At least it is certainly not obvious that they are the same. Nevertheless, the two forms are shown below to be equivalent. Equation (3.43) is found to be a more useful form for our present purposes of deducing the specific heat.

Let us now digress a bit to show that the above two expressions [(3.33) and (3.43)] for $\mathscr{E}_T$ are the same for the free-electron model wherein the total number of electrons in the system is conserved. Let N_{below} denote the total number of empty electronic states below the Fermi energy,

$$N_{\text{below}} = \int_0^{\mathscr{E}_F} 2\tilde{g}(\mathscr{E})[1-\tilde{f}(\mathscr{E})]\,d\mathscr{E}. \tag{3.44}$$

Let N_{above} denote the number of filled electronic states above the Fermi energy,

$$N_{\text{above}} = \int_{\mathscr{E}_F}^{\infty} 2\tilde{g}(\mathscr{E})\tilde{f}(\mathscr{E})\,d\mathscr{E}. \tag{3.45}$$

To the extent that the shift in $\mathscr{E}_F$ with temperature can be neglected (see §6), any filled states above $\mathscr{E}_F$ must be accompanied by an equal number of empty states below $\mathscr{E}_F$. This follows from the conservation of the total number of electrons and the fact that at 0°K all electronic states above $\mathscr{E}_F$ are empty and all those below $\mathscr{E}_F$ are filled. Thus

$$N_{\text{above}} = N_{\text{below}}. \tag{3.46}$$

Multiplying both sides by $\mathscr{E}_F$ leads to a relation

$$\mathscr{E}_F N_{\text{below}} - \mathscr{E}_F N_{\text{above}} = 0, \tag{3.47}$$

and substituting into the integrals (3.44) and (3.45) gives

$$\int_0^{\mathscr{E}_F} 2\tilde{g}(\mathscr{E})[1 - \tilde{f}(\mathscr{E})]\mathscr{E}_F \, d\mathscr{E} - \int_{\mathscr{E}_F}^{\infty} 2\tilde{g}(\mathscr{E})\tilde{f}(\mathscr{E})\mathscr{E}_F \, d\mathscr{E} = 0, \tag{3.48}$$

or equivalently,

$$\int_0^{\mathscr{E}_F} 2\tilde{g}(\mathscr{E})\mathscr{E}_F \, d\mathscr{E} - \int_0^{\infty} 2\tilde{g}(\mathscr{E})\tilde{f}(\mathscr{E})\mathscr{E}_F \, d\mathscr{E} = 0. \tag{3.49}$$

Adding this null expression to our initial expression (3.33) for $\mathscr{E}_T$ gives

$$\mathscr{E}_T = \int_0^{\infty} 2\tilde{g}(\mathscr{E})\tilde{f}(\mathscr{E})[\mathscr{E} - \mathscr{E}_F] \, d\mathscr{E} + \int_0^{\mathscr{E}_F} 2\tilde{g}(\mathscr{E})\mathscr{E}_F \, d\mathscr{E}, \tag{3.50}$$

which is identical to our second result (3.43) obtained from $\mathscr{E}_T^{(0)}$ and $\Delta\mathscr{E}_T$. Although the results for $\mathscr{E}_T$ are equivalent, the transformation effected in the integral makes a good approximate evaluation of the specific heat far easier. This we now proceed to do.

PROJECT 3.3 Pressure Exerted on Walls of a Container by Trapped Particles

Assume a particle is trapped in a three-dimensional square-well potential having edges of length L with infinitely high potential energy barriers at the edges (i.e., at the walls). Calculate the average force exerted on each wall and thereby determine the pressure within the container. [*Hint*: For the analogous one-dimensional problem, see ter Haar (1975).]

2.5 Specific Heat Evaluation

Considering $\mathscr{E}_F$ to be temperature independent, which is shown in §6 to be a reasonably good approximation, we obtain by differentiating Eq. (3.43) for $\mathscr{E}_T$ the result

$$C_{\text{el}} = \frac{d\mathscr{E}_T}{dT} = \int_0^{\infty} 2\tilde{g}(\mathscr{E}) \frac{d\tilde{f}}{dT} (\mathscr{E} - \mathscr{E}_F) \, d\mathscr{E}. \tag{3.51}$$

Substituting expression (3.35) for $d\tilde{f}/dT$, we obtain

$$C_{\text{el}} = \int_0^{\infty} \frac{2\tilde{g}(\mathscr{E})[(\mathscr{E} - \mathscr{E}_F)^2/k_B T^2]\exp[(\mathscr{E} - \mathscr{E}_F)/k_B T]}{\{\exp[(\mathscr{E} - \mathscr{E}_F)/k_B T] + 1\}^2} \, d\mathscr{E}. \tag{3.52}$$

Let us recall our previous discussion of the narrow range of energy over which $d\mathcal{f}/d\mathscr{E}$ is appreciably different from zero and the fact that $d\mathcal{f}/dT \propto d\mathcal{f}/d\mathscr{E}$, so that the slowly varying function $\tilde{g}(\mathscr{E})$ can be replaced by $\tilde{g}(\mathscr{E}_F)$ in the above integral. In addition, let us make the following change of variables in the integral,

$$\eta \equiv (\mathscr{E} - \mathscr{E}_F)/k_B T, \qquad d\eta = (k_B T)^{-1}\, d\mathscr{E}. \tag{3.53}$$

The lower limit becomes $-\mathscr{E}_F/k_B T$ and the upper limit is again infinity. Thus

$$C_{el} = 2k_B^2 T \tilde{g}(\mathscr{E}_F) I, \tag{3.54}$$

where

$$I \equiv \int_{-\mathscr{E}_F/k_B T}^{\infty} \frac{\eta^2 e^{\eta}}{(e^{\eta} + 1)^2}\, d\eta. \tag{3.55}$$

The integral I is temperature dependent, but this is not a large effect. The integrand takes on values greater than 0.1 for positive η, but for negative values of η in the range of $-\mathscr{E}_F/k_B T \simeq -40$ it has extremely small values of the order of $\eta^2 e^{\eta}$. Thus the range of integration can be extended to $-\infty$ without appreciable error, in which case I becomes a definite integral which has the value $\pi^2/3$. Therefore

$$C_{el} \simeq \tfrac{2}{3}\pi^2 k_B^2 \tilde{g}(\mathscr{E}_F) T \qquad \textbf{(electronic specific heat for both spin directions)}, \tag{3.56}$$

with $\tilde{g}(\mathscr{E}_F)$ denoting the density of states per direction of spin given by Eq. (3.22), assuming unit volume for the system. Writing this result as

$$C_{el} = \gamma_{el} T, \tag{3.57}$$

where

$$\gamma_{el} = \tfrac{2}{3}\pi^2 k_B^2 \tilde{g}(\mathscr{E}_F) \tag{3.58}$$

is a temperature-independent quantity, serves to emphasize the *linear* dependence of the specific heat on temperature predicted by the *quantum-mechanical* free-electron model. This contrasts with the *temperature-independent result* given by Eq. (3.31) which was deduced from the *classical* free-electron model.

Evaluating Eq. (3.22) at $\mathscr{E} = \mathscr{E}_F$ and substituting Eq. (3.29) gives

$$\begin{aligned}\tilde{g}(\mathscr{E}_F) &= (V/4\pi^2)(2m/\hbar^2)^{3/2}\mathscr{E}_F^{1/2} \\ &= (V/4\pi^2)(2m/\hbar^2)^{3/2}[(\hbar^2/2m)^{1/2}(3\pi^2 N/V)^{1/3}] \\ &= (V/4\pi^2)(2m/\hbar^2)(3\pi^2 N/V)^{1/3}.\end{aligned} \tag{3.59}$$

Substituting this result into Eq. (3.58) gives

$$\gamma_{el} = (mV/3\hbar^2)(3\pi^2 N/V)^{1/3} k_B^2. \tag{3.60}$$

EXERCISE Compute the electronic specific heat coefficient γ_{el} using Eq. (3.60) for several metals of your choice.

EXERCISE Compare your calculated values of the electronic specific heat to published experimental values, and attempt to draw some conclusions.

In the limit of unit volume, $V = 1$, the parameter γ_{el} is designated the *electronic specific heat coefficient*. If $V \neq 1$, then $C_{el} = \gamma_{el} T$ is the *total heat capacity* of the system of electrons. It can be seen that γ_{el} involves the electron density N/V and thus is predicted to differ from metal to metal. The other quantities such as m, $\hbar$, and k_B are physical constants which might be expected to be the same for different metals.

Typical values of the electronic specific heat coefficient for metals such as sodium, copper, and aluminum are in the range 0.6–1.4 mJ/mole deg^2. The agreement between experiment and the expression derived above using the quantum-mechanical free-electron model (FEM) is rather good, especially in view of the poor agreement of the classical result (3.31) with experiment. If we use the definition

$$\gamma_{el}^{(expl)} \equiv (m_{th}^* V/3\hbar^2)(3\pi^2 N/V)^{1/3} k_B^2, \tag{3.61}$$

where the parameter m_{th}^* is labeled the *thermal effective mass*, then the ratio

$$\gamma_{el}^{(expl)}/\gamma_{el}^{(FEM)} = m_{th}^*/m \tag{3.62}$$

provides a measure of the agreement of experiment with the quantum free-electron model. This ratio is found to be in the range 1.2–1.5 for the alkali metals, so it can be seen that the free-electron model is quite successful for this group of metals. The lack of perfect agreement is somewhat to be expected, considering that in the free-electron model the interactions of the conduction electrons with the lattice potential and with the thermal vibrations of the ions of the lattice are neglected. Likewise, the interactions of the conduction electrons among themselves have been neglected.

It should be pointed out explicitly that the electronic component of the specific heat deduced above is relatively important primarily at temperatures in the neighborhood of liquid helium temperatures ($\simeq 4°$K). At considerably higher temperatures, the specific heat of most solids is dominated by the lattice contribution. Quantized lattice vibrations mathematically analogous to the quantum results for the harmonic oscillator (cf. Chap. 1, §12.3) account for absorption of energy by the creation of "energy quanta" of lattice vibrational energy known as *phonons*. The reason that one can observe the *conduction electron* contribution at very low temperatures follows from the more rapid falloff of the lattice contribution (which varies as T^3) relative to the conduction electron contribution (which varies as T). [See, for example, R. A. Smith (1963).]

EXERCISE Numerically evaluate Eq. (3.55) for several temperatures and a typical set of parameter values to deduce the temperature dependence. (Note that this represents the departure of the specific heat from a strict linear dependence on temperature.)

PROJECT 3.4 Electronic Specific Heat for a Two-Dimensional Free-Electron System

Derive an expression for the electronic specific heat for a two-dimensional free-electron system. [*Hint*: Use the density of states derived in Project 3.2 (Chap. 3, §1.8) for the two-dimensional free-electron system.)

PROJECT 3.5 Sodium as a Free-Electron Metal

1. A macroscopic crystal of one mole of sodium occupies what volume?
2. Compute a numerical value for the density of states $\tilde{w}(\mathbf{k})$ versus wave vector $\mathbf{k}$ for this metal crystal.
3. Compute a value for the density of states $\mathscr{G}(\mathbf{p})$ versus momentum for this metal crystal.
4. Compute a value for the density of states $\tilde{g}(\mathscr{E})$ versus energy $\mathscr{E}$ for this metal crystal.
5. Calculate the value of the Fermi energy at 0°K.
6. Calculate the Fermi momentum p_F at 0°K.
7. Calculate the Fermi wave vector k_F at 0°K.
8. Calculate the Fermi wavelength λ_F at 0°K.
9. Calculate the difference between the Fermi energy at 300°K and the Fermi energy at 0°K.
10. Calculate the change in the Fermi momentum p_F between 0°K and 300°K.
11. Calculate the change in the Fermi wave vector k_F between 0°K and 300°K.
12. Calculate the change in the Fermi wavelength λ_F between 0°K and 300°K.
13. Calculate the numerical value of the electronic specific heat coefficient $\gamma_{él}$.

PROJECT 3.6 Free-Electron Model Computations

Choose any five real metals, and assume an appropriate valence for each in order to estimate the number of conduction electrons per atom. Compute values using the three-dimensional FEM for the Fermi energy $\mathscr{E}_F$, Fermi velocity v_F, Fermi momentum p_F, Fermi wave vector k_F, Fermi wavelength λ_F, Fermi temperature T_F, and electronic specific heat coefficient γ_{el}. Organize your final results in the form of a table so that conclusions may be inferred for the different metals. (*Note*: As an alternative procedure to using the valence, you may obtain the density of conduction electrons from appropriate literature data, such as Hall coefficient measurements.)

2.6 Ratio of Quantum to Classical Specific Heats

An informative ratio of the quantum and classical specific heats can be readily obtained which illustrates the large difference in the two predictions. Multiplying expression (3.59) for $\tilde{g}(\mathscr{E}_F)$ by the ratio

$$\mathscr{E}_F/k_B T_F = (\hbar^2/2m)(3\pi^2 N/V)^{2/3}/k_B T_F, \tag{3.63}$$

this ratio being unity by definition $T_F \equiv \mathscr{E}_F/k_B$ of the Fermi temperature, gives

$$\tilde{g}(\mathscr{E}_F) = 3N/4k_B T_F. \tag{3.64}$$

Substituting this expression into Eq. (3.58) for γ_{el} gives

$$\gamma_{el} = \tfrac{2}{3}\pi^2 k_B^2 \tilde{g}(\mathscr{E}_F) = N\pi^2 k_B/2T_F. \tag{3.65}$$

The corresponding classical result (3.31) is $C_{el}^{(class)} = 3Nk_B/2$, so the ratio of the quantum free-electron and the classical results is

$$C_{el}^{(FEM)}/C_{el}^{(class)} = \gamma_{el}T/(3Nk_B/2) = \tfrac{1}{3}\pi^2(T/T_F). \tag{3.66}$$

Considering that at room temperature the value of $k_B T$ is of the order of 1/40 eV, whereas the Fermi energy $\mathscr{E}_F = k_B T_F$ was deduced (see §1) to have values in the range 3–5 eV, the ratio thus is seen to have values of the order of 0.02 at room temperature. Correspondingly lower values are predicted at lower temperatures ($\lesssim$ 4°K) at which the experiments are performed. At 1°K, for example, the ratio given by Eq. (3.66) is of the order of 0.5×10^{-4}. This factor of T/T_F for the ratio of the quantum to the classical specific heats is in agreement with the qualitative discussion given in the early part of this section. This successful application of the quantum free-electron model represents one of its greatest triumphs and indeed, it represents one of the greatest triumphs of quantum mechanics in general, since there is apparently no way whatsoever to derive analogous results by means of a realistic classical model.

EXERCISE Compute the ratio of the quantum to the classical electronic specific heats at 1, 2, 3, and 4°K for several metals of your choice.

PROJECT 3.7 Density of States for Lattice Vibrational Modes

A density of states $\mathscr{D}(\omega)$ for lattice vibrational modes as a function of frequency ω can be deduced for simple harmonic oscillator models of coupled ions by employing periodic boundary conditions, somewhat analogously to the derivation of the electronic density of states $\tilde{g}(\mathscr{E})$ for the three-dimensional free-electron model. Carry through such a derivation, and point out the similarities and differences between the derivations for $\mathscr{D}(\omega)$ and $\tilde{g}(\mathscr{E})$. [*Hint*: First refer to a literature derivation of the lattice specific heat based on the Debye model, and then extend your considerations to deduce a more general result for $\mathscr{D}(\omega)$. See, for example, Kittel (1971).]

PROJECT 3.8 Lattice Specific Heat

Apply the Bose–Einstein distribution function [Eq. (2.152)] together with the concept of a single lattice vibrational mode to deduce the lattice vibrational contribution to the specific heat of a solid. [*Hint*: The vibrational mode can be considered to absorb energies $n\hbar\omega$ analogous to the harmonic oscillator (cf. Chap. 1, §12.3), this energy being due to population by individual quanta of energy $\hbar\omega$. These energy quanta can be treated somewhat like elemental particles (called *phonons*) which satisfy the same statistics as bosons. You may wish to refer to the *Einstein model*, which can be found in some standard solid state text, such as R. A. Smith (1963).]

3 Electrical Conductivity and the Derivation of Ohm's Law

3.1 Electrical Forces and Acceleration

The force **F** on an electron produced by an electric field **E** is $-e\mathbf{E}$. This force gives rise to a time dependence $d\mathbf{p}/dt$ of the linear momentum $\mathbf{p} = \hbar\mathbf{k}$ of the electron according to Newton's second law,

$$-e\mathbf{E} = \mathbf{F} = d\mathbf{p}/dt = \hbar\, d\mathbf{k}/dt. \tag{3.67}$$

Each filled electronic state **k** will thus be changed by an amount

$$\delta\mathbf{k} = -(e/\hbar)\mathbf{E}\,\delta t \tag{3.68}$$

by an applied electric field in a time increment δt. Since the amount $\delta\mathbf{k}$ is

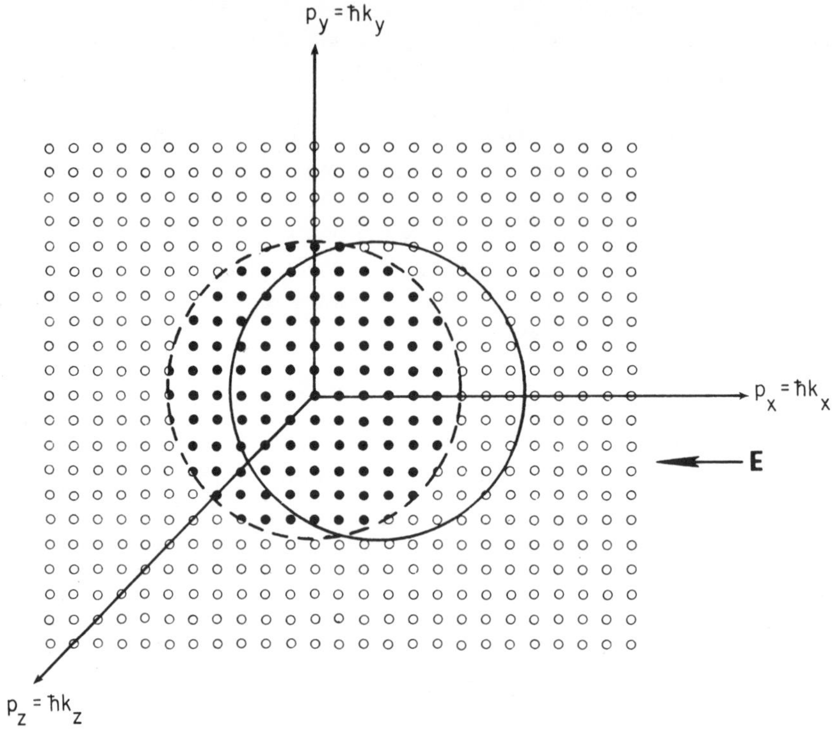

Fig. 3.7 Shift of the Fermi sphere in momentum space under the action of an externally applied electric field $\mathbf{E} = -\hat{\mathbf{x}}E_x$. (The allowed momentum states consistent with periodic boundary conditions are indicated by the small circles; the darkened areas represent those states which are occupied at 0°K in zero electric field. Note that the boundary (large dashed sphere) separating the occupied and unoccupied states in zero field is shifted to the right (large solid sphere) as the electrons are accelerated to the right by an applied electric field. Since $\mathbf{p} = \hbar\mathbf{k}$, a shift of the Fermi sphere in momentum space by $\delta\mathbf{p}$ is equivalent to a shift of the Fermi sphere in $\mathbf{k}$ space by $\delta\mathbf{k} = \delta\mathbf{p}/\hbar$.)

independent of the state $\mathbf{k}$, the entire Fermi sphere in $\mathbf{k}$ space is uniformly displaced by the field, as illustrated in Fig. 3.7.

Prior to the application of the field, the net momentum of the electrons is zero. This is true because at that point the Fermi sphere is symmetrical around the origin in $\mathbf{k}$ space (see Fig. 3.4); for every occupied state $\mathbf{k}$ there is an occupied state at $-\mathbf{k}$, with the net momentum $[\hbar\mathbf{k} + \hbar(-\mathbf{k})]$ of the pair of states being zero.

3.2 Collisions and the Relaxation Time Approximation

The acceleration of the electrons by the applied electric field results in an increased electron velocity antiparallel to the electric field direction. Such an increase in velocity in a specific direction gives rise to a net electric current in the solid. However, collisions of the electrons with impurities and defects and interaction with the vibrations of the lattice oppose this current by scattering the electrons into more or less random directions. The greater the electron velocity in a specific direction, the more frequently collisions occur which oppose this velocity. A steady state is achieved whenever the accelerating effects of the applied electric field are precisely compensated for (on the average) by the increased electron scattering, in which case the electron momentum reaches some constant average value. This is equivalent to saying that the displaced Fermi sphere will be maintained in some given position in **k** space consistent with the applied field and the collision processes. The net momentum will then be nonzero since the displaced sphere will not be symmetrical with respect to the origin of **k** space.

Suppose the electric field is removed after the Fermi sphere has reached its new steady-state position. Then the collisions which the electrons continuously undergo tend to relax the distribution back to the equilibrium state which existed before the field was applied. One simple model is based on the assumption that there is some characteristic collision time τ for scattering of the electrons by the imperfections of the lattice. The center of the displaced Fermi sphere can then be considered to return to the origin in **k** space in a time of the order of τ after the electric field is removed. The net momentum is reduced to zero by the effect of collisions in redistributing the occupied states to those bounded by a constant energy surface in **k** space, namely, a sphere centered at the origin.

Although elastic scattering alone is sufficient to reduce the net momentum to zero by redistributing the occupied states over constant energy surfaces, the fact that elastic scattering is characterized by no exchange of energy in the scattering event means that the electrons will still be in excited states after scattering. Inelastic phonon processes are therefore needed also to return the distribution to the ground state.

The characteristic relaxation time τ is important from the standpoint of estimating the displacement of the Fermi sphere with a given electric field. Since scattering randomizes the momentum within a time of the order of τ after each collision, the field can accelerate the electrons only over a time of the order of τ, on the average. Identifying τ with δt in Eq. (3.68) thus gives

$$\delta \mathbf{k}_\tau = -(e/\hbar)\mathbf{E}\tau. \tag{3.69}$$

The *average* increase is half this value. In the steady state, every electron is given (on the average) an additional incremental momentum $\delta \mathbf{p}_\tau = \frac{1}{2}\hbar\, \delta \mathbf{k}_\tau$ by the field, with a corresponding incremental velocity change of

$$\delta \mathbf{v}_\tau = \delta \mathbf{p}_\tau / m = -\tfrac{1}{2}(e\tau/m)\mathbf{E}. \tag{3.70}$$

3.3 Current Density and Electrical Resistivity

If there are n electrons of charge $q = -e$ per unit volume, the electric current density $\mathscr{J}$ is

$$\mathscr{J} = nq\, \delta \mathbf{v}_\tau = \tfrac{1}{2}(ne^2\tau/m)\mathbf{E}. \tag{3.71}$$

This has the form of Ohm's law

$$\mathscr{J} = \sigma \mathbf{E} \tag{3.72}$$

for an isotropic medium. The electrical conductivity σ just derived on the basis of the free-electron model is thus the scalar quantity

$$\sigma = ne^2\tau/2m. \tag{3.73}$$

The resistivity ρ is then

$$\rho = 1/\sigma = 2m/ne^2\tau. \tag{3.74}$$

Qualitatively the conductivity is given by the product of three factors: the charge density $-ne$, the ratio $-e/m$, and the time τ. The factor $-e/m$ determines the acceleration of the electron in a given electric field, and τ is the free time during which the field accelerates the carrier. It is quite significant that *all* of the conduction electrons participate in electrical transport, whereas statistically only a fraction of the conduction electrons participate in the specific heat (cf. §2).

3.4 Mean Free Path of Conduction Electrons

The use of an experimental value for the conductivity σ allows one to compute the relaxation time τ for a given metal for which the conduction electron density n is known. For copper at room temperature, τ as obtained by means of Eq. (3.74) is of the order of 10^{-14} sec. The Fermi velocity v_F for copper, as obtained by means of Eq. (3.29) and the nonrelativistic energy–momentum relation $\mathscr{E}_F = \frac{1}{2}mv_F^2$, is of the order of 10^8 cm/sec. Thus the mean free path $l_e = v_F\tau$ between scattering events is of the order of 10^{-6} cm, which represents 20 or 30 lattice spacings. At low temperatures, σ can be orders of magnitude larger if the metal is a very pure single crystal with few imperfections, and τ increases correspondingly. Since the Fermi velocity v_F does not change appreciably with temperature, the mean free path $l_e = v_F\tau$ at low temperatures thus can be enormous. Such long mean free paths cannot be explained by a classical model in which the ion cores must be considered effective in scattering. These long mean free paths, however, are readily understood on the basis of a quantum model, since in this model the periodic array of ion cores does not scatter the electrons randomly, on the average, and Fermi–Dirac statistics prevent all defect-related scattering events requiring the transfer of an electron into a momentum and spin state which is already occupied.

PROJECT 3.9 Dispersion Relation for Electromagnetic Wave Propagation in Metals

1. Write the general form of Maxwell's equations.
2. Reduce these equations to a simpler form, assuming Ohmic conduction currents and pointwise charge neutrality.
3. Solve for the **H** field in terms of the **E** field, assuming solutions of the plane wave form, namely, $\mathbf{E} = \mathbf{E}_0 \exp[i(\mathbf{k} \cdot \mathbf{r} - \omega t)]$.
4. Derive the dispersion relation.

5. Identify the portion of the dispersion relation due to displacement currents and the portion due to conduction currents.
6. Discuss the linearity of Maxwell's equations and the superposition of solutions corresponding to the simultaneous propagation of waves differing from one another in frequency and wavelength.
7. Consider the wave vector $\mathbf{k}$ to be given by $(\alpha + i\beta)\hat{\mathbf{n}}$, and solve for α and β.
8. Approximate α and β for the case of a good (but not perfect) dielectric material, and write down the resulting approximate expression for $\alpha + i\beta$.
9. Approximate α and β for the case of a good (but not perfect) metal, and write the resulting approximate expression for $\alpha + i\beta$.
10. Prove that the phase velocity for propagation in a material with nonzero conductivity is ω/α instead of ω/k as it is in a perfect dielectric.
11. Solve for the phase velocity in a good (but not perfect) dielectric.
12. Solve for the phase velocity in a good (but not perfect) metal.

4 Thermal Electron Emission from Metals

4.1 Work Function Barrier

The potential energy barriers at the interfaces separating a metal from the surrounding region of empty space are not infinitely high for electrons, as assumed in §1 to simplify the problem of computing wave functions for the free-electron model. The barrier height is only of the order of several electron-volts for most metals. This barrier, which constrains the electrons to remain within the metal, is known as the work function ϕ of the metal. This is illustrated in Fig. 3.8. The energy eigenfunctions which we deduced by assuming ϕ to be infinitely large do not differ very much from those which are deduced by using more realistic values for ϕ. [For a treatment of the *finite* square-well potential, see R. A. Smith (1963).]

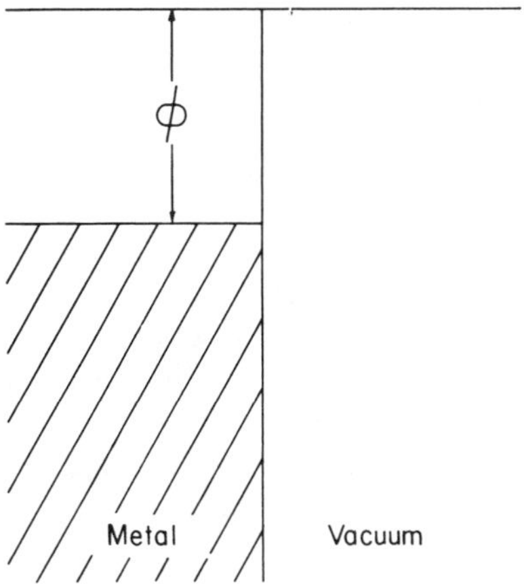

Fig. 3.8 Electron energy-level diagram illustrating the nearly filled states (crosshatched) in a metal; the quantum states are nearly empty over an energy region ϕ from the top of the nearly filled states to the vacuum energy.

4.2 Thermal Emission Picture

The Fermi–Dirac distribution function determines the statistical distribution of electrons as a function of energy. At nonzero temperatures, there is a finite (though small) probability that some of the energy levels in the metal above $\mathscr{E}_F + \phi$ will be occupied. These levels correspond to *unbound* electronic states, which means that the classical energy barrier is insufficiently high to confine such electrons within the metal. These electrons can cross the barrier with a finite kinetic energy and thereby escape from the metal. Such electrons are said to be *thermally emitted*. Because the occupation probability of the higher energy levels increases with increasing temperature, the thermal emission current increases accordingly. Let us now attempt to compute this emission current on the basis of the quantum free-electron model for metals.

4.3 Quantitative Details

In §1 the density of states $\mathscr{G}(\mathbf{p})$ in momentum space per direction of spin was found to be

$$\mathscr{G}(\mathbf{p}) = V/(2\pi\hbar)^3 = V/h^3, \tag{3.75}$$

where V is the volume of the metal specimen. For unit volume ($V = 1$) the density of electronic states for both spin directions becomes $2/h^3$. The Fermi–Dirac distribution function $\mathscr{f}(\mathscr{E})$ determines the occupation probability as a function of energy, and since $\mathscr{E} = \mathscr{E}(\mathbf{p})$, it also gives the occupation probability for a given momentum $\mathbf{p}$. Hence $(2/h^3)\mathscr{f}(\mathscr{E})\, d\mathbf{p}$ gives the number of electrons per unit volume of metal which have a momentum in the elemental volume $dp_x\, dp_y\, dp_z$ located at $\mathbf{p}$ in momentum space. These electrons have an x component of momentum between p_x and $p_x + dp_x$, with a corresponding velocity between v_x and $v_x + dv_x$. The number of electrons $[(2/h^3)\mathscr{f}(\mathscr{E})\, d\mathbf{p}]$ multiplied by v_x gives the number of electrons with momentum $\mathbf{p}$ striking unit area of a plane oriented perpendicular to the $\hat{\mathbf{x}}$ direction per unit time. If these electrons have an x component of momentum large enough so that

$$p_x^2/2m > \mathscr{E}_F + \phi, \tag{3.76}$$

we presume they leave the metal, whereas electrons with a lesser momentum in the x direction will be presumed to be reflected from the metal surface which is perpendicular to the x direction. The total emission current is obtained by summing all of the various contributions. This requires integrating over all x-momentum values satisfying the inequality (3.76) and over all possible values of the y and z components of momentum. The electron particle current density J_e emitted from the metal interface perpendicular to the x direction is therefore given by

$$J_e = \int_{-\infty}^{\infty} dp_z \int_{-\infty}^{\infty} dp_y \int_{p_x^{(\text{free})}}^{\infty} (2/h^3)\mathscr{f}(\mathscr{E}) v_x\, dp_x, \tag{3.77}$$

where

$$p_x^{(\text{free})} = [2m(\mathscr{E}_F + \phi)]^{1/2}. \tag{3.78}$$

In the free-electron model, the total energy $\mathscr{E}$ is all kinetic energy,

$$\mathscr{E} = (p_x^2 + p_y^2 + p_z^2)/2m, \tag{3.79}$$

since the potential energy is zero. In terms of the dimensionless variable η defined by

$$k_B T \eta \equiv [(p_x^2 + p_y^2 + p_z^2)/2m] - \mathscr{E}_F = \mathscr{E} - \mathscr{E}_F, \tag{3.80}$$

the Fermi–Dirac distribution function

$$\tilde{f}(\mathscr{E}) = \{\exp[(\mathscr{E} - \mathscr{E}_F)/k_B T] + 1\}^{-1} \tag{3.81}$$

becomes

$$\tilde{f}(\mathscr{E}) = [e^\eta + 1]^{-1}. \tag{3.82}$$

The dummy variable in the first integral in Eq. (3.77) is p_x, and the integration is carried out under conditions of constant p_y and constant p_z. If the variable of integration is changed from p_x to η in this first integral, then

$$k_B T \, d\eta = (p_x/m) \, dp_x = v_x \, dp_x. \tag{3.83}$$

The lower limit $p_x^{(\text{free})}$ is replaced by

$$\eta^{(\text{free})} = (k_B T)^{-1}[\phi + (p_y^2 + p_z^2)/2m]. \tag{3.84}$$

Hence

$$J_e = \int_{-\infty}^{\infty} dp_z \int_{-\infty}^{\infty} dp_y \int_{\eta^{(\text{free})}}^{\infty} (2/h^3)[k_B T \, d\eta/(e^\eta + 1)]. \tag{3.85}$$

Since

$$1/(e^\eta + 1) = [e^{-\eta}/(e^{-\eta} + 1)] = -(d/d\eta) \ln(e^{-\eta} + 1), \tag{3.86}$$

the first integration can be carried out immediately to give

$$\int_{\eta^{(\text{free})}}^{\infty} [d\eta/(e^\eta + 1)] = \ln\{1 + \exp[-\eta^{(\text{free})}]\}. \tag{3.87}$$

Since ϕ is of the order of 3 eV and $k_B T$ at room temperature is of the order of $\frac{1}{40}$ eV, it follows that $\phi/k_B T \gg 1$. Also $(p_y^2 + p_z^2)/2m$ has some positive value, so that Eq. (3.84) yields the inequality $\eta^{(\text{free})} \gg 1$. Denoting by δ the small exponential quantity involving $\eta^{(\text{free})}$,

$$\delta \equiv \exp[-\eta^{(\text{free})}] \ll 1, \tag{3.88}$$

the logarithm can be approximated by the first term in a series expansion,

$$\ln[1 + \delta] \simeq \delta, \tag{3.89}$$

to give

$$\ln\{1+\exp[-\eta^{(\text{free})}]\} \simeq \exp[-\eta^{(\text{free})}]$$
$$= \exp(-\phi/k_{\rm B}T)\exp(-p_y^2/2mk_{\rm B}T)\exp(-p_z^2/2mk_{\rm B}T). \tag{3.90}$$

It should be pointed out that this approximation in some way is equivalent to replacing Fermi–Dirac (quantum) statistics by Boltzmann (classical) statistics for those electrons which have sufficient thermal energy to surmount the work function barrier. The particle current (3.85) thereby takes the form

$$J_{\rm e} = (2k_{\rm B}T/h^3)e^{-\phi/k_{\rm B}T}\int_{-\infty}^{\infty} e^{-p_z^2/2mk_{\rm B}T}\,dp_z \int_{-\infty}^{\infty} e^{-p_y^2/2mk_{\rm B}T}\,dp_y. \tag{3.91}$$

The remaining integrals can be cast in terms of the definite integral $\mathscr{I}$,

$$\mathscr{I} \equiv \int_{-\infty}^{\infty} e^{-\zeta^2}\,d\zeta. \tag{3.92}$$

Let us change the variable p_y in the first integral in Eq. (3.91) to $\zeta \equiv p_y/(2mk_{\rm B}T)^{1/2}$, with

$$d\zeta = dp_y/(2mk_{\rm B}T)^{1/2}. \tag{3.93}$$

Then

$$\int_{-\infty}^{\infty} e^{-p_y^2/2mk_{\rm B}T}\,dp_y = (2mk_{\rm B}T)^{1/2}\mathscr{I}. \tag{3.94}$$

Similarly, for the second integral let

$$\zeta' \equiv p_z/(2mk_{\rm B}T)^{1/2} \tag{3.95}$$

to obtain

$$\int_{-\infty}^{\infty} e^{-p_z^2/2mk_{\rm B}T}\,dp_z = (2mk_{\rm B}T)^{1/2}\mathscr{I}. \tag{3.96}$$

Then Eq. (3.91) takes the form

$$J_{\rm e} = (4mk_{\rm B}^2T^2/h^3)\mathscr{I}^2 e^{-\phi/k_{\rm B}T}. \tag{3.97}$$

The value of $\mathscr{I}$ can be found in tables of definite integrals, or it can be evaluated by the following technique. Consider the quantity $\mathscr{I}^2$,

$$\mathscr{I}^2 = \int_{-\infty}^{\infty} e^{-x^2}\,dx \int_{-\infty}^{\infty} e^{-y^2}\,dy = \int_{-\infty}^{\infty}\int_{-\infty}^{\infty} dx\,dy\,e^{-(x^2+y^2)}. \tag{3.98}$$

A conversion from rectangular to polar coordinates can be effected by letting $x = r\cos\theta$ and $y = r\sin\theta$, in which case the elemental area $d\mathscr{A}$ is given by $(r\,d\theta)\,dr$. Thus $\mathscr{I}^2$ becomes

$$\mathscr{I}^2 = \int_0^{\infty}\int_0^{2\pi} e^{-r^2} r\,dr\,d\theta = \int_0^{\infty} 2\pi e^{-r^2} r\,dr. \tag{3.99}$$

Changing variables again by setting r^2 equal to the new dummy variable ρ gives

$$\mathscr{I}^2 = \int_0^\infty \pi e^{-\rho}\, d\rho = \pi. \tag{3.100}$$

The particle current (3.97) is therefore

$$J_{\mathrm{e}} = (4\pi m k_{\mathrm{B}}^2 T^2/h^3) e^{-\phi/k_{\mathrm{B}}T}, \tag{3.101}$$

and the corresponding charge current given by $\mathscr{J}_{\mathrm{e}} = -eJ_{\mathrm{e}}$ is

$$\mathscr{J}_{\mathrm{e}} = AT^2 e^{-\phi/k_{\mathrm{B}}T}, \tag{3.102}$$

with

$$A = -4\pi m e k_{\mathrm{B}}^2/h^3. \tag{3.103}$$

The minus sign means simply that the current is due to an electron flux. The result (3.102) is known as the *Richardson–Dushman equation.* The theoretical value of A deduced above in terms of the fundamental constants m, e, k_{B}, and h is 120 A/cm^2 deg^2. Experimentally, A has been found to have values extending from 32 for platinum ($\phi \simeq 5.3$ eV) to 160 for cesium ($\phi \simeq 1.8$ eV), with tungsten ($\phi \simeq 4.5$ eV) having an intermediate value of 75. This type of agreement between theory and experiment is remarkably good, considering the simplicity of the free-electron model and the experimental techniques frequently used in the past in making the measurements.

PROJECT 3.10 One-Dimensional Free-Electron System

Consider a one-dimensional free-electron model in which conduction electrons are confined to a line of length L. (It may aid your thinking to consider a very thin metal wire or a polymer chain.)

1. Solve the Schrödinger equation.
2. Apply periodic boundary conditions.
3. Obtain energy eigenfunctions.
4. Obtain energy eigenvalues.
5. Solve the momentum eigenvalue equation to find the momentum eigenfunctions.
6. Deduce the momentum eigenvalues.
7. Derive the density of states versus wave vector.
8. Derive the density of states versus momentum.
9. Derive the density of states versus energy.
10. Derive the electronic specific heat.
11. Derive the electrical conductivity.
12. Derive the thermal electron emission current.
(*Hint*: see Chap. 1, §10.)

5 General Method for Evaluating Statistical Quantities Involving Fermi–Dirac Statistics

The purpose of this section is to introduce a more general technique for dealing with integrals containing the Fermi–Dirac distribution function. Such integrals appear frequently in expressions for the statistical properties of a metal, such as the specific heat treated in §2. The idea behind the general approach is to convert the integral to a form containing the derivative of the Fermi function so that use can be made of the delta-function characteristics of this derivative. As a byproduct of the development, we deduce the temperature dependence of the Fermi energy $\mathscr{E}_F$ and show that it is relatively small, as we assumed in our previous development of the electronic specific heat.

Consider the integral

$$\mathscr{I} = \int_0^\infty \tilde{z}(\mathscr{E})\mathfrak{f}(\mathscr{E})\, d\mathscr{E}, \tag{3.104}$$

where $\tilde{z}(\mathscr{E})$ is some smooth slowly varying function of $\mathscr{E}$. An integration by parts with $u = \mathfrak{f}(\mathscr{E})$ and $dv = \tilde{z}(\mathscr{E})\, d\mathscr{E}$ gives

$$\mathscr{I} = \mathfrak{f}(\mathscr{E})\mathscr{Z}(\mathscr{E})|_0^\infty - \int_0^\infty \mathscr{Z}(\mathscr{E})[d\mathfrak{f}(\mathscr{E})/d\mathscr{E}]\, d\mathscr{E}, \tag{3.105}$$

where

$$\mathscr{Z}(\mathscr{E}) \equiv \int_0^{\mathscr{E}} \tilde{z}(\mathscr{E})\, d\mathscr{E}. \tag{3.106}$$

At the upper limit, $\mathfrak{f}(\mathscr{E}) = \mathfrak{f}(\infty) = 0$, while $\mathscr{Z}(\mathscr{E})$ might well diverge to ∞. At the lower limit, $\mathfrak{f}(\mathscr{E}) \simeq 1$ and $\mathscr{Z}(\mathscr{E}) = \mathscr{Z}(0) = 0$. Thus, while there is no question about the zero value of the product $\mathfrak{f}(\mathscr{E})\mathscr{Z}(\mathscr{E})$ at the lower limit, some additional examination of the value of the product at the upper limit is necessary before it can be judged to be zero. The easiest approach is to recognize that $\mathfrak{f}(\mathscr{E}) \propto e^{-\mathscr{E}/k_BT}$ as $\mathscr{E} \to \infty$, since the Fermi–Dirac distribution function in this limit is well approximated by the Boltzmann distribution function (cf. Chap. 2, §2.4.6). The integral $\mathscr{Z}(\mathscr{E})$ typically varies as $\mathscr{E}^{p/2}$, where p is some finite integer. The product $\mathfrak{f}(\mathscr{E})\mathscr{Z}(\mathscr{E})$ in such cases approaches zero as $\mathscr{E} \to \infty$ in accordance with l'Hôpital's rule [see Wylie (1951)] involving successive differentiation of numerator and denominator. Therefore Eq. (3.105) reduces to

$$\mathscr{I} = -\int_0^\infty \mathscr{Z}(\mathscr{E})[d\mathfrak{f}(\mathscr{E})/d\mathscr{E}]\, d\mathscr{E}. \tag{3.107}$$

This integral has the nice feature that it involves the derivative of the Fermi function. The derivative differs appreciably from zero only over the energy range within several k_BT of $\mathscr{E}_F$, as can be noted in Fig. 3.6. By hypothesis, $\tilde{z}(\mathscr{E})$ is a slowly varying smooth function of $\mathscr{E}$; it follows that $\mathscr{Z}(\mathscr{E})$ will likewise have these features. An approximation for $\mathscr{Z}(\mathscr{E})$ valid over a region within several k_BT of $\mathscr{E}_F$ can therefore be obtained by making a Taylor series expansion [see Wylie (1951)] of $\mathscr{Z}(\mathscr{E})$ about $\mathscr{E}_F$,

$$\mathscr{Z}(\mathscr{E}) = \mathscr{Z}(\mathscr{E}_F) + \left.\frac{d\mathscr{Z}(\mathscr{E})}{d\mathscr{E}}\right|_{\mathscr{E}=\mathscr{E}_F}(\mathscr{E}-\mathscr{E}_F) + \frac{1}{2!}\left.\frac{d^2\mathscr{Z}(\mathscr{E})}{d\mathscr{E}^2}\right|_{\mathscr{E}=\mathscr{E}_F}(\mathscr{E}-\mathscr{E}_F)^2$$

$$+ \cdots + \frac{1}{n!}\left.\frac{d^n\mathscr{Z}(\mathscr{E})}{d\mathscr{E}^n}\right|_{\mathscr{E}=\mathscr{E}_F}(\mathscr{E}-\mathscr{E}_F)^n + \cdots. \tag{3.108}$$

Abbreviating the constants $(d^n\mathscr{Z}(\mathscr{E})/d\mathscr{E}^n)|_{\mathscr{E}=\mathscr{E}_F}$ by $d^n\mathscr{Z}(\mathscr{E}_F)/d\mathscr{E}^n$ and substituting into Eq. (3.107) for $\mathscr{I}$ gives

$$-\mathscr{I} = \mathscr{Z}(\mathscr{E}_F)\int_0^\infty \frac{d\mathscr{f}(\mathscr{E})}{d\mathscr{E}}\,d\mathscr{E} + \frac{d\mathscr{Z}(\mathscr{E}_F)}{d\mathscr{E}}\int_0^\infty (\mathscr{E}-\mathscr{E}_F)\frac{d\mathscr{f}(\mathscr{E})}{d\mathscr{E}}\,d\mathscr{E}$$
$$+\frac{1}{2}\frac{d^2\mathscr{Z}(\mathscr{E}_F)}{d\mathscr{E}^2}\int_0^\infty (\mathscr{E}-\mathscr{E}_F)^2\frac{d\mathscr{f}(\mathscr{E})}{d\mathscr{E}}\,d\mathscr{E} + \cdots$$
$$+\frac{1}{n!}\frac{d^n\mathscr{Z}(\mathscr{E}_F)}{d\mathscr{E}^n}\int_0^\infty (\mathscr{E}-\mathscr{E}_F)^n\frac{d\mathscr{f}(\mathscr{E})}{d\mathscr{E}}\,d\mathscr{E} + \cdots. \tag{3.109}$$

The first integral on the right-hand side yields $\mathscr{f}(\infty) - \mathscr{f}(0) \simeq -1$. The remaining integrals have the characteristic form

$$\mathscr{H}_j = \int_0^\infty (\mathscr{E}-\mathscr{E}_F)^j[d\mathscr{f}(\mathscr{E})/d\mathscr{E}]\,d\mathscr{E} \qquad (j = 1, 2, \ldots, n, \ldots). \tag{3.110}$$

Furthermore, by definition of $\mathscr{Z}(\mathscr{E})$,

$$\mathscr{Z}(\mathscr{E}_F) = \int_0^{\mathscr{E}_F} \mathscr{z}(\mathscr{E})\,d\mathscr{E}. \tag{3.111}$$

Since in addition

$$d\mathscr{Z}(\mathscr{E})/d\mathscr{E} = \mathscr{z}(\mathscr{E}), \tag{3.112}$$

the derivatives of $\mathscr{Z}(\mathscr{E})$ appearing in $\mathscr{I}$ have the characteristic form,

$$\frac{d^j\mathscr{Z}(\mathscr{E}_F)}{d\mathscr{E}^j} = \left.\frac{d^{j-1}\mathscr{z}(\mathscr{E})}{d\mathscr{E}^{j-1}}\right|_{\mathscr{E}=\mathscr{E}_F} \equiv \frac{d^{j-1}\mathscr{z}(\mathscr{E}_F)}{d\mathscr{E}^{j-1}}. \tag{3.113}$$

Thus

$$\mathscr{I} = \int_0^{\mathscr{E}_F} \mathscr{z}(\mathscr{E})\,d\mathscr{E} - \mathscr{z}(\mathscr{E}_F)\mathscr{H}_1 - \frac{1}{2}\frac{d\mathscr{z}(\mathscr{E}_F)}{d\mathscr{E}}\mathscr{H}_2 - \cdots - \frac{1}{n!}\frac{d^{n-1}\mathscr{z}(\mathscr{E}_F)}{d\mathscr{E}^{n-1}}\mathscr{H}_n - \cdots$$
$$= \int_0^{\mathscr{E}_F} \mathscr{z}(\mathscr{E})\,d\mathscr{E} - \mathscr{z}(\mathscr{E}_F)\mathscr{H}_1 - \sum_{n=1}^{\infty}\frac{1}{(n+1)!}\mathscr{H}_{n+1}\frac{d^n\mathscr{z}(\mathscr{E}_F)}{d\mathscr{E}^n}. \tag{3.114}$$

At this point the reader must wonder whether any simplification at all has been achieved, since the integral $\mathscr{I}$ has been expressed as an infinite series of new integrals. We now show that the new integrals can be expressed in terms of tabulated functions, and the series converges so rapidly in practice that only the first several terms of the sum make any appreciable contribution.

Differentiation of

$$\mathscr{f}(\mathscr{E}) = \{\exp[(\mathscr{E}-\mathscr{E}_F)/k_BT] + 1\}^{-1} \tag{3.115}$$

gives

$$\frac{d\mathscr{f}(\mathscr{E})}{d\mathscr{E}} = -(k_BT)^{-1}\frac{\exp[(\mathscr{E}-\mathscr{E}_F)/k_BT]}{\{\exp[(\mathscr{E}-\mathscr{E}_F)/k_BT] + 1\}^2}. \tag{3.116}$$

Substituting this derivative into the integral in Eq. (3.110) for $\mathscr{H}_j$ and making the variable change

$$\eta = (\mathscr{E} - \mathscr{E}_{\mathrm{F}})/k_{\mathrm{B}}T, \qquad d\eta = (k_{\mathrm{B}}T)^{-1}\, d\mathscr{E}, \tag{3.117}$$

gives

$$\mathscr{H}_j = -(k_{\mathrm{B}}T)^j \int_{-\mathscr{E}_{\mathrm{F}}/k_{\mathrm{B}}T}^{\infty} \frac{\eta^j e^\eta}{(e^\eta + 1)^2}\, d\eta. \tag{3.118}$$

For values of η less than $-\mathscr{E}_{\mathrm{F}}/k_{\mathrm{B}}T$, which is of the order of -100 at room temperature, the integrand is negligibly small relative to its value when η is of the order of unity. Thus the dependence of the value of $\mathscr{H}_j$ on the lower limit is small, and it is a good approximation to replace the lower limit by $-\infty$. In addition, multiplying numerator and denominator of the integrand by $e^{-\eta}$ gives

$$\mathscr{H}_j \simeq -(k_{\mathrm{B}}T)^j \int_{-\infty}^{\infty} \frac{\eta^j\, d\eta}{(e^\eta + 1)(e^{-\eta} + 1)}, \tag{3.119}$$

which shows that the integrand is an odd function of η whenever j is odd and an even function of η whenever j is even. Thus

$$\mathscr{H}_j \simeq \begin{cases} 0 & (j \text{ odd}) \\ -2(k_{\mathrm{B}}T)^j \displaystyle\int_0^{\infty} \frac{\eta^j\, d\eta}{(e^\eta + 1)(e^{-\eta} + 1)} & (j \text{ even}), \end{cases} \tag{3.120}$$

so that only half of the terms contribute to the value of the series for $\mathscr{I}$ given in Eq. (3.114).

The definite integral in Eq. (3.120) for $j = 2$ has the value $\pi^2/6$, so that

$$\mathscr{H}_2 = -(\pi^2/3)(k_{\mathrm{B}}T)^2. \tag{3.121}$$

Since $\mathscr{H}_1 \simeq \mathscr{H}_3 \simeq 0$, $\mathscr{I}$ given by Eq. (3.114) can be approximated through terms of order $d^2\tilde{z}(\mathscr{E}_{\mathrm{F}})/d\mathscr{E}^2$ by

$$\mathscr{I} \simeq \int_0^{\mathscr{E}_{\mathrm{F}}} \tilde{z}(\mathscr{E})\, d\mathscr{E} + \frac{\pi^2}{6}(k_{\mathrm{B}}T)^2 \frac{d\tilde{z}(\mathscr{E}_{\mathrm{F}})}{d\mathscr{E}}. \tag{3.122}$$

It is informative to compare this result with the original definition for $\mathscr{I}$, namely,

$$\mathscr{I} = \int_0^{\infty} \tilde{z}(\mathscr{E}) \tilde{f}(\mathscr{E})\, d\mathscr{E}. \tag{3.123}$$

We have thus obtained an approximation to an integral *over all energies* in terms of an integral *over energies below the Fermi energy* and a correction term which is quadratic in the temperature.

The power and beauty of this approximation technique reside in its versatility, due to the fact that $\tilde{z}(\mathscr{E})$ can be chosen to be nearly any function which is physically meaningful, and thus the technique is applicable to the study of a wide range of statistical properties of metals. As one example of the application of this technique, let us now compute the temperature dependence of the Fermi energy.

6 The Temperature Dependence of the Fermi Energy and Other Applications of the General Approximation Technique

6.1 Fermi Energy

The technique of evaluating integrals involving the Fermi–Dirac distribution function which was developed in the preceding section can be employed to determine the temperature dependence of the Fermi energy $\mathscr{E}_F$. Let us choose $\tilde{z}(\mathscr{E})$ to be the density of states $2\tilde{g}(\mathscr{E})$, so that the integral $\mathscr{I}$ as defined by Eq. (3.123) and approximated by Eq. (3.122) becomes the total number N of electrons in the system,

$$N = \int_0^{\infty} 2\tilde{g}(\mathscr{E})\tilde{f}(\mathscr{E})\, d\mathscr{E} \simeq \int_0^{\mathscr{E}_F} 2\tilde{g}(\mathscr{E})\, d\mathscr{E} + \frac{\pi^2}{3}(k_B T)^2 \frac{d\tilde{g}(\mathscr{E}_F)}{d\mathscr{E}}. \tag{3.124}$$

At absolute zero, $\tilde{f}(\mathscr{E}) = 0$ above $\mathscr{E}_F(0)$ and $\tilde{f}(\mathscr{E}) = 1$ below $\mathscr{E}_F(0)$, so that another expression for N is

$$N = \int_0^{\mathscr{E}_F(0)} 2\tilde{g}(\mathscr{E})\, d\mathscr{E}. \tag{3.125}$$

Conservation of the total number of particles in the system is the condition which determines the temperature-dependent quantity $\mathscr{E}_F$. Subtracting the second expression (3.125) from the first expression (3.124) and dividing through by 2 gives

$$\int_0^{\mathscr{E}_F} \tilde{g}(\mathscr{E})\, d\mathscr{E} - \int_0^{\mathscr{E}_F(0)} \tilde{g}(\mathscr{E})\, d\mathscr{E} + \frac{\pi^2}{6}(k_B T)^2 \frac{d\tilde{g}(\mathscr{E}_F)}{d\mathscr{E}} \simeq 0, \tag{3.126}$$

or equivalently,

$$\int_{\mathscr{E}_F(0)}^{\mathscr{E}_F} \tilde{g}(\mathscr{E})\, d\mathscr{E} \simeq -\frac{\pi^2}{6}(k_B T)^2 \frac{d\tilde{g}(\mathscr{E}_F)}{d\mathscr{E}}. \tag{3.127}$$

For a density of states which increases with increasing energy, as in the three-dimensional free-electron model expression (3.22) for which $\tilde{g}(\mathscr{E}) \propto \mathscr{E}^{1/2}$, the right-hand side of Eq. (3.127) is negative. Thus for such cases the left-hand side of Eq. (3.127) shows that $\mathscr{E}_F < \mathscr{E}_F(0)$. That is, the Fermi energy *decreases* as the temperature is increased. (The variation of $\mathscr{E}_F$ with $\mathscr{E}$ would of course be in the opposite direction for the one-dimensional free-electron model, for which Eq. (1.296) gives $\tilde{g}(\mathscr{E}) \propto \mathscr{E}^{-1/2}$ so that $d\tilde{g}(\mathscr{E})/d\mathscr{E} < 0$.)

If we approximate the integral in Eq. (3.127) by considering $\tilde{g}(\mathscr{E}) \simeq \tilde{g}(\mathscr{E}_F(0))$ over the range between $\mathscr{E}_F(0)$ and $\mathscr{E}_F$, then we obtain

$$\tilde{g}[\mathscr{E}_F(0)][\mathscr{E}_F - \mathscr{E}_F(0)] \simeq -\frac{\pi^2}{6}(k_B T)^2 \frac{d\tilde{g}(\mathscr{E}_F)}{d\mathscr{E}}. \tag{3.128}$$

If we further assume that $d\tilde{g}(\mathscr{E})/d\mathscr{E}$ evaluated at $\mathscr{E}_F$ is approximately equal to this same derivative evaluated at $\mathscr{E}_F(0)$, then we obtain the following result from Eq.

(3.128) for the temperature dependence of $\mathscr{E}_F$,

$$\mathscr{E}_F \simeq \mathscr{E}_F(0) - \frac{\pi^2}{6}(k_B T)^2 \left\{ \frac{d}{d\mathscr{E}} [\ln \tilde{g}(\mathscr{E})] \right\}_{\mathscr{E} = \mathscr{E}_F(0)}. \tag{3.129}$$

Thus $\mathscr{E}_F(T)$ is predicted to vary with temperature as T^2, namely, in a *quadratic* manner.

It is of interest to ascertain the magnitude of the quadratic term with respect to the constant term $\mathscr{E}_F(0)$. Using Eq. (3.25) for the three-dimensional density of states gives

$$\ln \tilde{g}(\mathscr{E}) = \ln[(V/4\pi^2)(2m/\hbar^2)^{3/2}] + \tfrac{1}{2} \ln \mathscr{E}, \tag{3.130}$$

so that

$$\{(d/d\mathscr{E})[\ln \tilde{g}(\mathscr{E})]\}_{\mathscr{E}=\mathscr{E}_F(0)} = 1/[2\mathscr{E}_F(0)]. \tag{3.131}$$

Substituting into expression (3.129) for $\mathscr{E}_F$ gives our final result for the temperature dependence of the Fermi energy,

$$\mathscr{E}_F \simeq \mathscr{E}_F(0)[1 - (\pi^2/12)\{k_B T/\mathscr{E}_F(0)\}^2]. \tag{3.132}$$

Recalling that $k_B T/\mathscr{E}_F(0)$ is of the order of 0.01 at room temperature but even smaller at lower temperatures, one sees immediately that the quadratic temperature term is only of the order of 10^{-4} of the temperature independent term. Thus our assumption that the temperature dependence of the Fermi energy $\mathscr{E}_F$ could be neglected insofar as our previous electronic specific heat derivation is concerned seems to be reasonably good. Approximations such as these are frequently tricky, however, as will be illustrated shortly.

EXERCISE Compute and plot $\mathscr{E}_F(T)$ versus T for several metals.

EXERCISE Evaluate the next higher order term beyond the quadratic term T^2 in $\mathscr{E}_F(T)$ and estimate its magnitude relative to that of the quadratic term given by Eq. (3.132).

6.2 Chemical Potential

The matter of notation deserves a comment at this point. Our temperature-dependent Fermi energy $\mathscr{E}_F$ is called the *chemical potential* ζ by some authors, so everywhere we have used $\mathscr{E}_F$, one could substitute ζ. This is simple enough; the only confusing point is that the quantity $\zeta(0) = \mathscr{E}_F(0)$ is sometimes written (cf. Ziman [1964], for example) as simply $\mathscr{E}_F$, in which case $\mathscr{E}_F$ denotes the same quantity as our parameter $\mathscr{E}_F$ only in the limit of zero temperature. The terms *chemical potential* and *Fermi level* are generally considered to be synonyms, whereas the term *Fermi energy* is sometimes reserved for the constant quantity $\zeta(0) = \mathscr{E}_F(0)$. In actual usage, however, difficulty seldom arises. This is due to the fact that the temperature dependence of the chemical potential is so very small.

6.3 Electronic Specific Heat

Let us now deduce the temperature dependence of the electronic specific heat using our new approximation technique, and compare the results with those obtained in §2. We thus choose $\tilde{z}(\mathscr{E})$ to be the product of the energy $\mathscr{E}$ and twice the density of states $2\tilde{g}(\mathscr{E})$, where $\tilde{g}(\mathscr{E})$ is the density of states per spin direction. The integral $\mathscr{I}$ then becomes the total energy $\mathscr{E}_{\mathrm{T}}$ of the electron system,

$$\mathscr{E}_{\mathrm{T}} \simeq \int_0^{\mathscr{E}_{\mathrm{F}}} 2\mathscr{E}\tilde{g}(\mathscr{E})\, d\mathscr{E} + \frac{\pi^2}{3}(k_{\mathrm{B}}T)^2 \left\{ \frac{d}{d\mathscr{E}}[\mathscr{E}\tilde{g}(\mathscr{E})] \right\}_{\mathscr{E} = \mathscr{E}_{\mathrm{F}}} . \tag{3.133}$$

The electronic heat capacity C_{el} is given by

$$C_{\mathrm{el}} = d\mathscr{E}_{\mathrm{T}}/dT. \tag{3.134}$$

If we again neglect the temperature dependence of $\mathscr{E}_{\mathrm{F}}$, we obtain

$$C_{\mathrm{el}} \simeq \frac{2\pi^2 k_{\mathrm{B}}^2 T}{3} \left\{ \frac{d}{d\mathscr{E}} [\mathscr{E}\tilde{g}(\mathscr{E})] \right\}_{\mathscr{E} = \mathscr{E}_{\mathrm{F}}(0)} \tag{3.135}$$

This certainly has the linear temperature dependence. Since

$$(d/d\mathscr{E})[\mathscr{E}\tilde{g}(\mathscr{E})] = \tilde{g}(\mathscr{E}) + \mathscr{E}[d\tilde{g}(\mathscr{E})/d\mathscr{E}], \tag{3.136}$$

we further obtain

$$C_{\mathrm{el}} \simeq (2\pi^2 k_{\mathrm{B}}^2/3)\tilde{g}(\mathscr{E}_{\mathrm{F}}(0))T + (2\pi^2 k_{\mathrm{B}}^2/3)\mathscr{E}_{\mathrm{F}}(0)\, \frac{d\tilde{g}[\mathscr{E}_{\mathrm{F}}(0)]}{d\mathscr{E}}\, T \tag{3.137}$$

as an approximate value for the heat capacity. A comparison with Eq. (3.56), however, shows agreement only if we neglect the second term. The additional approximation of neglecting the second term is unacceptable, since the ratio of the second term to the first is

$$\mathscr{E}_{\mathrm{F}}(0)\{(d/d\mathscr{E})[\ln \tilde{g}(\mathscr{E})]\}_{\mathscr{E} = \mathscr{E}_{\mathrm{F}}(0)} = \tfrac{1}{2} \tag{3.138}$$

for the three-dimensional free-electron model. We thus obtain a specific heat coefficient that is apparently 50% too large.

To illustrate that the approximation technique can nevertheless yield a result more in conformity with our previous derivation, let us consider the contribution of the term

$$\frac{d}{dT}\int_0^{\mathscr{E}_{\mathrm{F}}} 2\mathscr{E}\tilde{g}(\mathscr{E})\, d\mathscr{E} \tag{3.139}$$

neglected above in $d\mathscr{E}_{\mathrm{T}}/dT$. Only the upper limit is temperature dependent. To differentiate a definite integral, we can use the *Leibniz rule*:

If

$$\phi(x) = \int_{a(x)}^{b(x)} f(t, x)\, dt, \tag{3.140}$$

where $a(x)$ and $b(x)$ are differentiable functions of x, and $f(t, x)$ and $\partial f(t, x)/\partial x$ are continuous functions of both x and t, then

$$\frac{d\phi}{dx} = \int_{a(x)}^{b(x)} \frac{\partial f(t, x)}{\partial x}\, dt + f[b(x), x]\frac{db(x)}{dx} - f[a(x), x]\frac{da(x)}{dx}. \tag{3.141}$$

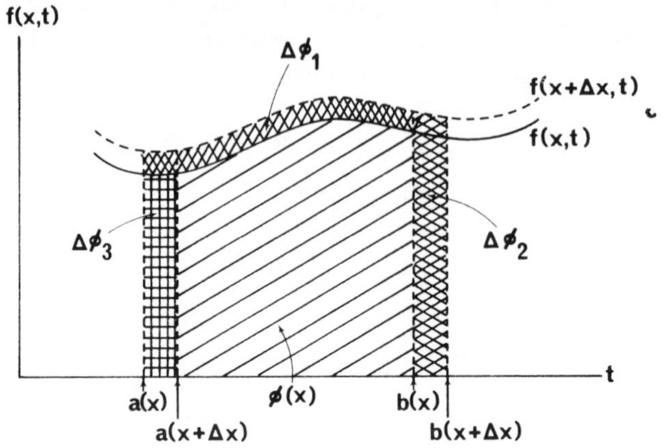

Fig. 3.9 Diagram representing the three contributions represented in the Leibniz rule for differentiation of an integral. [The area $\phi(x)$ given by $\phi(x) = \int_{a(x)}^{b(x)} f(x, t)\, dt$ under the curve $f(x, t)$ vs t will change when x is increased by an amount Δx because of (i) the change $\Delta\phi_1$ introduced by shifting the curve $f(x, t)$ to the new curve $f(x + \Delta x, t)$, (ii) the change $\Delta\phi_2$ introduced by shifting the upper limit from $b(x)$ to $b(x + \Delta x)$, and (iii) the change $\Delta\phi_3$ introduced by shifting the lower limit from $a(x)$ to $a(x + \Delta x)$.]

The contribution of each of the three terms on the right-hand side is indicated schematically in Fig. 3.9.

In our case, the lower limit in Eq. (3.139) is zero and the integrand is not a function of T, so that

$$\frac{d}{dT}\int_0^{\mathscr{E}_F} 2\mathscr{E}\tilde{g}(\mathscr{E})\, d\mathscr{E} = 2\mathscr{E}_F\tilde{g}(\mathscr{E}_F)\frac{d\mathscr{E}_F}{dT}. \tag{3.142}$$

Using expression (3.129)

$$\mathscr{E}_F \simeq \mathscr{E}_F(0) - \frac{\pi^2}{6}(k_B T)^2\left\{\frac{d}{d\mathscr{E}}[\ln \tilde{g}(\mathscr{E})]\right\}_{\mathscr{E}=\mathscr{E}_F(0)} \tag{3.143}$$

gives

$$\frac{d\mathscr{E}_F}{dT} \simeq -\frac{\pi^2}{3}k_B^2 T\left[\frac{1}{\tilde{g}(\mathscr{E})}\frac{d\tilde{g}(\mathscr{E})}{d\mathscr{E}}\right]_{\mathscr{E}=\mathscr{E}_F(0)}. \tag{3.144}$$

Substituting this result into Eq. (3.142) and assuming $\tilde{g}(\mathscr{E}_F)/\tilde{g}(\mathscr{E}_F(0)) \simeq 1$ and $\mathscr{E}_F \simeq \mathscr{E}_F(0)$ gives

$$\frac{d}{dT}\int_0^{\mathscr{E}_F} 2\mathscr{E}\tilde{g}(\mathscr{E})\, d\mathscr{E} \simeq -\tfrac{2}{3}\pi^2 k_B^2 T\mathscr{E}_F(0)\frac{d\tilde{g}[\mathscr{E}_F(0)]}{d\mathscr{E}}. \tag{3.145}$$

This contribution is equal in magnitude but opposite in sign to the second term in Eq. (3.137) for C_{el}, so that it cannot be neglected if the approximation is to be valid. Thus we are left with only the first term which agrees exactly with our earlier derivation. This perhaps is not too surprising when we remember that both terms in (3.133) involve the temperature dependence of the Fermi energy $\mathscr{E}_F$, which effect was neglected entirely in the first derivation (§2). This does serve to give us some confidence that our derived specific heat is accurate through terms quadratic in the temperature.

EXERCISE Evaluate the T^3 term in the electronic specific heat, and estimate its magnitude relative to the linear term.

PROJECT 3.11 Free-Electron Metal with a Twist

A hypothetical material, made up of long linear chains of silver atoms with a lattice constant $a_0 = 2.56$ Å, was used to construct a perfectly reflecting cubic cavity with walls one atomic layer in thickness. The separation between chains making up the walls was $3a_0$. The intrachain long-wavelength acoustic group-velocity v_{1t} for the transverse waves was found to have the value 1590 m/sec while the corresponding velocity v_{1l} for the longitudinal wave had the value 3600 m/sec. On the other hand, the interchain group velocities were found to be very nearly zero. Experimental measurements at the reststrahl frequency ν_0 of CsI (molecular weight = 259.83) showed the ratio $\mathscr{R}$ of the total lattice vibrational energy density at ν_0 in the walls of the cavity to the total radiant energy density at ν_0 in the cavity to be 1.182×10^9. The interatomic force constants in CsI can be considered as a first approximation to be the same as those in CsCl (molecular weight = 168.37), which has an experimentally measured absorption maximum at 102.0 microns. The ratio of the molecular weights of I_2 and Cl_2 is 3.58.

The cavity was then filled with a mole of lithium atoms, which condensed into a simple cubic free-electron metal occupying the entire cavity, each atom donating a single electron to the conduction band. A previous series of photoemission experiments showed the bottom of the conduction band to lie 7.03 eV below the vacuum level. An electric field of 10^6 volts/cm was maintained while extracting electrons from the metal at a temperature near 0°K. Assuming the electron tunnel current to be $\mathscr{I}(E_0) = I(E_0)\mathscr{T}$, where $I(E_0)$ is initially 120 A/cm^2 sec at $E_0 = 10^6$ V/cm, and $\mathscr{T}$ is the transmission coefficient, deduce the numerical value of the time t' necessary to deplete the metal of 30% of its electrons, assuming that an immobile negative charge replaces each conduction electron extracted from the metal.

Next, compute numerical values for the total electronic specific heat for this metal at 160°C and the corresponding change in the Fermi energy from its 0°K value. [*Hint*: See R. A. Smith (1963).]

7 The Boltzmann Equation

7.1 Derivation for Quantum Electron System

The simple approach used in §3.3 to derive Ohm's law and the electrical conductivity was remarkably successful. A more sophisticated approach to transport phenomena in general and electrical transport in particular is the *Boltzmann equation*. Since this equation is used widely in the literature and has great utility for many different applications, it is worthwhile for us to develop an understanding of it.

Let $\mathcal{f}(\mathbf{k}, \mathbf{r}, t)$ be the probability at time t of occupation of the state corresponding to the wave vector $\mathbf{k}$ at a point in the crystal given by the position vector $\mathbf{r}$. When we consider the variation of $\mathcal{f}$ with $\mathbf{r}$ we are concerned only with variations which are very small over distances of the order of a lattice spacing. Wave functions corresponding to a definite momentum $\mathbf{p}$ and a definite value of

$\mathbf{k} = \mathbf{p}/\hbar$ may then be specified locally in the crystal. The spatial variation allows for inhomogeneities in the crystal (such as impurity gradients) and also for temperature variations across the sample. Since the occupation probability $\mathcal{f}(\mathbf{k}, \mathbf{r}, t)$ considered herein is a more general quantity than the Fermi–Dirac function $\mathcal{f}(\mathscr{E})$ derived in Chap. 2 for a spatially uniform system of electrons in thermal equilibrium, $\mathcal{f}(\mathbf{k}, \mathbf{r}, t)$ will be referred to generally as the *distribution function*. Under conditions of thermal equilibrium (where there will be zero current), the distribution function will be written as $\mathcal{f}_0(\mathbf{k}, \mathbf{r}, t)$. It is well to keep in mind that for a spatially uniform system in thermal equilibrium, the occupation probability must be given by the Fermi–Dirac function $\mathcal{f}(\mathscr{E}) = \{\exp[(\mathscr{E} - \mathscr{E}_F)/k_B T] - 1\}^{-1}$ derived in Chap. 2, so that $\mathcal{f}(\mathbf{k}, \mathbf{r}, t) \to \mathcal{f}(\mathscr{E})$ in this limit. The dependence upon $\mathbf{k}$ reduces to a dependence on $\mathscr{E}$ in this case, in accordance with the free particle energy–momentum relation $\mathscr{E} = p^2/2m = \hbar^2 k^2/2m$.

In thermal equilibrium, the probability $\mathcal{f}(\mathscr{E}) = \mathcal{f}(\hbar^2 k^2/2m)$ that a state with wave vector $\mathbf{k}$ is occupied is the same as that of a state with wave vector $-\mathbf{k}$, so that no transport takes place. In order to have transport of charge (electrical conduction) or transport of energy (thermal conduction) the distribution must be modified to a nonequilibrium value by electric or magnetic fields or temperature gradients. We will see how this is done shortly.

The equation

$$\mathbf{F} = d\mathbf{p}/dt = \hbar\, d\mathbf{k}/dt \equiv \hbar\dot{\mathbf{k}} \tag{3.146}$$

was used in §3 in the derivation of Ohm's law; it tells us the effects produced on the $\mathbf{k}$ vector of a conduction electron by an externally applied force $\mathbf{F}$. An electron which at time t has the position $\mathbf{r}$ and momentum $\mathbf{p} = m\mathbf{v}$ will at the time $t - dt$ have had the position $\mathbf{r} - d\mathbf{r} = \mathbf{r} - \mathbf{v}\,dt$ and the momentum $\mathbf{p} - d\mathbf{p} = \hbar\mathbf{k} - \mathbf{F}\,dt$, and similarly at the time $t + dt$ it will have the position $\mathbf{r} + d\mathbf{r} = \mathbf{r} + \mathbf{v}\,dt$ and the momentum $\mathbf{p} + d\mathbf{p} = \hbar\mathbf{k} + \mathbf{F}\,dt$. Let us focus our attention upon the occupation of the energy levels in an element $d\Omega_{\text{phase space}} = d\mathbf{r}\,d\mathbf{p} \equiv (dx\,dy\,dz)(dp_x\,dp_y\,dp_z)$ of phase space. Using our assumption of slowly varying spatial variations in composition or temperature, we can conclude that the density of states, the Fermi energy and the equilibrium occupation probability $\mathcal{f}_0$ will vary insignificantly over the differential distances dx, dy, and dz in question. The time dependence of the occupation probability $\mathcal{f}(\mathbf{k}, \mathbf{r}, t)$ can therefore be deduced directly from the time dependence of the electron density in the element $d\Omega_{\text{phase space}}$. An *equation of continuity* can be invoked for the 6-dimensional phase space as readily as for the flow of particles and the buildup of particle densities in real space (Chap. 1, §9). The time rate of change of the occupation probability within the element $d\Omega_{\text{phase space}}$ at a given position $\mathbf{r}, \mathbf{p}$ of phase space is given by the difference between the inflow and the outflow of particles into this element. Consider first of all a particle density $n(x)$ which varies spatially in some manner, such as illustrated in Fig. 3.10.

We speak of particle densities because this is easier to visualize than occupation probabilities, but such a view is strictly correct only if all particles

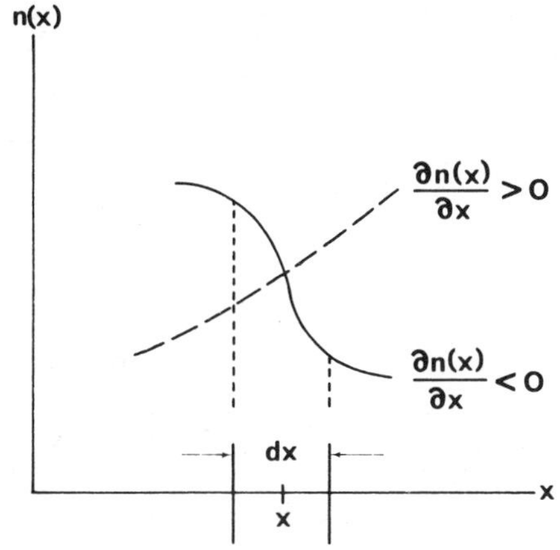

Fig. 3.10 Particle density variation with position x.

have the same velocity, i.e., a velocity corresponding to a given value of $\mathbf{k}$. In the more general case of a density of particles having a range of velocity vectors, the quantity $n(x)$ must therefore represent a particle density per unit volume per unit $\mathbf{k}$-vector range at a specific $\mathbf{k}$ vector and at a specific position $\mathbf{r}$. Thus we could write $n(x) = \tilde{f}(\mathbf{k}, \mathbf{r})\tilde{g}(\mathbf{k}, \mathbf{r})\, d\mathbf{k}$, where $\tilde{g}$ is the density of states at a specific $\mathbf{k}$ vector for the system at the spatial position $\mathbf{r}$. We then implicitly make the assumption that $\tilde{g}(\mathbf{k}, \mathbf{r})$ is at most a slowly varying function of $\mathbf{k}$ and $\mathbf{r}$ on a microscopic scale so that its variation can be neglected in setting up the equation for local particle balance. If $\partial n/\partial x < 0$, as indicated by the solid curve in Fig. 3.10, then for positive velocity v_x there will be more particles entering the region dx during a time increment dt than are leaving the region. This is due to the fact that there are more particles at distance $v_x\, dt$ to the left of the position x than at position x, so that in time dt more particles move into the region dx than move out of this region. Thus the time dependence of $n(x)$ due to the particle velocity v_x is

$$dn(x)|_{v_x \,\text{produced}} = -\left[\frac{\partial n(x)}{\partial x}\right] dx = -\left[\frac{\partial n(x)}{\partial x}\right] v_x\, dt, \qquad (3.147\text{a})$$

since the particles move a distance $dx = v_x\, dt$ during the time increment dt. The negative sign assures us that $dn(x)/dt$ is positive when v_x is positive and $\partial n(x)/\partial x$ is negative. On the other hand, if $\partial n(x)/\partial x > 0$, as indicated by the *dashed* curve in Fig. 3.10, then for positive velocity v_x there will be fewer particles entering the region dx during a time increment dt than leaving the region, corresponding to a negative value of $dn(x)$. Therefore the above expression (3.147a) is also applicable to this case. Similarly, a reversal in sign of v_x in Eq. (3.147a) reverses the sign of $dn(x)$, in agreement with what one expects physically. These considerations are independent of the specific choice of x, holding equally well for negative values of x as for positive values of x. This is evident mathematically from Eq. (3.147a) which involves the coordinate x only in the derivative [i.e., in the *slope* $\partial n(x)/\partial x$]. If we now consider the variation of the electron density in the y direction or the z direction also, we must substitute $n(\mathbf{r}) = n(x, y, z)$ for $n(x)$. It is evident that, aside from replacement of the coordinate x by the appropriate variable y or z, the same general results follow in the same straightforward way. Thus we can write a generalization of (3.147a),

$$dn(\mathbf{r})|_{\mathbf{v}\text{ produced}} = -\frac{\partial n(\mathbf{r})}{\partial x}\,dx - \frac{\partial n(\mathbf{r})}{\partial y}\,dy - \frac{\partial n(\mathbf{r})}{\partial z}\,dz = -[\nabla n(\mathbf{r})]\cdot d\mathbf{r}$$

$$= -[\nabla n(\mathbf{r})]\cdot \mathbf{v}\,dt = -\mathbf{v}\cdot[\nabla n(\mathbf{r})]\,dt, \tag{3.147b}$$

where $\nabla \equiv \hat{\mathbf{x}}\,\partial/\partial x + \hat{\mathbf{y}}\,\partial/\partial y + \hat{\mathbf{z}}\,\partial/\partial z$ is the gradient operator and $d\mathbf{r} \equiv \hat{\mathbf{x}}\,dx + \hat{\mathbf{y}}\,dy + \hat{\mathbf{z}}\,dz$ is the incremental vector distance.

Because the occupation probability changes directly as the electron density, as argued above, we can write analogously

$$d\mathcal{f}(\mathbf{k}, \mathbf{r}, t)|_{\mathbf{v}\text{ produced}} = -[\nabla \mathcal{f}(\mathbf{k}, \mathbf{r}, t)]\cdot \mathbf{v}\,dt = -\mathbf{v}\cdot[\nabla \mathcal{f}(\mathbf{k}, \mathbf{r}, t)]\,dt. \tag{3.147c}$$

In a somewhat analogous way, the number of electrons in $d\Omega_{\text{phase space}}$ having momentum in the range $\mathbf{p}$ to $\mathbf{p} + d\mathbf{p}$ is changing with time because of applied forces $\mathbf{F}$. Figure 3.11 constitutes a sketch of the number of electrons per unit volume per unit momentum range as a function of momentum p_x in the x direction, assuming fixed values for p_y and p_z. The sketch for positive p_x looks much like the plot of the Fermi–Dirac function (see Fig. 2.4) because $\mathcal{G}(\mathbf{p}) = 1/h^3$ which is independent of p_x, and the thermal equilibrium distribution function $\mathcal{f}_0$ varies as $\mathscr{E} = (p_x^2 + p_y^2 + p_z^2)/2m$ in the free-electron model. The negative p_x domain is a reflection of the positive domain since $\mathscr{E}(-p_x) = \mathscr{E}(p_x)$. At a particular value of p_x such as indicated in Fig. 3.11, note that the slope of $2\mathcal{G}(\mathbf{p})\mathcal{f}(p_x)$ is negative. A force F_x that tends to increase p_x in the positive direction takes the particles at momentum p_x to a new momentum value

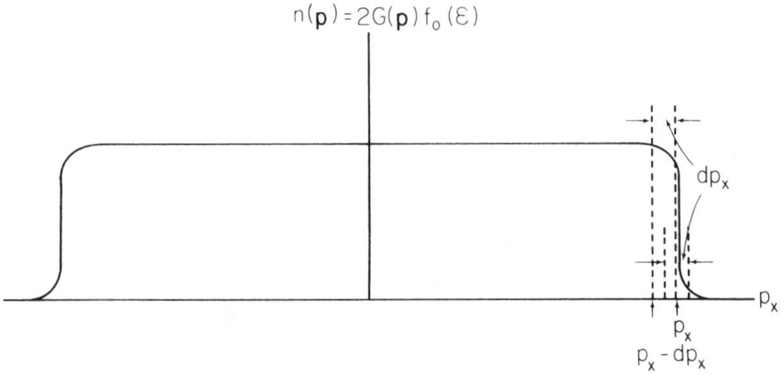

Fig. 3.11 Variation of the number of electrons $n(\mathbf{p})$ per unit volume with x momentum p_x, assuming fixed values for the y and z momentum components. (The electron density $n(\mathbf{p})$ is given by a product of the density of states $2\mathcal{G}(\mathbf{p})$ with the occupation probability $\mathcal{f}(\mathscr{E})$; the factor of 2 is due to the two allowed values of the electron spin quantum number.)

$p_x + dp_x = p_x + F_x\, dt$ in a time dt, thus depleting the region dp_x at p_x. However, the force simultaneously brings even more particles into the region dx, since Fig. 3.11 shows the density at $p_x - dp_x$ to be greater than the density at p_x. Thus

$$dn(p_x)|_{F_x \text{ produced}} = -\left[\frac{\partial n(p_x)}{\partial p_x}\right] dp_x = -\left[\frac{\partial n(p_x)}{\partial p_x}\right] F_x\, dt. \qquad (3.148a)$$

Considering the other momentum components p_y and p_z analogously, we arrive at the conclusion that

$$dn(\mathbf{p})|_{\mathbf{F} \text{ produced}} = -\left[\frac{\partial n(\mathbf{p})}{\partial p_x}\right] F_x\, dt - \left[\frac{\partial n(\mathbf{p})}{\partial p_y}\right] F_y\, dt - \left[\frac{\partial n(\mathbf{p})}{\partial p_z}\right] F_z\, dt$$

$$= -\,[\nabla_{\mathbf{p}} n(\mathbf{p})] \cdot \mathbf{F}\, dt = -\,\mathbf{F} \cdot [\nabla_{\mathbf{p}} n(\mathbf{p})]\, dt. \qquad (3.148b)$$

The symbol $\nabla_{\mathbf{p}}$ denotes the operation $\hat{\mathbf{x}}\, \partial/\partial p_x + \hat{\mathbf{y}}\, \partial/\partial p_y + \hat{\mathbf{z}}\, \partial/\partial p_z$, and since $\mathbf{p} = \hbar \mathbf{k}$, we can also write $\nabla_{\mathbf{p}} = \hbar^{-1} \nabla_{\mathbf{k}}$, where $\nabla_{\mathbf{k}} = \hat{\mathbf{x}}\, \partial/\partial k_x + \hat{\mathbf{y}}\, \partial/\partial k_y + \hat{\mathbf{z}}\, \partial/\partial k_z$. The occupation probability varies directly as the electron density because $\mathscr{G}(\mathbf{p})$ is uniform, so that we can write

$$d\mathscr{f}(\mathbf{k}, \mathbf{r}, t)|_{\mathbf{F} \text{ produced}} = -\,[\hbar^{-1} \nabla_{\mathbf{k}} \mathscr{f}(\mathbf{k}, \mathbf{r}, t)] \cdot \mathbf{F}\, dt = -\,\mathbf{F} \cdot [\hbar^{-1} \nabla_{\mathbf{k}} \mathscr{f}(\mathbf{k}, \mathbf{r}, t)]\, dt. \qquad (3.148c)$$

In addition to the forces and velocities considered explicitly above, there are collisions of the electrons with point defects, lattice vibrations, and other similar deviations from perfect lattice periodicity, and such collisions tend to bring any given nonequilibrium occupation probability $\mathscr{f}(\mathbf{k}, \mathbf{r}, t)$ towards its thermal equilibrium value which we denote as $\mathscr{f}_0(\mathbf{k}, \mathbf{r}, t)$. In the absence of either an applied force $\mathbf{F}$ or a spatial gradient $\nabla \mathscr{f}$ (such as could be produced by an electric field or a temperature gradient, respectively), the collisions will eventually return the electron system to its thermal equilibrium state. In addition, a nonzero $\mathbf{F}$ and/or a nonzero $\nabla \mathscr{f}$ can be counterbalanced continuously by the collisions such that macroscopically there will be no discernible time dependence of $\mathscr{f}$. This latter situation is denoted as the *steady state*. In any case, a contribution $d\mathscr{f}(\mathbf{k}, \mathbf{r}, t)|_{\text{coll}}$ must be included when writing the total variation of $\mathscr{f}(\mathbf{k}, \mathbf{r}, t)$ over a time increment dt.

The total time rate of change of the distribution function can therefore be written as the sum of the various contributions discussed above,

$$d\mathscr{f}(\mathbf{k}, \mathbf{r}, t)/dt = \partial \mathscr{f}(\mathbf{k}, \mathbf{r}, t)/\partial t|_{\text{coll}} - \hbar^{-1} \mathbf{F} \cdot \nabla_{\mathbf{k}} \mathscr{f}(\mathbf{k}, \mathbf{r}, t) - \mathbf{v} \cdot \nabla \mathscr{f}(\mathbf{k}, \mathbf{r}, t)$$

(the Boltzmann equation), (3.149)

where we have used Eqs. (3.147c) and (3.148c) and the collision term. This result, known as *Boltzmann's transport equation*, is the fundamental equation governing all transport phenomena.

7.2 Limiting Cases

In the *steady-state limit* there is no observable time dependence of $\tilde{f}$, in which case $d\tilde{f}/dt = 0$ and we obtain

$$(\partial \tilde{f}/\partial t)|_{\text{coll}} = \hbar^{-1}\mathbf{F} \cdot \nabla_{\mathbf{k}}\tilde{f} + \mathbf{v} \cdot \nabla\tilde{f} \qquad \text{(steady-state limit)}. \qquad (3.150)$$

If the external fields, temperature gradients, and similar spatial inhomogeneities are removed from the sample, the distribution will relax back to thermal equilibrium. The rate of relaxation will depend upon the efficiency of the collision mechanisms in bringing about an *equipartition* of kinetic energy of motion in the three spatial directions. As a simple case, we could assume that the rate of relaxation is directly proportional to the deviation from equilibrium,

$$(\partial \tilde{f}/\partial t)_{\text{coll}} = -(\tilde{f} - \tilde{f}_0)/\tau, \qquad (3.151)$$

corresponding to a time dependence given by $\tilde{f} = \tilde{f}_0 + [(\tilde{f}_{\text{initial}} - \tilde{f}_0)]\exp(-t/\tau)$ in the absence of forces. In this expression, $\tilde{f}_0$ represents $\tilde{f}$ at thermal equilibrium, and τ is some characteristic relaxation time. This approximation has some validity if the departure from thermal equilibrium is not too great. (In real nonisotropic crystals, however, there is little justification for assuming τ to be a *scalar* quantity.)

Under the approximation (3.151), the steady-state Boltzmann equation (3.150) becomes

$$-(\tilde{f} - \tilde{f}_0)/\tau = \hbar^{-1}\mathbf{F} \cdot \nabla_{\mathbf{k}}\tilde{f} + \mathbf{v} \cdot \nabla\tilde{f} \qquad (3.152)$$

or equivalently

$$\tilde{f} = \tilde{f}_0 - \hbar^{-1}\tau\mathbf{F} \cdot \nabla_{\mathbf{k}}\tilde{f} - \tau\mathbf{v} \cdot \nabla\tilde{f}. \qquad (3.153)$$

For a homogeneous crystal at constant temperature, $\tilde{f}$ is independent of $\mathbf{r}$, in which case

$$\tilde{f} = \tilde{f}_0 - \hbar^{-1}\tau\mathbf{F} \cdot \nabla_{\mathbf{k}}\tilde{f} \qquad \text{(homogeneous medium)}. \qquad (3.154)$$

For $\tilde{f}$ near enough to $\tilde{f}_0$, we can approximate Eq. (3.154) by replacing $\nabla_{\mathbf{k}}\tilde{f}$ by $\nabla_{\mathbf{k}}\tilde{f}_0$,

$$\tilde{f} = \tilde{f}_0 - \hbar^{-1}\tau\mathbf{F} \cdot \nabla_{\mathbf{k}}\tilde{f}_0. \qquad (3.155)$$

However, $\tilde{f}_0$ is a function of $\mathscr{E} = \mathscr{E}(k_x, k_y, k_z)$, so

$$\nabla_{\mathbf{k}}\tilde{f}_0 = \hat{\mathbf{x}}\frac{\partial \tilde{f}_0}{\partial k_x} + \hat{\mathbf{y}}\frac{\partial \tilde{f}_0}{\partial k_y} + \hat{\mathbf{z}}\frac{\partial \tilde{f}_0}{\partial k_z},$$

$$\nabla_{\mathbf{k}}\tilde{f}_0 = \hat{\mathbf{x}}\frac{d\tilde{f}_0}{d\mathscr{E}}\left(\frac{\partial \mathscr{E}}{\partial k_x}\right) + \hat{\mathbf{y}}\frac{d\tilde{f}_0}{d\mathscr{E}}\left(\frac{\partial \mathscr{E}}{\partial k_y}\right) + \hat{\mathbf{z}}\frac{d\tilde{f}_0}{d\mathscr{E}}\left(\frac{\partial \mathscr{E}}{\partial k_z}\right) = \left(\frac{d\tilde{f}_0}{d\mathscr{E}}\right)\nabla_{\mathbf{k}}\mathscr{E}. \qquad (3.156)$$

The dispersion relation $\mathscr{E}(\mathbf{k})$ for the free-electron model follows from $\mathscr{E} = p^2/2m = \hbar^2 k^2/2m$,

$$\mathscr{E} = (\hbar^2/2m)(k_x^2 + k_y^2 + k_z^2), \qquad (3.157)$$

in which case

$$\nabla_{\mathbf{k}}\mathscr{E} = \hbar^2\mathbf{k}/m = \hbar\mathbf{p}/m = \hbar\mathbf{v}, \tag{3.158}$$

where $\mathbf{v}$ is the electron velocity for a filled state characterized by the wave vector $\mathbf{k}$. Thus

$$\nabla_{\mathbf{k}}\tilde{f}_0 = (d\tilde{f}_0/d\mathscr{E})\hbar\mathbf{v}, \tag{3.159}$$

and substituting this result into the above expression (3.155) for $\tilde{f}$ gives

$$\tilde{f} \simeq f_0 - \tau(d\tilde{f}_0/d\mathscr{E})\mathbf{F} \cdot \mathbf{v}. \tag{3.160}$$

The Boltzmann equation thus has led us to an approximate expression for the occupation probability for a system of particles which obey a free electron dispersion relation and which are acted on by an applied force $\mathbf{F}$.

7.3   Electrical Conductivity

As a specific example, consider the force $\mathbf{F}$ to be due to an applied electric field in the x direction, $\mathbf{E} = E_x\hat{\mathbf{x}}$. Then $\mathbf{F} = -eE_x\hat{\mathbf{x}}$ gives the force acting upon each electron, and $\mathbf{F} \cdot \mathbf{v} = -eE_x v_x$. Substituting into the above expression (3.160) for $\tilde{f}$ gives

$$\tilde{f} \simeq \tilde{f}_0 + \tau eE_x v_x \, d\tilde{f}_0/d\mathscr{E}. \tag{3.161}$$

Considering the quantum free-electron model, the density of states $\tilde{w}(\mathbf{k})$ in momentum space per direction of spin per unit volume of the metal is $1/8\pi^3$ [Eq. (3.16)]. The particle current density J_x in the x direction is

$$J_x = \int 2\tilde{w}(\mathbf{k})\tilde{f}v_x \, d\mathbf{k}, \tag{3.162}$$

where the factor of 2 takes into account the two spin directions per allowed $\mathbf{k}$ vector. That is, $2\tilde{w}(\mathbf{k})\tilde{f}$ gives the density of occupied electronic states at $\mathbf{k}$, and $2\tilde{w}(\mathbf{k})\tilde{f} \, d\mathbf{k}$ gives the number of electrons in the element of volume $d\mathbf{k}$ in wave vector space. The product of this number and the corresponding velocity $v_x = p_x/m = \hbar k_x/m$ in the x direction gives the x component of the current density contribution due to this group of electrons, which when integrated over all of $\mathbf{k}$ space yields the total current density J_x in the x direction. Substituting our expression (3.161) for $\tilde{f}$ into this integral and using the value $1/4\pi^3$ for $2\tilde{w}(\mathbf{k})$ gives

$$J_x = (4\pi^3)^{-1}\int v_x(\tilde{f}_0 + \tau eE_x v_x \, d\tilde{f}_0/d\mathscr{E}) \, d\mathbf{k}. \tag{3.163}$$

For the free-electron parabolic dispersion relation $\mathscr{E}(\mathbf{k}) = \hbar^2 k^2/2m$, we have $\mathscr{E}(\mathbf{k}) = \mathscr{E}(-\mathbf{k})$. This follows from Eq. (3.11). For every allowed state $\mathbf{k}$, there is a degenerate state with wave vector $-\mathbf{k}$. (See Fig. 3.12.) The thermal equilibrium distribution function $\tilde{f}_0$ for electrons in a homogeneous metal is the Fermi–Dirac distribution function, which gives the same occupation probability for states

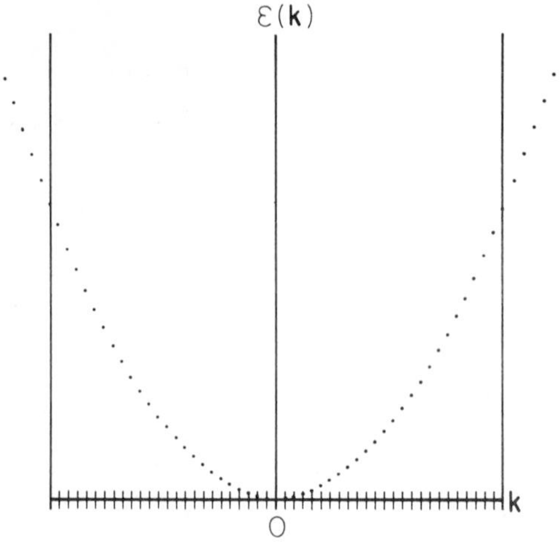

Fig. 3.12 Energy eigenvalues $\mathscr{E}(\mathbf{k})$ as a function of wave vector $\mathbf{k}$ for the free-electron model. [The discrete values of $\mathbf{k}$ follow from the application of periodic boundary conditions to the three-dimensional solid; the energy eigenvalues then follow from Eq. (3.11), namely, $\mathscr{E} = \hbar^2k^2/2m$.]

with the same energy. Thus $v_x\mathscr{f}_0 = (\hbar/m)\mathscr{f}_0k_x$ is an odd function of k_x, and this portion of the integrand gives no contribution to the value of the above integral. Hence

$$J_x = (4\pi^3)^{-1}\int \tau eE_xv_x^2(d\mathscr{f}_0/d\mathscr{E})\,d\mathbf{k}. \tag{3.164}$$

The factor eE_x is independent of $\mathbf{k}$, but the relaxation time τ in general will depend on $\mathbf{k}$. For an isotropic crystal, this will reduce to a dependence on $|\mathbf{k}|$, which in the free-electron model is equivalent to an energy dependence $\tau = \tau(\mathscr{E})$, since $\mathscr{E} = \hbar^2k^2/2m$ for this model.

The triple integral over $d\mathbf{k}$ in Eq. (3.164) can be simplified to a single integral over $d|\mathbf{k}|$ by expressing $d\mathbf{k}$ in spherical polar coordinates and performing the integrations over the θ and ϕ coordinates. That is, $d\mathbf{k} = k^2 \sin\theta\,dk\,d\theta\,d\phi$, where k indicates $|\mathbf{k}|$. This, however, is easily converted to an integral over the energy by using $\mathscr{E} = \hbar^2k^2/2m$.

Thus $k^2 = 2m\mathscr{E}/\hbar^2$, so that

$$d\mathscr{E} = (\hbar^2k/m)\,dk, \tag{3.165}$$

$$dk = d[\hbar^{-1}(2m\mathscr{E})^{1/2}] = \hbar^{-1}m(2m\mathscr{E})^{-1/2}\,d\mathscr{E}. \tag{3.166}$$

This leads to

$$\begin{aligned} d\mathbf{k} &= k^2\sin\theta\,dk\,d\theta\,d\phi = (2m\mathscr{E}/\hbar^2)\sin\theta\,(m/\hbar)(2m\mathscr{E})^{-1/2}\,d\mathscr{E}\,d\theta\,d\phi \\ &= (m/\hbar^3)(2m\mathscr{E})^{1/2}\sin\theta\,d\theta\,d\phi\,d\mathscr{E}. \end{aligned} \tag{3.167}$$

The velocity $v_x = p_x/m = \hbar k_x/m$ can likewise be obtained by resolving $\mathbf{k}$ into spherical polar coordinates,

$$v_x = (\hbar/m)k_x = (\hbar/m)k \sin\theta\cos\phi = m^{-1}(2m\mathscr{E})^{1/2}\sin\theta\cos\phi. \tag{3.168}$$

Substituting Eqs. (3.167) and (3.168) into Eq. (3.164) for J_x gives

$$\begin{aligned} J_x &= (4\pi^3)^{-1}eE_x\left(\frac{2}{m}\right)\left(\frac{m}{\hbar^3}\right)(2m)^{1/2} \\ &\quad\times\int\tau(\mathscr{E})\left(\frac{d\mathscr{f}_0}{d\mathscr{E}}\right)(\mathscr{E}\sin^2\theta\cos^2\phi)(\mathscr{E}^{1/2}\sin\theta\, d\mathscr{E}\, d\theta\, d\phi) \\ &= \left[\frac{(2m/\hbar^2)^{3/2}eE_x}{4\pi^3 m}\right]\int_0^\infty \tau(\mathscr{E})\mathscr{E}^{3/2}\left(\frac{d\mathscr{f}_0}{d\mathscr{E}}\right)d\mathscr{E}\int_0^\pi \sin^3\theta\, d\theta\int_0^{2\pi}\cos^2\phi\, d\phi \\ &= \left[\frac{(2m/\hbar^2)^{3/2}eE_x}{3\pi^2 m}\right]\int_0^\infty \tau(\mathscr{E})\mathscr{E}^{3/2}\left(\frac{d\mathscr{f}_0}{d\mathscr{E}}\right)d\mathscr{E}. \end{aligned} \tag{3.169}$$

The integrations over ϕ and θ in Eq. (3.169) were carried out to obtain factors of π and $\frac{4}{3}$, respectively. We note from Eq. (3.169) that the current density J_x is linearly related to the electric field E_x, so that the deduced dependence is in accordance with Ohm's law. Defining the conductivity σ by $\mathscr{J}_x = -eJ_x = \sigma E_x$, we thus obtain from Eq. (3.169) the following result for the electrical conductivity,

$$\sigma = \frac{-(2m)^{3/2}e^2}{3\pi^2 m\hbar^3}\int_0^\infty \tau(\mathscr{E})\mathscr{E}^{3/2}\left(\frac{d\mathscr{f}_0}{d\mathscr{E}}\right)d\mathscr{E}. \tag{3.170}$$

This expression can be simplified for metals by recalling that the energy derivative of the Fermi–Dirac distribution function is almost zero everywhere except within the narrow band of energies within several multiples of k_BT of the Fermi energy $\mathscr{E}_F$. Since $\tau(\mathscr{E})$ and $\mathscr{E}^{3/2}$ do not vary sharply over such a small energy range, the product $\tau(\mathscr{E})\mathscr{E}^{3/2}$ can be replaced by $\tau(\mathscr{E}_F)\mathscr{E}_F^{3/2}$ in Eq. (3.170) without modifying the value of the integral significantly. The product $\tau(\mathscr{E}_F)\mathscr{E}_F^{3/2}$ can then be removed from the integral. Since in addition

$$\int_0^\infty \left(\frac{d\mathscr{f}_0}{d\mathscr{E}}\right)d\mathscr{E} = \mathscr{f}_0(\infty) - \mathscr{f}_0(0) \simeq 0 - 1 = -1, \tag{3.171}$$

we thus obtain

$$\sigma \simeq [(2m)^{3/2}e^2\tau(\mathscr{E}_F)\mathscr{E}_F^{3/2}/3\pi^2 m\hbar^3]. \tag{3.172}$$

This result becomes more and more exact at lower temperatures, since $d\mathscr{f}_0/d\mathscr{E}$ approaches the negative of the Dirac delta function $\delta(\mathscr{E} - \mathscr{E}_F)$ as the temperature approaches zero.

PROJECT 3.12 Boltzmann Equation with a Magnetic Field

1. Apply the Boltzmann equation to the case of a homogeneous solid at a uniform temperature in the presence of stationary electric and magnetic fields. In particular, deduce the current for the

situation in which the magnetic field is along the z direction and the electric field is in the xy plane, and from this result derive the Hall coefficient.
2. Derive the *magnetoconductivity* tensor.

PROJECT 3.13 Boltzmann Equation with a Temperature Gradient

Apply the Boltzmann equation to the following situations:
1. Electrical transport in a spatially homogeneous metal at a uniform temperature.
2. Electrical and thermal transport in a spatially homogeneous metal with a uniform temperature gradient.
3. In Part 2, solve for the electronic contribution to the thermal conductivity for the case of blocking electrodes.
4. Solve for the Lorentz ratio, the Lorentz number, and justify the Wiedemann–Franz law.
5. Deduce expressions for the Thomson coefficient and the absolute thermoelectric power.
6. Give descriptions of the Ettingshausen, Nernst, and Righi–Leduc effects.
[*Hint*: See Smith (1963).]

7.4 Mean Free Path

The conductivity as derived above is not exactly the same expression which was previously obtained [Eq. (3.73)] using the simpler approach; however it can be converted to a closely analogous form. The electron density $n = N/V$ is related to the Fermi energy $\mathscr{E}_F$ by Eq. (3.29), which gives

$$\mathscr{E}_F^{3/2} = (\hbar^2/2m)^{3/2} 3\pi^2 n. \tag{3.173}$$

Substituting this evaluation of $\mathscr{E}_F^{3/2}$ into expression (3.172) for σ gives

$$\sigma = ne^2\tau(\mathscr{E}_F)/m, \tag{3.174}$$

which is identical to Eq. (3.73) with the exception that the relaxation time at the Fermi surface replaces the average (or effective) relaxation time introduced in the simpler approach. This shows that it is actually the relaxation time at the Fermi surface which is important in limiting the conductivity. The electrons deeper within the Fermi sphere have no adjacent empty states into which they can be scattered. The mean free path l_e between scattering events thus will be determined by $\tau(\mathscr{E}_F)$ in accordance with

$$l_e = v_F \tau(\mathscr{E}_F), \tag{3.175}$$

where v_F is the magnitude of the velocity at the Fermi surface.

This special example of the use of the Boltzmann transport equation is sufficient to illustrate the approach. It is a powerful and general technique which can be used for semiconductors as well as metals. It can be used to treat heat conduction as well as electrical conduction, and the effects of externally applied magnetic fields on thermal and electrical transport can be included. The Boltzmann transport equation is not restricted to the free-electron model, since it can be applied readily to a solid with an energy band structure.

PROBLEMS

1. Write a short essay on the quantum-mechanical free-electron model for metals. Describe the model, discuss the approximations involved relative to a real metal and list the successes of the model relative to classical theory.

2. A particle is confined to a three-dimensional cubic potential-energy well 1 m on a side with infinitely high barriers. Compute its ground-state energy and the energy of the first five excited states. (Choose a coordinate system in which the box lies inside the first quadrant, with the origin of the coordinate system at one corner of the box.)

3. In Problem 2, estimate the probability in each of the states that the particle is within a 1-cm^3 element of volume centered at position $\mathbf{r} = 0.25\hat{\mathbf{x}} + 0.25\hat{\mathbf{y}} + 0.25\hat{\mathbf{z}}$ in units of meters.

4. Determine the degeneracy of the five lowest energy levels for a particle in a three-dimensional cubic potential well with infinitely high walls.

5. Determine the energy expectation values for a particle in a three-dimensional well of infinite depth for the free-electron model.

6. (a) Determine the momentum expectation values for a particle in a three-dimensional potential well of infinite depth for the free-electron model. (b) Also determine $\langle \mathbf{p}^2 \rangle$.

7. Red light having $\lambda = 6300$ Å is found to be of the correct energy to promote an electron in a box (viz., infinite square-well potential) from the ground state to the first excited state. What is the largest dimension of the box?

8. (a) Give the differences between electron–electron collisions as predicted classically and quantum mechanically. (b) State and justify the reasons for these differences.

9. Explain how the Fermi–Dirac distribution function in quantum statistics leads directly to the prediction of electron speeds in metals within one or two orders of magnitude of the speed of light.

10. Compute the Fermi energy $\mathscr{E}_F$ for Cu, Ag, Au, Na, Li, K, Cs.

11. Compute the Fermi wavelength λ_F for Problem 10 above and compare with the lattice spacing d.

12. Show that the average energy per electron at 0°K is $\langle \mathscr{E} \rangle = \frac{3}{5}\mathscr{E}_F(0)$, where $\mathscr{E}_F(0)$ is the Fermi energy at 0°K.

13. Find the density of states per unit energy range at the Fermi energy for sodium. Assume that the free-electron model holds.

14. How many electrons per unit volume occupy energy states within 0.5 eV of the Fermi energy in sodium at 0°K?

15. Determine the electronic specific heat of copper at $T = 300$°K. (Assume one carrier per atom.)

16. Calculate the specific heat for an electron gas with Fermi energy 7 eV, and compare it with the specific heat for a corresponding system of particles obeying classical statistics.

17. What is the room temperature heat capacity of an electron cloud containing 6×10^{22} electrons which has an effective Fermi temperature of 37,000°K?

18. (a) Assume lithium follows the FEM closely. Calculate the Fermi energy of lithium at 27°C. Also calculate the Fermi velocity. (b) Calculate the electronic specific heat for lithium at 300°K. (c) Assuming the Fermi energy calculated in Part (a) for lithium is correct at absolute zero, obtain the Fermi energy at 300°K and 6000°K using the equation derived using the approximation technique.

19. Calculate $\tau(\mathscr{E}_F)$, the relaxation time at the Fermi surface, for (a) copper, (b) gold, and (c) zinc.

20. Use the relaxation time approximation to find the conductivity of n-type silicon doped with 1×10^{18} donor atoms/cm^3. Assume $\tau_n = 10^{-10}$ sec.

21. In the free-electron model how would you expect the resistance of a metal to change with respect to temperature? Why?

22. A filament in a vacuum tube is made of thoriated tungsten. In operation the filament reaches a temperature of 1600°K. Assume the electron emission is not space-charge limited and that no reflection occurs at the metal–vacuum interface. If the filament is 2 cm long and has a diameter of 0.02 cm, what electron emission current leaves the filament?

23. Deduce the thermionic current for tungsten wire with 0.1 cm^2 surface area at 1727°C.

24. The cathode of a diode has an area of 0.1 cm^2 and is operated at 1140°C. If a current of 1 A is emitted, what is the work function?

25. A certain metal when heated to 1000°K produces a current of 3.2×10^{-2} A/cm^2. Determine its electronic work function.

26. Evaluate the following integrals using the general approximation technique:

$$I = \int_0^\infty \tilde{z}(\mathscr{E})\tilde{f}(\mathscr{E})\, d\mathscr{E},$$

where $\tilde{f}(\mathscr{E})$ is the Fermi probability function, and (a) $\tilde{z}(\mathscr{E}) = A\mathscr{E}^{3/2}$ with $A = \text{const}$, (b) $\tilde{z}(\mathscr{E}) = Ae^{-\mathscr{E}/k_BT}$ with $A = \text{const}$.

27. (a) Does the Fermi energy increase or decrease with a temperature increase in the three-dimensional free-electron model? (b) What is the approximate fractional change of $\mathscr{E}_F$ as the temperature goes from 0°K to room temperature?

28. Assume the function $G(x) = (1/5\pi^2)(2m/\hbar^2)^{3/2}(k_BTx + \mathscr{E}_F)^{5/2}$. (a) Show that the integral obtained by substituting $dG(\mathscr{E})/d\mathscr{E}$ for $\tilde{z}(\mathscr{E})$ in Problem 26 gives a representation of $\langle\mathscr{E}\rangle$. (b) Compute $\langle\mathscr{E}\rangle$ in series form through two nonzero terms. (c) Compute the electronic specific heat C_{el} for $T > 0$°K.

29. How would one proceed to calculate the approximate energy necessary to raise the temperature of a neutron star by an order of magnitude? (Assume we know its mass and its very high density.)

30. For a one-dimensional FEM, do the following: (a) Set up the Hamiltonian. (b) Write the Schrödinger equation. (c) Solve for the eigenfunctions. (d) Solve for the eigenvalues, assuming periodic boundary conditions. (e) Compute the density of states $\tilde{w}(\mathbf{k})$ as a function of wave vector and the density of states $\mathscr{G}(\mathbf{p})$ as a function of momentum. (f) Compute the density of states $\tilde{g}(\mathscr{E})$ as a function of energy. (g) Compute the Fermi energy $\mathscr{E}_F$ as a function of electron density n. (h) Compute the electronic specific heat. (i) Compute the electrical conductivity. (j) Compute the thermal emission current.

31. Repeat Problem 30 above for the two-dimensional free-electron model. Consider unequal lengths L_x, L_y for the potential well in the x and y directions.

32. (a) Compare the derived results for the one- and two-dimensional free-electron models obtained in Problems 30 and 31. (b) Tabulate results for the one-, two-, and three-dimensional free-electron models, noting especially the differences in the functional dependence of the density of states $\tilde{g}(\mathscr{E})$ on energy $\mathscr{E}$.

33. (a) Derive the average energy per electron for a nondegenerate free-electron gas at a finite temperature. (b) Derive the formula for the energy per unit volume for electromagnetic radiation in a cavity as a function of frequency and temperature.

34. (a) List the two primary contributions to the specific heat in a metal. (b) What is the experimentally observed temperature dependence of each component?

35. (a) Derive the screened Coulomb potential appropriate for a test charge immersed in a metal. (b) Define: (i) Screening length, (ii) Bare Coulomb potential, (iii) Screened (or "dressed") Coulomb potential, (iv) Friedel oscillations.

36. (a) Show, starting with the Boltzmann transport equation and using the relaxation time approximation, that the electrical conductivity of a homogeneous semiconductor, viewed as a Boltzmann gas of free electrons and holes, can be written as $\sigma = e(n\mu_n + p\mu_p)$. The parameters μ_n, μ_p are the "mobilities," which represent the average drift velocity per unit applied electric field for electrons and holes. (b) Evaluate μ_n and μ_p in terms of the relaxation times τ_n and τ_p.

PART III

Approximation Techniques for the Schrödinger Equation

CHAPTER 4

THE WKB APPROXIMATION AND ELECTRON TUNNELING

Once an electron is represented by the wave function, it penetrates into a classically forbidden region, and can tunnel through a reasonably thin potential barrier without any real "tunnel." L. Esaki (1967)

1 Development of the WKB Approximation

The Schrödinger equation is difficult to solve unless the potential energy $U(x)$ has some particularly simple form, so approximation techniques are frequently useful. Perturbation techniques, variational techniques, and the WKB approximation are all well-known tools in quantum mechanics. This section is concerned with the development of the Wentzel–Kramers–Brillouin (WKB) technique.

Consider the one-dimensional time-independent Schrödinger equation as obtained from Eq. (1.130),

$$d^2\phi/dx^2 + (2m/\hbar^2)(\mathscr{E} - U)\phi = 0, \tag{4.1}$$

where $U = U(x)$ is the potential energy and $\mathscr{E}$ is the total energy of the particle in question. Let

$$k(x) = [(2m/\hbar^2)(\mathscr{E} - U)]^{1/2}. \tag{4.2}$$

Whenever $\mathscr{E} < U$, k is imaginary; in such cases we can replace k by $i\eta(x)$, where $\eta(x)$ is real. Regions for which $k(x)$ is real are called classically allowed regions, while regions for which $k(x)$ is imaginary are called classically forbidden regions. The following mathematical development is valid for both cases.

Whenever U is a constant the eigenfunctions are of the form $\phi(x) = A\exp(\pm ikx)$, with A equal to a constant, as can be verified immediately by substitution into the Schrödinger equation. Suppose U is not a constant, but instead varies slowly with x. Then the eigenfunctions should be nearly plane waves. Instead of the form $\exp(\pm ikx)$, let us choose $\exp[iw(x)]$, where $w(x)$

approaches $\pm kx$ in the limit where U is a constant. If

$$\phi(x) = A \exp[iw(x)], \tag{4.3}$$

then

$$d\phi/dx = i\phi \, dw/dx, \tag{4.4}$$

$$d^2\phi/dx^2 = i(i\phi(dw/dx)^2 + \phi \, d^2w/dx^2). \tag{4.5}$$

By substituting into the Schrödinger equation,

$$d^2\phi/dx^2 + k^2\phi = 0, \tag{4.6}$$

we obtain

$$-\phi(dw/dx)^2 + i\phi \, d^2w/dx^2 + k^2\phi = 0. \tag{4.7}$$

Dividing through by ϕ then gives us the following equation

$$i \, d^2w/dx^2 - (dw/dx)^2 + [k(x)]^2 = 0. \tag{4.8}$$

This alternate to the Schrödinger equation is nonlinear and therefore generally difficult to solve exactly; however, it provides a good basis for an approximation. In the limit in which U is a constant, $w = \pm kx$, with k independent of x. Then $dw/dx = \pm k$, and $d^2w/dx^2 = 0$. Therefore in cases for which $U(x)$ varies slowly with x, d^2w/dx^2 can be expected to be relatively small, though not zero. Equation (4.8) can then be approximated by

$$(dw/dx)^2 \simeq [k(x)]^2. \tag{4.9}$$

Thus

$$dw/dx \simeq \pm k(x), \tag{4.10}$$

$$d^2w/dx^2 \simeq \pm k'(x), \tag{4.11}$$

which must therefore be small. (The prime denotes differentiation with respect to x.) Integration of the above expression for dw/dx gives

$$w \simeq \pm \int^x k(x) \, dx + \Upsilon_0, \tag{4.12}$$

where Υ_0 is a constant of integration. Let this approximate value of w be designated by w_0,

$$w_0 = \pm \int^x k(x) \, dx + \Upsilon_0, \tag{4.13}$$

which represents the first in a sequence of successive approximations to w. Rearranging Eq. (4.8) in the form

$$(dw/dx)^2 = [k(x)]^2 + i \, d^2w/dx^2 \tag{4.14}$$

leads to

$$dw/dx = \pm\{[k(x)]^2 + i \, d^2w/dx^2\}^{1/2}, \tag{4.15}$$

so that

$$w(x) = \pm \int^x \{[k(x)]^2 + i\, d^2w/dx^2\}^{1/2}\, dx + \Upsilon, \tag{4.16}$$

where Υ is an appropriate constant of integration to be determined from normalization of the wave function. Using w_0 as the first approximation to w leads to

$$w_1(x) = \pm \int^x \{[k(x)]^2 + i\, d^2w_0/dx^2\}^{1/2}\, dx + \Upsilon_1. \tag{4.17}$$

This in turn can be used to obtain

$$w_2(x) = \pm \int^x \{[k(x)]^2 + i\, d^2w_1/dx^2\}^{1/2}\, dx + \Upsilon_2. \tag{4.18}$$

This can be continued as far as necessary; in general we can write

$$w_{n+1}(x) = \pm \int^x \{[k(x)]^2 + iw_n''(x)\}^{1/2}\, dx + \Upsilon_{n+1}, \tag{4.19}$$

where the prime means differentiation with respect to x.

Since

$$w_0' = \pm k(x), \tag{4.20}$$

$w_0'' = \pm k'(x)$, so that

$$w_1(x) = \pm \int^x \{[k(x)]^2 \pm ik'(x)\}^{1/2}\, dx + \Upsilon_1. \tag{4.21}$$

Convergence requires that each successive approximation must not deviate too markedly from the preceding approximation. From a comparison of $w_0(x)$ and $w_1(x)$, we see that this requires

$$|\pm k'(x)| \ll |[k(x)]^2|. \tag{4.22}$$

Thus, the binomial expansion can be used to approximate the square root in the integral in the following form of $w_1(x)$,

$$w_1(x) = \pm \int^x k(x)\left\{1 \pm i\frac{k'(x)}{[k(x)]^2}\right\}^{1/2} dx + \Upsilon_1. \tag{4.23}$$

Since $(1 \pm \delta)^{1/2} \simeq 1 \pm \frac{1}{2}\delta$ whenever $\delta \ll 1$, we obtain

$$\begin{aligned} w_1(x) &\simeq \pm \int^x k(x)\, dx + \frac{i}{2}\int^x \frac{k'(x)}{k(x)}\, dx + \Upsilon_1 \\ &\simeq +\left(\frac{i}{2}\right)\log_e[k(x)] \pm \int^x k(x)\, dx + \Upsilon_1. \end{aligned} \tag{4.24}$$

This expression is referred to as the WKB approximation. The stationary-state eigenfunction is given by

$$\phi = A \exp[iw(x)] \simeq A \exp[iw_1(x)], \tag{4.25}$$

so that

$$\begin{aligned}\phi &\simeq A \exp(i\Upsilon_1)\left\{\exp\left[\pm \int^x ik(x)\, dx\right]\right\} \exp\left\{i\left(\frac{i}{2}\right)\log_e[k(x)]\right\} \\ &\simeq A' \, k(x)^{-1/2} \exp\left[\pm i \int^x k(x)\, dx\right],\end{aligned} \tag{4.26}$$

where A' is determined from normalization of ϕ. The wave function $\psi(x, t)$ is given by the product of $\phi(x)$ and the usual time factor $\exp(-i\omega t)$ resulting from a separation of variables in the time-dependent Schrödinger equation. Thus

$$\psi(x, t) \simeq A'k(x)^{-1/2} \exp\left[\pm i \int^x k(x)\, dx - i\omega t\right] \quad \textbf{(WKB wave function)}, \tag{4.27}$$

where $\omega = \mathscr{E}/\hbar$. When $k(x)$ is real, the position dependence of this function is oscillatory; whenever $k(x) = i\eta(x)$ is imaginary, the position dependence of this function is exponential. It is worthwhile to note that the exponential portion represents the lowest order approximation to the wave function, whereas the pre-exponential factor $k(x)^{-1/2}$ is the first-order correction which involves the power series expansion of the square root in Eq. (4.23).

Consider the condition $|k'(x)| \ll |k(x)|^2$ stated above for the validity of this approximation. For the case where $k(x)$ is real, the momentum of the particle is

$$p = \hbar k = \hbar(2\pi/\lambda) = h/\lambda. \tag{4.28}$$

Then $k'(x)$ is $(d/dx)(p/\hbar)$ and $k(x)^2$ is $(p/\hbar)(2\pi/\lambda)$, and the condition can thus be written in the form

$$|dp/dx| \ll |p(2\pi/\lambda)|, \tag{4.29}$$

or equivalently,

$$|\lambda \, dp/dx| \ll |2\pi p|. \tag{4.30}$$

However, dp/dx represents in this case the change in momentum with x, so when multiplied by λ, the change in momentum over a wavelength is obtained. Therefore we can say that the change in momentum of the particle over a wavelength must be much less than the momentum itself in order for the approximation to be valid. Clearly p changes rapidly in regions where $|dU/dx|$ is large, so that the WKB approximation cannot be expected to hold at points where $U(x)$ varies rapidly. Analogous considerations for the case where k is imaginary lead to a similar conclusion.

Let us now consider the case of classical turning points, defined as points in space where $U = \mathscr{E}$ so that $k = 0$ and therefore $p = \hbar k = 0$. Clearly if $p \simeq 0$, the above inequality cannot be satisfied even for very small changes of potential with position, so the above approximation does not hold. Therefore even for cases in which the WKB approximation is valid on both sides of a classical turning point, it is necessary in general to employ some method to extend the wave function through the classical turning point. This becomes rather involved from a mathematical standpoint, and will not be treated here. Our primary application of the WKB technique is to the phenomenon of barrier penetration, for which connection formulas are not needed in the simplest approximation. More exact treatments of barrier penetration [cf. Merzbacher (1970)] do make use of connection formulas, and they are necessary in order to apply the WKB method to the problem of finding the bound energy levels for an arbitrary potential well.

PROJECT 4.1 Mathematics of the WKB Method

1. Work out and present all mathematical details of the following article on the WKB method: "Two Notes on Phase-Integral Methods" by W. H. Furry, *Phys. Rev.* **71**, 360 (1947).
2. Work out and present all mathematical details of the following article on the WKB method: "On the Connection Formulas and the Solutions of the Wave Equation" by R. E. Langer, *Phys. Rev.* **51**, 669 (1937).

2 Application of the WKB Technique to Barrier Penetration

Consider a case in which the wave function is strongly attenuated in a classically forbidden region extending from $x = x_1$ to $x = x_2$. The ratio $\psi(x_2)/\psi(x_1)$ evaluated according to the above WKB approximation for ψ is

$$\frac{\psi(x_2)}{\psi(x_1)} = \frac{A'k(x_2)^{-1/2}\exp\left[i\int^{x_2} k(x)\,dx - i\omega t\right]}{A'k(x_1)^{-1/2}\exp\left[i\int^{x_1} k(x)\,dx - i\omega t\right]}$$

$$= \left[\frac{k(x_1)}{k(x_2)}\right]^{1/2}\exp\left[i\int_{x_1}^{x_2} k(x)\,dx\right] = \left[\frac{\eta(x_1)}{\eta(x_2)}\right]^{1/2}\exp\left[-\int_{x_1}^{x_2}\eta(x)\,dx\right], \tag{4.31}$$

where

$$\eta(x) = \{(2m/\hbar^2)[U(x) - \mathscr{E}]\}^{1/2}. \tag{4.32}$$

The corresponding transmission coefficient given by $|\psi(x_2)/\psi(x_1)|^2(v_2/v_1)$, where v_2 and v_1 are the velocities, is thus approximately

$$\mathscr{T} \simeq \exp\left[-2\int_{x_1}^{x_2}\eta(x)\,dx\right]. \tag{4.33}$$

This shows that the critical factor in tunneling is the barrier segment at energies above the particle energy $\mathscr{E}$.

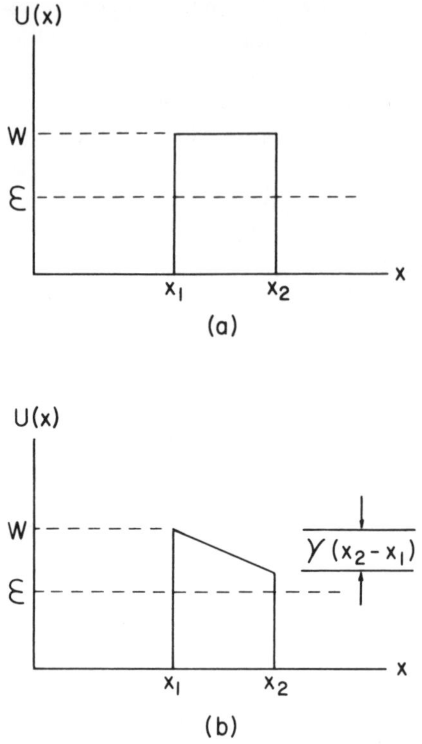

Fig. 4.1 (a) One-dimensional square energy barrier W impeding the motion of a particle of kinetic energy $\mathscr{E}$ in the x direction. (b) Trapezoidal energy barrier. (Such a barrier can result from the application of a uniform applied field in the forward direction to the square barrier.)

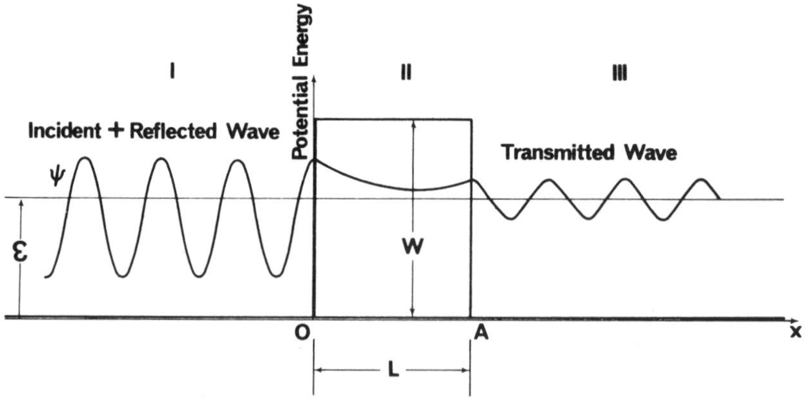

Fig. 4.2 Incident, reflected, and transmitted wave function components for a particle having kinetic energy $\mathscr{E}$ incident from Region I on a potential barrier (Region II) of height W and width L. (The transmission coefficient $\mathscr{T}$ in this case of zero of potential energy in Region III is determined from $\mathscr{T} = |\psi_{\text{III}}/\psi_{\text{I}}|^2$, where ψ_{III} is the transmitted wave and ψ_{I} is the incident wave.)

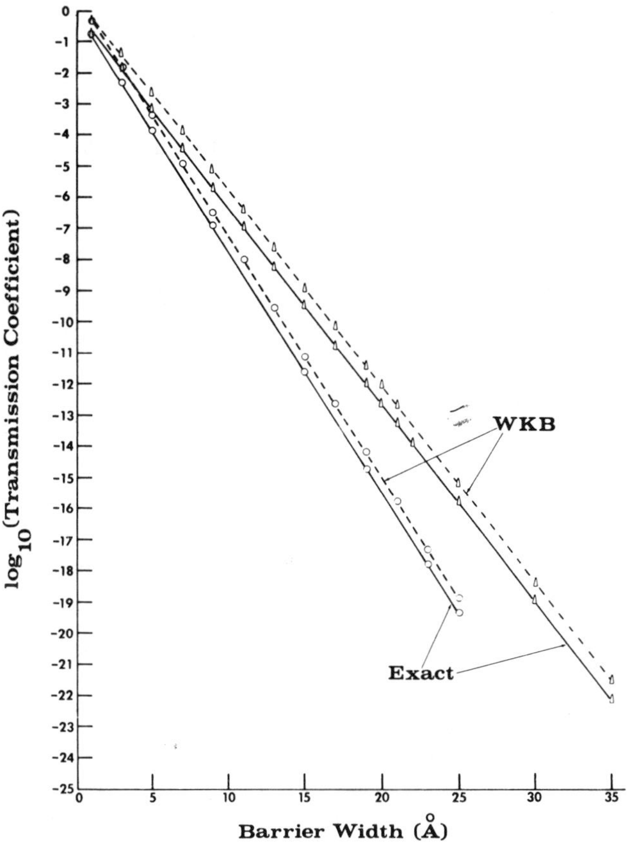

Fig. 4.3 Transmission coefficient versus barrier width for the rectangular energy barrier of Fig. 4.2. [The dashed curves illustrate results using the WKB approximation (4.34), the upper curve (triangles) representing an energy barrier exceeding the kinetic energy of the incident electron by 2 eV and the lower curve (circles) representing an energy barrier exceeding the kinetic energy of the incident electron by 3 eV. The corresponding solid curves were computed from the exact result deduced in Chap. 1, §11.3.2, using an incident electron kinetic energy of 1 eV.]

Consider now the limiting case of a rectangular ("square") barrier [$U(x) = W$ for $x_1 \leqslant x \leqslant x_2$; $U(x) = 0$ for $x < x_1$ and for $x > x_2$], as sketched in Fig. 4.1a. We designate the corresponding transmission coefficient $\mathscr{T}_{\text{Sq}}$. Since η is independent of x over the region $x_1 \leqslant x \leqslant x_2$, we immediately deduce the result

$$\mathscr{T}_{\text{Sq}} \simeq \exp\{-2[(2m/\hbar^2)(W - \mathscr{E})]^{1/2}(x_2 - x_1)\}. \tag{4.34}$$

This result is illustrated in Figs. 4.2 and 4.3.

Consider now the case of a trapezoidal barrier [$U(x) = 0$ for $x < x_1$; $U(x) = W - \gamma(x - x_1)$ for $x_1 \leqslant x \leqslant x_2$, where γ is a constant; $U(x) = 0$ for $x > x_2$], as sketched in Fig. 4.1b. We consider the case in which the entire region $x_1 \leqslant x \leqslant x_2$ is classically forbidden, and designate the corresponding transmission coefficient $\mathscr{T}_{\text{trap}}$,

$$\mathscr{T}_{\text{trap}} = \exp\left[-2\int_{x_1}^{x_2} \{(2m/\hbar^2)[W - \gamma(x - x_1) - \mathscr{E}]\}^{1/2}\right] dx. \tag{4.35}$$

If we designate the integrand as $\zeta(x)^{1/2}$, so that over the barrier region

$$\zeta(x) = (2m/\hbar^2)[W - \gamma(x - x_1) - \mathscr{E}], \tag{4.36}$$

then

$$d\zeta = -(2m\gamma/\hbar^2)\,dx, \tag{4.37}$$

$$\begin{aligned}
\mathscr{T}_{\text{trap}} &= \exp\left[-2\int_{\zeta(x_1)}^{\zeta(x_2)} \zeta^{1/2}(-\hbar^2/2m\gamma)\,d\zeta\right] \\
&= \exp\{(2\hbar^2/3m\gamma)[\zeta(x_2)^{3/2} - \zeta(x_1)^{3/2}]\} \\
&= \exp\left\{-\frac{4(2m)^{1/2}}{3\gamma\hbar}[(W - \mathscr{E})^{3/2} - [W - \mathscr{E} - \gamma(x_2 - x_1)]^{3/2}]\right\}.
\end{aligned} \tag{4.38}$$

If γ is considered to be very small, then the binomial expansion can be used,

$$\begin{aligned}
[W - \mathscr{E} - \gamma(x_2 - x_1)]^{3/2} &= (W - \mathscr{E})^{3/2}\left(1 - \frac{\gamma(x_2 - x_1)}{(W - \mathscr{E})}\right)^{3/2} \\
&\simeq (W - \mathscr{E})^{3/2}\left(1 - \frac{3}{2}\frac{\gamma(x_2 - x_1)}{(W - \mathscr{E})}\right. \\
&\qquad \left. + \left(\frac{1}{2}\right)\left(\frac{3}{2}\right)\left(\frac{1}{2}\right)\frac{\gamma^2(x_2 - x_1)^2}{(W - \mathscr{E})^2} + \cdots\right),
\end{aligned} \tag{4.39}$$

or equivalently,

$$\begin{aligned}
[W - \mathscr{E} - \gamma(x_2 - x_1)]^{3/2} &\simeq (W - \mathscr{E})^{3/2} - \tfrac{3}{2}\gamma(x_2 - x_1)(W - \mathscr{E})^{1/2} \\
&\quad + \tfrac{3}{8}\gamma^2(x_2 - x_1)^2/(W - \mathscr{E})^{1/2} + \cdots.
\end{aligned} \tag{4.40}$$

Therefore we obtain for this limit of small γ,

$$\mathscr{T}_{\text{trap}} \simeq \exp\left[-\frac{4(2m)^{1/2}}{3\hbar}\left[\tfrac{3}{2}(x_2 - x_1)(W - \mathscr{E})^{1/2} - \frac{3}{8}\frac{\gamma(x_2 - x_1)^2}{(W - \mathscr{E})^{1/2}}\right]\right]. \tag{4.41}$$

In the limit in which $\gamma \to 0$, this reduces to the result (4.34) previously obtained for the square barrier, as expected. For $\gamma \neq 0$, we see that the transmission probability increases exponentially with γ. A positive γ can represent a forward electric field bias for current transport; this leads to exponential increases in the current with electric field, in agreement with experimental measurements on electron tunneling.

Finally, we consider the other extreme limiting case, namely, that in which γ is so large that $\gamma(x_2 - x_1)$ is equal to $W - \mathscr{E}$. The barrier is triangular in this limit. Both x_1 and x_2 are classical turning points, so the WKB approximation is not actually justified. The exponential form of $\mathscr{T}_{\text{trap}}$ still turns out to give a meaningful indication of the transmission probability for this triangular barrier. Thus we obtain from Eq. (4.38) the result

$$\mathscr{T}_{\text{tri}} \simeq \exp[-4(2m)^{1/2}(W-\mathscr{E})^{3/2}/3\gamma\hbar]. \tag{4.42}$$

In this limit of large γ, the transmission probability again increases with increasing γ, but the functional dependence is somewhat different from that in the previously considered small γ limit. This latter form is useful in the study of *field emission* of electrons from metals, in which case $W-\mathscr{E}$ is the metal work function in units of energy per electron, and γ is the product $|eE_0|$, where e is the magnitude of the electronic charge and E_0 is the electric field which aids the emission of electrons. The current produced in this limit is referred to as "Fowler–Nordheim tunneling."

The electrical current is given by the product of the incident electron flux and the transmission coefficient. An integration over the various filled states is generally required, as indicated in the next section.

PROJECT 4.2 Alpha-Particle Tunneling

An alpha particle, once it is formed inside a nucleus, cannot escape unless it penetrates or surmounts the surrounding Coulomb barrier. Kelvin originally suggested that the particles emitted by a radioactive element are "evaporated" by the nuclei from within a potential crater. Rutherford's scattering experiments together with this classical explanation led to a paradox involving the law of conservation of energy. This paradox is completely resolved by quantum mechanics, which includes the possibility of tunneling through classically disallowed regions. This leads to what is known as the *Gamow factor*. On the basis of this information, formulate a simple theory of alphy decay.

Even long before the advent of quantum mechanics, Geiger and Nuttall formulated an empirical rule between the half-lives of alpha-particle emitters and the corresponding velocities of the emitted alpha particles. What qualitative prediction for this relationship is given by your theory? State other qualitative features of alpha decay which you can deduce from analogy with our treatment of electron tunneling through potential barriers.

PROJECT 4.3 Ramsauer Effect

A strong electric field is capable of removing electrons from individual atoms (or ions) in a gas (or plasma). This phenomenon can be easily explained in terms of the potential energy of an electron in the atom in the presence of an external electric field. Give two qualitative theories for this effect, one based solely on classical mechanics and the other including the modifications and additions introduced by quantum mechanics. Be sure to give the relevant fundamental equations for each of the theories. [*Hint*: See Merzbacher (1970).]

PROJECT 4.4 Triangular Barrier

Consider a triangular energy barrier extending from $x = 0$ to $x = L$, with the potential energy given as follows:

Region I:	$U(x) = 0$	$(-\infty < x < 0)$
Region II:	$U(x) = \alpha x$	$(0 < x < L)$
Region III:	$U(x) = U_0$	$(L < x < \infty)$

where α is a positive constant given by $\alpha = U_{\max}/L$ and U_0 is a constant energy less than the energy $U_{\max}$.

1. Solve for the transmission and reflection coefficients, using the WKB approximation. (Assume $\mathscr{E} < U_{max}$.)
2. Attempt to evaluate the transmission coefficient exactly by solving the Schrödinger equation for the potential energy function given above.
3. Compare the exact results with the results obtained using the WKB approximation. [*Hint*: For literature references to the triangular barrier, see C. B. Duke (1969).]

PROJECT 4.5 Fowler–Nordheim Tunneling

Choose the maximum energy U_{max} in the problem of a triangular barrier to be 1, 2, 3, 4, 5, or 10 eV, and choose the barrier thickness L to be 1 Å. Plot the WKB transmission coefficient $\mathscr{T}$ and the corresponding reflection coefficient $\mathscr{R} = 1 - \mathscr{T}$ as a function of incident particle energy $\mathscr{E}$, scanning the range $0 < \mathscr{E} < 2W$. Assume the particle density in the incident beam to be 10^{18} particles/cm^3.

3 Tunneling in Metal–Insulator–Metal Structures

3.1 Formulation of Tunnel Current Expression

A diagram of a typical metal–insulator–metal structure is shown in Fig. 4.4. With an applied electric field, this energy level diagram is modified to that of Fig. 4.5. Of importance are the electronic density of states $\tilde{g}(\mathscr{E})$ in the two metals, the Fermi–Dirac distribution functions for the metals, and the potential energy barrier for electrons in the layer of insulator separating the two metals. The insulator layer may be quite thin so that electron tunneling can be a major current transport mechanism when an external voltage is applied between the two metals.

The analysis for current through the layer as a function of applied voltage proceeds by considering the electron fluxes incident at the interfaces and the quantum mechanical transmission coefficient. Consider x to be the distance in

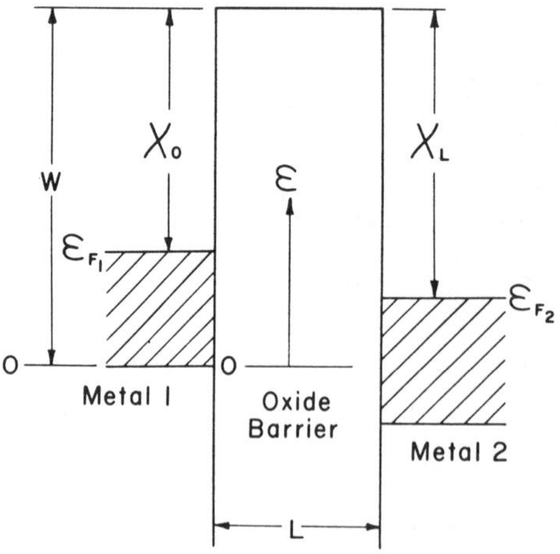

Fig. 4.4 Electron energy-level diagram for two metals with the filled electronic energy levels (crosshatched portion) of the two metals separated by an energy barrier of thickness L due to an insulating oxide.

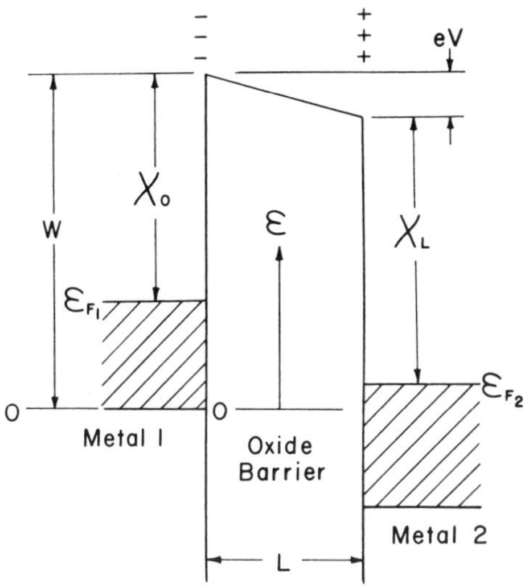

Fig. 4.5 Electron energy-level diagram for a metal–oxide–metal (MOM) configuration with a uniform applied electric field leading to a voltage difference V across the insulating oxide.

the insulator perpendicular to one of the parallel metal electrode interfaces. As a standard convention, we will usually measure all energies with respect to the bottom of the conduction band in metal 1. The final results are independent of choice of this reference point for the zero of energy. Let us consider the density of states $\tilde{g}(\mathscr{E})$ to be resolved into a spectrum $\tilde{g}(\mathscr{E}, p_x)$ characterized by the value of the x component of the momentum, such that

$$\tilde{g}(\mathscr{E}) = \int_{-\infty}^{\infty} \tilde{g}(\mathscr{E}, p_x)\, dp_x. \tag{4.43}$$

Then $2\tilde{g}(\mathscr{E}, p_x)\, dp_x\, d\mathscr{E}$ represents the number of *electronic* states with total energy in the range $\mathscr{E}$ to $\mathscr{E} + d\mathscr{E}$ but restricted to have the x component of momentum in the range p_x to $p_x + dp_x$. (The factor of 2 takes care of spin degeneracy.) Since the occupation probability is given by the Fermi–Dirac function $\mathcal{J}(\mathscr{E})$, the corresponding number of occupied states is $2\mathcal{J}(\mathscr{E})\tilde{g}(\mathscr{E}, p_x)\, dp_x\, d\mathscr{E}$. The product of this number and the x component of the velocity $v_x = p_x/m$ gives the flux of electrons incident on the barrier with x component of momentum between p_x and $p_x + dp_x$,

$$\text{Flux} = 2(p_x/m)\tilde{g}(\mathscr{E}, p_x)\mathcal{J}(\mathscr{E})\, dp_x\, d\mathscr{E}. \tag{4.44}$$

The probability that an incident electron will penetrate the barrier will vary with the x component of the momentum. The y and z components of momentum cannot be expected to contribute directly to overcoming or penetrating an energy barrier $U(x)$ for motion along the x direction. Thus for the present problem Eq. (4.32) takes the form

$$\eta(x) = ((2m/\hbar^2)[U(x) - (p_x^2/2m)])^{1/2}, \tag{4.45}$$

where $\eta(x)$ determines the WKB transmission coefficient $\mathscr{T}$ according to Eq. (4.33). Thus

$$\mathscr{T} = \mathscr{T}(p_x), \tag{4.46}$$

and the maximum number of electrons per unit time which could cross the insulator in the energy and momentum ranges considered is

$$(\text{Flux})\mathscr{T}(p_x) = 2(p_x/m)\mathscr{T}(p_x)\tilde{g}(\mathscr{E}, p_x)\tilde{f}(\mathscr{E})\, dp_x\, d\mathscr{E}. \tag{4.47}$$

An electron can cross the insulator, however, only if there is a corresponding unfilled allowable state in the opposite metal electrode. To obtain the current, then, we must weigh the above expression with the probability $\mathscr{A}$ of finding such an empty state. If conservation of transverse momentum is required and the free electron model is used for the quasi-continuum of electronic states in the two metals, then it is reasonable to assume that there will be one electronic state in the conduction band of the second metal corresponding to each tunneling electron proceeding from the first metal. The occupation probability of the final state will be given by the Fermi–Dirac function, so that $\mathscr{A}$ will simply be equal to $1 - \tilde{f}(\mathscr{E})$ for the metal in question. On the basis of this approximation, then $\mathscr{A} = \mathscr{A}(\mathscr{E})$ only.

We could raise the question as to whether or not the y and z components of momentum are conserved in this transition of the electron from a given state in one metal to a corresponding state in the other metal, as would be necessary if there were no possibility of elastic scattering processes accompanying the transition. Since the interfaces of actual metals are not atomically smooth, there is the possibility of elastic scattering processes both as the electron enters and as it leaves the insulator. Such transitions are neglected in the present simple treatment, however.

Likewise there is the possibility of transitions which may involve inelastic scattering processes, such as those involving electron–phonon scattering and electron–impurity scattering. Although such inelastic processes can give important physical information about the system [Duke (1969)], the contribution to the overall tunnel current is found to be relatively minor. A comprehensive treatment must of course include all such possible "channels" for the current through the insulator; it is not in the interest of simplicity, however, to attempt such a treatment here.

If we let $J(\mathscr{E}, p_x)\, dp_x\, d\mathscr{E}$ denote the partial electron-particle current due to electrons which are in the momentum range p_x to $p_x + dp_x$ and in the energy range $\mathscr{E}$ to $\mathscr{E} + d\mathscr{E}$ in one metal electrode which penetrate the barrier and enter the opposite metal electrode, then

$$J(\mathscr{E}, p_x)\, dp_x\, d\mathscr{E} = 2(p_x/m)\mathscr{T}(p_x)\tilde{g}(\mathscr{E}, p_x)\tilde{f}(\mathscr{E})\mathscr{A}(\mathscr{E})\, dp_x\, d\mathscr{E}. \tag{4.48}$$

There is the possibility of a reverse current also, so the *net* partial current will be determined by the difference between the two. If the metal on the left in Fig. 4.4 is denoted by 1 and the metal on the right is denoted by 2, then the net partial current between metals 1 and 2 is

$$J^{(\text{net})}(\mathscr{E}, p_x)\, dp_x\, d\mathscr{E} = 2(p_x/m)\mathscr{T}(p_x)[\tilde{g}_1(\mathscr{E}, p_x)\tilde{f}_1(\mathscr{E})\mathscr{A}_2(\mathscr{E}) - \tilde{g}_2(\mathscr{E}, p_x)\tilde{f}_2(\mathscr{E})\mathscr{A}_1(\mathscr{E})]\, dp_x\, d\mathscr{E}, \tag{4.49}$$

where for our simple calculation

$$\mathscr{A}_1(\mathscr{E}) = 1 - \tilde{f}_1(\mathscr{E}), \tag{4.50}$$

$$\mathscr{A}_2(\mathscr{E}) = 1 - \tilde{f}_2(\mathscr{E}). \tag{4.51}$$

It is helpful to recall our stated convention of measuring energies with respect to the bottom of the conduction band of metal 1. The Fermi–Dirac functions $\tilde{f}_1(\mathscr{E})$ and $\tilde{f}_2(\mathscr{E})$ evaluated at a given energy $\mathscr{E}$ then differ because the Fermi energies $\mathscr{E}_{F_1}$ and $\mathscr{E}_{F_2}$ for the two metals are different with respect to the zero of energy, as can be seen from Fig. 4.5. The transmission coefficient $\mathscr{T}(p_x)$ depends principally upon $U(x) - (p_x^2/2m)$, as can be seen from Eqs. (4.33) and (4.45). It is therefore essentially the same for forward and reverse penetration; this fact has been used explicitly in formulating the expression (4.49). The total electron particle current $J_{\text{tot}}^{(\text{net})}$ is obtained by integrating $J^{(\text{net})}(\mathscr{E}, p_x)\, dp_x\, d\mathscr{E}$ over all possible values of p_x and $\mathscr{E}$,

$$J_{\text{tot}}^{(\text{net})} = \iint J^{(\text{net})}(\mathscr{E}, p_x)\, dp_x\, d\mathscr{E}. \tag{4.52}$$

3.2 Appropriate Density of States

Let us now calculate the quantity $\tilde{g}(\mathscr{E}, p_x)$ on the basis of the free-electron model for use in the above expression. This quantity is defined to be the subset of those states of energy in the range $\mathscr{E}$ to $\mathscr{E} + d\mathscr{E}$ which have momentum in the range p_x to $p_x + dp_x$. Suppose that we take the density of states

$$\mathscr{G}(\mathbf{p}) = (2\pi\hbar)^{-3} \tag{4.53}$$

in momentum space [Eq. (3.19)] per direction of spin for unit volume of metal and multiply it by the volume of momentum space $\frac{4}{3}\pi p^3$ containing states with energy equal to or less than some arbitrary value $\mathscr{E}$. This gives

$$n(\mathscr{E}) = (2\pi\hbar)^{-3}(\tfrac{4}{3}\pi p^3) = (2\pi\hbar)^{-3}[\tfrac{4}{3}\pi(2m\mathscr{E})^{3/2}] \tag{4.54}$$

allowed levels, and a differentiation with respect to energy gives the density $\tilde{g}(\mathscr{E})$ of levels per direction of spin per unit range at this energy [Eq. (3.22)]. Our problem is somewhat different since we wish only the subset of these levels corresponding to a momentum in the x direction in the range p_x to $p_x + dp_x$. Thus instead of using the entire spherical volume of momentum space containing states with energy equal to or less than $\mathscr{E}$ in our derivation, we must use only the volume of the constant energy sphere which lies between two planes located at p_x and $p_x + dp_x$ which are perpendicular to the p_x axis. (See Fig. 4.6.) Thus we must consider the constant energy sphere to be divided into elemental

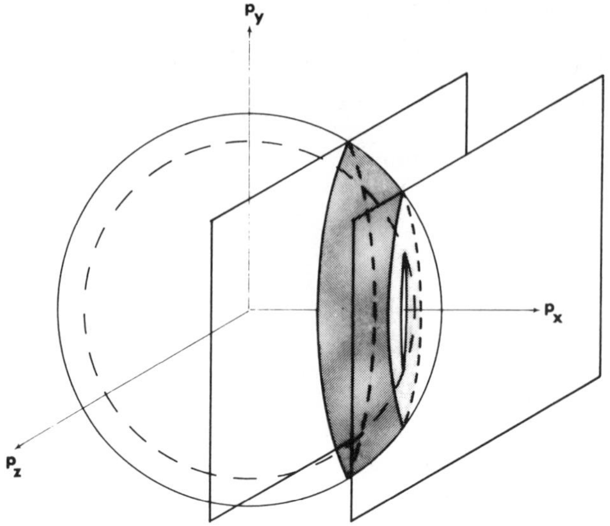

Fig. 4.6 Segment of the Fermi sphere containing electronic states having momentum between p_x and $p_x + dp_x$ and having energy between $\mathscr{E}$ and $\mathscr{E} + d\mathscr{E}$. (This diagram is useful for deriving the density of states versus energy for a specific x-momentum value.)

slices of width dp_x, the slices being bounded by planes perpendicular to the p_x axis. The radius of an elemental slice bounded by a plane at p_x is $(2m\mathscr{E} - p_x^2)^{1/2}$, with a corresponding area of $\pi(2m\mathscr{E} - p_x^2)$. The volume of the elemental slice bounded by the planes at p_x and $p_x + dp_x$ is $\pi(2m\mathscr{E} - p_x^2)\,dp_x$, and the elemental number $n(\mathscr{E}, p_x)\,dp_x$ of states contained within the slice is

$$n(\mathscr{E}, p_x)\,dp_x = (2\pi\hbar)^{-3}\pi(2m\mathscr{E} - p_x^2)\,dp_x. \tag{4.55}$$

It is clear that this elemental slice contains all allowed states with energy equal to or less than $\mathscr{E}$ and with momentum values in the range p_x to $p_x + dp_x$. The number $n(\mathscr{E}, p_x)\,dp_x$ can be looked upon as the product of a density and an interval dp_x. A differentiation of $n(\mathscr{E}, p_x)$ with respect to energy yields the density of such states per unit energy range at energy $\mathscr{E}$, which is the required quantity $\tilde{g}(\mathscr{E}, p_x)$,

$$\tilde{g}(\mathscr{E}, p_x) = \frac{dn(\mathscr{E}, p_x)}{d\mathscr{E}} = \frac{2\pi m}{(2\pi\hbar)^3} = \frac{2\pi m}{h^3}. \tag{4.56}$$

The fact that this quantity is independent of both $\mathscr{E}$ and p_x is both interesting and helpful.

A similar way of obtaining the quantity $n(\mathscr{E}, p_x)$ is to consider an integration of the density of states in momentum space over all values of p_y and p_z subject to the conditions that the energy $(p_x^2 + p_y^2 + p_z^2)/2m$ be equal to or less than $\mathscr{E}$ and the x component of momentum be p_x. Thus

$$n(\mathscr{E}, p_x) = \int_{-p_z^{(\max)}}^{p_z^{(\max)}} \int_{-p_y^{(\max)}}^{p_y^{(\max)}} (2\pi\hbar)^{-3}\,dp_y\,dp_z. \tag{4.57}$$

Since the maximum values of p_y and p_z for a fixed p_x value are determined from $\mathscr{E} = (p_x^2 + p_y^2 + p_z^2)/2m$, the value of the double integral above is simply the product of $(2\pi\hbar)^{-3}$ and the area of a circle of radius $(2m\mathscr{E} - p_x^2)^{1/2}$. Thus

$$n(\mathscr{E}, p_x) = (2\pi\hbar)^{-3}\pi(2m\mathscr{E} - p_x^2), \tag{4.58}$$

as previously deduced. Multiplying this area by the elemental quantity dp_x would give the elemental volume of the slice of the constant energy surface bounded by planes perpendicular to p_x at p_x and $p_x + dp_x$. An integration of these elemental volumes from $p_x = -(2m\mathscr{E})^{1/2}$ to $(2m\mathscr{E})^{1/2}$ would give the total number $n(\mathscr{E})$ of states contained in the constant energy sphere,

$$n(\mathscr{E}) = \int_{-(2m\mathscr{E})^{1/2}}^{(2m\mathscr{E})^{1/2}} n(\mathscr{E}, p_x)\, dp_x = \int_{-(2m\mathscr{E})^{1/2}}^{(2m\mathscr{E})^{1/2}} (2\pi\hbar)^{-3}\pi(2m\mathscr{E} - p_x^2)\, dp_x$$

$$= \pi(2\pi\hbar)^{-3}\, 2(2m)^{3/2}\, \mathscr{E}^{3/2}[1 - \tfrac{1}{3}], \tag{4.59}$$

which is equal to the value previously given [Eq. (4.54)]. Thus we have confidence that the quantity $n(\mathscr{E}, p_x)$ is the proper one to have used in deducing $\tilde{g}(\mathscr{E}, p_x)$.

3.3 Integration of Tunnel Current Expression

Substituting Eqs. (4.50), (4.51), and (4.56) into Eq. (4.49) yields

$$J^{(\mathrm{net})}(\mathscr{E}, p_x)\, dp_x\, d\mathscr{E} = (2\pi^2\hbar^3)^{-1}[\mathscr{f}_1(\mathscr{E}) - \mathscr{f}_2(\mathscr{E})]\mathscr{T}(p_x)p_x\, dp_x\, d\mathscr{E}, \tag{4.60}$$

which in turn can be used in Eq. (4.52) to obtain the total current. In carrying out the integration over energy before integrating over the x component of momentum, we must be careful to add only the contributions from energies greater than $p_x^2/2m$, since energies less than this would require imaginary momentum components in either the y or z directions. Thus the lower limit on the energy integral will be $p_x^2/2m$, and Eq. (4.52) gives

$$J_{\mathrm{tot}}^{(\mathrm{net})} = (2\pi^2\hbar^3)^{-1} \int_0^\infty \mathscr{T}(p_x)p_x\, dp_x \int_{p_x^2/2m}^\infty [\mathscr{f}_1(\mathscr{E}) - \mathscr{f}_2(\mathscr{E})]\, d\mathscr{E}. \tag{4.61}$$

The Fermi–Dirac function (2.151) is easy to integrate if the numerator and denominator are first multiplied by $\exp[-(\mathscr{E} - \mathscr{E}_{\mathrm{F}})/k_{\mathrm{B}}T]$,

$$\mathscr{I} \equiv \int_{p_x^2/2m}^\infty \mathscr{f}(\mathscr{E})\, d\mathscr{E} = \int_{p_x^2/2m}^\infty \frac{\exp[-(\mathscr{E} - \mathscr{E}_{\mathrm{F}})/k_{\mathrm{B}}T]\, d\mathscr{E}}{1 + \exp[-(\mathscr{E} - \mathscr{E}_{\mathrm{F}})/k_{\mathrm{B}}T]}. \tag{4.62}$$

Changing variables as follows

$$\zeta = (\mathscr{E} - \mathscr{E}_{\mathrm{F}})/k_{\mathrm{B}}T, \qquad d\zeta = (k_{\mathrm{B}}T)^{-1}\, d\mathscr{E}, \tag{4.63}$$

we obtain

$$\mathscr{I} = \int_{\zeta_0}^\infty \frac{\exp(-\zeta)k_{\mathrm{B}}T\, d\zeta}{1 + \exp(-\zeta)}, \tag{4.64}$$

where

$$\zeta_0 \equiv [(p_x^2/2m) - \mathscr{E}_{\mathrm{F}}]/k_{\mathrm{B}}T. \tag{4.65}$$

Integration gives

$$\mathscr{I} = k_B T \ln[1 + \exp(-\zeta_0)]. \tag{4.66}$$

Thus Eq. (4.61) takes the form

$$J_{\text{tot}}^{(\text{net})} = (k_B T/2\pi^2\hbar^3) \int_0^\infty dp_x\, p_x \mathscr{T}(p_x) \ln\left[\frac{1 + \exp\{[\mathscr{E}_{F_1} - (p_x^2/2m)]/k_B T\}}{1 + \exp\{[\mathscr{E}_{F_2} - (p_x^2/2m)]/k_B T\}}\right]. \tag{4.67}$$

This expression is quite fundamental in tunneling theory as applied to metal–oxide–metal thin film structures.

The Fermi–Dirac functions in Eq. (4.61) vary with $(\mathscr{E} - \mathscr{E}_F)/k_B T$ for the metal in question according to Eq. (2.151). Different metals will have different Fermi energies, and in addition any externally applied voltage V_{ext} will also adjust the relative positions of the Fermi levels of the two metals. For different metals and zero applied voltage, there will occur a tunneling of electrons until at equilibrium the two Fermi levels will be aligned with a resultant potential difference developed across the insulating layer.

Because the development of conduction bands and the meaning of energy gaps in insulators is not presented in detail until Chap. 7, it is worthwhile as a preliminary to expound upon these concepts a bit and show how they are intimately involved in the phenomenon of tunneling in metal–insulator–metal structures.

3.4 Metal–Insulator–Metal Energy Diagrams

The Fermi energy $\mathscr{E}_F$ for a metal relative to the bottom of the conduction band is determined to large extent by the conduction electron density (see Chap. 3, §1.9), and this varies from metal to metal. The depth $\mathscr{E}_F + \phi$ of the "square-well potential" in the free electron approximation, constituting as it does some measure of the energy required to abstract a conduction electron from the lowest energy level in the conduction band of a metal at zero potential and remove it to infinity, is also characteristic of the metal in question. The electrostatic potential of the metal relative to some reference potential must likewise be considered. Thus two different metals, labeled 1 and 2, may have potential energy diagrams as illustrated in Fig. 4.7a with zero voltage between the metals, but will have the energy for each electron shifted by an amount eV for a difference in potential $V = V_2 - V_1$ between the two metals, as illustrated in Fig. 4.7b. (The parameter e is the electronic charge magnitude.) It can be noted from the diagrams in Fig. 4.7 that the difference in electron energies $\Delta\mathscr{E}_B^{(0)}$ between the bottoms of the conduction bands for *the case of zero applied voltage* is

$$\Delta\mathscr{E}_B^{(0)} = (\mathscr{E}_{F_2} - \mathscr{E}_{F_1}) + (\phi_2 - \phi_1),$$

where the quantities $\mathscr{E}_{F_1}$ and $\mathscr{E}_{F_2}$ are the Fermi energies as computed in the usual manner considering the electron densities in the metals, and the quantities ϕ_1 and ϕ_2 are the vacuum work functions for the metals. Likewise it can be seen

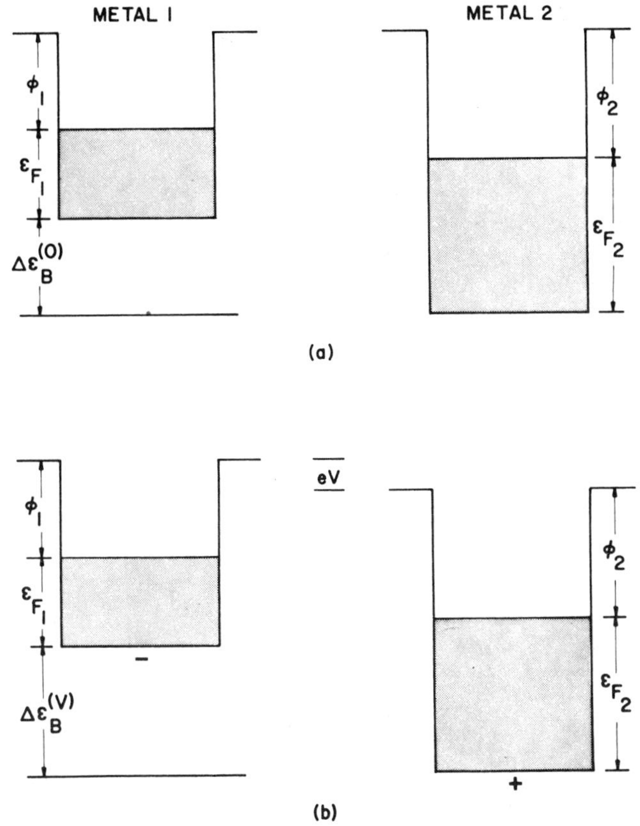

Fig. 4.7 Energy-level diagram for two metals separated in free space. (a) Zero potential difference between the metals. (b) Applied voltage difference V between the two metals. (The dark areas denote the filled portions of the conduction bands.)

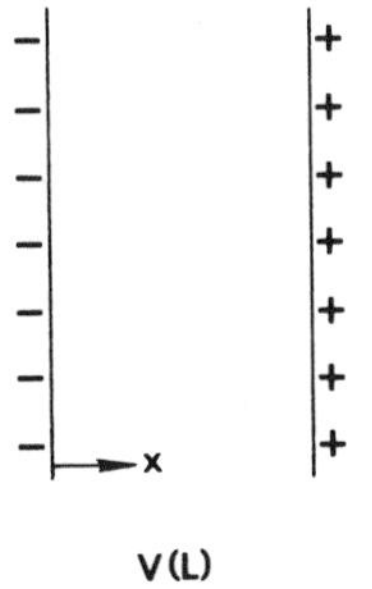

Fig. 4.8 The electric field produced by a negative surface charge at $x = 0$ and an equal but positive surface charge at $x = L$ is negative, thus yielding a positive electrostatic potential difference in accordance with the relation $V(x) = -\int_0^x E(x)\,dx$. The *voltage* is often taken to be the magnitude $|V(L)|$.

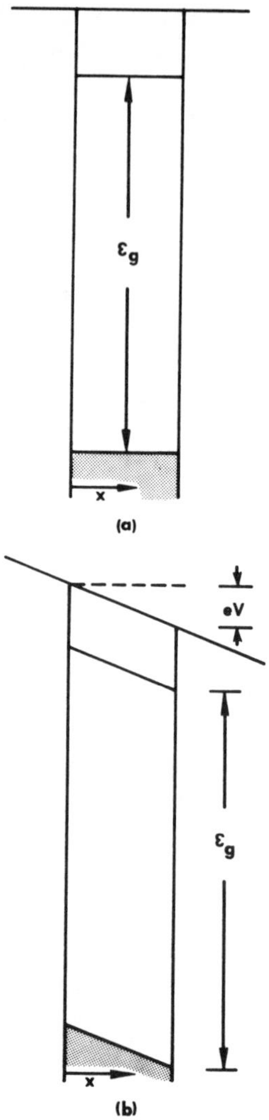

Fig. 4.9 Energy-level diagram for an insulator. (a) Zero applied voltage. (b) Applied voltage V. (The dark area denotes the completely filled valence band of the insulator, and the energy gap $\mathscr{E}_g$ is the difference in energy between the top of the filled valence band and the bottom of the empty conduction band.)

from Fig. 4.7 that the difference in electron energies $\Delta\mathscr{E}_B^{(V)}$ between the bottoms of the conduction bands *for the case of an applied voltage* V between the two metals is

$$\Delta\mathscr{E}_B^{(V)} = \Delta\mathscr{E}_B^{(0)} + eV,$$

where a positive voltage constitutes the case of excess negative charge on metal 1, relative to metal 2, which represents the forward bias condition for electrons in the sense of electron transfer from metal 1 to metal 2.

The voltage sign convention follows from basic electrostatic theory, where the electrostatic potential difference between positions $\mathbf{r}_1$ and $\mathbf{r}_2$ is defined as the negative line integral of the electric field vector along an arbitrary path extending from $\mathbf{r}_1$ to $\mathbf{r}_2$, $\Delta V = V_2 - V_1 = -\int_{\mathbf{r}_1}^{\mathbf{r}_2} E(\mathbf{r}) \cdot d\mathbf{r}$. The sign of the electric field follows from the application of Gauss's law. For metal 1 to the left and metal 2 to the right, with the net charge on metal 2 being equal in magnitude but opposite in sign to that on metal 1 (Fig. 4.8), the sign of the field is the same as the sign of the net charge on metal 1. The potential energy of a positive charge is thus lowered as it travels from position $\mathbf{r}_1$ to position $\mathbf{r}_2$ in the direction of a positive field. The change in the electrostatic energy $U(\mathbf{r})$ of a charge q in moving from $\mathbf{r}_1$ to $\mathbf{r}_2$ is given by $\Delta U(\mathbf{r}) = q\,\Delta V = q(V_2 - V_1)$.

Let us now consider separating metal 1 from metal 2 in Fig. 4.7 by an insulator having a band gap $\mathscr{E}_g$, as shown in Fig. 4.9. The conduction levels in the insulator lie in the energy region below the vacuum level but above the forbidden gap $\mathscr{E}_g$.

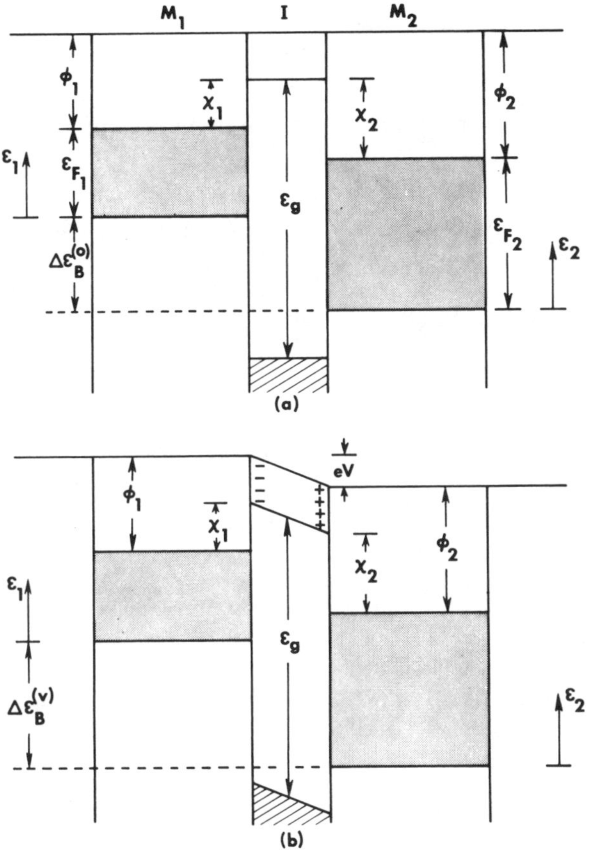

Fig. 4.10 Energy-level diagram for a combined metal–insulator–metal system such as a practical tunnel current device. (a) Zero voltage. (b) Applied voltage V. (Note how this figure is a combination of Figs. 4.7 and 4.9.)

No conduction is possible for electrons in the entirely filled band illustrated as the crosshatched area below the energy gap region $\mathscr{E}_g$, due to the essential restrictions enunciated in the Pauli exclusion principle.

If metals 1 and 2 are placed on either side of, and very near to, the insulator, the electron potential energy diagram has the appearance of that illustrated in Fig. 4.10. It can be noted from Fig. 4.10 that an electron at the Fermi level of metal 1 need only surmount a barrier of height χ_1 before entering metal 2 for the case of two metals separated by a solid state insulator, whereas the barrier height would be ϕ_1 for the same two metals separated by a vacuum region. Analogous to the terminology "metal–vacuum work functions" for ϕ_1 and ϕ_2, we use the terminology "metal–dielectric work functions" for χ_1 and χ_2, and in cases for which the dielectric is an oxide formed by chemical reaction of one of the metals with ambient oxygen, the term "metal–oxide work function" is used for χ. It can be noted from Fig. 4.10 that $\phi_2 - \phi_1 = \chi_2 - \chi_1$, so that we can also write

$$\Delta\mathscr{E}_{\mathrm{B}}^{(0)} = (\mathscr{E}_{\mathrm{F}_2} - \mathscr{E}_{\mathrm{F}_1}) + (\chi_2 - \chi_1),$$

with a nonzero voltage again yielding

$$\Delta\mathscr{E}_{\mathrm{B}}^{(V)} = \mathscr{E}_{\mathrm{B}}^{(0)} + eV$$

for the energy difference between the bottoms of the conduction bands in the two metals.

The situation in which a semiconductor replaces the dielectric between the two metals is quite similar, although the energy gap $\mathscr{E}_g$ is then relatively small. Typically the energy gap in an intrinsic semiconductor is of the order of a few multiples of the thermal energy ($k_B T \simeq 0.025$ eV at $T = 300°$K), instead of having values exceeding an electron volt or so as is the case for good electrical insulators.

PROJECT 4.6 Electron Tunneling Data

Examine carefully the literature data for single-particle tunneling. Can you conclude unambiguously that the quantum phenomenon of electron tunneling has indeed been observed in solids? Summarize the extent to which there is essential agreement between present theory and experiment, and the extent to which there is a lack of agreement. What critical experiments may yet be needed to test the theory? [*Hint*: Keep in mind the possibility of experimental error, the existence of alternate transport mechanisms, and the extent to which one should trust the predictions of a highly simplified idealistic model of any real physical system. See Gundlach and Simmons (1969).]

PROJECT 4.7 Electron Tunneling Theory

Work out the mathematical details of a theoretical development of tunneling through space charge regions. [*Hint*: See Conley *et al.* (1966).]

4 Tunnel Current at 0°K between Two Metals Separated by a Rectangular Barrier

It is worthwhile to evaluate Eq. (4.67) for the specific situation of a rectangular ("square") barrier and very low temperatures. Consider the case in which the Fermi energy of metal 1 is higher than the Fermi energy of metal 2, with the metals separated by a square barrier of thickness L and of height W relative to the bottom of the conduction band of metal 1. (See the energy level diagram in Fig. 4.4 for the definition of the microsopic parameters.) By dividing the range of integration of Eq. (4.67) into segments I, II, and III corresponding to the momentum intervals $(0, \sqrt{2m\mathscr{E}_{F_2}})$, $(\sqrt{2m\mathscr{E}_{F_2}}, \sqrt{2m\mathscr{E}_{F_1}})$, $(\sqrt{2m\mathscr{E}_{F_1}}, \infty)$ and making use of the WKB result (4.34) for the transmission probability, we obtain

$$J_{\text{tot}} = J_{\text{I}} + J_{\text{II}} + J_{\text{III}}, \tag{4.68}$$

where

$$J_{\text{I}} = \frac{k_B T}{2\pi^2\hbar^3}\int_0^{\sqrt{2m\mathscr{E}_{F_2}}} \ln\left[\frac{1 + \exp\{[\mathscr{E}_{F_1} - (p_x^2/2m)]/k_B T\}}{1 + \exp\{[\mathscr{E}_{F_2} - (p_x^2/2m)]/k_B T\}}\right]$$
$$\times \exp\{-(2L/\hbar)[2mW - p_x^2]^{1/2}\}p_x\, dp_x, \tag{4.69}$$

with the expressions for J_{II} and J_{III} being the same except for the limits of integration.

The exponentials involving the Fermi energies approach either zero or infinity as the temperature approaches zero. In momentum region I, $p_x^2/2m$ is less than $\mathscr{E}_{F_1}$ and less than $\mathscr{E}_{F_2}$, so the exponentials diverge as $T \to 0°\text{K}$. The logarithmic factor thus has the limiting value $(\mathscr{E}_{F_1} - \mathscr{E}_{F_2})/k_B T$, so that

$$J_{\text{I}}(0°\text{K}) = (2\pi^2\hbar^3)^{-1}(\mathscr{E}_{F_1} - \mathscr{E}_{F_2})\int_0^{\sqrt{2m\mathscr{E}_{F_2}}} \exp\{-(2L/\hbar)[2mW - p_x^2]^{1/2}\}p_x\, dp_x. \tag{4.70}$$

In momentum region II, $p_x^2/2m$ is less than $\mathscr{E}_{F_1}$ but greater than $\mathscr{E}_{F_2}$, so that the exponential in the denominator of Eq. (4.69) approaches zero and the exponential in the numerator diverges as $T \to 0°\text{K}$. The logarithmic factor thus has the limiting value $[\mathscr{E}_{F_1} - (p_x^2/2m)]/k_B T$, so that

$$J_{\text{II}}(0°\text{K}) = (2\pi^2\hbar^3)^{-1}\int_{\sqrt{2m\mathscr{E}_{F_2}}}^{\sqrt{2m\mathscr{E}_{F_1}}} \exp\{-(2L/\hbar)[2mW - p_x^2]^{1/2}\}$$
$$\times [\mathscr{E}_{F_1} - (p_x^2/2m)]p_x\, dp_x. \tag{4.71}$$

In momentum region III, $p_x^2/2m$ is greater than $\mathscr{E}_{F_1}$ and greater than $\mathscr{E}_{F_2}$, so that the exponentials in the numerator and in the denominator of Eq. (4.69) approach zero as $T \to 0°\text{K}$. The logarithmic factor thus has the limiting value of zero, so that

$$J_{\text{III}}(0°\text{K}) = 0. \tag{4.72}$$

Needless to say, J_{III} is not zero at higher temperatures, since J_{III} contains the entire thermionic emission current (Chap. 3, §4), as well as the tunnel current from filled states above $\mathscr{E}_{F_1}$.

Let us change variables in Eqs. (4.70) and (4.71) by defining the dimensionless quantity

$$\zeta \equiv (2L/\hbar)(2mW - p_x^2)^{1/2}, \tag{4.73}$$

consistent with

$$p_x^2 = 2mW - (\hbar^2/4L^2)\zeta^2, \tag{4.74}$$

$$p_x\, dp_x = -(\hbar^2/4L^2)\zeta\, d\zeta. \tag{4.75}$$

Equation (4.70) and (4.71) thus become

$$J_{\text{I}}(0^\circ\text{K}) = (2\pi^2\hbar^3)^{-1}(\chi_{\text{L}} - \chi_0)(\hbar^2/4L^2)\int_{\alpha_{\text{L}}L}^{\alpha_{\text{W}}L} e^{-\zeta}\zeta\, d\zeta, \tag{4.76}$$

$$J_{\text{II}}(0^\circ\text{K}) = (2\pi^2\hbar^3)^{-1}(\hbar^2/4L^2)\int_{\alpha_0 L}^{\alpha_{\text{L}}L} e^{-\zeta}\zeta[(\hbar^2/8mL^2)\zeta^2 - \chi_0]\, d\zeta, \tag{4.77}$$

with

$$\chi_0 \equiv W - \mathscr{E}_{\text{F}_1}, \tag{4.78}$$

$$\chi_{\text{L}} \equiv W - \mathscr{E}_{\text{F}_2}, \tag{4.79}$$

$$\chi_{\text{L}} - \chi_0 = \mathscr{E}_{\text{F}_1} - \mathscr{E}_{\text{F}_2}, \tag{4.80}$$

$$\alpha_{\text{W}} \equiv (2/\hbar)(2mW)^{1/2}, \tag{4.81}$$

$$\alpha_0 \equiv (2/\hbar)[2m(W - \mathscr{E}_{\text{F}_1})]^{1/2} = (2/\hbar)(2m\chi_0)^{1/2}, \tag{4.82}$$

$$\alpha_{\text{L}} \equiv (2/\hbar)[2m(W - \mathscr{E}_{\text{F}_2})]^{1/2} = (2/\hbar)(2m\chi_{\text{L}})^{1/2}. \tag{4.83}$$

The quantities χ_0 and χ_{L} can be noted from Fig. 4.4 to be the work functions of metals 1 and 2 with respect to the barrier material. From standard integral tables,

$$\int \eta e^{-\eta}\, d\eta = -(\eta + 1)e^{-\eta}, \tag{4.84}$$

$$\int \eta^3 e^{-\eta}\, d\eta = -(\eta^3 + 3\eta^2 + 6\eta + 6)e^{-\eta}, \tag{4.85}$$

so that evaluation of Eqs. (4.76) and (4.77) can be carried out immediately to give

$$J_{\text{I}}(0^\circ\text{K}) = (4\pi^2 m\hbar^3)^{-1}(\hbar^2/4L^2)^2(\alpha_{\text{L}}^2L^2 - \alpha_0^2L^2) \times \{e^{-\alpha_{\text{L}}L}(\alpha_{\text{L}}L + 1) - e^{-\alpha_{\text{W}}L}(\alpha_{\text{W}}L + 1)\}, \tag{4.86}$$

$$J_{\text{II}}(0^\circ\text{K}) = (4\pi^2 m\hbar^3)^{-1}(\hbar^2/4L^2)^2\{e^{-\alpha_0 L}[2\alpha_0^2L^2 + 6\alpha_0 L + 6] - e^{-\alpha_{\text{L}}L}[(\alpha_{\text{L}}^3L^3 + 3\alpha_{\text{L}}^2L^2 + 6\alpha_{\text{L}}L + 6) - (\alpha_0^2\alpha_{\text{L}}L^3 + \alpha_0^2L^2)]\}. \tag{4.87}$$

Combining Eqs. (4.72), (4.86), and (4.87) according to Eq. (4.68) gives

$$J_{\text{tot}}(0^\circ\text{K}) = (4\pi^2 m\hbar^3)^{-1}(\hbar^2/4L^2)^2[e^{-\alpha_{\text{W}}L}(\alpha_0^2L^2 - \alpha_{\text{L}}^2L^2)(1 + \alpha_{\text{W}}L) + 2e^{-\alpha_0 L}(\alpha_0^2L^2 + 3\alpha_0 L + 3) - 2e^{-\alpha_{\text{L}}L}(\alpha_{\text{L}}^2L^2 + 3\alpha_{\text{L}}L + 3)]. \tag{4.88}$$

Approximations can be made which reduce this equation to a simpler form. If the ratio of the effective electron mass m to the free-electron mass m_e is designated as ι,

$$\iota \equiv m/m_e, \tag{4.89}$$

and energies are expressed in units of electron volts, then in units of reciprocal angstoms $\alpha_W \simeq 1.025\sqrt{\iota W}$ Å^{-1}, $\alpha_0 \simeq 1.025\sqrt{\iota\chi_0}$ Å^{-1}, and $\alpha_L \simeq 1.025\sqrt{\iota\chi_L}$ Å^{-1}. For $L \gtrsim 10$ Å, we thus expect the dimensionless quantities $\alpha_W L$, $\alpha_0 L$, and $\alpha_L L$ to have values of the order of or greater than 10, since W, χ_0, and χ_L will ordinarily be in the range 1–5 eV and ι should ordinarily have values in the range 0.1–1. For these values, the quadratic term $\alpha_0^2 L^2$ predominates over $3\alpha_0 L$ and 3, and likewise $\alpha_L^2 L^2$ predominates over $3\alpha_L L$ and 3. In addition, it is often the case that W is a factor of 2 or so larger than either χ_0 or χ_L, in which case the terms in $\exp(-\alpha_W L)$ can be neglected. Thus we obtain the approximate expression

$$J_{\text{tot}}(0^\circ\text{K}) \simeq 2(4\pi^2 m\hbar^3)^{-1}(\hbar^2/4L^2)^2[\alpha_0^2 L^2 e^{-\alpha_0 L} - \alpha_L^2 L^2 e^{-\alpha_L L}]. \tag{4.90}$$

Substituting the definitions (4.82) and (4.83) for α_0 and α_L gives

$$J_{\text{tot}}(0^\circ\text{K}) \simeq (4\pi^2\hbar L^2)^{-1}\{\chi_0 \exp[-(2L/\hbar)(2m\chi_0)^{1/2}] - \chi_L \exp[-(2L/\hbar)(2m\chi_L)^{1/2}]\}. \tag{4.91}$$

Whenever χ_L is very large, the current depends essentially on χ_0, with the term in χ_L being unimportant. The electrons then travel almost exclusively from metal 1 to metal 2, with very few going in the reverse direction. It is useful to view the two terms in this equation semantically as constituting "forward" and "reverse" currents. Except for the sign associated with the direction of travel, the expression for the current can be noted to be symmetrical in χ_0 and χ_L.

The derivation for the alternant case in which the Fermi level of metal 2 is higher than the Fermi level of metal 1 proceeds in the same manner as the derivation given above. In fact, the physical situation is exactly the same from the viewpoint of metal 2 as it was from that of metal 1 in the above derivation, with the exception that the direction of travel is reversed. Thus an interchange of the subscripts 1 and 2 and a replacement of $J_{\text{tot}}(0^\circ\text{K})$ by $-J_{\text{tot}}(0^\circ\text{K})$ converts the above result to one appropriate for metal 2. It can be seen that the equation obtained by such a conversion is identical to the equation before conversion, so the above expressions are valid for either case.

It can be seen from Eq. (4.88) that J_{tot} is zero whenever $\alpha_0 = \alpha_L$, corresponding to $\chi_0 = \chi_L$. This is in accord with the physical picture of a symmetrical structure with the energies $\mathscr{E}_{F_1}$ and $\mathscr{E}_{F_2}$ aligned on the energy level diagram in Fig. 4.4. In such a case there is no reason to expect a net electron tunnel current.

The decrease in current with the thickness L of the barrier is exponential; this can be seen from Eq. (4.90). Since the values of the parameters α_0 and α_L in Eq. (4.90) have been shown to be of the order of one reciprocal angstrom, the current falls by a factor of approximately 1/2.7183 with each angstrom increase in thickness, corresponding to a decrease in the current by a factor of 4.54×10^{-5} for each 10 Å increase in thickness. The only reason that there can be an appreciable current for barriers a few monolayers or more in thickness is the fact that the pre-exponential factor in the tunnel current expressions is quite large. For example, the factor $\chi_0/(4\pi^2\hbar L^2)$ in Eq. (4.91) has a value of approximately 3.85×10^{27} electrons/cm^2 sec whenever $L = 10$ Å and $\chi_0 = 1$ eV, corresponding to 6.17×10^8 A/cm^2. The electron tunnel current for these values of the parameters would therefore be of the order of $(6.17 \times 10^8)(4.54 \times 10^{-5}) = 2.80 \times 10^4$ A/cm^2, which is very large. An increase in thickness to 20 Å would decrease this by an additional factor of $\frac{1}{4}(4.54 \times 10^{-5})$ to give 0.318 A/cm^2.

PROJECT 4.8 Tunnel Current Development

Instead of using the WKB approximation for the transmission coefficient $\mathscr{T}$ in the integrals for the forward and reverse tunnel currents through a rectangular barrier, carry out (analytically or numerically) an analogous development using the exact result for the transmission coefficient for a rectangular barrier. Delineate specifically the physical differences between the two results, and compare (numerically) the results of the two approaches for a realistic set of parameter values.

PROJECT 4.9 Voltage Dependence of the Electron Tunnel Current

Choose a realistic set of numerical values for the relevant parameters, and compute the voltage dependence of the electron tunnel current from Eq. (4.88). Then compute the voltage dependence of the tunnel current from the approximate expression (4.91), and plot the results of both computations on the same graph (semilogarithmic plot). Draw logical conclusions from your comparison. [*Hint*: See §3.4.]

5 Tunnel Current at 0°K for Barriers of Arbitrary Shape

5.1 Taylor Series Expansion Technique

An approximate WKB expression for the transmission coefficient for a trapezoidal barrier such as that in Fig. 4.1b is given by Eq. (4.41). The parameter γ is a measure of the slope of the upper portion of the barrier, which can depend, for example, on the value of an applied electric field. For $\gamma = 0$, Eq. (4.41) reduces to the transmission coefficient (4.34) for a square barrier. It can be seen from Eq. (4.41) that the lowest term in γ involves $-\frac{3}{8}\gamma(x_2 - x_1)^2/(W - \mathscr{E})^{1/2}$ as the argument of the exponential function, which becomes for tunneling in the x direction $-\frac{3}{8}\gamma L^2/[W - (p_x^2/2m)]^{1/2}$ in the present notation. The 0°K tunnel current expressions analogous to Eq. (4.70) for J_{I} and the corresponding expressions for J_{II} and J_{III} have the additional factor

$$\exp\{(m/2\hbar^2)^{1/2}\gamma L^2/[W - (p_x^2/2m)]^{1/2}\}.$$

In terms of the dimensionless quantity ζ defined by Eq. (4.73), this additional factor is $\exp(2m\gamma L^3/\hbar^2\zeta)$, which of course will appear in the expressions for $J_{\text{I}}(0°\text{K})$ and $J_{\text{II}}(0°\text{K})$ analogous to Eqs. (4.76) and (4.77). The integrals then become more difficult to evaluate, and the subsequent development becomes more involved than that for a square barrier. An excellent development is given by Good and Müller (1956) for the trapezoidal barrier as modified by the electron image potential in the metal electrode.

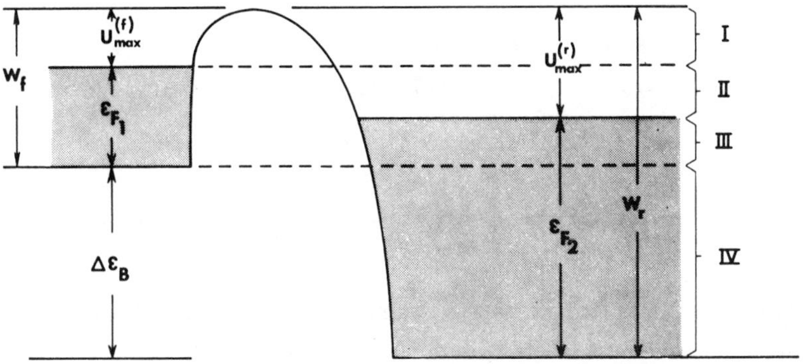

Fig. 4.11 Representation of an arbitrary potential energy barrier separating two metals. (Such a barrier could be derived, for example, from a rectangular barrier such as that illustrated in Fig. 4.10 by adding the effects of the electron image potentials in the two metals.)

For purposes of approximation and insight it is useful to have available simpler expressions for obtaining the tunnel current through a barrier of arbitrary shape. (See Fig. 4.11, for example.) The only difficulty in evaluating the current for the limiting case of 0°K is due to the integral of the transmission coefficient. The development leading to Eqs. (4.70) and (4.71) shows that

$$J_{\mathrm{I}}(0^\circ\mathrm{K}) = (2\pi^2\hbar^3)^{-1}(\mathscr{E}_{\mathrm{F}_1} - \mathscr{E}_{\mathrm{F}_2}) \int_0^{\sqrt{2m\mathscr{E}_{\mathrm{F}_2}}} \mathscr{T}(p_x)p_x\, dp_x, \tag{4.92}$$

$$J_{\mathrm{II}}(0^\circ\mathrm{K}) = (2\pi^2\hbar^3)^{-1} \int_{\sqrt{2m\mathscr{E}_{\mathrm{F}_2}}}^{\sqrt{2m\mathscr{E}_{\mathrm{F}_1}}} [\mathscr{E}_{\mathrm{F}_1} - (p_x^2/2m)]\mathscr{T}(p_x)p_x\, dp_x, \tag{4.93}$$

$$J_{\mathrm{III}}(0^\circ\mathrm{K}) = 0, \tag{4.94}$$

where $\mathscr{T}(p_x)$ is the appropriate transmission coefficient for the arbitrary barrier under consideration. The key to evaluating $J_{\mathrm{tot}} = J_{\mathrm{I}} + J_{\mathrm{II}} + J_{\mathrm{III}}$ for a barrier of arbitrary shape is the development of a good approximate expression for $\mathscr{T}(p_x)$. One method [cf. Stratton (1962)] is to assume that ln $\mathscr{T}(p_x)$ can be obtained by an exact solution of the Schrödinger equation (4.1) or by an approximate technique such as the WKB approximation (§1) and then carry out a Taylor series expansion of ln $\mathscr{T}(p_x)$ about the Fermi energy of the metal with the higher Fermi energy, since the primary contribution to the tunnel current comes from this energy region. This is due to the fact that very few states are populated above this energy, while the rapid decrease of the transmission coefficient with decreasing energy causes severe attenuation of the current from filled states at lower energies. The Taylor series for ln $\mathscr{T}(p_x)$ around a given energy $\mathscr{E}_0$ will be a power series in p_x^2,

$$\ln \mathscr{T}(p_x) \simeq \mathscr{P}_0 + \mathscr{P}_1(W_x - \mathscr{E}_0) + \tfrac{1}{2}\mathscr{P}_2(W_x - \mathscr{E}_0)^2 + \cdots, \tag{4.95}$$

where

$$W_x \equiv p_x^2/2m, \tag{4.96}$$

$$\mathscr{P}_0 \equiv [\ln \mathscr{T}(p_x)]_{W_x = \mathscr{E}_0}, \tag{4.97}$$

$$\mathscr{P}_1 \equiv \left[\frac{d \ln \mathscr{T}(p_x)}{dW_x}\right]_{W_x = \mathscr{E}_0}, \tag{4.98}$$

$$\mathscr{P}_2 \equiv \left[\frac{d^2 \ln \mathscr{T}(p_x)}{dW_x^2}\right]_{W_x = \mathscr{E}_0}, \tag{4.99}$$

etc., are constants which are barrier-shape dependent. Substituting Eq. (4.95) into Eqs. (4.92) and (4.93) and changing variables according to Eq. (4.96) gives

$$J_{\rm I}(0^\circ{\rm K}) = (2\pi^2\hbar^3)^{-1}(\mathscr{E}_{{\rm F}_1} - \mathscr{E}_{{\rm F}_2})m \int_0^{\mathscr{E}_{{\rm F}_2}} \exp[\mathscr{P}_0 + \mathscr{P}_1(W_x - \mathscr{E}_0) + \tfrac{1}{2}\mathscr{P}_2(W_x - \mathscr{E}_0)^2 + \cdots]\, dW_x, \tag{4.100}$$

$$J_{\rm II}(0^\circ{\rm K}) = (2\pi^2\hbar^3)^{-1}m \int_{\mathscr{E}_{{\rm F}_2}}^{\mathscr{E}_{{\rm F}_1}} (\mathscr{E}_{{\rm F}_1} - W_x)\exp[\mathscr{P}_0 + \mathscr{P}_1(W_x - \mathscr{E}_0) + \tfrac{1}{2}\mathscr{P}_2(W_x - \mathscr{E}_0)^2 + \cdots]\, dW_x. \tag{4.101}$$

The terms involving $\mathscr{P}_0$ and $\mathscr{P}_1$ will provide a suitable approximation if

$$|\tfrac{1}{2}(\mathscr{P}_2/\mathscr{P}_1)(W_x - \mathscr{E}_0)| \ll 1. \tag{4.102}$$

Whenever this is the case and the expansion is with respect to the Fermi energy of metal 1, $\mathscr{E}_0 = \mathscr{E}_{{\rm F}_1}$, Eqs. (4.100) and (4.101) lead to the following approximate result:

$$J_{\rm tot}(0^\circ{\rm K}) = J_{\rm I}(^\circ{\rm K}) + J_{\rm II}(0^\circ{\rm K}) \simeq (2\pi^2\hbar^3)^{-1}m\mathscr{P}_1^{-1}e^{\mathscr{P}_0}\{\mathscr{P}_1^{-1}[1 - e^{-\mathscr{P}_1(\mathscr{E}_{{\rm F}_1} - \mathscr{E}_{{\rm F}_2})}] - (\mathscr{E}_{{\rm F}_1} - \mathscr{E}_{{\rm F}_2})e^{-\mathscr{P}_1\mathscr{E}_{{\rm F}_1}}\}. \tag{4.103}$$

Let us evaluate the parameters in Eq. (4.103) for square and trapezoidal barriers. Using the expression

$$\ln \mathscr{T}(p_x) = -(2L/\hbar)[2m(W - W_x)]^{1/2} \tag{4.104}$$

obtained from Eq. (4.34) as applied to penetration of a square barrier in the x direction, together with the definitions (4.97)–(4.99) with $\mathscr{E}_0 = \mathscr{E}_{{\rm F}_1}$, gives

$$\mathscr{P}_0 = -(2L/\hbar)[2m(W - \mathscr{E}_{{\rm F}_1})]^{1/2}, \tag{4.105}$$

$$\mathscr{P}_1 = (2mL/\hbar)[2m(W - \mathscr{E}_{{\rm F}_1})]^{-1/2}, \tag{4.106}$$

$$\mathscr{P}_2 = (2m^2L/\hbar)[2m(W - \mathscr{E}_{{\rm F}_1})]^{-3/2}. \tag{4.107}$$

The condition (4.102) required for the validity of Eq. (4.103) is

$$|(W_x - \mathscr{E}_{{\rm F}_1})| \ll 4|(W - \mathscr{E}_{{\rm F}_1})|. \tag{4.108}$$

Since the result (4.103) for 0°K contains a zero contribution for $W_x > \mathscr{E}_{{\rm F}_1}$, the inequality (4.108) is required to hold only for $W_x \leqslant \mathscr{E}_{{\rm F}_1}$. The left-hand side has its largest value for $W_x = 0$, so the inequality will be met whenever the Fermi energy in metal 1 is much less than the barrier height W. (In fact, the contribution to the current from states considerably lower in energy than $\mathscr{E}_{F_1}$ is negligible due to the sharp decrease in the transmission coefficient with increasing barrier height, so the current will not be affected very much even if Eq. (4.108) is not well met for values of W_x near zero.) The definitions (4.78)–(4.80) can be used to convert Eq. (4.103) to a form involving χ_0 and χ_L instead of $\mathscr{E}_{F_1}$ and $\mathscr{E}_{F_2}$.

Using the expression

$$\ln \mathscr{T}(p_x) = -(2/\hbar)(2m)^{1/2}[L(W - W_x)^{1/2} - \tfrac{1}{4}\gamma L^2(W - W_x)^{-1/2}] \quad (4.109)$$

obtained from approximating Eq. (4.41) as applied to penetration of a trapezoidal barrier in the x direction (with γ representing a small slope in the top of the barrier), together with the definitions (4.97) and (4.98), gives

$$\mathscr{P}_0 = -2L(2m/\hbar^2)^{1/2}(W - \mathscr{E}_{F_1})^{1/2}[1 - \tfrac{1}{4}\gamma L(W - \mathscr{E}_{F_1})^{-1}], \quad (4.110)$$

$$\mathscr{P}_1 = L(2m/\hbar^2)^{1/2}(W - \mathscr{E}_{F_1})^{-1/2}[1 + \tfrac{1}{4}\gamma L(W - \mathscr{E}_{F_1})^{-1}]. \quad (4.111)$$

For the case of electron tunneling with the trapezoidal barrier produced by the application of a small uniform electric field E_0 to a square barrier,

$$\gamma = -eE_0. \quad (4.112)$$

Thus $\mathscr{P}_0$ and $\mathscr{P}_1$ contain terms linear in the electric field, so that the current given by Eq. (4.103) contains exponential terms with arguments linear in the electric field. The electrostatic potential $-E_0L$ associated with E_0 will also shift the relative positions of the Fermi levels $\mathscr{E}_{F_1}$ and $\mathscr{E}_{F_2}$ of the two metals by $|eE_0L|$. If χ_L is the usual parameter associated with the barrier in zero field, as illustrated in Fig. 4.4, then with a field E_0 applied,

$$\mathscr{E}_{F_2} = W - (\chi_L - eE_0L). \quad (4.113)$$

Equation (4.78) is unmodified by the field,

$$\mathscr{E}_{F_1} = W - \chi_0, \quad (4.114)$$

so we obtain

$$\mathscr{E}_{F_1} - \mathscr{E}_{F_2} = (W - \chi_0) - [W - (\chi_L - eE_0L)] = (\chi_L - \chi_0) - eE_0L. \quad (4.115)$$

The electron energy barrier is lowered by a negative field, so that $(\mathscr{E}_{F_1} - \mathscr{E}_{F_2})$ is greater than $\chi_L - \chi_0$ for a negative applied field. The electron tunnel current will of course be increased by a field with this polarity. Equations (4.110)–(4.115) can be substituted into Eq. (4.103) to give $J_{tot}(0°K)$ as a function of applied field. The resulting expression is restricted to the small field limit since this was the assumption made in deriving $\mathscr{T}_{trap}$ in Eq. (4.41) utilized in the present development.

For the large field limit, the expression

$$\ln \mathscr{T}(p_x) = -[4(2m)^{1/2}/3\gamma\hbar](W - W_x)^{3/2} \quad (4.116)$$

obtained from Eq. (4.42) as applied to penetration of a triangular barrier in the x

direction, together with the definitions (4.97) and (4.98) with $\mathscr{E}_0 = \mathscr{E}_{F_1}$, gives

$$\mathscr{P}_0 = -[4(2m)^{1/2}/3\gamma\hbar](W - \mathscr{E}_{F_1})^{3/2}, \tag{4.117}$$

$$\mathscr{P}_1 = [2(2m)^{1/2}/\gamma\hbar](W - \mathscr{E}_{F_1})^{1/2}. \tag{4.118}$$

Substitution of Eqs. (4.112)–(4.115) and Eqs. (4.117)–(4.118) into Eq. (4.103) then gives $J_{\text{tot}}(0°\text{K})$ as a function of applied field in the large field limit. Note that Eq. (4.103) will contain exponentials with arguments varying as $\gamma^{-1} \propto E_0^{-1}$. The electron current for this limit in which the field dependence is of the form $\exp(-E_0'/E_0)$ with E_0 being the homogeneous electric field and E_0' being a constant parameter, is referred to as *field emission*, *cold cathode emission*, or *Fowler–Nordheim tunneling*.

It is often useful to write the electric field E_0 in a tunnel current device as

$$E_0 = E_{\text{applied}} + E_{\text{built-in}}, \tag{4.119}$$

where $E_{\text{built-in}}$ is the electric field in the absence of an applied field which is established by spontaneous tunneling between the two metal electrodes to achieve equilibrium. That is, when two electrodes are placed near enough together so that the transmission coefficient is not prohibitively small, tunneling will occur until the metal Fermi levels are aligned. According to Eq. (4.115), this will give

$$V_{\text{built-in}} \equiv -E_{\text{built-in}}L = e^{-1}(\chi_0 - \chi_L). \tag{4.120}$$

Thus

$$E_{\text{built-in}} = -e^{-1}(\chi_0 - \chi_L)/L, \tag{4.121}$$

so that

$$E_0 = E_{\text{applied}} + e^{-1}(\chi_L - \chi_0)/L. \tag{4.122}$$

This transformation can be made throughout the equations of the present section whenever the tunnel current is required as a function of the applied electric field instead of in terms of the net electric field. More detailed predictions of the tunnel current as a function of applied electric field for the various field regions can be found in the literature [cf. Good and Müller (1956), Stratton (1962)].

5.2 Equivalent Square Barrier Technique

There is another useful approximation technique for obtaining the tunnel current through arbitrary barriers (such as illustrated in Fig. 4.11) which yields simpler equations than those just derived. This technique [cf. Simmons (1963a,b)] involves the determination of an equivalent square (or rectangular) barrier for the arbitrary barrier in question with subsequent utilization of expressions for the currents such as those given by Eqs. (4.88), (4.90), and (4.91) which are derived explicitly for a square (or rectangular) barrier. The *equivalent square barrier* $\langle U(x) \rangle$ corresponding to an arbitrary barrier $U(x)$ over the region $x_1 \leqslant x \leqslant x_2$ is defined to be the average value of $U(x)$ over this region,

$$\langle U(x) \rangle \equiv (x_2 - x_1)^{-1} \int_{x_1}^{x_2} U(x)\, dx. \tag{4.123}$$

As a specific example, consider the trapezoidal barrier $U(x) = W - \gamma x$ for $0 \leqslant x \leqslant L$,

$$\langle U(x) \rangle = L^{-1} \int_0^L (W - \gamma x)\, dx = W - \tfrac{1}{2}\gamma L. \tag{4.124}$$

Substituting $\gamma = -eE_0$ as given by Eq. (4.112) gives $\langle U(x) \rangle = W + \frac{1}{2}eE_0 L = W - \frac{1}{2}eV(L)$, where $V(L)$ is the electrostatic potential across the barrier. In an analogous fashion, the barrier heights χ_0 and χ_L as viewed from the forward and reverse directions, respectively, are replaced by

$$\langle \chi_0 + eE_0 x \rangle = \chi_0 + \tfrac{1}{2}eE_0 L = \chi_0 - \tfrac{1}{2}eV(L), \tag{4.125}$$

$$\langle \chi_L - eE_0(L - x) \rangle = \chi_L - \tfrac{1}{2}eE_0 L = \chi_L + \tfrac{1}{2}eV(L), \tag{4.126}$$

respectively. Substituting these two expressions for χ_0 and χ_L into Eq. (4.91), for example, gives

$$J_{tot}(0°\mathrm{K}) \simeq (8\pi^2\hbar L^2)^{-1}\{(2\chi_0 + eE_0 L) \exp[-2m^{1/2}\hbar^{-1}L(2\chi_0 + eE_0 L)^{1/2}] - (2\chi_L - eE_0 L) \exp[-2m^{1/2}\hbar^{-1}L(2\chi_L - eE_0 L)^{1/2}]\}. \tag{4.127}$$

The accuracy of results obtained by the equivalent square barrier technique relative to corresponding results obtained by the Taylor series expansion technique has been the subject of a numerical evaluation by Hartman (1964). The two techniques generally give consistent results, although the accuracy of one or the other may be somewhat better over a limited range of electric fields for any given set of values of the microscopic parameters. The simplicity of the equivalent square barrier technique is very appealing, and the equations derived by this method turn out to give a surprisingly good measure of the exact electron tunnel current.

PROJECT 4.10 Transmission Coefficient for Penetration of a Parabolic Potential Barrier

1. Use the WKB approximation to derive the transmission coefficient $\mathscr{T}$ for the parabolic potential energy barrier defined by

$$U(x) = \begin{cases} U_0[1 - (x/L)^2] & (-L \leqslant x \leqslant L), \\ 0 & (x < -L; x > L). \end{cases}$$

[*Hint*: Use Eqs. (4.32) and (4.33).]

2. Numerically evaluate your result in Part 1 as a function of incident energy $\mathscr{E}$ for some realistic choice of numerical values for the parameters, such as the electronic mass for m, 0.1–5 eV for U_0, and 1–20 Å for L.

3. Use the equivalent square barrier technique (cf. §5.2) to obtain an alternate numerical evaluation for $\mathscr{T}$ for each of the computations in Part 2.

4. Use the Taylor series expansion technique (cf. §5.1) to obtain yet another alternate numerical evaluation for $\mathscr{T}$ for each of the computations in Part 2.

5. Plot the corresponding results for Parts 1, 2, 3 in a single graph for $\mathscr{T}$ vs $\mathscr{E}$, and attempt to draw conclusions from your results.

6 Temperature Dependence of the Electron Tunnel Current

At elevated temperatures the Fermi–Dirac distribution functions are not discontinuous at the Fermi energy. The temperature dependence of the logarithmic function in Eq. (4.67) then gives rise to a modest temperature variation of the tunnel current.

Making the change of variables $W_x \equiv p_x^2/2m$ in Eq. (4.67) and introducing the Taylor series approximation (4.95) for $\ln \mathscr{T}(p_x)$ with respect to the energy $\mathscr{E}_{F_1}$ gives

$$J_{tot}(T) = (k_B T/2\pi^2\hbar^3) m e^{\mathscr{P}_0 - \mathscr{P}_1 \mathscr{E}_{F_1}} \int_0^\infty e^{\mathscr{P}_1 W_x} \ln\left[\frac{1+\exp[(\mathscr{E}_{F_1} - W_x)/k_B T]}{1+\exp[(\mathscr{E}_{F_2} - W_x)/k_B T]}\right] dW_x. \tag{4.128}$$

It is true that the Taylor series approximation is not valid for $W_x \gtrsim W$, but the contribution to the integral will be small at such energies due to the fact that the states will be unpopulated. Thus the upper limit of ∞ does not lead to any appreciable error if indeed the temperature is not so high as to cause the thermionic emission current (Chap. 3, §4) to become significant with respect to the electron tunnel current.

The assumption made in using the Taylor series approximation (4.95) is that $\mathscr{E}_{F_1} > \mathscr{E}_{F_2}$, with the primary contribution to the tunnel current arising from filled states in metal 1 in the region $W_x \simeq \mathscr{E}_{F_1}$. The quantity $(\mathscr{E}_{F_2} - W_x)/k_B T$ will be negative and very large in magnitude for $W_x \simeq \mathscr{E}_{F_1}$, so that $\exp[(\mathscr{E}_{F_2} - W_x)/k_B T]$ can be neglected with respect to unity in the denominator of the logarithmic factor in Eq. (4.128). If desired, the lower limit in the integral in Eq. (4.128) can be extended to $-\infty$ without modifying the value of $J_{tot}(T)$ since the integrand is essentially zero over this domain. Making these approximations, and employing the additional variables change $\exp[-(\mathscr{E}_{F_1} - W_x)/k_B T] \equiv \zeta$ leads to

$$J_{tot}(T) = [m(k_B T)^2/(2\pi^2\hbar^3)] e^{\mathscr{P}_0} \int_{\zeta_{lower}}^\infty \zeta^{\mathscr{P}_1 k_B T - 1} \ln(1 + \zeta^{-1})\, d\zeta, \tag{4.129}$$

where $\zeta_{lower} = \exp(-\mathscr{E}_{F_1}/k_B T)$. Let us now integrate by parts, letting $u = \ln(1 + \zeta^{-1})$ and $dv = \zeta^{\mathscr{P}_1 k_B T - 1} d\zeta$. Since $du = -[\zeta(\zeta + 1)]^{-1} d\zeta$ and $v = (\mathscr{P}_1 k_B T)^{-1} \zeta^{\mathscr{P}_1 k_B T}$, the product uv is an indeterminant at the upper limit, which can be readily shown to have the value zero provided $\mathscr{P}_1 k_B T < 1$. The product evaluated at the lower limit yields the approximate value

$$-\mathscr{E}_{F_1}/[\mathscr{P}_1 (k_B T)^2]\, e^{-\mathscr{P}_1 \mathscr{E}_{F_1}}, \tag{4.130}$$

assuming that $\exp(\mathscr{E}_{F_1}/k_B T)$ is much greater than unity. The integral resulting from the parts integration is

$$-\int v\, du = \int_{\zeta_{lower}}^\infty (\mathscr{P}_1 k_B T)^{-1} (\zeta + 1)^{-1} \zeta^{\mathscr{P}_1 k_B T - 1}\, d\zeta$$

$$= (\mathscr{P}_1 k_B T)^{-1} \left[\int_0^\infty \frac{\zeta^{\mathscr{P}_1 k_B T - 1}\, d\zeta}{\zeta + 1} - \int_0^{\zeta_{lower}} \frac{\zeta^{\mathscr{P}_1 k_B T - 1}\, d\zeta}{\zeta + 1}\right]. \tag{4.131}$$

Assuming $0 < \mathscr{P}_1 k_B T < 1$, the definite integral can be found in standard integral tables,

$$\int_0^\infty \frac{\zeta^{\mathscr{P}_1 k_B T - 1}\, d\zeta}{\zeta + 1} = \frac{\pi}{\sin(\mathscr{P}_1 \pi k_B T)}. \tag{4.132}$$

It is a good approximation over the region of integration to replace $\zeta + 1$ by unity in the denominator of the second integral, so that

$$\int_0^{\zeta_{\text{lower}}} \frac{\zeta^{\mathscr{P}_1 k_B T - 1}\, d\zeta}{\zeta + 1} \simeq \int_0^{\zeta_{\text{lower}}} \zeta^{\mathscr{P}_1 k_B T - 1}\, d\zeta$$

$$= (\mathscr{P}_1 k_B T)^{-1} \zeta^{\mathscr{P}_1 k_B T} \Big|_0^{\zeta_{\text{lower}}} = (\mathscr{P}_1 k_B T)^{-1} e^{-\mathscr{P}_1 \mathscr{E}_{F_1}}. \tag{4.133}$$

Substituting these evaluations into Eq. (4.129) thus gives

$$J_{\text{tot}}(T) = [m(k_B T)^2/(2\pi^2\hbar^3)] e^{\mathscr{P}_0} \{[-\mathscr{E}_{F_1}/\mathscr{P}_1(k_B T)^2] e^{-\mathscr{P}_1 \mathscr{E}_{F_1}}$$
$$+ (\mathscr{P}_1 k_B T)^{-1}[\pi \csc(\mathscr{P}_1 \pi k_B T) - (\mathscr{P}_1 k_B T)^{-1} e^{-\mathscr{P}_1 \mathscr{E}_{F_1}}]\}. \tag{4.134}$$

If the denominator of the logarithmic factor in Eq. (4.128) is not considered to be approximately unity, its contribution can be evaluated in exactly the same way as the contribution from the numerator. The result will have a negative sign attached and will have $\mathscr{E}_{F_1}$ replaced by $\mathscr{E}_{F_2}$, so that $e^{\mathscr{P}_0}$ will be replaced by $\exp(\mathscr{P}_0 - \mathscr{P}_1\mathscr{E}_{F_1} + \mathscr{P}_1\mathscr{E}_{F_2})$. This contribution will therefore be a factor of $\exp[-\mathscr{P}_1(\mathscr{E}_{F_1} - \mathscr{E}_{F_2})]$ lower than the contribution above, and therefore can be neglected whenever $\mathscr{P}_1(\mathscr{E}_{F_1} - \mathscr{E}_{F_2}) \gg 1$. An examination of $\mathscr{P}_1$ given by Eq. (4.106) for the case of a square barrier shows that it has values of the order of 5 $(\text{eV})^{-1}$ whenever $W - \mathscr{E}_{F_1} \simeq 1$ eV, so that the approximation is valid whenever $\mathscr{E}_{F_1} - \mathscr{E}_{F_2} \gtrsim 1$ eV. The quantity $\mathscr{P}_1 k_B T \simeq \frac{1}{8}$ at room temperature for this value of $\mathscr{P}_1$, so that the condition $0 < \mathscr{P}_1 k_B T < 1$ used in the above derivation is satisfied. Typical values of $\mathscr{P}_1 \pi k_B T$ will be of the order of $\frac{3}{8}$, so that $\pi \csc(\mathscr{P}_1 \pi k_B T)$ will generally be of the order of or greater than unity. Since $\mathscr{P}_1 \mathscr{E}_{F_1}$ will be the order of 10, the terms involving $e^{-\mathscr{P}_1 \mathscr{E}_{F_1}}$ will be a factor of 100 or so smaller than the term involving $\csc(\mathscr{P}_1 \pi k_B T)$ and thus can be neglected. Therefore we obtain

$$J_{\text{tot}}(T) \simeq \frac{m}{2\pi \mathscr{P}_1 \hbar^3} \frac{k_B T}{\sin(\mathscr{P}_1 \pi k_B T)} e^{\mathscr{P}_0}. \tag{4.135}$$

In the limit $T \to 0$, the sine function can be expanded to give

$$J_{\text{tot}}(0°\text{K}) \simeq (m/2\pi^2 \mathscr{P}_1^2 \hbar^3)\, e^{\mathscr{P}_0}. \tag{4.136}$$

Taking a ratio gives

$$J_{\text{tot}}(T)/J_{\text{tot}}(0°\text{K}) = (\mathscr{P}_1 \pi k_B T)/\sin(\mathscr{P}_1 \pi k_B T). \tag{4.137}$$

For small values of $\mathscr{P}_1 \pi k_B T$, the first two terms in the sine expansion lead to the approximation

$$J_{\text{tot}}(T)/J_{\text{tot}}(0°\text{K}) \simeq [1 - \tfrac{1}{6}(\mathscr{P}_1 \pi k_B T)^2]^{-1} \simeq 1 + \tfrac{1}{6}(\mathscr{P}_1 \pi k_B T)^2, \tag{4.138}$$

in which case the lowest order temperature dependence will be quadratic. The coefficient of the temperature dependence is proportional to $\mathscr{P}_1^2$ and is therefore dependent upon the barrier shape.

Comparing the 0°K result (4.136) to our former result (4.103) at 0°K shows agreement provided the terms in $\exp[-\mathscr{P}_1(\mathscr{E}_{F_1} - \mathscr{E}_{F_2})]$ and $\exp(-\mathscr{P}_1\mathscr{E}_{F_1})$ in Eq. (4.103) can be neglected. Since these approximations are of the same order as those made in deriving Eq. (4.136), the results of the two derivations are consistent, as expected.

PROJECT 4.11 Temperature Dependence of the Electron Tunnel Current

Choose a specific barrier shape and a realistic set of numerical values for the relevant parameters, and compute the temperature dependence of the electron tunnel current from Eq. (4.135). Then compute the temperature dependence of the tunnel current from the approximate expression (4.138), and plot the results of both computations on the same graph for comparison purposes. What conclusions can you draw?

7 Applications of Electron Tunneling

7.1 Solid State Research

The theory of interband electron tunneling was first developed by Zener in 1934. From an experimentalist's point of view, the field of electron tunneling dates from the discovery of the *p–n* tunnel diode by Esaki in 1957 and the discovery of tunneling through oxide layers by Fisher and Giaever in 1960. Tunneling has been observed between normal metals and also between superconducting metals, and exciting experiments have been carried out for each.

The observation by Giaever in 1960 of electron tunneling through an oxide layer separating metal films in the superconducting state and the theoretical prediction and observation of coupled pairs of superconducting electrons (*Cooper pairs*) by Josephson and Powell in 1962 were major milestones in the field of superconducting tunneling. In fact, electron tunneling is one of the most sensitive probes of the superconducting state. It provides detailed information about the phonon density of states and electron–phonon interactions.

The experiments using normal metals are affected by the nature of the tunnel barrier, which is frequently only a few atomic layers thick, and by the metal electrodes within a screening depth from the metal–oxide interfaces. Tunneling thus is a problem which involves much surface physics.

The overall conductance versus voltage dependence is fairly well described by the single-particle tunneling theory, the rudiments of which are the subject of this chapter, although many-body effects [March, Young, and Sampanthar (1967)] are also considered to be important. Experimental *zero-bias anomalies* have captured the imagination of theorists, and there has been much consideration of the interaction of the tunneling electron with the optical and acoustical phonon spectrums and with localized vibrational modes and energy levels due to magnetic and nonmagnetic impurities within the oxide barrier. Experiments by Jaklevic and Lambe (1966) on tunneling between metals have revealed evidence of the vibrational excitation of molecular species in the barrier. Since the vibrational frequencies are characteristic of the source (such as molecular species, electrode surfaces, oxide barrier), electron tunneling can be considered to be a mode of spectroscopy. The interested reader who wishes to study electron tunneling in greater depth and learn in greater detail about some of the applications of tunneling spectroscopy is referred to the monograph by Duke (1969) and to the excellent compilation edited by Burstein and Lundqvist (1969). Tunneling in metal–oxide–semiconductor structures has been used to measure semiconductor band gaps and the effect of temperature and pressure on the gap. Tunneling also promises to be an important tool for the study of band structure and Fermi surfaces.

7.2 The Esaki Diode

As one practical example of a tunnel device, let us examine how a negative resistance arises in the *Esaki tunnel diode*. The tunnel diode is a semiconductor device discovered by Leo Esaki in 1957 while he was working as a physicist at the Sony research laboratory in Japan. Tunneling is the dominant mode of current

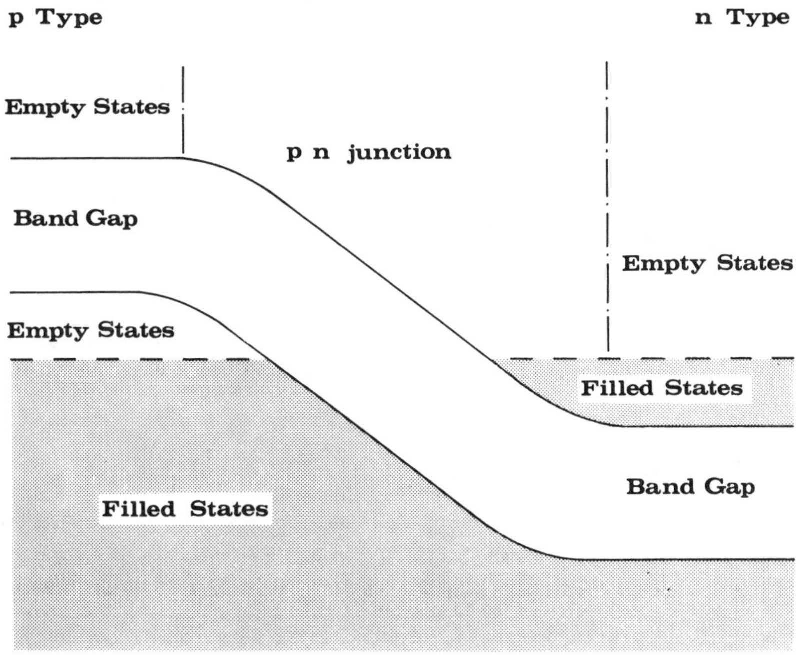

Fig. 4.12 Semiconductor *p–n* junction. (The varying electrostatic potential due to space charge within the region of contact between the *n*-type material containing electron carriers and the *p*-type material containing electron hole carriers causes the gradual bending of the energy bands which one notes in the energy level diagram for the *p–n* junction.

transport in the device, and the device exhibits the phenomenon of "negative resistance" which causes it to be attractive for use in a wide variety of high frequency oscillators, amplifiers, and similar electronic equipment.

The tunnel diode is basically a p–n semiconductor junction, which for present purposes can be viewed simply as two electronically dissimilar substances (see §3.4) separated by a region sufficiently narrow for tunneling to take place. The allowed energy bands separated by energy gaps occur at different locations on the energy level diagram for the p-type material and the n-type material (see Fig. 4.12) because of band bending produced by the voltage due to space charge within the junction, so the conduction bands are not horizontally matched when

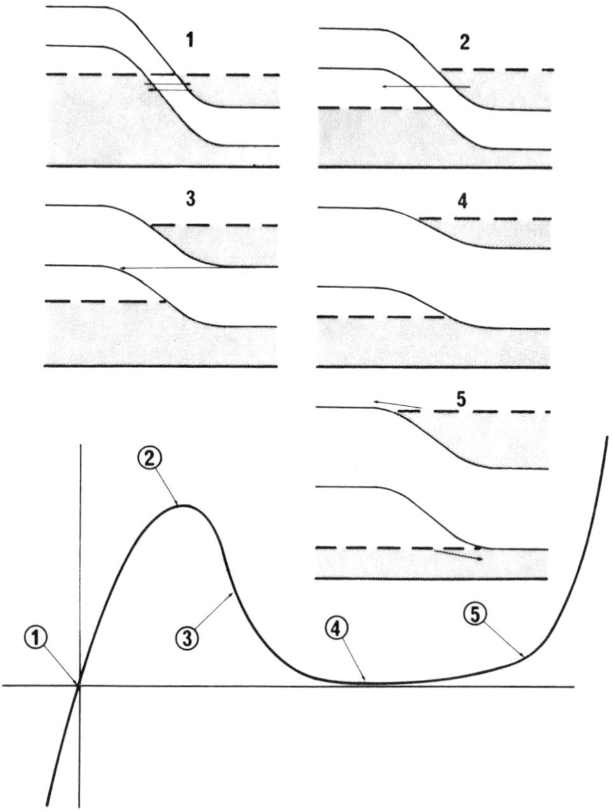

Fig. 4.13 Energy-level diagrams for various bias voltages applied to the Esaki diode and the resulting current–voltage characteristic. (1. Zero bias voltage gives zero current. 2. An applied voltage sufficient to match the filled energy levels in the conduction band on the n side of the p–n junction with empty energy levels in the conduction band on the p side results in a large tunnel current through the intervening energy gap. 3. An increase in the voltage such that some of the filled energy levels in the conduction band on the n side of the p–n junction are raised to the level of the energy gap on the p side results in a decrease in the tunnel current. 4. The tunnel current drops to a minimum value when there are no available states in the p side at the energies of the filled states on the n side of the junction. 5. The current increases again when the voltage is increased sufficiently to allow thermal emission of electrons and electron holes across the junction.)

there is a zero *applied* voltage. At equilibrium there is no net current, but with an applied voltage the bands are shifted relative to each other on either side of the junction, and tunneling can occur from the higher filled levels through the energy gap threading the *p–n* junction and into the empty levels on the opposite side. Thus an initial voltage results in a current. However, if the voltage V is increased sufficiently so that the uppermost filled levels are aligned with an energy *gap* in the material on the opposite side of the junction, then no tunneling can occur, and the voltage increment which initiates this situation is accompanied by a *decrease* in current I. Since dI/dV is negative for this situation, we say that we have a situation of *negative* resistance. If the voltage is increased still further, the filled levels on one side of the junction are raised further, and after continuous increases in the voltage the filled levels on the one side will be raised to the top region of the energy gap on the opposite side, in which case electrons can be thermally excited (see Chap. 3, §4) into the empty levels of the material on the opposite side, the current increases, and dI/dV is positive corresponding to a *positive* resistance. The sketches in Fig. 4.13 attempt to illustrate in a crude sort of way the current–voltage characteristic and the energy-level diagram which explains the observations.

In an analogous way, it is possible by means of tunneling to establish the location of a semiconductor band edge relative to the Fermi level of a metal. In the following section, a description is given of the use of tunneling in determining energy barriers between metals and semiconductors.

EXERCISE Qualitatively show how tunneling can lead to a determination of the energy for a semiconductor band edge relative to the Fermi level of a metal. (*Hint*: Draw an energy-level diagram analogous to that shown for the Esaki diode, and deduce the effects of increases in voltage on the current.)

PROJECT 4.12 Semiconductor *p–n* Junction Theory

Define and give background theory for the following:

1. intrinsic semiconductor **2.** extrinsic semiconductor
3. *n*-type semiconductor **4.** *p*-type semiconductor
5. *p–n* junction.

7.3 Determination of Metal–Semiconductor Barrier Heights

It is possible by means of tunneling to measure the height of the energy barrier between a metal and a semiconductor (or a dielectric material). A thin-film "sandwich" made up of two electrodes of a given metal separated by a very thin film of the semiconductor (or dielectric material) approximates a rectangular energy barrier for conduction electrons (cf. Fig. 4.14a), which under forward bias, assumes a trapezoidal shape. (See Fig. 4.1.) A tunnel current is produced by the forward bias if the barrier is sufficiently thin. The tunnel current increases with increasing bias voltage. At the critical voltage at which the trapezoidal barrier takes on a triangular shape at energies above the Fermi energy, the current–voltage curve undergoes a discontinuity in slope because subsequent

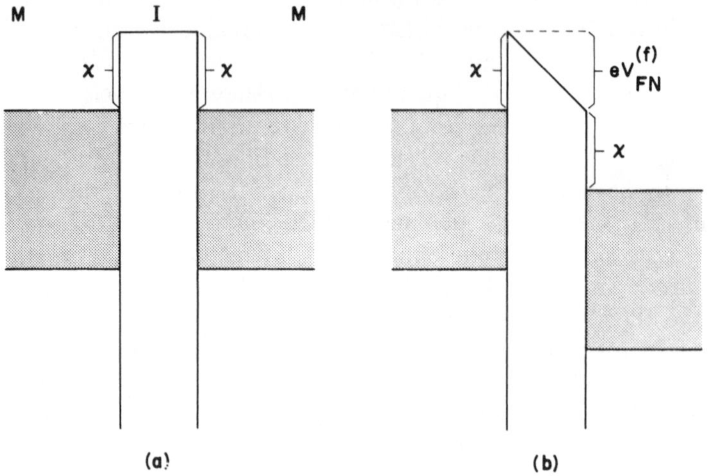

Fig. 4.14 Energy-level diagram for thin film "sandwich" consisting of two electrodes of the *same* metal separated by a very thin film of dielectric (or semiconducting) material. (a) Rectangular potential energy barrier under zero bias voltage. (b) Forward bias voltage sufficient to yield a triangular potential energy barrier which results in Fowler–Nordheim tunneling.

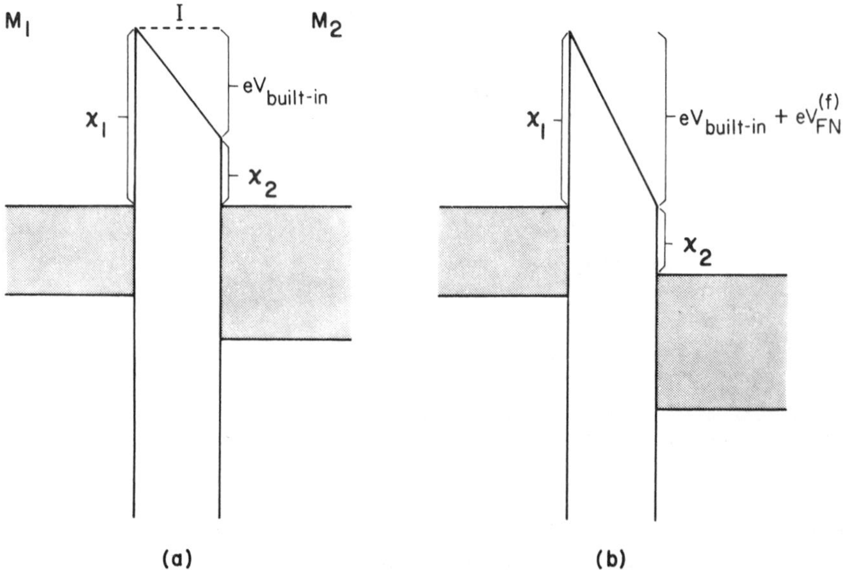

Fig. 4.15 Energy-level diagram for thin film "sandwich" consisting of two electrodes of *different* metal separated by a very thin film of dielectric material. (a) Tunneling occurs until equilibrium is established, which is achieved when the Fermi levels of the two metals are at the same energy. The charge transferred during tunneling sets up an electric field, thus resulting in a "built-in" voltage across the insulator under zero bias conditions. (b) With the addition of a forward bias, the triangular barrier needed for Fowler–Nordheim tunneling in the forward direction is achieved.

increases in voltage *narrow* the effective barrier seen by the electrons *in addition to* lowering it. This is clarified by the diagram in Fig. 4.14b. The forward bias voltage V_{FN} at this critical point is the energy per electron of the original barrier ($\chi = eV_{FN}$), thus giving an experimental measurement of the barrier height χ. The measurement can be repeated under reverse bias conditions. The tunnel current produced under voltages large enough to yield triangular barriers is called *Fowler–Nordheim tunneling* (§5.1).

If the two electrodes are of different metals, the work functions will be different, so the barriers χ_1 and χ_2 at the two interfaces of the thin film semiconductor (or dielectric material) will be different, corresponding to a difference in Fermi levels of the two metals. Tunneling will occur spontaneously until a sufficient voltage difference $V_{\text{built-in}}$ is established to equalize the Fermi levels. The resulting barrier will be trapezoidal, as illustrated in Fig. 4.15a. The application of an externally applied voltage $V_{FN}^{(f)}$ constituting a forward bias sufficient to initiate Fowler–Nordheim tunneling then gives an experimental measurement of χ_2, while the application of an externally applied voltage $V_{FN}^{(r)}$ constituting a reverse bias sufficient to initiate Fowler–Nordheim tunneling in the reverse direction gives an experimental measurement of χ_1. The difference $V_{FN}^{(r)} - V_{FN}^{(f)}$ gives $\chi_1 - \chi_2$, which in turn is equal to $eV_{\text{built-in}}$. Thus electron tunneling provides a powerful experimental technique for obtaining the microscopic physical parameters of solids. Such microscopic parameters are extremely important, for example, in determining the rate of oxide film growth on metals [Fromhold (1976)].

7.4 Tunneling between Superconductors

The Bardeen–Cooper–Schrieffer (BCS) theory of superconductivity provides a detailed microscopic theory for resistanceless currents based on the postulate that there exists a binding force between pairs of electrons which comes about in the following way. One electron, moving through the metal crystal, attracts the positively charged ion cores in the crystal lattice, with the consequence that the crystal lattice is temporarily distorted in such a way that it represents a wake of positive charge which in turn attracts a second electron. (See Fig. 4.16.) The resulting attractive force between the two electrons is quite weak; it binds pairs of electrons with equal but oppositely directed momenta and oppositely directed electronic spin with a binding energy which is characteristic of the metal in question. The pair binding corresponds to a forbidden energy gap spanning an energy range which is forbidden to unpaired electrons in the superconducting metal, the unpaired electrons resulting from thermal disruption of some of the electron pairs at temperatures above 0°K. A current–voltage measurement of the tunnel current between two superconductors separated by a thin insulating layer exhibits very little current until the applied voltage is equivalent to enough energy per electron to break up the superconducting electron pairs, at which point the current increases sharply. A schematic representation of such a current–voltage characteristic is shown in Fig. 4.17. Thus tunneling measurements can yield values for the superconducting energy gap, which in turn can be

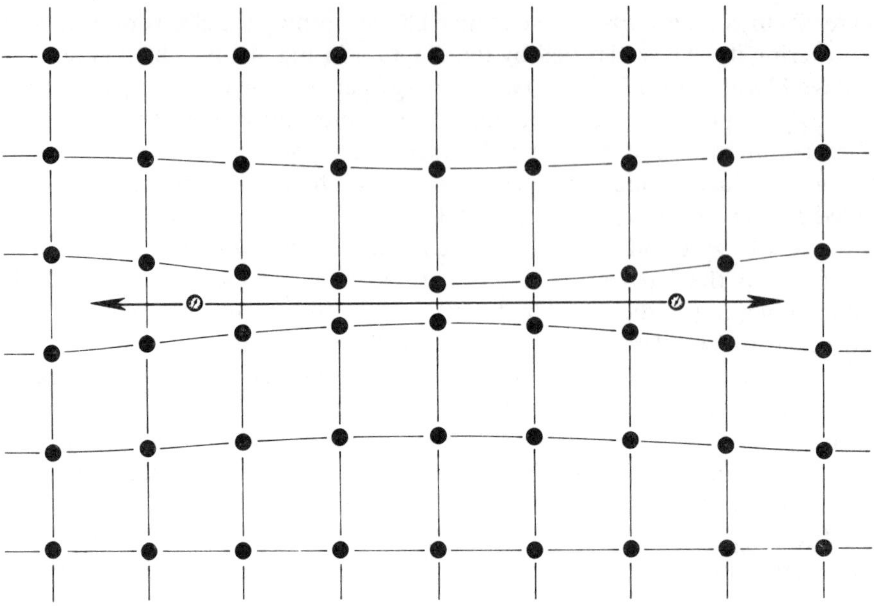

Fig. 4.16 An "electron pair" in a superconductor consists of two electrons having opposite momentum and spin which interact via the polarization of the charged lattice of ions. (The ionic polarization, which is equivalent to a "wake" of positive charge, results in a net reduction in energy for the correlated pair relative to a corresponding electron pair with uncorrelated momenta and spin; thus, the condensation of the ordinary conduction electrons in a normal metal into electron pairs in a superconductor results in a drop in energy (representing an "energy gap") of a few milli-electron-volts.)

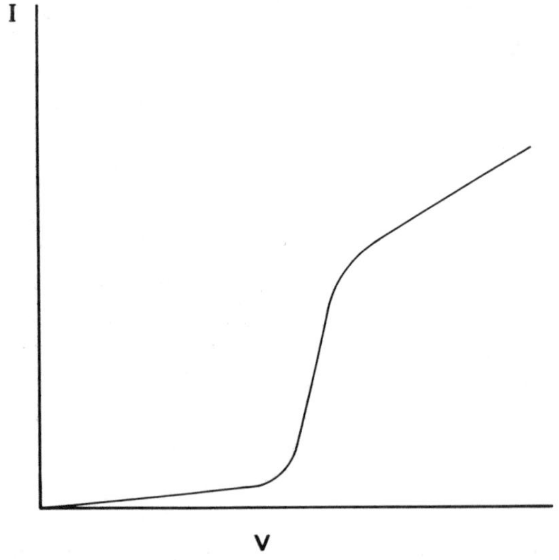

Fig. 4.17 Typical shape of the current–voltage characteristic for a tunnel current between two superconductors separated by a thin insulating layer. (The current remains very low until the applied voltage exceeds the superconductivity energy gap, but thereafter the current rises sharply because the applied voltage represents an energy sufficient to break the electron pairs and reduce them to normal conduction electrons.)

used to check some of the basic relationships appearing in the BCS theory. Such measurements were first carried out by Giaever (1960) at the General Electric Research Laboratory.

7.5 Josephson Tunnel Junctions

The BCS theory of superconductivity leads to the conclusion that the motions of all of the pairs of electrons are correlated, such that the pair centers-of-mass move with the same momentum when a superconducting current is flowing. A de Broglie wavelength can therefore be ascribed to the electron pairs. Thus the superconducting electrons behave as a macroscopic quantum fluid which has the property of phase coherence between the de Broglie waves for the electron pairs.

The English physicist Brian Josephson predicted in 1962 that the phase coherence of electron pairs could extend through a sufficiently thin insulator separating two superconducting metals so that the insulator itself would behave as a weak superconductor. In this situation, an electrical current produced by electron pair tunneling could flow through the barrier even in the absence of a voltage across the barrier; it would only be necessary to start the current flowing in a particular direction through the barrier with a voltage, and thereafter the voltage could be reduced to zero (taking care only to maintain a closed superconducting loop for a complete circuit path). This prediction of a zero voltage current (the *Josephson tunnel current*) was verified experimentally by Rowell (1963).

Another aspect of Josephson's work was a study of the effects of externally applied magnetic fields on the electron pair tunnel current. A magnetic field which threads the barrier will induce oppositely directed superconducting eddy currents in the superconducting electrodes in order to expel the magnetic field

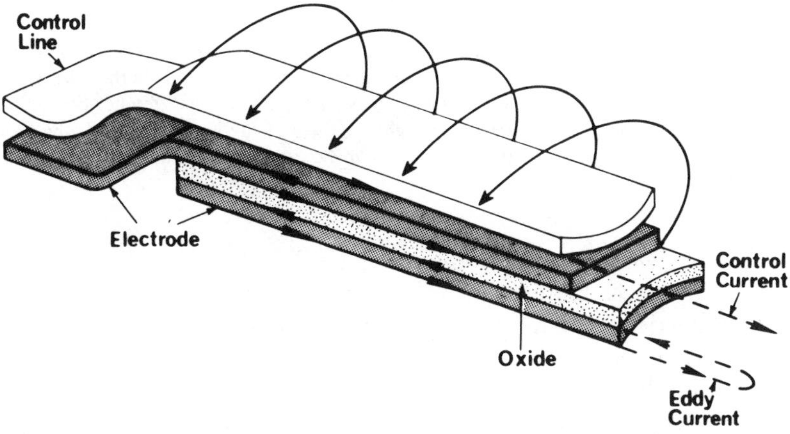

Fig. 4.18 A current through a "control line" over a thin film Josephson tunnel junction "sandwich" consisting of two superconducting electrodes separated by a few monolayers of oxide (or similar insulating material) produces curved magnetic lines of flux which thread the oxide and thereby set up oppositely directed eddy currents in the adjacent electrodes.

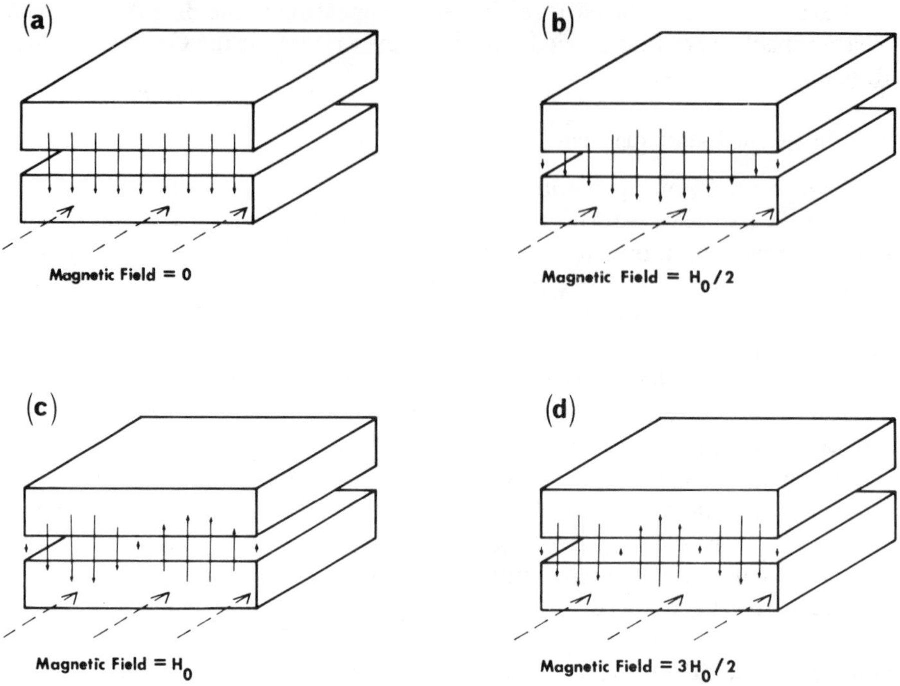

Fig. 4.19 Schematic representation of the change in phase of the wavefunction for two superconductors produced by magnetic flux threading the thin film insulator separating the metals. (Phase matching is a very important factor in determining the magnitude and direction of the current due to electron pair tunneling; thus, the electron pair tunnel current is position dependent along the junction because of the accumulated change in electron phase along the junction due to the magnetic flux.) (a) No magnetic field and no spatial dependence of the electron pair tunnel current; the net Josephson tunnel current through the device is a maximum under these conditions. (b) A magnetic field is impressed sufficient to yield an accumulated phase difference of π along the junction; the net electron pair tunnel current is decreased accordingly. (c) The magnetic field is doubled to give a phase difference of 2π along the junction; the net electron pair tunnel current is zero, since as much current flows in one direction in half of the junction as flows in the opposite direction in the other half of the junction. (d) The magnetic field is increased still further to give a phase difference of 3π along the junction; the electron pair tunnel current through one third of the junction cancels that due to a second third so the net Josephson tunnel current can be attributed to that current through the remaining third of the junction.

(see Fig. 4.18), which in turn produce a phase difference (cf. Fig. 4.19) for the electron pairs along the Josephson junction. Because electron pairs tunnel through the barrier only when their phase matches the phase of the electron pairs in the superconductor on the opposite side, the magnetic field causes a modulation of the flow of electron pairs across the junction with respect to position along the junction. The larger the magnetic field, the more rapid the modulation. The effect is a net reduction in the Josephson tunnel current through the junction, as illustrated in Fig. 4.20. With certain values of the magnetic field, the Josephson tunnel current is zero, as illustrated in Fig. 4.21.

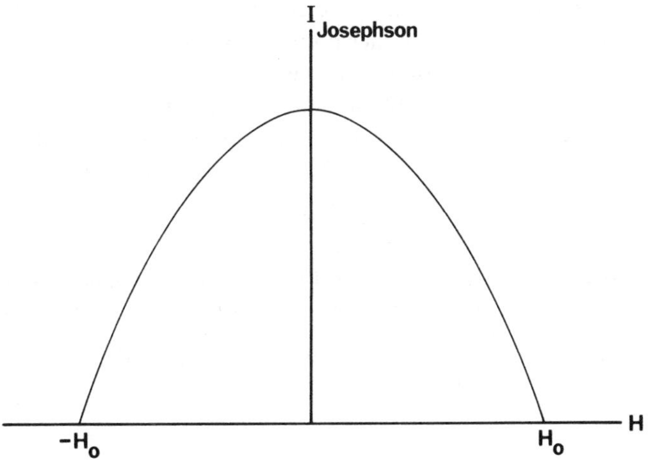

Fig. 4.20 Dependence of the maximum Josephson tunnel current on magnetic field in the junction. (The H_0 in this plot corresponds to the H_0 in Fig. 4.19.)

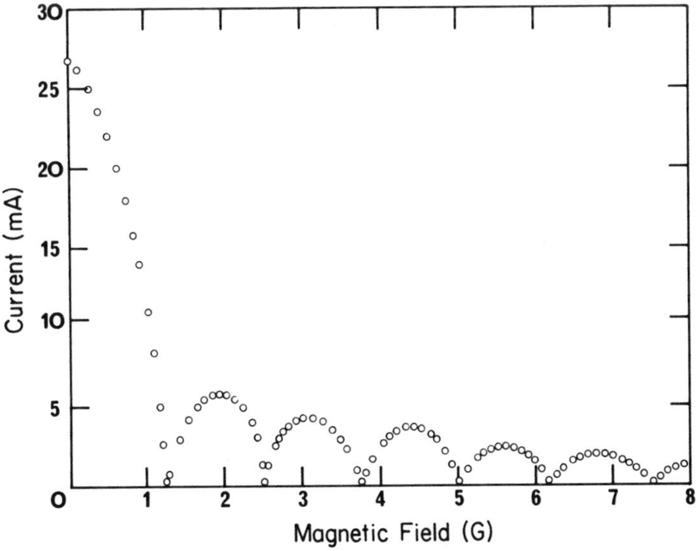

Fig. 4.21 Representation of the Josephson tunnel current versus applied magnetic field [Langenberg *et al.* (1966)].

A current source set for a fixed current through the junction can thus drive current through the junction under two possible conditions, namely, either a zero voltage drop across the junction (whenever the current is the Josephson tunnel current) or else a finite voltage drop across the junction (whenever the current is due to single-particle tunneling). This is readily understood by visualizing a horizontal line to be drawn in Fig. 4.17. These two voltage states provide the basis for memory and logic circuits which may someday supplant transistors in the highest performance computers. This would possibly mark the first major application of coherent matter waves in technology, analogous to the technical applications following the advent of lasers which are based on the use of coherent light waves. Giaever, Josephson, and Esaki received the 1973 Nobel Prize in physics for their research work on electron tunneling.

PROJECT 4.13 Applications of the Josephson Effect

1. Explain from your study of the literature how the Josephson effect and the concept of quantized magnetic flux has been used to obtain the most accurate available measurements to date of the ratio h/e (i.e., of Planck's constant to the electronic charge magnitude).
2. Explain the physical principles underlying the SQUID magnetometer.
3. Explain the operation of the SLUG voltmeter.
[*Hint*: See Tinkham (1975).]

CHAPTER 5

PERTURBATION THEORY, DIFFRACTION OF VALENCE ELECTRONS, AND THE NEARLY-FREE-ELECTRON MODEL

Light brings us the news of the Universe. Sir William Bragg (1931)

1 Stationary-State Perturbation Theory

1.1 Background

In Chap. 4 we considered the approximation technique known as the WKB method for solving the time-independent Schrödinger equation [Eq. (1.130)]. There are in addition the possibilities of direct numerical integration [Eisberg (1967)] of the Schrödinger equation, variational methods [Schiff (1968)], and the currently considered perturbation techniques. In the present chapter we develop approximation techniques based on small potential energy perturbations for solving this important equation.

Stationary-state perturbation theory is a technique for obtaining the modifications of the discrete energy eigenvalues and corresponding eigenfunctions when a relatively small change is made in the Hamiltonian of a time-independent Schrödinger equation for which an exact analytical solution is available. The Hamiltonian $\mathscr{H}$ is written as the sum of two parts, a large part $\mathscr{H}_0$ for which the Schrödinger equation can be solved exactly, and a small part $\mathscr{V}$ denoting the difference between $\mathscr{H}$ and $\mathscr{H}_0$. Both $\mathscr{H}$ and $\mathscr{H}_0$ are considered to represent Hamiltonians for a physical system with real eigenvalues, so that $\mathscr{H}$, $\mathscr{H}_0$, and $\mathscr{V}$ will be Hermitian operators. We introduce a dimensionless parameter λ having values between 0 and 1 to order the terms in our perturbation series; the Hamiltonian

$$\mathscr{H} = \mathscr{H}_0 + \lambda\mathscr{V} \tag{5.1}$$

thus ranges from the completely unperturbed Hamiltonian $\mathscr{H}_0$ to the completely perturbed Hamiltonian $\mathscr{H}_0 + \mathscr{V}$ as λ is increased from zero to one. This can be considered to be a gradual "turning on" of the perturbation in which we shift the energy eigenvalues and corresponding eigenfunctions from the initial to the final stationary-state values. It is assumed that the perturbation expansion does not diverge in the limit $\lambda = 1$.

Let the symbols $\phi_n^{(0)}$ and $\mathscr{E}_n^{(0)}$ represent normalized eigenfunctions and eigenvalues respectively of the unperturbed Hamiltonian $\mathscr{H}_0$, and let ϕ_n and $\mathscr{E}_n$ represent the perturbed eigenfunctions and eigenvalues. Thus we have the following equations,

$$\mathscr{H}_0\phi_n^{(0)} = \mathscr{E}_n^{(0)}\phi_n^{(0)}, \tag{5.2}$$

$$\mathscr{H}\phi_n = \mathscr{E}_n\phi_n. \tag{5.3}$$

1.2 Nondegenerate Case

We first treat the case in which the unperturbed eigenfunctions $\phi_n^{(0)}$ are nondegenerate. We assume that both the eigenvalues $\mathscr{E}_n$ and the eigenfunctions ϕ_n of $\mathscr{H}$ can be expanded in powers of the perturbation parameter λ, and seek to determine coefficients occurring in the perturbation expansion,

$$\mathscr{E}_n = \mathscr{E}_n^{(0)} + \lambda\mathscr{E}_n^{(1)} + \lambda^2\mathscr{E}_n^{(2)} + \cdots, \tag{5.4}$$

$$\phi_n = \phi_n^{(0)} + \lambda\phi_n^{(1)} + \lambda^2\phi_n^{(2)} + \cdots. \tag{5.5}$$

We say that $\lambda^j\mathscr{E}_n^{(j)}$ is the jth-order correction to the energy eigenvalue, and $\lambda^j\phi_n^{(j)}$ is the jth-order correction to the corresponding eigenfunction. Substituting these expansions into Eq. (5.3) yields

$$\begin{aligned}(\mathscr{H}_0 + \lambda\mathscr{V})[\phi_n^{(0)} + \lambda\phi_n^{(1)} + \lambda^2\phi_n^{(2)} + \cdots]\\ = [\mathscr{E}_n^{(0)} + \lambda\mathscr{E}_n^{(1)} + \lambda^2\mathscr{E}_n^{(2)} + \cdots][\phi_n^{(0)} + \lambda\phi_n^{(1)} + \lambda^2\phi_n^{(2)} + \cdots].\end{aligned} \tag{5.6}$$

This gives

$$\begin{aligned}&\mathscr{H}_0\phi_n^{(0)} + \mathscr{H}_0\lambda\phi_n^{(1)} + \mathscr{H}_0\lambda^2\phi_n^{(2)} + \cdots + \lambda\mathscr{V}\phi_n^{(0)} + \lambda^2\mathscr{V}\phi_n^{(1)} + \lambda^3\mathscr{V}\phi_n^{(2)} + \cdots\\ &\quad = \mathscr{E}_n^{(0)}\phi_n^{(0)} + \mathscr{E}_n^{(0)}\lambda\phi_n^{(1)} + \mathscr{E}_n^{(0)}\lambda^2\phi_n^{(2)} + \cdots\\ &\qquad + \lambda\mathscr{E}_n^{(1)}\phi_n^{(0)} + \lambda^2\mathscr{E}_n^{(1)}\phi_n^{(1)} + \lambda^3\mathscr{E}_n^{(1)}\phi_n^{(2)} + \cdots\\ &\qquad + \lambda^2\mathscr{E}_n^{(2)}\phi_n^{(0)} + \lambda^3\mathscr{E}_n^{(2)}\phi_n^{(1)} + \lambda^4\mathscr{E}_n^{(2)}\phi_n^{(2)} + \cdots.\end{aligned} \tag{5.7}$$

The different orders of the perturbation approximation are given by the coefficients of the corresponding powers of λ. Collecting terms in the same power of λ, we obtain

$$\begin{aligned}&\mathscr{H}_0\phi_n^{(0)} + \lambda[\mathscr{H}_0\phi_n^{(1)} + \mathscr{V}\phi_n^{(0)}] + \lambda^2[\mathscr{H}_0\phi_n^{(2)} + \mathscr{V}\phi_n^{(1)}]\\ &+ \lambda^3[\mathscr{H}_0\phi_n^{(3)} + \mathscr{V}\phi_n^{(2)}] + \cdots\\ &\quad = \mathscr{E}_n^{(0)}\phi_n^{(0)} + \lambda[\mathscr{E}_n^{(0)}\phi_n^{(1)} + \mathscr{E}_n^{(1)}\phi_n^{(0)}] + \lambda^2[\mathscr{E}_n^{(0)}\phi_n^{(2)} + \mathscr{E}_n^{(1)}\phi_n^{(1)} + \mathscr{E}_n^{(2)}\phi_n^{(0)}]\\ &\qquad + \lambda^3[\mathscr{E}_n^{(0)}\phi_n^{(3)} + \mathscr{E}_n^{(1)}\phi_n^{(2)} + \mathscr{E}_n^{(2)}\phi_n^{(1)} + \mathscr{E}_n^{(3)}\phi_n^{(0)}] + \cdots.\end{aligned} \tag{5.8}$$

Since this equation is supposedly valid for a continuous range of λ, we can equate the coefficients of equal powers of λ on both sides to obtain a series of equations that represent successively higher orders of the perturbation. Thus we obtain

$$\mathscr{H}_0\phi_n^{(0)} = \mathscr{E}_n^{(0)}\phi_n^{(0)}, \tag{5.9}$$

$$\mathscr{H}_0\phi_n^{(1)} + \mathscr{V}\phi_n^{(0)} = \mathscr{E}_n^{(0)}\phi_n^{(1)} + \mathscr{E}_n^{(1)}\phi_n^{(0)}, \tag{5.10}$$

$$\mathscr{H}_0\phi_n^{(2)} + \mathscr{V}\phi_n^{(1)} = \mathscr{E}_n^{(0)}\phi_n^{(2)} + \mathscr{E}_n^{(1)}\phi_n^{(1)} + \mathscr{E}_n^{(2)}\phi_n^{(0)}, \tag{5.11}$$

$$\vdots$$

The solution to the first of these equations is already known to us according to our initial assumption. To solve the second of these equations, we expand $\phi_n^{(1)}$ in terms of the complete set of unperturbed eigenfunctions $\phi_l^{(0)}$,

$$\phi_n^{(1)} = \sum_l a_l\phi_l^{(0)}, \tag{5.12}$$

where the summation sign denotes a summation over the discrete set of eigenfunctions and an integration over the continuous set of eigenfunctions. Clearly the set of coefficients a_l will be dependent upon the index n characterizing the energy level in question, although we do not designate this explicitly by adding the additional index n to a_l. We substitute Eq. (5.12) into Eq. (5.10) to obtain

$$\mathscr{H}_0 \sum_l a_l\phi_l^{(0)} + \mathscr{V}\phi_n^{(0)} = \mathscr{E}_n^{(0)} \sum_l a_l\phi_l^{(0)} + \mathscr{E}_n^{(1)}\phi_n^{(0)}. \tag{5.13}$$

However, the first term on the left can be replaced by $\sum_l a_l\mathscr{E}_l^{(0)}\phi_l^{(0)}$. We multiply by $\phi_s^{(0)*}$ and integrate over all space, making use of the orthonormality of the $\phi_s^{(0)}$,

$$\sum_l a_l\mathscr{E}_l^{(0)}\,\delta_{sl} + \mathscr{V}_{sn} = \mathscr{E}_n^{(0)} \sum_l a_l\,\delta_{sl} + \mathscr{E}_n^{(1)}\,\delta_{sn}, \tag{5.14}$$

where

$$\mathscr{V}_{sn} = \int \phi_s^{(0)*}\mathscr{V}\phi_n^{(0)}\,d\mathbf{r}. \tag{5.15}$$

The quantity $\mathscr{V}_{sn}$ is the sn matrix element of the perturbation energy $\mathscr{V}$ in the representation in which the unperturbed Hamiltonian $\mathscr{H}_0$ is diagonal. (That is, the $\phi_l^{(0)}$ have been chosen to be the set of orthogonal normalized basis states which are the eigenfunctions of $\mathscr{H}_0$.) The δ_{sl} reduce each of the above sums to a single term, so we obtain

$$a_s\mathscr{E}_s^{(0)} + \mathscr{V}_{sn} = a_s\mathscr{E}_n^{(0)} + \mathscr{E}_n^{(1)}\,\delta_{sn}. \tag{5.16}$$

For $s = n$, this gives

$$a_n\mathscr{E}_n^{(0)} + \mathscr{V}_{nn} = a_n\mathscr{E}_n^{(0)} + \mathscr{E}_n^{(1)},$$

or equivalently

$$\mathscr{E}_n^{(1)} = \mathscr{V}_{nn}. \tag{5.17}$$

The first-order correction to the energy thus is the expectation value of the perturbation operator $\mathscr{V}$ in the unperturbed state $\phi_n^{(0)}$; hence it depends only on the zero-order wave functions. No information is given regarding the coefficient a_n by this equation, but later this coefficient will be shown to be determined by the normalization condition. For $s \neq n$, the above equation (5.16) gives

$$a_s \mathscr{E}_s^{(0)} + \mathscr{V}_{sn} = a_s \mathscr{E}_n^{(0)},$$

or equivalently

$$a_s = \mathscr{V}_{sn}/(\mathscr{E}_n^{(0)} - \mathscr{E}_s^{(0)}) \qquad (s \neq n). \tag{5.18}$$

The a_s determine the first-order corrections to the wave function. Note from the energy denominator that the coefficient a_s, which determines the degree to which the state $\phi_s^{(0)}$ has been *admixed* into the wave function ϕ_n by the perturbation, is relatively larger for those states which are nearby in energy. If the matrix elements $\mathscr{V}_{sn}$ $(s, n = 1, 2, \ldots)$ are considered to constitute a matrix, then we can see from the above results that *the diagonal elements of the matrix give the first-order corrections to the energy levels while the off-diagonal elements determine the first-order correction to the wave functions.* We must still evaluate the remaining coefficient a_n. This coefficient is determined from the normalization of the wave function (5.5), as will now be shown. Integrating the product $\phi_n^* \phi_n$ over all space to terms linear in λ gives

$$\int \phi_n^* \phi_n \, d\mathbf{r} = 1 + \lambda \left[\int \phi_n^{(0)*} \phi_n^{(1)} \, d\mathbf{r} + \int \phi_n^{(1)*} \phi_n^{(0)} \, d\mathbf{r} \right]. \tag{5.19}$$

But we can use the expansion (5.12) for $\phi_n^{(1)}$ to give

$$\int \phi_n^{(0)*} \phi_n^{(1)} \, d\mathbf{r} = \sum_l a_l \int \phi_n^{(0)*} \phi_l^{(0)} \, d\mathbf{r} = \sum_l a_l \, \delta_{nl} = a_n. \tag{5.20}$$

Taking the complex conjugate of Eq. (5.20) gives

$$a_n^* = \int \phi_n^{(1)*} \phi_n^{(0)} \, d\mathbf{r}. \tag{5.21}$$

Therefore we obtain from Eq. (5.19),

$$\int \phi_n^* \phi_n \, d\mathbf{r} = 1 + \lambda(a_n + a_n^*), \tag{5.22}$$

so that as a requirement for normalization,

$$a_n = -a_n^*. \tag{5.23}$$

Hence a_n is purely imaginary. Let $a_n = i\gamma_n$, where γ_n is real. Thus to first order,

$$\phi_n = \phi_n^{(0)} + \lambda \phi_n^{(1)}, \tag{5.24}$$

where

$$\phi_n^{(1)} = i\gamma_n \phi_n^{(0)} + \sum_{l \neq n} a_l \phi_l^{(0)}. \tag{5.25}$$

Thus

$$\phi_n = \phi_n^{(0)}(1 + i\lambda\gamma_n) + \lambda \sum_{l \neq n} a_l \phi_l^{(0)}, \tag{5.26}$$

with

$$a_l = \mathscr{V}_{ln}/(\mathscr{E}_n^{(0)} - \mathscr{E}_l^{(0)}) \qquad (l \neq n). \tag{5.27}$$

But to first order in λ,

$$1 + i\lambda\gamma_n \simeq \exp[i\lambda\gamma_n], \tag{5.28}$$

so the first term can be written $\exp(i\lambda\gamma_n)\phi_n^{(0)}$. The choice of γ_n corresponds to a particular choice of phase factor of the unperturbed wave function $\phi_n^{(0)}$. The specific value is unimportant because it does not affect the value of the probability density $\phi_n^{(0)*}\phi_n^{(0)}$. Choosing the value to be zero,

$$a_n = i\gamma_n = 0, \tag{5.29}$$

the first-order wave function becomes

$$\phi_n = \phi_n^{(0)} + \lambda \sum_{l \neq n} \frac{\mathscr{V}_{ln}}{\mathscr{E}_n^{(0)} - \mathscr{E}_l^{(0)}} \phi_l^{(0)}. \tag{5.30}$$

PROJECT 5.1 Stationary-State Perturbation of the Harmonic Oscillator

1. Calculate the first-order perturbation theory corrections to the one-dimensional harmonic oscillator when a perturbation $V(x) = Ax^3$ is applied, where A is a constant. (*Hint*: Look up the ground-state wave functions and relationships between these functions in standard reference works on quantum mechanics.)
2. Calculate the first-order perturbation theory corrections to the one-dimensional harmonic oscillator when a perturbation $V(x) = Bx^4$ is applied, where B is a constant.

PROJECT 5.2 Stationary-State Perturbation due to Electron–Electron Interaction in the Helium Atom

Use first-order perturbation theory to compute the ground-state energy correction due to the Coulomb interaction between the two electrons in the helium atom, using as a starting point the ground-state wave function

$$\phi^{(0)}(\mathbf{r}_1, \mathbf{r}_2) = (8a_0^{-3}/\pi)\exp[-2a_0^{-1}(r_1 + r_2)]$$

deduced by neglecting the electron–electron Coulomb energy. In the above wave function the symbols $\mathbf{r}_1$ and $\mathbf{r}_2$ denote the position vectors of the two electrons relative to the nucleus, and a_0 is the Bohr radius of the hydrogen atom. [*Hint*: The corresponding ground-state energy is $8(-\frac{1}{2}e^2/4\pi\varepsilon_0 a_0)$.]

1.3 Degenerate Case

We must now consider the case in which the unperturbed energy eigenvalue $\mathscr{E}_n^{(0)}$ is degenerate. In this case, $\mathscr{E}_l^{(0)} = \mathscr{E}_n^{(0)}$ for some $l \neq n$, and the denominator vanishes in the above expansion (5.30). We can avoid the difficulty raised by this situation if it is possible to arrange to have the $\mathscr{V}_{ln}$ to be zero whenever $\mathscr{E}_l^{(0)} = \mathscr{E}_n^{(0)}$. That this is in fact a requirement for the validity of the treatment follows by writing Eq. (5.16) in the form

$$a_s[\mathscr{E}_n^{(0)} - \mathscr{E}_s^{(0)}] + \mathscr{E}_n^{(1)}\delta_{sn} = \mathscr{V}_{sn} \tag{5.31}$$

and examining the situation in which $\mathscr{E}_s^{(0)} = \mathscr{E}_n^{(0)}$ for some $s \neq n$. *Thus we can say that the perturbation treatment fails in first order if there is degeneracy of the unperturbed state in zero order and the perturbation "connects" the degenerate states in first order.* (The latter phrase involving the word "connect" means simply that $\mathscr{V}_{sn} \neq 0$, so that in a certain sense the states $\phi_s^{(0)}$ and $\phi_n^{(0)}$ are linked through the perturbation operator $\mathscr{V}$.)

Consider the case where the eigenvalue $\mathscr{E}_n^{(0)}$ is $(p+1)$-fold degenerate, so that

$$\mathscr{E}_n^{(0)} = \mathscr{E}_{n+1}^{(0)} = \cdots = \mathscr{E}_{n+p}^{(0)}. \tag{5.32}$$

The corresponding orthonormal eigenfunctions $\phi_n^{(0)}, \phi_{n+1}^{(0)}, \ldots, \phi_{n+p}^{(0)}$ of $\mathscr{H}_0$ can be considered to span a $(p+1)$-dimensional *linear manifold* or *subspace*. Since the $\phi_i^{(0)}$ $(i = 1, 2, \ldots, n, n+1, \ldots, n+p, n+p+1, \ldots)$ represent a complete set of functions, and the subset designated by indices $n, n+1, \ldots, n+p$ are the only ones corresponding to the eigenvalue $\mathscr{E}_n^{(0)}$, then any eigenfunction of $\mathscr{H}_0$ corresponding to the eigenvalue $\mathscr{E}_n^{(0)}$ can be constructed from a linear superposition of the *basis vectors* (i.e., the basis functions) in this $(p+1)$-dimensional manifold. Let us designate this initial set of functions $\phi_{n+j}^{(0)}$ $(j = 0, 1, 2, \ldots, p)$ by ξ_j $(j = 0, 1, 2, \ldots, p)$ to emphasize that these can be considered as a set of basis vectors for the manifold. Because the basis vectors of a manifold are not unique, it is possible to replace the set ξ_j by any other orthonormal set η_j $(j = 0, 1, 2, \ldots, p)$ constructed from a linear superposition of the ξ_j. Each of the functions η_j, being a linear combination of eigenfunctions of $\mathscr{H}_0$ corresponding to the eigenvalue $\mathscr{E}_n^{(0)}$, will also be an eigenfunction of $\mathscr{H}_0$ corresponding to this same eigenvalue. Thus, the set of η_j can be substituted for the ξ_j in the set $\phi_{n+j}^{(0)}$ $(j = 0, 1, 2, \ldots)$.

Suppose now that there exists a particular set η_j, the members of which are *in addition* orthonormal eigenfunctions of the perturbation operator $\mathscr{V}$,

$$\mathscr{V}\eta_j = \tilde{v}_j\eta_j \qquad (j = 0, 1, 2, \ldots, p). \tag{5.33}$$

Then the matrix elements

$$\int \eta_i^*\mathscr{V}\eta_j\, d\mathbf{r} = \int \eta_i^*\tilde{v}_j\eta_j\, d\mathbf{r} = \tilde{v}_j \int \eta_i^*\eta_j\, d\mathbf{r} = \tilde{v}_j\, \delta_{ij} \tag{5.34}$$

are noted to satisfy the requirements imposed by Eq. (5.31) for validity of the perturbation expansion for the case of degeneracy. The act of constructing linear superpositions of the ξ_j $(j = 0, 1, 2, \ldots, p)$ which meet this criterion that the new orthonormal set η_j $(j = 0, 1, 2, \ldots, p)$ be eigenfunctions of the operator $\mathscr{V}$ is called *diagonalization* of the operator $\mathscr{V}$ in the $(p+1)$-dimensional manifold in question. With respect to this new basis set, the matrix $\mathscr{V}_{n+j, n+k}$ is a diagonal matrix. If both sets of basis functions are orthonormal, the transformation from the ξ_j to the η_j is said to be *unitary*. Therefore, following the diagonalization process, we replace the ξ_j by the η_j for the set $\phi^{(0)}_{n+j}$ $(j = 0, 1, \ldots, p)$. The set η_j are referred to as a "*good*" set of zero-order wave functions corresponding to the degenerate eigenvalue $\mathscr{E}^{(0)}_n$.

Since $\mathscr{E}^{(1)}_n = \mathscr{V}_{nn}$, the first-order corrections to the energy $\mathscr{E}^{(0)}_n$ are simply

$$\mathscr{E}^{(1)}_{n+j} = \int \eta_j^* \mathscr{V} \eta_j \, d\mathbf{r} = \int \eta_j^* \tilde{v}_j \eta_j \, d\mathbf{r} = \tilde{v}_j. \tag{5.35}$$

Hence to first order,

$$\mathscr{E}_{n+j} = \mathscr{E}^{(0)}_n + \lambda \tilde{v}_j \qquad (j = 0, 1, 2, \ldots, p). \tag{5.36}$$

The eigenvalues $\tilde{v}_0, \tilde{v}_1, \ldots, \tilde{v}_p$ of $\mathscr{V}$ in the $(p + 1)$-dimensional manifold representing the degenerate zero-order eigenfunctions of $\mathscr{H}_0$ may be nondegenerate, degenerate, or partially degenerate. Thus we can say that *the perturbation splits the degeneracy* into a maximum of $(p + 1)$ levels. The first-order eigenfunctions corresponding to each of these levels, whether degenerate or nondegenerate, follow immediately from our previous treatment if we set the undetermined coefficients of the degenerate zero-order wave functions in Eq. (5.12) to zero.

The coefficients a_s for the degenerate set (i.e., $s = n, n + 1, n + 2, \ldots, n + p$) are left undetermined by Eq. (5.31), and only one undetermined parameter can be determined from the previously illustrated normalization procedure. The reason for including only one function η_k in each new eigenfunction ϕ_{n+k} and omitting the other functions η_j is simply that a given value of $\tilde{v}_j$ is associated with each perturbed energy $\mathscr{E}_{n+j}$; it would not do to admix functions η_j $(j \neq k)$ corresponding to *different* values of the energy level modifications $\tilde{v}_j$ into a given wave function, since then we would no longer have a wave function characteristic of a *single* energy level. This assumes of course that the perturbation splits the degeneracy; some additional flexibility is available in those cases having *degenerate* values for $\tilde{v}_j$, but even in those cases there can be no harm in choosing the undetermined coefficients to be zero and thereby avoid admixing the various degenerate basis states $\phi^{(0)}_{n+k}$ $(k = 1, 2, \ldots, p)$ into each wave function which is constructed for the perturbed levels which remain degenerate.

The eigenfunctions are then given by Eq. (5.30), where we specifically indicate by a prime on the sum that we omit any terms involving $\mathscr{V}_{ln}$ whenever $\mathscr{E}^{(0)}_l = \mathscr{E}^{(0)}_n$,

$$\phi_{n+k} = \eta_k + \lambda \sum_{l \neq n}{}' \frac{\mathscr{V}_{ln}}{\mathscr{E}^{(0)}_n - \mathscr{E}^{(0)}_l} \phi^{(0)}_l \qquad (k = 0, 1, 2, \ldots, p). \tag{5.37}$$

The first-order eigenfunctions are modified relative to the zero-order eigenfunctions even for cases where the degeneracy is not lifted by the perturbation. Whether or not the degeneracy will be lifted in any specific case depends on the symmetry of the eigenfunctions and the symmetry of the perturbation potential, a problem which can be approached fruitfully with the techniques of group theory.

PROJECT 5.3 Stationary-State Square-Well Perturbation Acting on Particle in a Box

An otherwise free particle moving inside a three-dimensional infinite square-well potential with sides of length L experiences a small uniform potential energy U_1 when it is within the central cubical region having dimensions equal to $\frac{1}{2}L$. Compute the perturbation corrections to the eigenfunctions and eigenvalues. (*Hint*: First work out the corresponding one-dimensional problem.)

PROJECT 5.4 Stationary-State Gravitational Perturbation Acting on Particle in a Box

An otherwise free particle moving inside a three-dimensional infinite square-well potential with sides of length L experiences a uniform vertical force $-mg$ resulting in a position-dependent gravitational potential energy $U = mgz$ within the box, where z represents the vertical distance. Use stationary-state perturbation theory to compute the gravitational modifications to the eigenfunctions and energy eigenvalues.

2 Elementary Treatment of Diagonalization

It remains for us to present convincing arguments that the diagonalization of the perturbation operator $\mathscr{V}$ in the $(p + 1)$-dimensional manifold containing the $p + 1$ degenerate eigenfunctions of $\mathscr{H}_0$ corresponding to the eigenvalue $\mathscr{E}_n^{(0)}$ can indeed be carried out. We first choose the special case of a twofold degeneracy $(p = 1)$ as a specific example. It will be shown that it is not difficult to extend the procedure to larger manifolds. Let

$$\eta_j = a_{j1}\xi_1 + a_{j2}\xi_2 \qquad (j = 1, 2), \tag{5.38}$$

with the a_{ij} representing constants to be determined. We require

$$\mathscr{V}\eta_j = \tilde{v}_j\eta_j \qquad (j = 1, 2), \tag{5.39}$$

where we must yet determine $\tilde{v}_1$ and $\tilde{v}_2$. Thus we have the equation

$$\mathscr{V}(a_{j1}\xi_1 + a_{j2}\xi_2) = \tilde{v}_j(a_{j1}\xi_1 + a_{j2}\xi_2) \qquad (j = 1, 2). \tag{5.40}$$

Now either $\tilde{v}_1 = \tilde{v}_2$ or $\tilde{v}_1 \neq \tilde{v}_2$. If $\tilde{v}_1 = \tilde{v}_2 = \tilde{v}$, the two equations given by $j = 1, 2$ are

$$\mathscr{V}(a_{11}\xi_1 + a_{12}\xi_2) = \tilde{v}(a_{11}\xi_1 + a_{12}\xi_2), \tag{5.41}$$

$$\mathscr{V}(a_{21}\xi_1 + a_{22}\xi_2) = \tilde{v}(a_{21}\xi_1 + a_{22}\xi_2). \tag{5.42}$$

Multiplying the first by $1/a_{12}$ and the second by $1/a_{22}$ gives

$$\mathscr{V}[(a_{11}/a_{12})\xi_1 + \xi_2] = \tilde{v}[(a_{11}/a_{12})\xi_1 + \xi_2], \tag{5.43}$$

$$\mathscr{V}[(a_{21}/a_{22})\xi_1 + \xi_2] = \tilde{v}[(a_{21}/a_{22})\xi_1 + \xi_2]. \tag{5.44}$$

Subtracting the two equations gives

$$\mathscr{V}[(a_{11}/a_{12}) - (a_{21}/a_{22})]\xi_1 = \tilde{v}[(a_{11}/a_{12}) - (a_{21}/a_{22})]\xi_1. \tag{5.45}$$

Either $a_{11}/a_{12} = a_{21}/a_{22}$ or else

$$\mathscr{V}\xi_1 = \tilde{v}\xi_1. \tag{5.46}$$

The former cannot be true if η_1 and η_2 are to represent linearly independent eigenfunctions, so the latter must be true, which means that ξ_1 itself is an eigenfunction of $\mathscr{V}$. Clearly the analogous procedure of multiplying the first equation by $1/a_{11}$ and the second by $1/a_{21}$ and subtracting gives the corresponding result

$$\mathscr{V}\xi_2 = \tilde{v}\xi_2. \tag{5.47}$$

Therefore we have no need to diagonalize further for this particular case.

For $\tilde{v}_1 \neq \tilde{v}_2$, we must devise a general procedure for diagonalization. Multiplying Eq. (5.40) by ξ_k^* and integrating over all space gives

$$a_{j1}\int \xi_k^* \mathscr{V}\xi_1\, d\mathbf{r} + a_{j2}\int \xi_k^* \mathscr{V}\xi_2\, d\mathbf{r} = \tilde{v}_j a_{j1}\,\delta_{k1} + \tilde{v}_j a_{j2}\,\delta_{k2}, \tag{5.48}$$

where we have used the orthonormality of the ξ_j. This equation can be written more compactly by adopting the matrix element notation

$$\mathscr{M}_{ij} \equiv \int \xi_i^* \mathscr{V}\xi_j\, d\mathbf{r}. \tag{5.49}$$

Equation (5.48) becomes

$$a_{j1}(\mathscr{M}_{k1} - \tilde{v}_j\delta_{k1}) + a_{j2}(\mathscr{M}_{k2} - \tilde{v}_j\delta_{k2}) = 0 \qquad (j, k = 1, 2). \tag{5.50}$$

For $j = 1$, $k = 1, 2$ leads to the following pair of homogeneous algebraic equations

$$a_{11}(\mathscr{M}_{11} - \tilde{v}_1) + a_{12}\mathscr{M}_{12} = 0, \tag{5.51}$$

$$a_{11}\mathscr{M}_{21} + a_{12}(\mathscr{M}_{22} - \tilde{v}_1) = 0. \tag{5.52}$$

A nontrivial solution exists only if the condition

$$\begin{vmatrix} \mathscr{M}_{11} - \tilde{v}_1 & \mathscr{M}_{12} \\ \mathscr{M}_{21} & \mathscr{M}_{22} - \tilde{v}_1 \end{vmatrix} = 0 \tag{5.53}$$

is satisfied. For $j = 2$, $k = 1, 2$ leads to the corresponding pair of equations

$$a_{21}(\mathscr{M}_{11} - \tilde{v}_2) + a_{22}\mathscr{M}_{12} = 0, \tag{5.54}$$

$$a_{21}\mathscr{M}_{21} + a_{22}(\mathscr{M}_{22} - \tilde{v}_2) = 0. \tag{5.55}$$

The existence of a nontrivial solution for a_{21} and a_{22} then requires

$$\begin{vmatrix} \mathscr{M}_{11} - \tilde{v}_2 & \mathscr{M}_{12} \\ \mathscr{M}_{21} & \mathscr{M}_{22} - \tilde{v}_2 \end{vmatrix} = 0. \tag{5.56}$$

The two determinants in Eqs. (5.53) and (5.56) give identical second-degree algebraic equations (called *secular equations*) for $\tilde{v}_1$ and $\tilde{v}_2$. Solving the secular equation then gives both roots; one root is arbitrarily designated $\tilde{v}_1$ and the other root, provided it is distinct from the first, is then designated $\tilde{v}_2$. Using $\tilde{v}_1$ and $\tilde{v}_2$ as obtained in this way, each pair of homogeneous equations [(5.51) and

(5.52); (5.54) and (5.55)] can then be individually solved for the ratio of the two unknown coefficients involved in Eq. (5.38). Normalization of each of the η_j then completes the determination of the a_{ij} except for an arbitrary phase factor (having no physical content) for each η_j ($j = 1, 2$).

For the case now considered in which $\tilde{v}_1 \neq \tilde{v}_2$, the eigenfunctions η_1 and η_2 can be shown to be orthogonal, as follows. Multiply the first of the pair of equations

$$\mathscr{V}\eta_1 = \tilde{v}_1\eta_1, \tag{5.57}$$

$$\mathscr{V}\eta_2 = \tilde{v}_2\eta_2 \tag{5.58}$$

by η_2^* and the second by η_1^* and integrate over all space. This gives

$$\int \eta_2^*\mathscr{V}\eta_1 \, d\mathbf{r} = \tilde{v}_1 \int \eta_2^* \eta_1 \, d\mathbf{r}, \tag{5.59}$$

$$\int \eta_1^*\mathscr{V}\eta_2 \, d\mathbf{r} = \tilde{v}_2 \int \eta_1^* \eta_2 \, d\mathbf{r}. \tag{5.60}$$

However,

$$\begin{aligned} \langle\eta_1|\mathscr{V}|\eta_2\rangle \equiv \int \eta_1^*\mathscr{V}\eta_2 \, d\mathbf{r} &= \int (\mathscr{V}\eta_1)^*\eta_2 \, d\mathbf{r} \\ &= \left[\int \eta_2^*\mathscr{V}\eta_1 \, d\mathbf{r}\right]^* = \langle\eta_2|\mathscr{V}|\eta_1\rangle^*, \end{aligned} \tag{5.61}$$

since $\mathscr{V}$ is a Hermitian operator [see Eq. (1.231)]. In addition, $\tilde{v}_2 = \tilde{v}_2^*$ since $\tilde{v}_2$ is the eigenvalue of a Hermitian operator. Substituting this result into Eq. (5.60), taking the complex conjugate, and subtracting from Eq. (5.59) gives

$$0 = (\tilde{v}_1 - \tilde{v}_2) \int \eta_2^* \eta_1 \, d\mathbf{r}. \tag{5.62}$$

Since by hypothesis $\tilde{v}_1 \neq \tilde{v}_2$ in this particular case, then

$$\int \eta_2^*\eta_1 \, d\mathbf{r} = 0, \tag{5.63}$$

which proves the orthogonality of η_1 and η_2. This summarizes for our diagonalized perturbation operator the proof of one of the general theorems given in Chap. 1, §8, viz., that eigenfunctions of a Hermitian operator corresponding to different eigenvalues are orthogonal. Thus the new set of basis functions η_k which we construct is indeed an orthogonal set.

If we had a manifold of three eigenfunctions ξ_1, ξ_2, ξ_3 instead of only two, we would have nine coefficients a_{ij}. The determinantal equation would yield three values of $\tilde{v}_j$. Each such value would yield ratios for a_{2j}/a_{1j} and a_{3j}/a_{1j}, which together with normalization would serve to determine one eigenfunction of $\mathscr{V}$

within an arbitrary phase factor. The procedure is readily generalized to even larger manifolds of functions as is now shown.

Consider the general case of a $(p + 1)$-dimensional manifold. First we separate out all those ξ_k which are already eigenfunctions of $\mathscr{V}$, set them equal to an equivalent number of the η_j, and construct the remaining η_j from linear combinations of the remaining ξ_k. We assume that there are r remaining ξ_k, so that we must construct r additional η_j. Thus

$$\eta_j = \sum_k a_{jk}\xi_k \qquad (j = 1, 2, \ldots, r), \tag{5.64}$$

with

$$\mathscr{V}\eta_j = \tilde{v}_j\eta_j \qquad (j = 1, 2, \ldots, r). \tag{5.65}$$

Combining Eq. (5.64) with Eq. (5.65) gives

$$\mathscr{V} \sum_k a_{jk}\xi_k = \tilde{v}_j \sum_k a_{jk}\xi_k \qquad (j = 1, 2, \ldots, r). \tag{5.66}$$

Multiplying by ξ_l^* and integrating over all space gives

$$\sum_k a_{jk}\mathscr{M}_{lk} = \tilde{v}_j \sum_k a_{jk}\,\delta_{lk} = \tilde{v}_j a_{jl} \qquad (j, l = 1, 2, \ldots, r), \tag{5.67}$$

or equivalently

$$\sum_k a_{jk}(\mathscr{M}_{lk} - \tilde{v}_j\,\delta_{lk}) = 0 \qquad (j, l = 1, 2, \ldots, r). \tag{5.68}$$

The determinantal equation is obtained from

$$\begin{vmatrix} \mathscr{M}_{11} - \tilde{v}_j & \mathscr{M}_{12} & \mathscr{M}_{13} & \cdots & \mathscr{M}_{1r} \\ \mathscr{M}_{21} & \mathscr{M}_{22} - \tilde{v}_j & \mathscr{M}_{23} & \cdots & \mathscr{M}_{2r} \\ \mathscr{M}_{31} & \mathscr{M}_{32} & \mathscr{M}_{33} - \tilde{v}_j & \cdots & \mathscr{M}_{3r} \\ \vdots & \vdots & \vdots & & \vdots \\ \mathscr{M}_{r1} & \mathscr{M}_{r2} & \mathscr{M}_{r3} & & \mathscr{M}_{rr} - \tilde{v}_j \end{vmatrix} = 0. \tag{5.69}$$

Thus we obtain values for $\tilde{v}_j$, each of which gives through the set of homogeneous equations the ratios a_{j2}/a_{j1}, $a_{j3}/a_{j1}, \ldots,$ a_{jr}/a_{j1} necessary to determine the corresponding eigenfunction within the requirements of normalization and an arbitrary phase factor.

PROJECT 5.5 Commuting Operators Have a Complete Set of Simultaneous Eigenfunctions Also in Cases of Degeneracy

Considering the fact that an operator can be diagonalized within a manifold of functions, reexamine the theorem "Commuting Operators have a complete set of simultaneous eigenfunctions" (which was proven in Chap. 1, §8.8 for *nondegenerate* eigenfunctions) for the case of *degenerate* eigenfunctions. Attempt to construct a rigorous proof of the theorem for this more general case.

3 Higher-Order Perturbations and Applications

3.1 Second-Order Treatment for Nondegenerate Case

We can now derive expressions for the higher-order perturbations. These are most useful when there is no first-order shift in the energy. Consider the expansions

$$\phi_n^{(1)} = \sum_l a_l^{(1)} \phi_l^{(0)}, \tag{5.70}$$

$$\phi_n^{(2)} = \sum_l a_l^{(2)} \phi_l^{(0)}. \tag{5.71}$$

The sets of coefficients $a_l^{(1)}$ and $a_l^{(2)}$ will be characteristic of the index n, although this is not indicated explicitly by a subscript. The $a_l^{(1)}$ are in actuality the same coefficients a_l which we introduced in the first-order treatment. We substitute into Eq. (5.11) and obtain

$$\begin{aligned}\mathscr{H}_0 \sum_l a_l^{(2)} \phi_l^{(0)} &+ \mathscr{V} \sum_l a_l^{(1)} \phi_l^{(0)} \\ &= \mathscr{E}_n^{(0)} \sum_l a_l^{(2)} \phi_l^{(0)} + \mathscr{E}_n^{(1)} \sum_l a_l^{(1)} \phi_l^{(0)} + \mathscr{E}_n^{(2)} \phi_n^{(0)}.\end{aligned} \tag{5.72}$$

The first sum is

$$\mathscr{H}_0 \sum_l a_l^{(2)} \phi_l^{(0)} = \sum_l a_l^{(2)} \mathscr{H}_0 \phi_l^{(0)} = \sum_l a_l^{(2)} \mathscr{E}_l^{(0)} \phi_l^{(0)}. \tag{5.73}$$

We substitute this into Eq. (5.72), multiply through by $\phi_s^{(0)*}$ and integrate over all space making use of the orthonormality of the $\phi_s^{(0)}$. Thus

$$\sum_l a_l^{(2)} \mathscr{E}_l^{(0)} \delta_{sl} + \sum_l a_l^{(1)} \mathscr{V}_{sl} = \mathscr{E}_n^{(0)} \sum_l a_l^{(2)} \delta_{sl} + \mathscr{E}_n^{(1)} \sum_l a_l^{(1)} \delta_{sl} + \mathscr{E}_n^{(2)} \delta_{sn}, \tag{5.74}$$

which gives

$$a_s^{(2)} \mathscr{E}_s^{(0)} + \sum_l a_l^{(1)} \mathscr{V}_{sl} = \mathscr{E}_n^{(0)} a_s^{(2)} + \mathscr{E}_n^{(1)} a_s^{(1)} + \mathscr{E}_n^{(2)} \delta_{sn} \tag{5.75}$$

or equivalently

$$a_s^{(2)} [\mathscr{E}_n^{(0)} - \mathscr{E}_s^{(0)}] = \sum_l a_l^{(1)} \mathscr{V}_{sl} - \mathscr{E}_n^{(1)} a_s^{(1)} - \mathscr{E}_n^{(2)} \delta_{sn}. \tag{5.76}$$

From our first-order treatment we have $\mathscr{E}_n^{(1)} = \mathscr{V}_{nn}$ according to Eq. (5.17), and for $l \neq n$,

$$a_l^{(1)} = \mathscr{V}_{ln} / (\mathscr{E}_n^{(0)} - \mathscr{E}_l^{(0)})$$

according to Eq. (5.18). Recall that $a_n^{(1)}$ is determined by the normalization requirement on the first-order wave function, and we found that it could be chosen to be zero. This choice is also acceptable in the second-order treatment, since we have an additional coefficient $a_n^{(2)}$ which can provide normalization of the second-order wave function.

If $s = n$ in the above equation (5.76) for the second-order treatment, we obtain

$$\begin{aligned}\mathscr{E}_n^{(2)} &= \sum_l a_l^{(1)}\mathscr{V}_{sl} - \mathscr{E}_n^{(1)}a_s^{(1)} = \sum_l a_l^{(1)}\mathscr{V}_{nl} - \mathscr{V}_{nn}a_n^{(1)} \\ &= \sum_{l\neq n} a_l^{(1)}\mathscr{V}_{nl} = \sum_{l\neq n} [\mathscr{V}_{ln}\mathscr{V}_{nl}/(\mathscr{E}_n^{(0)} - \mathscr{E}_l^{(0)})]. \qquad (5.77)\end{aligned}$$

The product $\mathscr{V}_{ln}\mathscr{V}_{nl}$ is equal to $\mathscr{V}_{nl}^*\mathscr{V}_{nl} = |\mathscr{V}_{nl}|^2$, or alternatively, to $\mathscr{V}_{ln}\mathscr{V}_{ln}^* = |\mathscr{V}_{ln}|^2$; these relations follow from the fact that $\mathscr{V}$ is Hermitian [see Eq. (1.231)].

If $s \neq n$ in Eq. (5.76), we obtain

$$\begin{aligned}a_s^{(2)}[\mathscr{E}_n^{(0)} - \mathscr{E}_s^{(0)}] &= \sum_l a_l^{(1)}\mathscr{V}_{sl} - \mathscr{E}_n^{(1)}a_s^{(1)} \\ &= a_n^{(1)}\mathscr{V}_{sn} + \sum_{l\neq n} \frac{\mathscr{V}_{sl}\mathscr{V}_{ln}}{\mathscr{E}_n^{(0)} - \mathscr{E}_l^{(0)}} - \frac{\mathscr{V}_{sn}\mathscr{V}_{nn}}{\mathscr{E}_n^{(0)} - \mathscr{E}_s^{(0)}}. \qquad (5.78)\end{aligned}$$

Therefore

$$a_s^{(2)} = \frac{a_n^{(1)}\mathscr{V}_{sn}}{\mathscr{E}_n^{(0)} - \mathscr{E}_s^{(0)}} - \frac{\mathscr{V}_{sn}\mathscr{V}_{nn}}{[\mathscr{E}_n^{(0)} - \mathscr{E}_s^{(0)}]^2} + \sum_{l\neq n} \frac{\mathscr{V}_{sl}\mathscr{V}_{ln}}{[\mathscr{E}_n^{(0)} - \mathscr{E}_s^{(0)}][\mathscr{E}_n^{(0)} - \mathscr{E}_l^{(0)}]} \quad (s \neq n). \qquad (5.79)$$

The coefficient $a_n^{(1)}$ can be chosen to be zero, as previously discussed. The coefficient $a_n^{(2)}$ is obtained from normalization of the second-order wave function.

Therefore, to second order

$$\begin{aligned}\mathscr{E}_n = \mathscr{E}_n^{(0)} + \lambda\mathscr{E}_n^{(1)} + \lambda^2\mathscr{E}_n^{(2)} &= \mathscr{E}_n^{(0)} + \lambda\mathscr{V}_{nn} \\ &\quad + \lambda^2 \sum_{l\neq n} [|\mathscr{V}_{ln}|^2/(\mathscr{E}_n^{(0)} - \mathscr{E}_l^{(0)})], \qquad (5.80)\end{aligned}$$

$$\begin{aligned}\phi_n &= \phi_n^{(0)} + \lambda\phi_n^{(1)} + \lambda^2\phi_n^{(2)} \\ &= \phi_n^{(0)} + \lambda \sum_{l\neq n} \frac{\mathscr{V}_{ln}}{\mathscr{E}_n^{(0)} - \mathscr{E}_l^{(0)}}\phi_l^{(0)} + \lambda^2 a_n^{(2)}\phi_n^{(0)} \\ &\quad + \lambda^2 \sum_{l\neq n}\left[\sum_{m\neq n} \frac{\mathscr{V}_{lm}\mathscr{V}_{mn}}{[\mathscr{E}_n^{(0)} - \mathscr{E}_l^{(0)}][\mathscr{E}_n^{(0)} - \mathscr{E}_m^{(0)}]} - \frac{\mathscr{V}_{ln}\mathscr{V}_{nn}}{[\mathscr{E}_n^{(0)} - \mathscr{E}_l^{(0)}]^2}\right]\phi_l^{(0)}. \qquad (5.81)\end{aligned}$$

EXERCISE Develop the equations needed to evaluate the third-order perturbation effects.

3.2 Example Illustrating How to Apply Stationary-State Theory

As an example of the use of stationary-state perturbation theory, let us refer back to §10 of Chap. 1 to the problem of the one-dimensional potential well having length L with infinitely high potential walls at the edges. Suppose that a uniform potential energy U_0 is added as a perturbation over the length γL $(0 \leqslant \gamma \leqslant 1)$ in the central region of the potential well (see Fig. 5.1). We specify that U_0 is much less than the ground-state energy $\mathscr{E}_1 = \pi^2\hbar^2/2mL^2$ (assuming

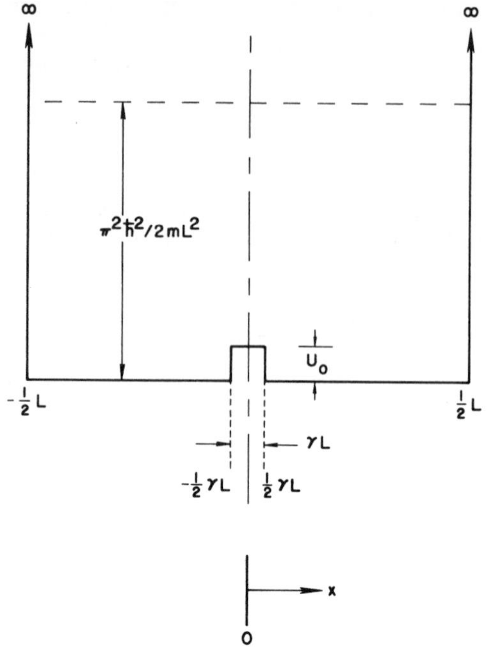

Fig. 5.1 Perturbation on the square-well potential. [The perturbation energy U_0 symmetrically spans a fraction of the potential well width L, as determined by the value of the parameter γ $(0 \leqslant \gamma \leqslant 1)$.]

fixed boundary conditions) in order to assure the validity of our use of the perturbation treatment results. The eigenfunctions of the unperturbed Hamiltonian given by Eq. (1.284) are

$$\phi_n(x) = \begin{cases} A_n \sin(n\pi x/L) & (n \text{ even}) \\ B_n \cos(n\pi x/L) & (n \text{ odd}), \end{cases}$$

with $|A_n| = |B_n| = (2/L)^{1/2}$ as given by Eq. (1.289). The corresponding unperturbed energy eigenvalues $\mathscr{E}_n^{(0)}$ are nondegenerate, and are given by Eq. (1.285),

$$\mathscr{E}_n^{(0)} = (\hbar^2/2m)(n\pi/L)^2.$$

The perturbed energy levels $\mathscr{E}_n$ can be obtained from Eq. (5.4) (for the limit $\lambda \to 1$), with Eq. (5.17) used to compute the first-order correction $\mathscr{E}_n^{(1)}$ and Eq. (5.77) used to compute the second-order correction $\mathscr{E}_n^{(2)}$. We thus can write the first-order energy corrections as

$$\mathscr{E}_n^{(1)} = \int_{-\frac{1}{2}\gamma L}^{\frac{1}{2}\gamma L} [A_n \sin(n\pi x/L)]^* U_0 [A_n \sin(n\pi x/L)]\, dx \qquad (n \text{ even}),$$

$$\mathscr{E}_n^{(1)} = \int_{-\frac{1}{2}\gamma L}^{\frac{1}{2}\gamma L} [B_n \cos(n\pi x/L)]^* U_0 [B_n \cos(n\pi x/L)]\, dx \qquad (n \text{ odd}).$$

The limits of the integrals have the values indicated because $U(x)$ is zero outside of this domain, therefore rendering the integrands zero and thereby eliminating any contribution over the remainder of the potential well. The above first-order corrections readily reduce to

$$\begin{aligned}\mathscr{E}_n^{(1)} &= \frac{2}{L} U_0 \int_{-\frac{1}{2}\gamma L}^{\frac{1}{2}\gamma L} \sin^2(n\pi x/L)\, dx \\ &= \frac{2U_0}{n\pi} \int_{-\frac{1}{2}\gamma n\pi}^{\frac{1}{2}\gamma n\pi} \sin^2\eta\, d\eta = \frac{2U_0}{n\pi}\left(\tfrac{1}{2}\eta - \tfrac{1}{4}\sin 2\eta\right)\Big|_{-\frac{1}{2}\gamma n\pi}^{\frac{1}{2}\gamma n\pi} \\ &= (U_0/n\pi)[\gamma n\pi - \sin(\gamma n\pi)] \qquad (n \text{ even}),\end{aligned}$$

$$\begin{aligned}\mathscr{E}_n^{(1)} &= \frac{2}{L} U_0 \int_{-\frac{1}{2}\gamma L}^{\frac{1}{2}\gamma L} \cos^2(n\pi x/L)\, dx \\ &= \frac{2U_0}{n\pi} \int_{-\frac{1}{2}\gamma n\pi}^{\frac{1}{2}\gamma n\pi} \cos^2\eta\, d\eta = \frac{2U_0}{n\pi}\left(\tfrac{1}{2}\eta + \tfrac{1}{4}\sin 2\eta\right)\Big|_{-\frac{1}{2}\gamma n\pi}^{\frac{1}{2}\gamma n\pi} \\ &= (U_0/n\pi)[\gamma n\pi + \sin(\gamma n\pi)] \qquad (n \text{ odd}).\end{aligned}$$

Note from these results that the energy level corrections reduce to zero in the limit $\gamma \to 0$, as they must, since this eliminates the perturbation. Likewise, the energy-level perturbations reduce to zero as $U_0 \to 0$, which is again to be expected. In the limit $\gamma \to 1$, the perturbation energy U_0 extends across the entire potential well, and the energy corrections have the single value U_0 for all levels, which is correct from a physical standpoint. If $\gamma \ll 1$, then for the lower-lying energy levels (n small), $\sin(\gamma n\pi) \simeq \gamma n\pi$ and we obtain $\mathscr{E}_n^{(1)} \simeq 0$ for n even and $\mathscr{E}_n^{(1)} \simeq 2\gamma U_0$ for n odd. This can be understood physically by noting that the *even* n solutions represent sine functions [see Eq. (1.284)], which have nodes at $x = 0$; the perturbation has little effect since for small γ the perturbation potential energy is thus localized in the neighborhood of a very small probability density $\phi^*\phi$. On the other hand, the *odd* n solutions represent cosine functions [see Eq. (1.284)], which have maxima at $x = 0$; the perturbation has a maximum effect since for small γ the perturbation potential energy is thus localized in the neighborhood of a maximum in the probability density $\phi^*\phi$.

EXERCISE Evaluate the first-order corrections to the energy eigenfunctions for the above considered perturbed particle in a box. [*Hint*: See Eq. (5.30).]

EXERCISE Evaluate the second-order corrections to the energy eigenvalues for the above considered perturbed particle in a box. [*Hint*: See Eq. (5.77).]

EXERCISE Evaluate the second-order corrections to the energy eigenfunctions for the above considered perturbed particle in a box. [*Hint*: See Eq. (5.79).]

EXERCISE Use reasonable numerical values for the parameters, such as $L = 1$–10 Å, $U_0 = 0.01$ eV, and $\gamma = 0.1$, to evaluate the first-order corrections to the ground-state energy and the energies of the first four excited states produced by the perturbation considered above for the infinite one-dimensional square-well potential.

EXERCISE Repeat the above problem with the exception that U_0 extends from a to b, where a and b are two arbitrary points within the square-well potential.

4 Degenerate Case for Second-Order Treatment

The above results are immediately applicable if the $\phi_n^{(0)}$ are nondegenerate. Some care is needed in the degenerate case because the energy denominator can be zero. Consider the case in which an eigenfunction $\phi_p^{(0)}$ is degenerate with $\phi_n^{(0)}$, so that $\mathscr{E}_p^{(0)} = \mathscr{E}_n^{(0)}$. The double sum in Eq. (5.81) contains the factors $\mathscr{E}_n^{(0)} - \mathscr{E}_l^{(0)}$ and $\mathscr{E}_n^{(0)} - \mathscr{E}_m^{(0)}$, with l and m representing summation indices, so the energy denominator goes to zero whenever $l = p$ or $m = p$. To avoid divergence to infinity, the product $\mathscr{V}_{lm}\mathscr{V}_{mn}$ appearing in the numerator must be chosen to be zero whenever $l = p$ or $m = p$. For the case $l = p$, we require $\mathscr{V}_{pm}\mathscr{V}_{mn} = 0$. For the case $m = p$, we require $\mathscr{V}_{lp}\mathscr{V}_{pn} = 0$. Our previous diagonalization of $\mathscr{V}$ for the first-order treatment was performed in order that $\mathscr{V}_{pn} = 0$ for any state $\phi_p^{(0)}$ which is degenerate with $\phi_n^{(0)}$, so the second of these requirements is equivalent to the requirement for the validity of the first-order treatment. The first requirement, however, is an additional stipulation characteristic of the second-order treatment; it demands that either $\mathscr{V}_{pm} = 0$ or $\mathscr{V}_{mn} = 0$. (This requirement is sometimes expressed as follows: *There can exist no state $\phi_m^{(0)}$ which connects the degenerate states $\phi_p^{(0)}$ and $\phi_n^{(0)}$ through the perturbation.*) To summarize, our second-order treatment is valid for the degenerate case if we arrange to have $\mathscr{V}_{pn} = 0$ and $\mathscr{V}_{mn} = 0$, or alternately, if we arrange to have $\mathscr{V}_{pn} = 0$ and $\mathscr{V}_{pm} = 0$, where the index p represents any state $\phi_p^{(0)}$ which is degenerate with $\phi_n^{(0)}$. Therefore we should arrange to have all matrix elements $\mathscr{V}_{mn} = 0$ ($m = 1, 2, \ldots, p, \ldots$) or else arrange to have all matrix elements $\mathscr{V}_{pm} = 0$ ($m = 1, 2, \ldots, n, \ldots$). We see that this may require a more comprehensive diagonalization of the perturbation operator than was necessary for the first-order treatment. It is sufficient to diagonalize the submatrix including all rows and columns labeled by the subscript m for which either $\mathscr{V}_{pm}$ or $\mathscr{V}_{mn}$ is not zero. (Of course the eigenfunctions can be ordered in such a way that the row and columns in question are brought together.) This procedure is more complicated analytically than is necessary since in actuality it is only required that either $\mathscr{V}_{mn} = 0$ or $\mathscr{V}_{pm} = 0$, but not necessarily both, as stated above. An alternate procedure is to expand the exact eigenfunctions in powers of λ, as was done previously, but include both degenerate functions $\phi_n^{(0)}$ and $\phi_p^{(0)}$ in zero order. This procedure is illustrated below.

5 Removal of Degeneracy in Second Order

Assume we have two degenerate states $\phi_n^{(0)}$ and $\phi_p^{(0)}$, so that $\mathscr{E}_p^{(0)} = \mathscr{E}_n^{(0)}$, and let us consider the problem of finding an approximation for the exact eigenfunction ϕ_n correct to terms through second order. Let us try the expansions

$$\phi_n = a_n\phi_n^{(0)} + a_p\phi_p^{(0)} + \lambda \sum_l{}' a_l^{(1)}\phi_l^{(0)} + \lambda^2 \sum_l{}' a_l^{(2)}\phi_l^{(0)}, \tag{5.82}$$

$$\phi_p = b_n\phi_n^{(0)} + b_p\phi_p^{(0)} + \lambda \sum_l{}' b_l^{(1)}\phi_l^{(0)} + \lambda^2 \sum_l{}' b_l^{(2)}\phi_l^{(0)}, \tag{5.83}$$

$$\mathscr{E}_n = \mathscr{E}_n^{(0)} + \lambda\mathscr{E}_n^{(1)} + \lambda^2\mathscr{E}_n^{(2)}, \tag{5.84}$$

where the prime associated with the summation symbol means in this case that the terms corresponding to the indices n and p are specifically not to be included in the summations since these terms represent the degenerate eigenfunctions which are already included in zero order. Substituting the above expressions for ϕ_n and $\mathscr{E}_n$ into the equation

$$(\mathscr{H}_0 + \lambda\mathscr{V})\phi_n = \mathscr{E}_n\phi_n \tag{5.85}$$

and keeping only terms through second order in λ gives

$$\begin{aligned}
&a_n\mathscr{H}_0\phi_n^{(0)} + a_p\mathscr{H}_0\phi_p^{(0)} + \lambda \sum_l{}' a_l^{(1)}\mathscr{H}_0\phi_l^{(0)} + \lambda^2 \sum_l{}' a_l^{(2)}\mathscr{H}_0\phi_l^{(0)} \\
&\quad + \lambda a_n\mathscr{V}\phi_n^{(0)} + \lambda a_p\mathscr{V}\phi_p^{(0)} + \lambda^2 \sum_l{}' a_l^{(1)}\mathscr{V}\phi_l^{(0)} + \mathscr{O}(\lambda^3) \\
&\quad = a_n\mathscr{E}_n^{(0)}\phi_n^{(0)} + a_p\mathscr{E}_n^{(0)}\phi_p^{(0)} + \lambda \sum_l{}' a_l^{(1)}\mathscr{E}_n^{(0)}\phi_l^{(0)} + \lambda^2 \sum_l{}' a_l^{(2)}\mathscr{E}_n^{(0)}\phi_l^{(0)} \\
&\qquad + \lambda a_n\mathscr{E}_n^{(1)}\phi_n^{(0)} + \lambda a_p\mathscr{E}_n^{(1)}\phi_p^{(0)} + \lambda^2 \sum_l{}' a_l^{(1)}\mathscr{E}_n^{(1)}\phi_l^{(0)} + \mathscr{O}(\lambda^3) \\
&\qquad + \lambda^2 a_n\mathscr{E}_n^{(2)}\phi_n^{(0)} + \lambda^2 a_p\mathscr{E}_n^{(2)}\phi_p^{(0)} + \mathscr{O}(\lambda^3) + \mathscr{O}(\lambda^4),
\end{aligned} \tag{5.86}$$

where the symbol $\mathscr{O}(\lambda^n)$ means we have neglected terms of order n in λ. We now substitute the relation

$$\mathscr{H}_0\phi_l^{(0)} = \mathscr{E}_l^{(0)}\phi_l^{(0)} \tag{5.87}$$

into the first four terms in the above expression. Using the fact that $\mathscr{E}_p^{(0)} = \mathscr{E}_n^{(0)}$, we then see that the first two terms on the left-hand side of the above equation reduce to the first two terms on the right-hand side, so these four terms can be eliminated. Multiplying the remaining terms by $\phi_n^{(0)*}$ and integrating over all space gives one equation; repeating except for use of the factor $\phi_p^{(0)*}$ instead of $\phi_n^{(0)*}$ gives a second equation. Taking into consideration the orthonormality of the $\phi_l^{(0)}$, we obtain for the first equation

$$\begin{aligned}
&\lambda \sum_l{}' a_l^{(1)}\mathscr{E}_l^{(0)}\,\delta_{nl} + \lambda^2 \sum_l{}' a_l^{(2)}\mathscr{E}_l^{(0)}\,\delta_{nl} + \lambda a_n\mathscr{V}_{nn} + \lambda a_p\mathscr{V}_{np} + \lambda^2 \sum_l{}' a_l^{(1)}\mathscr{V}_{nl} \\
&\quad = \lambda \sum_l{}' a_l^{(1)}\mathscr{E}_n^{(0)}\,\delta_{nl} + \lambda^2 \sum_l{}' a_l^{(2)}\mathscr{E}_n^{(0)}\,\delta_{nl} \\
&\qquad + \lambda a_n\mathscr{E}_n^{(1)} + \lambda^2 \sum_l{}' a_l^{(1)}\mathscr{E}_n^{(1)}\,\delta_{nl} + \lambda^2 a_n\mathscr{E}_n^{(2)},
\end{aligned} \tag{5.88}$$

or equivalently, since all terms involving δ_{nl} in the sums are zero due to the fact that $l \neq n$,

$$\lambda a_n\mathscr{V}_{nn} + \lambda a_p\mathscr{V}_{np} + \lambda^2 \sum_l{}' a_l^{(1)}\mathscr{V}_{nl} = \lambda a_n\mathscr{E}_n^{(1)} + \lambda^2 a_n\mathscr{E}_n^{(2)}. \tag{5.89}$$

The second equation is

$$\lambda a_n\mathscr{V}_{pn} + \lambda a_p\mathscr{V}_{pp} + \lambda^2 \sum_l{}' a_l^{(1)}\mathscr{V}_{pl} = \lambda a_p\mathscr{E}_n^{(1)} + \lambda^2 a_p\mathscr{E}_n^{(2)}. \tag{5.90}$$

We can obtain a third equation in a similar manner. Instead of using the complex conjugate of one of the degenerate eigenfunctions $\phi_n^{(0)}$ and $\phi_p^{(0)}$, we multiply by an arbitrary zero-order eigenfunction $\phi_m^{(0)*}$ ($m \neq n$, $m \neq p$) and again integrate over all space. The resulting equation is

$$\lambda \sum_l{}' a_l^{(1)}\mathscr{E}_l^{(0)}\,\delta_{ml} + \lambda^2 \sum_l{}' a_l^{(2)}\mathscr{E}_l^{(0)}\,\delta_{ml} + \lambda a_n \mathscr{V}_{mn} + \lambda a_p \mathscr{V}_{mp} + \lambda^2 \sum_l{}' a_l^{(1)}\mathscr{V}_{ml}$$

$$= \lambda \sum_l{}' a_l^{(1)}\mathscr{E}_n^{(0)}\,\delta_{ml} + \lambda^2 \sum_l{}' a_l^{(2)}\mathscr{E}_n^{(0)}\,\delta_{ml} + \lambda^2 \sum_l{}' a_l^{(1)}\mathscr{E}_n^{(1)}\,\delta_{ml}, \tag{5.91}$$

or equivalently

$$\lambda a_m^{(1)}\mathscr{E}_m^{(0)} + \lambda^2 a_m^{(2)}\mathscr{E}_m^{(0)} + \lambda a_n \mathscr{V}_{mn} + \lambda a_p \mathscr{V}_{mp} + \lambda^2 \sum_l{}' a_l^{(1)}\mathscr{V}_{ml}$$

$$= \lambda a_m^{(1)}\mathscr{E}_n^{(0)} + \lambda^2 a_m^{(2)}\mathscr{E}_n^{(0)} + \lambda^2 a_m^{(1)}\mathscr{E}_n^{(1)}. \tag{5.92}$$

Since λ is a variable parameter, we separate the equations into first-order terms and second-order terms. We note that the first-order terms in λ in the first two equations give the pair of equations

$$a_n[\mathscr{V}_{nn} - \mathscr{E}_n^{(1)}] + a_p \mathscr{V}_{np} = 0, \tag{5.93}$$

$$a_n \mathscr{V}_{pn} + a_p[\mathscr{V}_{pp} - \mathscr{E}_n^{(1)}] = 0. \tag{5.94}$$

The requirement for a nontrivial solution for the a_n and a_p is

$$\begin{vmatrix} \mathscr{V}_{nn} - \mathscr{E}_n^{(1)} & \mathscr{V}_{np} \\ \mathscr{V}_{pn} & \mathscr{V}_{pp} - \mathscr{E}_n^{(1)} \end{vmatrix} = 0. \tag{5.95}$$

This determinant leads to a second-degree algebraic equation in $\mathscr{E}_n^{(1)}$. One of the roots may be identified as $\mathscr{E}_n^{(1)}$ and the other as $\mathscr{E}_p^{(1)}$, since the same determinantal equation must arise when the corresponding perturbation expansion is carried through for ϕ_p. The corresponding two sets of coefficients for a_n, a_p obtained when each of the roots is used in turn and the pair of homogeneous equations solved can likewise be identified with the two pairs of coefficients (a_n, a_p) and (b_n, b_p). Although the homogeneous equations give only the ratios a_n/a_p and b_n/b_p, normalization completes the determination of each coefficient to within an arbitrary phase factor for the eigenfunction. The first-order terms in λ in Eq. (5.92) yield an evaluation of $a_m^{(1)}$,

$$a_m^{(1)} = (a_n \mathscr{V}_{mn} + a_p \mathscr{V}_{mp})/(\mathscr{E}_n^{(0)} - \mathscr{E}_m^{(0)}) \qquad (m \neq n,\ p). \tag{5.96}$$

Since $\mathscr{E}_m^{(0)} \neq \mathscr{E}_n^{(0)}$ for any index m, due to the fact that the only zero-order eigenfunction degenerate with $\phi_n^{(0)}$ was assumed to be $\phi_p^{(0)}$, then $a_m^{(1)}$ is well-behaved. Thus we have avoided the difficulty of zero denominators by the present technique.

If $\mathscr{E}_n^{(1)}$ happens to be a double root then the perturbation fails to split the degeneracy in first order, in which case we find that it has been worthwhile to carry along the second-order terms. As we have shown previously, whenever the degeneracy remains in first order then $\phi_n^{(0)}$ and $\phi_p^{(0)}$ must both be eigenfunctions of the perturbation operator $\mathscr{V}$ with the common eigenvalue $\mathscr{E}_n^{(1)}$, so that

$$\mathscr{V}_{nn} = \int \phi_n^{(0)*}\mathscr{V}\phi_n^{(0)}\, d\mathbf{r} = \int \phi_n^{(0)*}\mathscr{E}_n^{(1)}\phi_n^{(0)}\, d\mathbf{r} = \mathscr{E}_n^{(1)}, \tag{5.97}$$

$$\mathscr{V}_{pp} = \int \phi_p^{(0)*}\mathscr{V}\phi_p^{(0)}\, d\mathbf{r} = \int \phi_p^{(0)*}\mathscr{E}_n^{(1)}\phi_p^{(0)}\, d\mathbf{r} = \mathscr{E}_n^{(1)}, \tag{5.98}$$

$$\mathscr{V}_{np} = \int \phi_n^{(0)*}\mathscr{V}\phi_p^{(0)}\, d\mathbf{r} = \int \phi_n^{(0)*}\mathscr{E}_n^{(1)}\phi_p^{(0)}\, d\mathbf{r} = 0, \tag{5.99}$$

$$\mathscr{V}_{pn} = \int \phi_p^{(0)*}\mathscr{V}\phi_n^{(0)}\, d\mathbf{r} = \int \phi_p^{(0)*}\mathscr{E}_n^{(1)}\phi_n^{(0)}\, d\mathbf{r} = 0. \tag{5.100}$$

Returning now to the contribution of the second-order terms in λ, Eqs. (5.89) and (5.90) yield the following results:

$$a_n\mathscr{E}_n^{(2)} = \sum_l{}' a_l^{(1)}\mathscr{V}_{nl}, \tag{5.101}$$

$$a_p\mathscr{E}_n^{(2)} = \sum_l{}' a_l^{(1)}\mathscr{V}_{pl}. \tag{5.102}$$

Equation (5.92) gives

$$a_m^{(2)}[\mathscr{E}_m^{(0)} - \mathscr{E}_n^{(0)}] = a_m^{(1)}\mathscr{E}_n^{(1)} - \sum_l{}' a_l^{(1)}\mathscr{V}_{ml}. \tag{5.103}$$

Substituting the first-order coefficients (5.96) into the first two of these equations gives

$$a_n\mathscr{E}_n^{(2)} = \sum_l{}' \mathscr{V}_{nl}[(a_n\mathscr{V}_{ln} + a_p\mathscr{V}_{lp})/(\mathscr{E}_n^{(0)} - \mathscr{E}_l^{(0)})], \tag{5.104}$$

$$a_p\mathscr{E}_n^{(2)} = \sum_l{}' \mathscr{V}_{pl}[(a_n\mathscr{V}_{ln} + a_p\mathscr{V}_{lp})/(\mathscr{E}_n^{(0)} - \mathscr{E}_l^{(0)})]. \tag{5.105}$$

We can rewrite these in the form of two homogeneous algebraic equations for a_n and a_p,

$$a_n\left[\sum_l{}'\left[\frac{\mathscr{V}_{nl}\mathscr{V}_{ln}}{\mathscr{E}_n^{(0)} + \mathscr{E}_l^{(0)}}\right] - \mathscr{E}_n^{(2)}\right] + a_p\sum_l{}'\left[\frac{\mathscr{V}_{nl}\mathscr{V}_{lp}}{\mathscr{E}_n^{(0)} - \mathscr{E}_l^{(0)}}\right] = 0, \tag{5.106}$$

$$a_n\sum_l{}'\left[\frac{\mathscr{V}_{pl}\mathscr{V}_{ln}}{\mathscr{E}_n^{(0)} - \mathscr{E}_l^{(0)}}\right] + a_p\left[\sum_l{}'\left[\frac{\mathscr{V}_{pl}\mathscr{V}_{lp}}{\mathscr{E}_n^{(0)} - \mathscr{E}_l^{(0)}}\right] - \mathscr{E}_n^{(2)}\right] = 0. \tag{5.107}$$

The determinant of the coefficients of the a_n and a_p gives a secular equation which is of the second degree in $\mathscr{E}_n^{(2)}$. Either of the roots can be designated as $\mathscr{E}_n^{(2)}$, with the alternate root constituting $\mathscr{E}_p^{(2)}$. We thus have the possibility of removing the degeneracy in second order. The pairs of coefficients a_n, a_p then resulting from solution of the homogeneous equations (5.106) and (5.107) with subsequent normalization for each of the values of $\mathscr{E}_n^{(2)}$ are identified as a_n, a_p and

b_n, b_p respectively, and then used in the expansions (5.82) and (5.83) for ϕ_n and ϕ_p. These serve to determine the proper linear combinations of the unperturbed degenerate functions $\phi_n^{(0)}$ and $\phi_p^{(0)}$ to use as zero-order basis functions.

The coefficients needed for the second-order corrections to ϕ_n and ϕ_p are obtained from Eq. (5.103) following substitution of Eq. (5.96) for $a_l^{(1)}$,

$$a_m^{(2)} = \sum_l{}' \frac{(a_n\mathscr{V}_{ln} + a_p\mathscr{V}_{lp})\mathscr{V}_{ml}}{[\mathscr{E}_n^{(0)} - \mathscr{E}_l^{(0)}][\mathscr{E}_n^{(0)} - \mathscr{E}_m^{(0)}]} - \frac{(a_n\mathscr{V}_{mn} + a_p\mathscr{V}_{mp})\mathscr{E}_n^{(1)}}{[\mathscr{E}_n^{(0)} - \mathscr{E}_m^{(0)}]^2}. \quad (5.108)$$

This completes our treatment of stationary-state perturbation theory. In the next section, we develop time-dependent perturbation theory.

6 Time-Dependent Perturbation Theory

Let us consider the case for which the perturbation $\mathscr{V}$ is time dependent. The total Hamiltonian $\mathscr{H} = \mathscr{H}_0 + \mathscr{V}$ is then time dependent; there are no exact stationary states for the system, and the energy of the system is not conserved with time. A charged particle passing through a system of other charged particles is one example of this type of perturbation. A transient electromagnetic field applied to a system of particles is a second example. If we consider the initial state of the system to be some given eigenstate of $\mathscr{H}_0$, then our problem can be considered to be that of determining the probabilities with which the system will be found in the various eigenstates of $\mathscr{H}_0$ following the application of the perturbation. It is characteristic of many perturbations that the interaction with the system is limited in time. This is the case of an incoming charged particle being deflected by a fixed charged scattering center, such as a conduction electron in a metal being scattered by an ionized impurity in the lattice.

Consider the unperturbed case, for which

$$\mathscr{H}_0\phi_n^{(0)} = \mathscr{E}_n^{(0)}\phi_n^{(0)}. \quad (5.109)$$

The corresponding time-dependent eigenfunctions are

$$\psi_n^{(0)} = \phi_n^{(0)} \exp[-(i/\hbar)\mathscr{E}_n^{(0)}t] = \phi_n^{(0)} \exp(-i\omega_n t), \quad (5.110)$$

where $\omega_n \equiv \mathscr{E}_n^{(0)}/\hbar$. If the arbitrary linear combination of the $\phi_n^{(0)}$

$$\Phi_0(\mathbf{r}) = \sum_n c_n\phi_n^{(0)} \quad (5.111)$$

represents the general solution to the unperturbed problem at time $t = 0$, then

$$\Psi_0(\mathbf{r}, t) = \sum_n c_n\psi_n^{(0)} = \sum_n c_n\phi_n^{(0)}e^{-i\omega_n t} \quad (5.112)$$

represents the general solution to the unperturbed problem as a function of time. Each term in this superposition oscillates with its characteristic frequency ω_n, where

$$\omega_n = \mathscr{E}_n^{(0)}/\hbar. \quad (5.113)$$

If a perturbation $\mathscr{V}$ is now applied, the wave function can still be expanded in terms of the complete set of orthonormal eigenfunctions $\phi_n^{(0)}$ of the unperturbed problem at any instant of time, but the coefficients will change with time because of the perturbation. If we allow for the time dependence of the coefficients by permitting c_n to be $c_n(t)$, then

$$\Psi(\mathbf{r}, t) = \sum_n c_n(t)\phi_n^{(0)}e^{-i\omega_n t} \tag{5.114}$$

represents a valid solution to the perturbed problem for which $\mathscr{H} = \mathscr{H}_0 + \mathscr{V}$. Our problem thus reduces to the determination of the $c_n(t)$ for the perturbed problem, for which the equation of motion is

$$\mathscr{H}\Psi = (\mathscr{H}_0 + \mathscr{V})\Psi = i\hbar\, \partial\Psi/\partial t. \tag{5.115}$$

Substituting the above expansion for Ψ into this equation of motion gives

$$(\mathscr{H}_0 + \mathscr{V}) \sum_n c_n(t)\phi_n^{(0)}e^{-i\omega_n t} = i\hbar\frac{\partial}{\partial t} \sum_n c_n(t)\phi_n^{(0)}e^{-i\omega_n t}. \tag{5.116}$$

Using the fact that

$$\mathscr{H}_0\phi_n^{(0)} = \mathscr{E}_n^{(0)}\phi_n^{(0)} \tag{5.117}$$

and employing the usual rules for differentiation of a product, we obtain

$$\sum_n \mathscr{E}_n^{(0)}c_n(t)\phi_n^{(0)}e^{-i\omega_n t} + \sum_n \mathscr{V} c_n(t)\phi_n^{(0)}e^{-i\omega_n t}$$

$$= i\hbar \sum_n c_n(t)\phi_n^{(0)}(-i\omega_n)e^{-i\omega_n t} + i\hbar \sum_n \frac{dc_n(t)}{dt}\, \phi_n^{(0)}e^{-i\omega_n t}. \tag{5.118}$$

The first sum on the right is equal to the first sum on the left because by definition $\mathscr{E}_n^{(0)} = \hbar\omega_n$. Multiplying the remaining terms by the arbitrary term $\psi_j^{(0)*} = \phi_j^{(0)*}\exp(+i\omega_j t)$ and integrating over all space gives

$$\sum_n c_n(t)\mathscr{V}_{jn}e^{i(\omega_j-\omega_n)t} = i\hbar \sum_n \frac{dc_n(t)}{dt}\, \delta_{jn}e^{i(\omega_j-\omega_n)t}, \tag{5.119}$$

where

$$\mathscr{V}_{jn} \equiv \int \phi_j^{(0)*}\mathscr{V}\phi_n^{(0)}\, d\mathbf{r} \tag{5.120}$$

and we have used the orthonormality of the $\phi_j^{(0)}$. (Recall that $\mathscr{V}$ can be time dependent in this development.) The Kronecker delta function eliminates all terms but the one for which $n = j$ on the right side, so we obtain

$$i\hbar\frac{dc_j(t)}{dt} = \sum_n c_n(t)\mathscr{V}_{jn}e^{i(\omega_j-\omega_n)t} \qquad (j = 1, 2, 3, \ldots). \tag{5.121}$$

Thus we have replaced the Schrödinger equation by a system of simultaneous linear homogeneous differential equations. As in the case of time-independent perturbation theory, to develop a perturbation expansion we introduce the parameter λ such that

$$\mathscr{H} = \mathscr{H}_0 + \lambda \mathscr{V}, \tag{5.122}$$

$$c_j(t) = c_j^{(0)}(t) + \lambda c_j^{(1)}(t) + \lambda^2 c_j^{(2)}(t) + \cdots. \tag{5.123}$$

The above equation becomes

$$i\hbar \frac{dc_j^{(0)}}{dt} + i\hbar\lambda \frac{dc_j^{(1)}}{dt} + i\hbar\lambda^2 \frac{dc_j^{(2)}}{dt} + \cdots$$
$$= \lambda \sum_n c_n^{(0)} \mathscr{V}_{jn} e^{i(\omega_j - \omega_n)t} + \lambda^2 \sum_n c_n^{(1)} \mathscr{V}_{jn} e^{i(\omega_j - \omega_n)t} + \cdots. \tag{5.124}$$

The three relations obtained by equating terms which are respectively of zero, first, and second order in λ are

$$i\hbar \frac{dc_j^{(0)}}{dt} = 0, \tag{5.125}$$

$$i\hbar \frac{dc_j^{(1)}}{dt} = \sum_n c_n^{(0)} \mathscr{V}_{jn} e^{i(\omega_j - \omega_n)t}, \tag{5.126}$$

$$i\hbar \frac{dc_j^{(2)}}{dt} = \sum_n c_n^{(1)} \mathscr{V}_{jn} e^{i(\omega_j - \omega_n)t}. \tag{5.127}$$

Higher-order terms can of course be obtained in the same manner. The procedure to solve this sequence of equations is that of successive integration, since the equation for the derivative of any given coefficient involves only the lower-order coefficients. The integration of Eq. (5.125) gives immediately,

$$c_j^{(0)} = \text{const.} \tag{5.128}$$

Thus the $c_j^{(0)}$ $(j = 1, 2, \ldots)$ are the initial values of the problem which describe the state before application of the perturbation.

EXERCISE Deduce the third-order time-dependent perturbation equation analogous to the zero-, first-, and second-order perturbation equations (5.125)–(5.127).

For our further development we consider the state to be initially in the eigenstate

$$\psi_s^{(0)} = \phi_s^{(0)} \exp(-i\omega_s t). \tag{5.129}$$

Then

$$c_j^{(0)} = \delta_{js}, \tag{5.130}$$

where δ_{js} is unity only if $j = s$ and is zero otherwise. Then the first-order equation [i.e., (5.126)] becomes

$$i\hbar \frac{dc_j^{(1)}}{dt} = \sum_n \delta_{ns} \mathscr{V}_{jn} e^{i(\omega_j - \omega_n)t} = \mathscr{V}_{js} e^{i(\omega_j - \omega_s)t}. \tag{5.131}$$

Therefore

$$c_j^{(1)}(t) = (i\hbar)^{-1} \int_{-\infty}^{t} \mathscr{V}_{js} e^{i(\omega_j - \omega_s)t}\, dt \qquad (j \neq s). \tag{5.132}$$

We consider $c_s(t)$ to continue to have values which do not deviate too much from unity; otherwise, the perturbation approach begins to become inexact if we restrict ourselves to only the low-order approximations. Thus the limiting case is considered in which the wave function does not deviate too much from the unperturbed value, so the coefficients c_j for $j \neq s$ remain quite small relative to c_s.

PROJECT 5.6 Time-Dependent Perturbation of a Particle in a Box by the Application of a Uniform Electric Field

Consider a particle of mass m and charge q which is initially in its ground state in a one-dimensional potential well of length L with infinitely high potential energy barriers (Chap. 1, §10). At time $t = 0$, a uniform electric field E_0 is switched on and maintained at a fixed value until time t' at which it is switched off. Compute the first-order perturbation coefficients $c_j^{(1)}(t')$ for the first five excited states following the perturbation.

7 Example: Harmonic Perturbation

7.1 Basic Equations

Consider the specific case for which the perturbation is harmonic in time starting from $t = 0$, when it is initially switched on, and lasts for a time t',

$$\mathscr{V}(\mathbf{r}, t) = 0 \qquad (t < 0), \tag{5.133}$$

$$\mathscr{V}(\mathbf{r}, t) = \mathscr{U}(\mathbf{r}) \sin \omega t \qquad (0 \leqslant t \leqslant t'), \tag{5.134}$$

$$\mathscr{V}(\mathbf{r}, t) = 0 \qquad (t > t'). \tag{5.135}$$

Then during times t for which $0 \leqslant t < t'$ is satisfied,

$$\mathscr{V}_{js} = \mathscr{U}_{js} \sin \omega t, \tag{5.136}$$

where

$$\mathscr{U}_{js} \equiv \int \phi_j^{(0)*} \mathscr{U}(\mathbf{r}) \phi_s^{(0)}\, d\mathbf{r}. \tag{5.137}$$

The corresponding integral expression (5.132) for $c_j^{(1)}$ becomes

$$c_j^{(1)} = (i\hbar)^{-1} \mathscr{U}_{js} \int_0^t e^{i(\omega_j - \omega_s)t} \sin \omega t\, dt. \tag{5.138}$$

Using the Euler identity, $\sin y = (2i)^{-1}(e^{iy} - e^{-iy})$, we obtain

$$\begin{aligned} c_j^{(1)} &= -(2\hbar)^{-1} \mathscr{U}_{js} \int_0^t [e^{i(\omega_j - \omega_s + \omega)t} - e^{i(\omega_j - \omega_s - \omega)t}]\, dt \\ &= -\frac{\mathscr{U}_{js}}{2\hbar} \frac{e^{i(\omega_j - \omega_s + \omega)t} - 1}{i(\omega_j - \omega_s + \omega)} + \frac{\mathscr{U}_{js}}{2\hbar} \frac{e^{i(\omega_j - \omega_s - \omega)t} - 1}{i(\omega_j - \omega_s - \omega)}. \end{aligned} \tag{5.139}$$

The factors $\exp[i(\omega_j - \omega_s \pm \omega)]$ have magnitudes of unity; $c_j^{(1)}$ has a resonance character because it takes on relatively large values whenever one or the other of the denominators approaches zero, which requires

$$\mathscr{E}_j^{(0)} \simeq \mathscr{E}_s^{(0)} \pm \hbar\omega. \tag{5.140}$$

The energy input of the perturbation can be considered positive for an energy absorption process and negative for an energy emission process, corresponding respectively to the source of the perturbation giving quanta of energy to the physical system or receiving such energy quanta from the physical system (Fig. 1.5). Thus the effect of a harmonic perturbation is to change the total energy of the system, provided the matrix element for the transition is nonzero.

Since the system is initially considered to be in state s, the population of state j requires energy absorption if $E_s^{(0)} < E_j^{(0)}$ and energy emission if $E_s^{(0)} > E_j^{(0)}$.

As one nears resonance, the alternate term in the above expression will be relatively small and can be neglected since the factor $(\omega_j - \omega_s \pm \omega)$ in that denominator will approach $\mp 2\omega$. Application of l'Hôpital's rule [Wylie (1951)] then gives the result that $c_j^{(1)}$ approaches $\pm(\mathscr{U}_{js}/2\hbar)t$. The assumption that the energy levels $\mathscr{E}_j^{(0)}$ and $\mathscr{E}_s^{(0)}$ are perfectly sharp is not a good one for real systems in the laboratory, because broadening mechanisms are always present. In addition, the uncertainty relation $\Delta\omega\ \Delta t \gtrsim 1$ (Chap. 1, §7.8) applied to the harmonic perturbation requires some spread in ω due to the fact that the perturbation is switched on at $t = 0$. Thus we must examine the situation for values of ω somewhat off resonance. If we define

$$\delta\omega \equiv \omega_j - \omega_s \mp \omega, \tag{5.141}$$

where the sign is chosen so that $\delta\omega \to 0$ as ω passes through the resonance frequency, then the predominant term in $c_j^{(1)}$ can be written in the form

$$\pm \frac{\mathscr{U}_{js}}{2\hbar} \frac{e^{i(\delta\omega)t} - 1}{i(\delta\omega)} = \frac{\mathscr{U}_{js}\, e^{i\frac{1}{2}(\delta\omega)t} \sin \frac{1}{2}(\delta\omega)t}{\hbar(\delta\omega)}, \tag{5.142}$$

so that the probability for finding the system in the state j at time t for a perturbation near the resonance frequency is

$$|c_j^{(1)}(t)|^2 \simeq [|\mathscr{U}_{js}|^2 \sin^2 \tfrac{1}{2}(\delta\omega)t]/\hbar^2(\delta\omega)^2. \tag{5.143}$$

This would seem to predict a periodic oscillation of the system into and out of the state j during application of the perturbation for long periods of time, but it must be remembered that the result is valid only in the limit in which the initial state is depleted only negligibly by the perturbation. Thus for a system involving only a single particle, we would be well advised to take the small time limit of Eq. (5.143). If $\frac{1}{2}(\delta\omega)t \ll 1$, Eq. (5.143) reduces to the short time limiting expression

$$|c_j^{(1)}(t)|^2 \simeq (|\mathscr{U}_{js}|^2/4\hbar^2)t^2.$$

If the quadratic time dependence is bothersome to the reader, then he should skip ahead to §10 to find out how this anomaly is eliminated in the derivation of the Golden Rule.

7.2 Application to a Particle Trapped in a Square-Well Potential

To apply our results to a specific problem, let us consider the application of a harmonic electric field (such as would be produced by an electromagnetic wave of light) to a particle of mass m and charge q initially in its ground state in a three-dimensional square-well potential. Furthermore, let us consider the electric field to be polarized in the x direction and to have a wavelength λ which is long compared to the dimensions L_x, L_y, and L_z of the potential well. The potential well (i.e., the "box") can be considered to extend over the region $(0, L_x)$, $(0, L_y)$, and $(0, L_z)$ in the x, y, and z directions, respectively. We consider the case of infinitely high potential walls at the edges of the box together with fixed boundary conditions. The unperturbed eigenfunctions thus given by Eq. (3.6) are

$$\phi_{\mathbf{n}}(\mathbf{r}) = (8/V)^{1/2} \sin(n_x\pi x/L_x) \sin(n_y\pi y/L_y) \sin(n_z\pi z/L_z),$$

with $\mathbf{n}$ representing any triplet (n_x, n_y, n_z) of positive integers. The corresponding eigenvalues are

$$\mathscr{E}_{\mathbf{n}} = (\hbar^2\pi^2/2m)[(n_x/L_x)^2 + (n_y/L_y)^2 + (n_z/L_z)^2].$$

The potential energy is $\mathscr{U} = qV$, where V is the position-dependent electrostatic potential due to the electric field of the light wave. Since the wavelength λ of the light is considered to be much greater than the largest dimension of the box, the electric field $\mathbf{E} = \hat{\mathbf{x}}E_x$ can for all practical purposes be considered to be uniform over the dimensions of the box. The field is in the x direction due to the choice of the polarization direction of the electromagnetic field. The electrostatic potential

$$V(\mathbf{r}) = -\int_0^{\mathbf{r}} \mathbf{E}(\mathbf{r}') \cdot d\mathbf{r}',$$

obtained by taking the line integral of the electric field from the origin to the arbitrary position $\mathbf{r}$, thus reduces to

$$V(x) = -\int_0^x E_x\, dx = -E_x x,$$

where E_x is independent of position x, y, and z over the box. The electric field E_x is, by hypothesis, a harmonic function of time with some specific frequency which can be labeled ω, so that we can write $E_x = E_0 \sin \omega t$. The potential energy $\mathscr{U}$ of interaction of the charge q with the harmonic light wave can thus be written $\mathscr{U} = qV = -qxE_0 \sin \omega t$. This is the *time-dependent perturbation* which acts on the system.

Let us designate the states of the system in Dirac notation so that $|\mathbf{n}\rangle \equiv \phi_{\mathbf{n}}(\mathbf{r})$ is the state corresponding to the energy eigenvalue $\mathscr{E}_{\mathbf{n}}$. The matrix elements needed for transitions from the ground state to the excited states are thus

$$\langle \mathbf{1}|\mathscr{U}|\mathbf{n}\rangle = \int_{\Omega_{\mathbf{r}}} \phi_{\mathbf{1}}(\mathbf{r})(-qxE_0 \sin \omega t)\, \phi_{\mathbf{n}}(\mathbf{r})\, d\mathbf{r}.$$

Factoring out the constant factors, we obtain

$$\langle \mathbf{1}|\mathscr{U}|\mathbf{n}\rangle = -(qE_0 \sin \omega t)\langle \phi_\mathbf{1}(\mathbf{r})|x|\phi_\mathbf{n}(\mathbf{r})\rangle.$$

The triple integral can be written as the product of three integrals since the integrand is separable; due to orthogonality of the sine functions, the y and z integrals give zero unless the quantum numbers n_y and n_z are the same for the initial and final states,

$$\int_0^{L_y} \sin\left(\frac{n_y \pi y}{L_y}\right) \sin\left(\frac{n_y' \pi y}{L_y}\right) dy = \tfrac{1}{2}L_y \delta_{n_y, n_y'},$$

$$\int_0^{L_z} \sin\left(\frac{n_z \pi z}{L_z}\right) \sin\left(\frac{n_z' \pi z}{L_z}\right) dz = \tfrac{1}{2}L_z \delta_{n_z, n_z'}.$$

The matrix element thus reduces to

$$\langle \mathbf{1}|\mathscr{U}|\mathbf{n}\rangle = (8/L_x L_y L_z)(L_y L_z/4)[-qE_0 \sin \omega t]\, \delta_{1,n_y}\delta_{1,n_z}\mathscr{I}_{1,n_x},$$

where

$$\mathscr{I}_{1,n_x} = \int_0^{L_x} x \sin\left(\frac{\pi x}{L_x}\right) \sin\left(\frac{n_x \pi x}{L_x}\right) dx.$$

This latter integral is easily evaluated by using the Euler identities and a subsequent integration by parts. The result can be written in the form

$$\mathscr{I}_{1,n_x} = -\frac{1}{2}L_x^2\left[\left(\frac{1-\cos(n_x-1)\pi}{(n_x-1)^2\pi^2}\right) - \left(\frac{1-\cos(n_x+1)\pi}{(n_x+1)^2\pi^2}\right)\right].$$

If n_x is an odd integer, the right-hand side is zero; if n_x is an even integer, we obtain

$$\mathscr{I}_{1,n_x} = -\frac{1}{2}L_x^2\left[\frac{2}{(n_x-1)^2\pi^2} - \frac{2}{(n_x+1)^2\pi^2}\right] = \frac{-4n_x L_x^2}{(n_x^2-1)^2\pi^2}.$$

EXERCISE Verify this final result by direct integration of $\mathscr{I}_{1,n_x}$ using the Euler identities and integrating the resulting expression by parts.

We therefore obtain

$$\langle \mathbf{1}|\mathscr{U}|\mathbf{n}\rangle = 0 \qquad (n_x \text{ odd}),$$

$$\langle \mathbf{1}|\mathscr{U}|\mathbf{n}\rangle = -\left(\frac{4n_x L_x^2}{(n_x^2-1)^2\pi^2}\right)\left(\frac{2}{L_x}\right)[-qE_0 \sin \omega t]\delta_{1,n_y}\delta_{1,n_z}$$

$$= 8qE_0 L_x\left(\frac{n_x}{(n_x^2-1)^2\pi^2}\right)\delta_{1,n_y}\delta_{1,n_z} \sin \omega t \qquad (n_x \text{ even}).$$

These results can be written in the form

$$\langle \mathbf{1}|\mathscr{U}|\mathbf{n}\rangle = \mathscr{U}_{\mathbf{1},\mathbf{n}} \sin \omega t,$$

where

$$\mathscr{U}_{\mathbf{1},\mathbf{n}} \equiv 8qE_0L_x\left(\frac{n_x}{(n_x^2-1)^2\pi^2}\right)\delta_{1,n_y}\delta_{1,n_z}\Delta_{n_x}^{(\text{even})},$$

with $\Delta_{n_x}^{(\text{even})}$ defined to be unity if n_x is an even integer but zero otherwise, namely,

$$\Delta_{n_x}^{(\text{even})} \equiv \begin{cases} 1 & (n_x = \text{even integer}) \\ 0 & (n_x \neq \text{even integer}). \end{cases}$$

Several interesting features of the present result can be noted immediately. First of all, the transition probability is zero unless the quantum numbers in the initial and final states are the *same* for directions *perpendicular* to the electric field; the transition probability is similarly zero unless the quantum number of the final state is *different* from that of the ground state for the direction *parallel* to the field. Furthermore, the ground state has an *odd* integer quantum number ($n_x = 1$), whereas the excited state must have an *even* integer quantum number n_x in order to have a nonzero transition probability. Next, it can be noted that the matrix element is directly proportional to the electric field strength E_0 and directly proportional to the "electric dipole" product qL_x. Finally, it can be noted that for $n_x \gg 1$, the matrix element is nearly inversely proportional to n_x^3.

By substituting our result for the matrix element of the perturbation directly into Eq. (5.143), we can obtain the probability of finding the particle in the state $|\mathbf{n}\rangle$ at time t when the frequency of the perturbing electromagnetic wave is very near the resonance frequency $\omega_{\mathbf{1},\mathbf{n}} = (\mathscr{E}_{\mathbf{n}} - \mathscr{E}_{\mathbf{1}})/\hbar$ corresponding to the energy difference between the states in question. In our example, $\delta\omega = \hbar^{-1}(\mathscr{E}_{\mathbf{n}} - \mathscr{E}_{\mathbf{1}}) - \omega$, and we obtain for the probability,

$$|c^{(1)}(t)|^2 \simeq [\hbar(\delta\omega)]^{-2}|\mathscr{U}_{\mathbf{1},\mathbf{n}}|^2 \sin^2[\tfrac{1}{2}(\delta\omega)t],$$

where the evaluation of $\mathscr{U}_{\mathbf{1},\mathbf{n}}$ has been explicitly carried out above. The fact that we obtain *selection rules*, as evidenced by the delta functions which arise in the evaluation of the matrix elements, is especially to be noted.

PROJECT 5.7 Time-Dependent Perturbation of a Harmonic Oscillator by Uniform and Oscillating Electric Fields

For a harmonic oscillator of mass m, charge q, and frequency $\nu_0 = \omega_0/2\pi$ which is initially in its ground state, compute the effect of the following perturbations in exciting the oscillator to the nth excited state.

1. A spatially uniform electric field which is turned on at $t = 0$, held at a constant value E_0 over the time period $(0 \leqslant t \leqslant T_0)$, and turned off at $t = T_0$.

2. A spatially uniform but sinusoidally varying electric field $E_0 \cos \omega t$ which is switched on at $t = 0$ and off at $t = T_0$. (Consider that in general $\omega \neq \omega_0$.)

3. For Part 2, plot and interpret your results for a range of values of ω above and below ω_0, using reasonable values of ω_0, q, T_0, and E_0 chosen such that the conditions for validity of the perturbation treatment are not violated.

4. Physically interpret the results which you obtained in Part 3.

PROJECT 5.8 Time-Dependent Perturbation of the Hydrogen Atom

Consider a beam of hydrogen atoms, each atom having momentum p with its electron being in the ground state; the atoms enter, pass through, and exit from a large but finite (length L) parallel plate capacitor, the beam axis being *parallel* to the plates. Assume there is an alternating electric field within the capacitor due to an applied sinusoidal voltage of frequency ω_0, where $\hbar\omega_0 > me^4/32\pi^2\varepsilon_0^2\hbar^2$. Compute the probability per unit time for each electron inside the capacitor to make a transition to a plane-wave state, thus ionizing the atom. What is the corresponding total probability per atom for the transition to occur?

8 Example: Constant Perturbation in First Order

Consider now the specific case where the perturbation $\mathscr{V}$ is a constant except for being switched on at a time which we again designate to be zero and switched off at time t',

$$\mathscr{V} = 0 \qquad (t < 0), \tag{5.144}$$

$$\mathscr{V} = \mathscr{U}(\mathbf{r}) \qquad (0 \leqslant t \leqslant t'), \tag{5.145}$$

$$\mathscr{V} = 0 \qquad (t > t'). \tag{5.146}$$

Then the integral expression (5.132) for $c_j^{(1)}$ for $0 \leqslant t \leqslant t'$ is

$$c_j^{(1)} = (i\hbar)^{-1}\,\mathscr{U}_{js}\int_0^t e^{i(\omega_j - \omega_s)t}\,dt = -\frac{\mathscr{U}_{js}[e^{i(\omega_j - \omega_s)t} - 1]}{\hbar(\omega_j - \omega_s)}$$

$$= -\frac{2i\mathscr{U}_{js}}{\hbar(\omega_j - \omega_s)}\,e^{i\frac{1}{2}(\omega_j - \omega_s)t}\sin[\tfrac{1}{2}(\omega_j - \omega_s)t]. \tag{5.147}$$

The probability for finding the system in the state j at time t is given by

$$|c_j^{(1)}(t)|^2 = \frac{4|\mathscr{U}_{js}|^2\sin^2[\frac{1}{2}(\omega_j - \omega_s)t]}{\hbar^2(\omega_j - \omega_s)^2} \qquad (0 \leqslant t \leqslant t'). \tag{5.148}$$

For $t > t'$, $c_j^{(1)}(t)$ and $|c_j^{(1)}(t)|^2$ are simply time-independent constants given by $c_j^{(1)}(t')$ and $|c_j^{(1)}(t')|^2$ because the integrand is zero for $t > t'$.

It is of interest to repeat this calculation under the somewhat more general stipulation that the perturbation is applied gradually to the system over some time interval prior to $t = 0$ instead of the sudden switching at $t = 0$. Let us designate the positive quantity α as a switching parameter, and introduce the following perturbation,

$$\mathscr{V} = e^{\alpha t}\,\mathscr{U}(\mathbf{r}) \qquad (t < 0), \tag{5.149}$$

$$\mathscr{V} = \mathscr{U}(\mathbf{r}) \qquad (0 \leqslant t \leqslant t'), \tag{5.150}$$

$$\mathscr{V} = 0 \qquad (t > t'). \tag{5.151}$$

As $\alpha \to \infty$, this reduces to the preceding case which is referred to as the sudden approximation. The integral for $c_j^{(1)}$ has the additional contribution $\delta c_j^{(1)}$,

$$\delta c_j^{(1)} = (i\hbar)^{-1}\mathscr{U}_{js}\int_{-\infty}^{0} e^{i(\omega_j - \omega_s - i\alpha)t}\,dt = \frac{-\mathscr{U}_{js}}{\hbar(\omega_j - \omega_s - i\alpha)}. \tag{5.152}$$

This contribution approaches zero as $\alpha \to \infty$, as expected. (This is shown more explicitly by multiplying numerator and denominator by the complex conjugate of the denominator before taking the limit $\alpha \to \infty$.) As $\alpha \to 0$, corresponding to a very slow switching on of the perturbation, we obtain

$$c_j^{(1)} + \delta c_j^{(1)} \to -\, \mathscr{U}_{js}\, e^{i(\omega_j - \omega_s)t}/\hbar(\omega_j - \omega_s), \tag{5.153}$$

$$|c_j^{(1)} + \delta c_j^{(1)}|^2 \to |\mathscr{U}_{js}|^2/\hbar^2(\omega_j - \omega_s)^2. \tag{5.154}$$

This time-independent result seems peculiar until we remember that the perturbation has been applied effectively for an infinite time. This result is therefore not physically realistic for cases in which we must consider the effect of perturbations in inducing transitions between eigenstates, as for example, the scattering of a conduction electron in a metal by the Coulomb potential of a charged impurity. The quantity $|c_j^{(1)} + \delta c_j^{(1)}|^2$ given in Eq. (5.154) is equal to the corresponding value $|a_j|^2$ obtained from the coefficient a_j which we derived in time-independent perturbation theory [Eq. (5.18)], so that it represents a result more closely related to the shift in stationary-state values by a time-independent perturbation than a result which is useful for predicting transition probabilities between stationary-state eigenstates.

A more appropriate way to study the effects introduced by switching would perhaps be to maintain the integral of $\mathscr{V}$ over the range $-\infty$ to t a constant while adjusting the sharpness of switching. We do not explore this matter further at the present time, although this should be done before applying formulas derived on the basis of the sudden approximation to an entirely different physical situation.

EXERCISE Examine the problem of switching on a perturbation under the assumption of a constant energy output from the source for $t < 0$. (*Hint*: Energy is the time integral of the power.)

PROJECT 5.9 Sudden Perturbation of Particle in a Box by Application of a Decaying Voltage

A charged particle trapped in a one-dimensional square-well potential of length L with infinitely high potential barriers at the edges is initially in its ground state (Chap. 1, §10). At time $t = 0$, a voltage V_0 is suddenly applied which subsequently decays exponentially with a time constant t'. Assuming that the voltage produces a spatially uniform field V/L over the potential well, compute the probabilities for excitation into the first, second, and third excited states after the passage of a very long time interval.

PROJECT 5.10 Sudden Perturbation for Particle in a Box by Wall Motion

Consider the case of a particle trapped in a one-dimensional square-well potential of width L having infinitely high potential energy barriers at the edges (Chap. 1, §10). Initially (i.e., for $t < 0$), the particle may be considered to be in the ground state. Suppose that suddenly (at $t = 0$) the infinitely high barrier at $x = L$ is displaced to $x = 4L$. For time $t > 0$, calculate the probability that the energy of the particle is less than it was initially.

9 Example: Constant Perturbation in Second Order

Let us first extend the treatment to second order before examining the implications of the first-order result. Substituting the above result (5.147) for $c_j^{(1)}$,

$$c_j^{(1)} = -\, \mathscr{U}_{js}[e^{i(\omega_j - \omega_s)t} - 1]/\hbar(\omega_j - \omega_s), \tag{5.155}$$

obtained on the basis of the sudden approximation for t between zero and t' into Eq. (5.127) for $c_j^{(2)}$ gives

$$i\hbar\frac{dc_j^{(2)}}{dt} = -\sum_n \frac{\mathcal{U}_{jn}\mathcal{U}_{ns}}{\hbar(\omega_n - \omega_s)}\,[e^{i(\omega_n - \omega_s)t} - 1]\, e^{i(\omega_j - \omega_n)t}$$

$$= -\sum_n \frac{\mathcal{U}_{jn}\mathcal{U}_{ns}}{\hbar(\omega_n - \omega_s)}\,[e^{i(\omega_j - \omega_s)t} - e^{i(\omega_j - \omega_n)t}]. \tag{5.156}$$

Integration from zero to t, for t between zero and t', yields

$$c_j^{(2)} = \sum_n \frac{\mathcal{U}_{jn}\mathcal{U}_{ns}}{\hbar^2(\omega_n - \omega_s)(\omega_j - \omega_s)}[e^{i(\omega_j - \omega_s)t} - 1]$$

$$-\sum_n \frac{\mathcal{U}_{jn}\mathcal{U}_{ns}}{\hbar^2(\omega_n - \omega_s)(\omega_j - \omega_n)}[e^{i(\omega_j - \omega_n)t} - 1]. \tag{5.157}$$

It is readily seen that the entire second sum would be missing if we had used Eq. (5.153) for $c_j^{(1)} + \delta c_j^{(1)}$ for the limit $\alpha \to 0$ in the integral instead of $c_j^{(1)}$. In addition, we would have to add on the contribution given by

$$\delta c_j^{(2)} = (i\hbar)^{-1} \sum_n \int_{-\infty}^{0} [c_n^{(1)} + \delta c_n^{(1)}]\,\mathcal{U}_{jn} e^{i(\omega_j - \omega_n - i\alpha)t}\, dt \tag{5.158}$$

and evaluate it in the limit $\alpha \to 0$, and this contribution simply subtracts the term associated with the -1 in the brackets in the remaining expression for $c_j^{(2)}$, so that we obtain finally

$$c_j^{(2)} \to \sum_n \frac{\mathcal{U}_{jn}\mathcal{U}_{ns}}{\hbar^2(\omega_n - \omega_s)(\omega_j - \omega_s)} e^{i(\omega_j - \omega_s)t}. \tag{5.159}$$

EXERCISE Calculate the third-order contribution to the time-dependent effect of a constant perturbation suddenly switched on at $t = 0$.

10 Transition Probability and Fermi's Golden Rule

The preceding results for a constant perturbation even in first order are seemingly not in accord with the concept of a transition probability P per unit time, because such a concept requires the total probability for a transition to be proportional to the time t during which the perturbation acts on the system. To obtain a transition probability P per unit time, we must consider transitions to a spectrum of final states which are closely spaced in energy and grouped around some state which we denote by m. This leads to a well-known result referred to as *Fermi's Golden Rule of time-dependent perturbation theory.*

Consider the particular case of a macroscopic system for which periodic boundary conditions (Chap. 1, §3.4 and Chap. 3, §1.7) are applicable, so that we are dealing with discrete eigenfunctions closely spaced in energy which are normalized to the volume of the system. The density of states as a function of the

$\mathbf{k}$ vector (Chap. 3, §1.8) is denoted by $\tilde{w}(\mathbf{k})$ and the density of states as a function of energy *for the particular group of states in question* is denoted by $\Theta(\mathscr{E})$. For example, the group of states in question might be bounded by two constant energy surfaces and some increment of solid angle $d\Omega$ in $\mathbf{k}$ space. For a free electron metal, $\Theta(\mathscr{E})$ could be expressed in terms of the total density of states $\tilde{g}(\mathscr{E})$, viz. $\Theta(\mathscr{E}) = \tilde{g}(\mathscr{E})(d\Omega/4\pi)$, where $\tilde{g}(\mathscr{E})$ is given by Eq. (3.25). We assume that $\Theta(\mathscr{E})$ is a relatively slowly varying function of $\mathscr{E}(\mathbf{k})$ in the neighborhood of the state signified by the subscript m, so that $\Theta(\mathscr{E}(\mathbf{k}_j))$ for the group of states in question denoted by wavevectors $\mathbf{k}_j$ can be considered to be equal to $\Theta(\mathscr{E}(\mathbf{k}_m))$.

Let us denote the transition probability per unit time from the initial state $\mathbf{k}_s$ to a given final state $\mathbf{k}_j$ by $P(\mathbf{k}_s, \mathbf{k}_j)$, so that the transition probability P_m per unit time for a transition to one or another of the group of final states j is

$$P_m = \sum_j P(\mathbf{k}_s, \mathbf{k}_j). \tag{5.160}$$

Since the states are closely spaced, this sum can be replaced by an integral

$$P_m \simeq \int \tilde{w}(\mathbf{k}_j)P(\mathbf{k}_s, \mathbf{k}_j)\, d\mathbf{k}_j \tag{5.161}$$

over the group of states in question. The transition probability $P(\mathbf{k}_s, \mathbf{k}_j)$ per unit time can be considered in the limit of small time t to be the square of the magnitude of the first-order coefficient $c_j^{(1)}(t)$ divided by the time t during which the perturbation has been applied,

$$P(\mathbf{k}_s, \mathbf{k}_j) = |c_j^{(1)}|^2/t = \frac{4|\mathscr{U}_{js}|^2 \sin^2[\frac{1}{2}(\omega_j - \omega_s)t]}{\hbar^2(\omega_j - \omega_s)^2 t}, \tag{5.162}$$

where we have used the result (5.147) of the sudden approximation for $c_j^{(1)}$. The matrix elements $\mathscr{U}_{js}$ can be considered to be independent of j over the group of states in question as a first approximation,

$$\mathscr{U}_{js} \simeq \mathscr{U}_{ms}, \tag{5.163}$$

because the states $\mathbf{k}_j$ are considered to be grouped in the neighborhood of $\mathbf{k}_m$. The expression for $P(\mathbf{k}_s, \mathbf{k}_j)$ is thus seen to depend on the energy difference $\hbar(\omega_j - \omega_s)$ but not especially on the particular value of $\mathbf{k}_j$ for the group of states in question. Thus

$$P_m \simeq \int \frac{4|\mathscr{U}_{ms}|^2 \sin^2[\frac{1}{2}(\omega_j - \omega_s)t]}{\hbar^2(\omega_j - \omega_s)^2 t}\tilde{w}(\mathbf{k}_j)\, d\mathbf{k}_j, \tag{5.164}$$

where the integration is to be carried out over the states in question. Since

$$\tilde{w}(\mathbf{k}_j)\, d\mathbf{k}_j = \Theta(\mathscr{E}(\mathbf{k}_j))\, d\mathscr{E}(\mathbf{k}_j) \simeq \Theta(\mathscr{E}(\mathbf{k}_m))\, d\mathscr{E}(\mathbf{k}_j), \tag{5.165}$$

we obtain

$$P_m \simeq \frac{4|\mathscr{U}_{ms}|^2\Theta[\mathscr{E}(\mathbf{k}_m)]}{t}\int \frac{\sin^2 \frac{1}{2}(\omega_j - \omega_s)t}{\hbar^2(\omega_j - \omega_s)^2}\, d\mathscr{E}(\mathbf{k}_j), \tag{5.166}$$

where the integral is over energies which are appropriate for the group of states $\mathbf{k}_j$. Due to the nature of the integrand, the maximum contribution can be expected to arise from regions of $\mathscr{E}(\mathbf{k}_j)$ for which $\omega_j \simeq \omega_s$. This represents essentially a *selection rule* requiring conservation of energy. If we agree to consider final states k_j which meet this criterion, i.e.,

$$\mathscr{E}(\mathbf{k}_m) \simeq \mathscr{E}(\mathbf{k}_s), \tag{5.167}$$

and furthermore are able to consider $\mathscr{U}_{js}$ and $\Theta(\mathscr{E}(\mathbf{k}_j))$ to be approximately equal to $\mathscr{U}_{ms}$ and $\Theta(\mathscr{E}(\mathbf{k}_m))$ over a region in $\mathbf{k}$ space which corresponds to an energy width which is broad with respect to the energy width of the function constituting the integrand, then we can extend the limits of integration from $-\infty$ to $+\infty$ without appreciable modification in the results. Making the variable change

$$y \equiv \tfrac{1}{2}(\omega_j - \omega_s)t = [\mathscr{E}(\mathbf{k}_j) - \mathscr{E}(\mathbf{k}_s)]t/2\hbar, \qquad dy = (t/2\hbar)\, d\mathscr{E}(\mathbf{k}_j) \tag{5.168}$$

in the integral then gives

$$\begin{aligned} P_m &\simeq \frac{4|\mathscr{U}_{ms}|^2\Theta(\mathscr{E}(\mathbf{k}_m))}{t}\int_{-\infty}^{\infty}\frac{\sin^2 y}{\hbar^2(2y/t)^2}(2\hbar/t)\,dy \\ &\simeq (2/\hbar)|\mathscr{U}_{ms}|^2\,\Theta(\mathscr{E}(\mathbf{k}_m))\int_{-\infty}^{\infty} y^{-2}\sin^2 y\,dy. \end{aligned} \tag{5.169}$$

The definite integral has the value π, so we obtain

$$P_m \simeq (2\pi/\hbar)|\mathscr{U}_{ms}|^2\,\Theta(\mathscr{E}(\mathbf{k}_m)). \tag{5.170}$$

This is called **Fermi's Golden Rule** of time-dependent perturbation theory. The transition probability thus obtained is, within the limits of validity of our treatment, independent of time.

The matrix element $\mathscr{U}_{ms}$ can be expected to depend upon the vectors $\mathbf{k}_s$ and $\mathbf{k}_m$ characterizing the initial and final states, even though $\mathscr{E}(\mathbf{k}_m) \simeq \mathscr{E}(\mathbf{k}_s)$. For example, consider the case of a conduction electron initially in a plane-wave state characterized by $\mathbf{k}_s$ to be scattered by the Coulomb potential of an ionized impurity into the plane-wave state characterized by $\mathbf{k}_m$. It would not be physically realistic to expect isotropic scattering, so P_m would be expected to depend at the very least upon the angle between $\mathbf{k}_s$ and $\mathbf{k}_m$. In this respect, P_m represents a quantity more closely related to the differential scattering cross section (Chap. 5, §11) than to the total scattering cross section. To obtain the quantity P_{total} corresponding to the total scattering cross section, the quantity P_m must be integrated over all groups of final states which contribute to the process.

PROJECT 5.11 Application of Fermi's Golden Rule to a Particle Trapped in a Square-Well Potential

Carry through an analogous treatment to that of §7.2 for the particle trapped in the three-dimensional square-well potential using Fermi's Golden Rule [Eq. (5.170)]. Explain the similarities and differences in the results obtained by the two methods.

11 Differential Cross Section for Scattering

Consider the case of transitions from one plane-wave state characterized by an initial **k** vector $\mathbf{k}_i$

$$\psi_i(\mathbf{r}, t) = A_i e^{i(\mathbf{k}_i \cdot \mathbf{r} - \omega_i t)} \tag{5.171}$$

and energy $\mathscr{E}(\mathbf{k}_i)$ to another plane-wave state of the same energy characterized by a final **k** vector $\mathbf{k}_f$,

$$\psi_f(\mathbf{r}, t) = A_f e^{i(\mathbf{k}_f \cdot \mathbf{r} - \omega_f t)}. \tag{5.172}$$

Let us normalize these plane-wave states to the volume V of our system, which for present purposes we consider to have linear dimensions L in each of the three orthogonal directions x, y, and z. Then $\psi^*\psi$ integrated from $-L/2$ to $+L/2$ in each of the three directions gives A^2L^3, which must be equal to unity. Therefore $A_i = A_f = L^{-3/2} = V^{-1/2}$. Consider the case where the perturbation inducing the transitions is time independent, as for example, the case where the Coulomb potential of a charged scattering center induces transitions of conduction electrons in a metal from one plane-wave eigenstate to another plane-wave eigenstate. Let us use Fermi's Golden Rule (5.170) to obtain an expression for the transition probability per unit time in a given direction in **k** space defined by the differential solid angle $d\Omega = \sin\theta \, d\theta \, d\phi$,

$$P_m \simeq (2\pi/\hbar)|\mathscr{U}_{fi}|^2 \Theta[\mathscr{E}(\mathbf{k}_f)]. \tag{5.173}$$

The quantity $\Theta[\mathscr{E}(\mathbf{k}_f)]$ in this case is the density of states of the system at the energy $\mathscr{E}(\mathbf{k}_f)$ corresponding to the element of volume $k_f^2 \sin\theta \, d\theta \, d\phi$ in momentum space, where we consider only a single direction of spin. (We ignore the possibility of additional final states for multiple directions of electron spin.) The matrix element $\mathscr{U}_{fi}$ is that of the scattering potential $\mathscr{U}(\mathbf{r})$ with respect to the initial and final states $\mathbf{k}_i$ and $\mathbf{k}_f$. Since we are considering collisions for which the energy of the particle is conserved in the scattering process,

$$\mathscr{E}(\mathbf{k}_f) = \mathscr{E}(\mathbf{k}_i), \tag{5.174}$$

then $\omega_f = \omega_i$, so

$$\mathscr{U}_{fi} = \int \psi_f^* \mathscr{U}(\mathbf{r}) \psi_i \, d\mathbf{r} = V^{-1} \int \mathscr{U}(\mathbf{r}) e^{i(\mathbf{k}_i - \mathbf{k}_f) \cdot \mathbf{r}} \, d\mathbf{r}. \tag{5.175}$$

The integration is to be carried out over the volume L^3 of the system.

The density of states $\tilde{w}(\mathbf{k})$ in **k** space is given by Eq. (3.16) for a free-electron metal,

$$\tilde{w}(\mathbf{k}_f) = V/8\pi^3, \tag{5.176}$$

which is independent of $\mathbf{k}_f$. Consider the vector $\mathbf{k}_f$ to be given in spherical polar coordinates by k_f, θ, ϕ. Then

$$d\mathbf{k}_f = k_f^2 \sin\theta \, dk_f \, d\theta \, d\phi. \tag{5.177}$$

If we assume spherical constant energy surfaces, then

$$\mathscr{E}(\mathbf{k}_f) = \hbar^2 k_f^2/2m^*, \tag{5.178}$$

$$d\mathscr{E}(\mathbf{k}_f) = (\hbar^2 k_f/m^*)\, dk_f, \tag{5.179}$$

where m^* is the "effective mass" of the particle undergoing the scattering. (The concept of *effective mass* arises in energy band theory (cf. Chap. 7). It includes the effect of the ion cores on the inertia of an electron in a periodic solid. For the moment, m^* can be considered to be analogous to the ordinary mass of a particle.) The substitution of Eqs. (5.176), (5.177), and (5.179) into the general relation

$$\Theta[\mathscr{E}(\mathbf{k}_f)]\, d\mathscr{E}(\mathbf{k}_f) = \tilde{w}(\mathbf{k}_f)\, d\mathbf{k}_f \tag{5.180}$$

gives the following expression for $\Theta[\mathscr{E}(\mathbf{k}_f)]$,

$$\Theta[\mathscr{E}(\mathbf{k}_f)] = V(m^*/8\pi^3\hbar^2)k_f \sin\theta\, d\theta\, d\phi. \tag{5.181}$$

By substituting the above results (5.175) and (5.181) for $\mathscr{U}_{fi}$ and $\Theta[\mathscr{E}(\mathbf{k}_f)]$ into P_m given by Eq. (5.173), we obtain the following expression

$$P_m \simeq (m^*/4\pi^2\hbar^3 V)k_f \sin\theta\, d\theta\, d\phi \left| \int \mathscr{U}(\mathbf{r})e^{i(\mathbf{k}_i - \mathbf{k}_f)\cdot\mathbf{r}}\, d\mathbf{r} \right|^2. \tag{5.182}$$

The *differential scattering cross section* $\sigma(\theta, \phi)$, however, *is defined as the transition probability per unit solid angle per unit incidence flux*,

$$\sigma(\theta, \phi) = P_m/(Cv\, d\Omega), \tag{5.183}$$

where v is the velocity $\hbar k/m^*$ of the particles being scattered which have a concentration C, and $d\Omega = \sin\theta\, d\theta\, d\phi$. The velocity of the particle in the initial and final states is the same in the approximation of spherical constant energy surfaces (i.e., $\mathscr{E} \propto k^2$), so $v = \hbar k_f/m^*$. We have considered in the development of expression (5.182) for P_m the scattering of a single particle initially in state $\mathbf{k}_i$, so that the concentration of the particles being scattered is simply $C = 1/V = 1/L^3$. Thus we obtain

$$\sigma(\theta, \phi) = Vm^* P_m/(\hbar k_f \sin\theta\, d\theta\, d\phi). \tag{5.184}$$

Substituting our above result (5.182) for P_m, we obtain

$$\sigma(\theta, \phi) = (m^*/2\pi\hbar^2)^2 \left| \int \mathscr{U}(\mathbf{r})e^{i(\mathbf{k}_i - \mathbf{k}_f)\cdot\mathbf{r}}\, d\mathbf{r} \right|^2. \tag{5.185}$$

This important result for arbitrary potentials will be used below for two specific problems:

(a) the coherent scattering of electrons by the periodic lattice potential to yield electron diffraction and the energy band structure of periodic solids; and

(b) the contribution to the electrical resistivity of a solid caused by the scattering of conduction electrons by randomly situated ionized impurities in the lattice.

12 Diffraction of Electrons by the Periodic Potential of a Crystal

An electron-diffraction experiment consists basically of directing a monoenergetic beam of electrons onto a crystal and searching with a detector for any orientations at which the outcoming electrons have intensity maxima. The crystal orientation and the energy per electron in the incident beam are the major experimental variables. The beam intensity must be sufficient for observation of the diffracted electrons, but otherwise is unimportant insofar as the physical phenomena are concerned. The question we ask is specifically what are the conditions necessary to produce a diffracted beam in a certain direction relative to the incident beam. Consider the incident beam to be directed along the $+z$ direction, and consider the possibility of a scattered beam along the direction denoted by θ, ϕ in spherical polar coordinates. Expression (5.182) for the transition probability then tells us that the condition for diffraction in a given direction is that the matrix element

$$\mathscr{U}_{\mathrm{fi}} = V^{-1} \int \mathscr{U}(\mathbf{r}) e^{i(\mathbf{k}_{\mathrm{i}} - \mathbf{k}_{\mathrm{f}})\cdot\mathbf{r}}\, d\mathbf{r} \tag{5.186}$$

be nonzero. Furthermore we can conclude that a diffraction maximum requires that the value of this integral be an extremum.

The potential $\mathscr{U}(\mathbf{r})$ for this situation is the periodic lattice potential $\mathscr{U}_0(\mathbf{r})$, which can be expressed in terms of a three-dimensional complex Fourier series,

$$\mathscr{U}_0(\mathbf{r}) = \sum_{\mathbf{G}} \mathscr{U}_{\mathbf{G}} e^{i\mathbf{G}\cdot\mathbf{r}}. \tag{5.187}$$

The vectors $\mathbf{G}$ are known as *reciprocal lattice vectors*; they are developed in detail in Chap. 6. Substituting Eq. (5.187) into $\mathscr{U}_{\mathrm{fi}}$ gives

$$\mathscr{U}_{\mathrm{fi}} = V^{-1} \sum_{\mathbf{G}} \mathscr{U}_{\mathbf{G}} \int e^{i(\mathbf{k}_{\mathrm{i}} - \mathbf{k}_{\mathrm{f}} + \mathbf{G})\cdot\mathbf{r}}\, d\mathbf{r}. \tag{5.188}$$

Due to the oscillatory nature of each integrand, the integrals will be practically zero except for cases in which the condition

$$\mathbf{k}_{\mathrm{f}} \simeq \mathbf{k}_{\mathrm{i}} + \mathbf{G} \tag{5.189}$$

is met for some reciprocal lattice vector $\mathbf{G}$. When this occurs, the integral has the value V, and the corresponding value of $\mathscr{U}_{\mathrm{fi}}$ is $\mathscr{U}_{\mathbf{G}}$. The matrix element is in this case equal to the coefficient of the Fourier component of the periodic potential corresponding to the reciprocal lattice vector satisfying the above condition, so the transition probability will be proportional to the square of the magnitude of this coefficient. In general, we can state the result as

$$\mathscr{U}_{\mathrm{fi}} = \sum_{\mathbf{G}} \mathscr{U}_{\mathbf{G}}\, \delta_{\mathbf{G}+\mathbf{k}_{\mathrm{i}},\, \mathbf{k}_{\mathrm{f}}}. \tag{5.190}$$

Equation (5.190) predicts that for a given incident beam denoted by $\mathbf{k}_{\mathrm{i}}$ we will have a number of diffracted beams in different directions $\mathbf{k}_{\mathrm{f}}$ because there are a large number of reciprocal lattice vectors **G** available for satisfying the condition contained in the Kronecker delta function. If we confine our attention to a single final state $\mathbf{k}_f$ instead of scanning over a solid angle of 4π, then Eq. (5.190) can be written as

$$\mathscr{U}_{fi} = \mathscr{U}_{\mathbf{G}}\, \delta_{\mathbf{G}+\mathbf{k}_i, \mathbf{k}_f}. \tag{5.191}$$

Since other effects due to the nature of the atomic scattering centers constituting the periodic lattice are also important, we can only conclude at this point that the condition for diffraction which we have deduced above is required but not necessarily sufficient. Certain diffraction maxima predicted by the above condition are not found experimentally due to other mutual cancellation effects.

The above requirement (5.189) that $\mathbf{k}_f - \mathbf{k}_i \simeq \mathbf{G}$ for diffraction will now be shown to lead to the well-known Bragg condition (1.89), namely $n\lambda = 2d \sin\theta$, where θ is the angle which the incident beam makes with a set of lattice planes of spacing d, and n is an integer denoting the order of diffraction. For spherical energy bands, the energy conservation requirement $\mathscr{E}(\mathbf{k}_f) = \mathscr{E}(\mathbf{k}_i)$ for elastic scattering means that $\mathbf{k}_f$ and $\mathbf{k}_i$ must be equal in magnitude; thus $k_f = k_i = 2\pi/\lambda$. Figure 5.2 illustrates the relationship between a set of lattice planes (viewed as being perpendicular to the surface of the diagram), the corresponding set of **G** vectors (equal in magnitude to $2n\pi/d$ $(n = 1, 2, \ldots)$ and perpendicular to the set of lattice planes at the point of incidence of the beam), the **k** vector $\mathbf{k}_i$ of the incident beam (viewed as lying in the surface of the page and making an angle θ with the lattice planes), and the **k** vector $\mathbf{k}_f$ of the diffracted beam. In analogy with electromagnetic radiation, the wave vector of the diffracted beam is assumed to lie in the plane of incidence (i.e., the plane defined by **G** and $\mathbf{k}_i$) at the

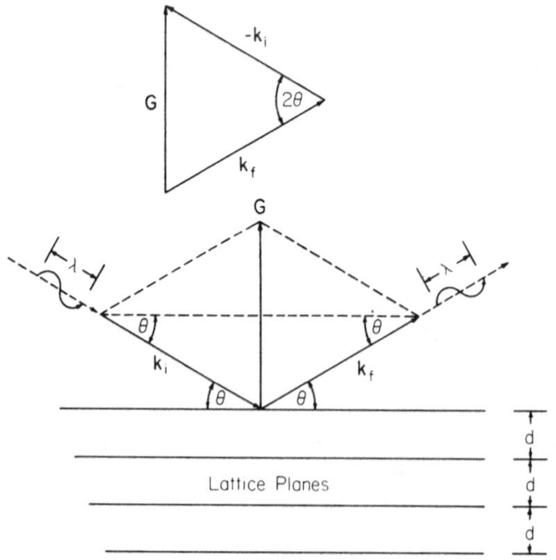

Fig. 5.2 Diffraction of waves with wave vector $\mathbf{k}_i$ by a sequence of parallel equally spaced lattice planes characterized by the reciprocal lattice vector **G**. (The wave vector of the diffracted beam is given by $\mathbf{k}_f = \mathbf{k}_i + \mathbf{G}$.)

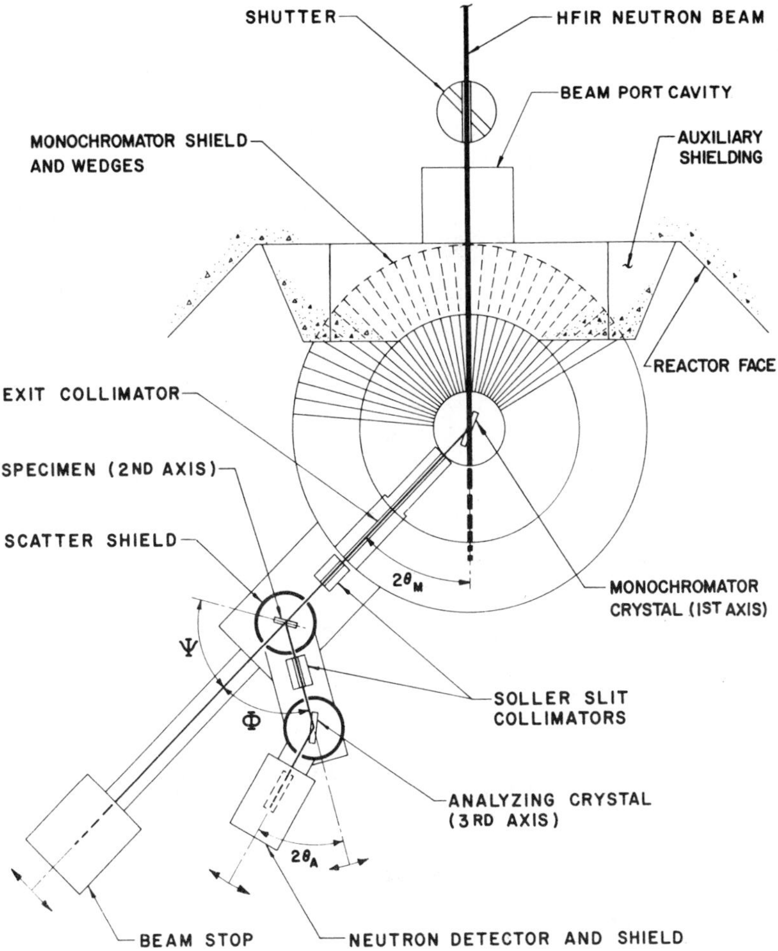

Fig. 5.3 Schematic diagram of a triple-axis neutron spectrometer installed at the High Flux Isotope Reactor (HFIR) at Oak Ridge National Laboratory. (The initial neutron energy is determined by Bragg reflection from the monochromator; the final energy, after scattering by the specimen, is determined by Bragg reflection from the analyzing crystal. Knowledge of the initial and final energies and the scattering angle Φ allows the determination of the energy and momentum transfer. This instrument is used primarily for measurement of phonon and magnon dispersion curves. The photograph was provided through the courtesy of Dr. R. M. Moon and Dr. M. K. Wilkinson of the Solid State Division of Oak Ridge National Laboratory, which is operated by the Nuclear Division of Union Carbide Corporation.)

same angle θ with respect to the lattice planes. From the isosceles triangle in the diagram, we deduce that

$$|\mathbf{G}|/2 = |\mathbf{k}_i| \sin \theta = |\mathbf{k}_f| \sin \theta. \tag{5.192}$$

Substituting $|\mathbf{G}| = 2n\pi/d$ and $|\mathbf{k}_i| = |\mathbf{k}_f| = 2\pi/\lambda$ gives immediately the Bragg condition (1.89),

$$\lambda = (2d/n) \sin \theta. \tag{5.193}$$

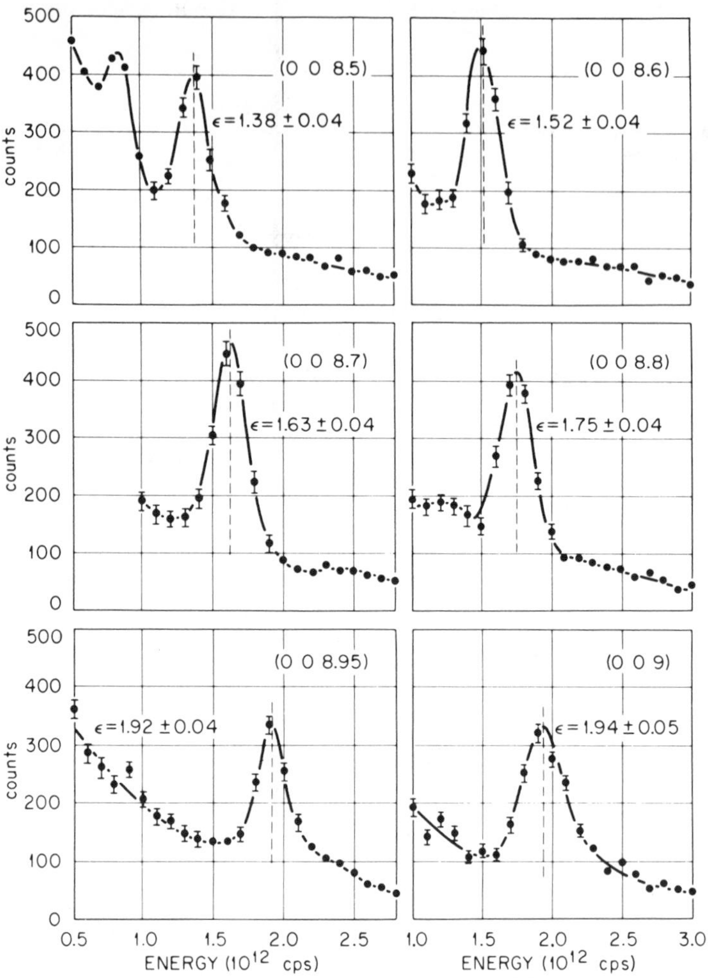

Fig. 5.4 Phonon peaks observed in iodine at 78°K with the neutron diffraction apparatus illustrated in Fig. 5.3. (Each graph shows neutron intensity versus energy transfer at a constant momentum transfer. The position of each peak determines a point on the phonon dispersion curve. This photograph was provided by Dr. R. M. Moon and Dr. M. K. Wilkinson of the Oak Ridge National Laboratory.)

The experimental observation of electron diffraction and its one-to-one correspondence with the results of x-ray diffraction seem to constitute irrefutable evidence for the wave nature of matter. Although electron diffraction and x-ray diffraction are found to be entirely alike in the general principles involved, experimental diffraction patterns reflect the fact that the lattice potential effective in the scattering of electrons differs from the corresponding lattice potential effective in the scattering of x rays. Neutron diffraction (Figs. 5.3 and 5.4) again follows the same general theory, but neutron scattering is influenced more strongly by the location of the atomic nuclei in the solid than is x-ray or electron diffraction, which depend to a large extent on the electron distribution within the solid.

13 Diffraction of Conduction Electrons and the Nearly-Free-Electron Model

13.1 Lattice Perturbation of States within the Brillouin Zone

The conduction electrons at the Fermi surface in a metal have wavelengths comparable to the lattice spacing, so that diffraction effects by the periodic lattice potential should perturb the results which we derived in Chap. 3 on the basis of the quantum free-electron model. Considering the conduction electrons to replace the role of the incident beam of electrons in the diffraction experiment described in the preceding section, we can immediately use the conclusion that whenever the matrix element

$$\mathscr{U}_{\mathrm{fi}} = \sum_{\mathbf{G}} \mathscr{U}_{\mathbf{G}}\, \delta_{\mathbf{G}+\mathbf{k}_{\mathrm{i}},\mathbf{k}_{\mathrm{f}}} \tag{5.194}$$

is nonzero, a conduction electron in the plane wave state $\mathbf{k}_{\mathrm{i}}$ will tend to be diffracted into the plane-wave state $\mathbf{k}_{\mathrm{f}}$. The rate of occurrence of this transition is proportional to the square of $|\mathscr{U}_{\mathbf{G}}|$, where $\mathscr{U}_{\mathbf{G}}$ is the particular coefficient of the complex Fourier component $e^{i\mathbf{G}\cdot\mathbf{r}}$ of the periodic potential $\mathscr{U}_0(\mathbf{r})$ for which

$$\mathbf{G} \simeq \mathbf{k}_{\mathrm{f}} - \mathbf{k}_{\mathrm{i}}. \tag{5.195}$$

Therefore we can conclude that the periodic lattice potential induces conduction electron transitions from one plane-wave state to another plane-wave state differing in $\mathbf{k}$ vector by a reciprocal lattice vector. The perturbing periodic potential thus mixes plane-wave states differing by reciprocal lattice vectors. The free-electron states and energy levels can thus be considered to be perturbed by the lattice potential, and in the limit where the perturbation is small, the system can be said to be described by a *nearly-free-electron (NFE) model.*

The perturbed energy through second order can be obtained by using our previously derived formulas (5.17) and (5.77) from time-independent perturbation theory

$$\mathscr{E}_i^{(1)} = \mathscr{V}_{ii}, \tag{5.196}$$

$$\mathscr{E}_i^{(2)} = \sum_{l \neq i} \frac{\mathscr{V}_{li}\mathscr{V}_{il}}{\mathscr{E}_i^{(0)} - \mathscr{E}_l^{(0)}}, \tag{5.197}$$

where the perturbed energy $\mathscr{E}_i$ is given by Eq. (5.4),

$$\mathscr{E}_i \simeq \mathscr{E}_i^{(0)} + \lambda \mathscr{E}_i^{(1)} + \lambda^2 \mathscr{E}_i^{(2)}. \tag{5.198}$$

The unperturbed state corresponding to a plane wave with wave vector $\mathbf{k}_{\mathrm{i}}$ has an energy, according to Eq. (3.11) for the free-electron model, of

$$\mathscr{E}_{\mathrm{i}}^{(0)} = \hbar^2 k_{\mathrm{i}}^2/2m, \tag{5.199}$$

where m is the electronic mass. According to the result (5.194) quoted above for

$\mathscr{U}_{\text{fi}}$, the diagonal matrix element corresponds to the reciprocal lattice vector $\mathbf{G} = 0$; the matrix element $\mathscr{V}_{ii}$ in Eq. (5.196) therefore has the value $\mathscr{U}_0$, and hence

$$\mathscr{E}_i^{(1)} = \mathscr{U}_0. \tag{5.200}$$

Thus the first-order correction to the energy level is simply the uniform potential energy (i.e., infinite wavelength Fourier component) contribution introduced by the perturbation. Similarly, Eq. (5.194) shows us that the off-diagonal matrix elements in Eq. (5.197) are given by

$$\mathscr{V}_{li} = \mathscr{U}_{\mathbf{G}}\, \delta_{\mathbf{G}+\mathbf{k}_i, \mathbf{k}_l} = \mathscr{U}_{\mathbf{G}}\, \delta_{\mathbf{G}, \mathbf{k}_l - \mathbf{k}_i}, \tag{5.201}$$

$$\mathscr{V}_{il} = \mathscr{U}_{\mathbf{G}'}\, \delta_{\mathbf{G}'+\mathbf{k}_l, \mathbf{k}_i} = \mathscr{U}_{\mathbf{G}'}\, \delta_{\mathbf{G}', \mathbf{k}_i - \mathbf{k}_l}. \tag{5.202}$$

The reason $\mathbf{G} \neq \mathbf{G}'$ is that $\mathbf{k}_l - \mathbf{k}_i$ is a different vector from $\mathbf{k}_i - \mathbf{k}_l$; in fact, it is the negative of $\mathbf{k}_i - \mathbf{k}_l$. We perhaps may not say that $\mathscr{V}_{li}^* = \mathscr{V}_{il}$, as we did in Eq. (5.77); the difference here is that we have written a real potential as a complex Fourier series with complex Fourier coefficients, and thus each term in the resulting matrix element is not real. The fact that the lattice potential must be real still proves useful to us in obtaining a compact result, as presently will be shown.

For every vector $\mathbf{G}$, there must be a vector $-\mathbf{G}$ in the Fourier expansion since the Fourier components $\exp(i\mathbf{G}\cdot\mathbf{r})$ and $\exp(-i\mathbf{G}\cdot\mathbf{r})$ are linearly independent. Hence if the *selection rule* (5.201) is met with $\mathbf{G}$ for $\mathbf{k}_l - \mathbf{k}_i$, then the *selection rule* (5.202) is met with $\mathbf{G}' = -\mathbf{G}$ for $\mathbf{k}_i - \mathbf{k}_l$. Therefore Eq. (5.202) can be written

$$\mathscr{V}_{il} = \mathscr{U}_{-\mathbf{G}}\, \delta_{-\mathbf{G}, \mathbf{k}_i - \mathbf{k}_l} = \mathscr{U}_{-\mathbf{G}}\, \delta_{\mathbf{G}, \mathbf{k}_l - \mathbf{k}_i}. \tag{5.203}$$

Equations (5.201) and (5.203) yield

$$\mathscr{V}_{li}\mathscr{V}_{il} = \mathscr{U}_{\mathbf{G}}\mathscr{U}_{-\mathbf{G}}(\delta_{\mathbf{G}, \mathbf{k}_l - \mathbf{k}_i})^2 = \mathscr{U}_{\mathbf{G}}\mathscr{U}_{-\mathbf{G}}\, \delta_{\mathbf{G}, \mathbf{k}_l - \mathbf{k}_i}. \tag{5.204}$$

Let us pause to examine the nature of the coefficients $\mathscr{U}_{\mathbf{G}}$. The lattice potential $\mathscr{U}_0(\mathbf{r})$ is real, so that

$$\mathscr{U}_0(\mathbf{r}) = \mathscr{U}_0^*(\mathbf{r}). \tag{5.205}$$

Substituting the expansion (5.187) into (5.205) gives

$$\sum_{\mathbf{G}} \mathscr{U}_{\mathbf{G}} e^{i\mathbf{G}\cdot\mathbf{r}} = \sum_{\mathbf{G}} \mathscr{U}_{\mathbf{G}}^* e^{-i\mathbf{G}\cdot\mathbf{r}}. \tag{5.206}$$

For every $\mathbf{G}$, there is a corresponding vector $-\mathbf{G}$ in the reciprocal lattice, as mentioned above and as shown explicitly in Chap. 6. Thus we can substitute $-\mathbf{G}$ for $\mathbf{G}$ in every term in the sum and reorder the terms,

$$\sum_{\mathbf{G}} \mathscr{U}_{\mathbf{G}}^* e^{-i\mathbf{G}\cdot\mathbf{r}} = \sum_{-\mathbf{G}} \mathscr{U}_{-\mathbf{G}}^* e^{i\mathbf{G}\cdot\mathbf{r}} = \sum_{\mathbf{G}} \mathscr{U}_{-\mathbf{G}}^* e^{i\mathbf{G}\cdot\mathbf{r}}. \tag{5.207}$$

Substituting this result into Eq. (5.206) gives

$$\sum_{\mathbf{G}} \mathscr{U}_{\mathbf{G}} e^{i\mathbf{G}\cdot\mathbf{r}} = \sum_{\mathbf{G}} \mathscr{U}_{-\mathbf{G}}^* e^{i\mathbf{G}\cdot\mathbf{r}}, \tag{5.208}$$

or equivalently,

$$\sum_{\mathbf{G}} (\mathscr{U}_{\mathbf{G}} - \mathscr{U}^*_{-\mathbf{G}})e^{i\mathbf{G}\cdot\mathbf{r}} = 0. \tag{5.209}$$

Since the Fourier components $e^{i\mathbf{G}\cdot\mathbf{r}}$ are linearly independent of one another, the coefficient $\mathscr{U}_{\mathbf{G}} - \mathscr{U}^*_{-\mathbf{G}}$ must be zero, so that

$$\mathscr{U}_{\mathbf{G}} = \mathscr{U}^*_{-\mathbf{G}}, \tag{5.210}$$

or equivalently,

$$\mathscr{U}^*_{\mathbf{G}} = \mathscr{U}_{-\mathbf{G}}. \tag{5.211}$$

Utilizing this result in Eq. (5.204) gives

$$\mathscr{V}_{li}\mathscr{V}_{il} = \mathscr{U}_{\mathbf{G}}\mathscr{U}_{-\mathbf{G}}\,\delta_{\mathbf{G},\mathbf{k}_l-\mathbf{k}_i} = \mathscr{U}_{\mathbf{G}}\mathscr{U}^*_{\mathbf{G}}\,\delta_{\mathbf{G},\mathbf{k}_l-\mathbf{k}_i} = |\mathscr{U}_{\mathbf{G}}|^2\,\delta_{\mathbf{G},\mathbf{k}_l-\mathbf{k}_i}. \tag{5.212}$$

Substituting Eq. (5.212) into the expression (5.197), we thus obtain

$$\mathscr{E}_i^{(2)} = \sum_{l\neq i} \frac{|\mathscr{U}_{\mathbf{G}}|^2\,\delta_{\mathbf{G},\mathbf{k}_l-\mathbf{k}_i}}{\mathscr{E}_i^{(0)} - \mathscr{E}_l^{(0)}}. \tag{5.213}$$

Denoting $\mathscr{E}_i^{(0)}$ by $\mathscr{E}(\mathbf{k}_i)$ and $\mathscr{E}_l^{(0)}$ by $\mathscr{E}(\mathbf{k}_l)$,

$$\mathscr{E}_i^{(2)} = \sum_{\mathscr{E}(\mathbf{k}_l)\neq\mathscr{E}(\mathbf{k}_i)} \frac{|\mathscr{U}_{\mathbf{G}}|^2\,\delta_{\mathbf{G},\mathbf{k}_l-\mathbf{k}_i}}{\mathscr{E}(\mathbf{k}_i) - \mathscr{E}(\mathbf{k}_l)}. \tag{5.214}$$

As the sum over the index l is carried out, the factor $\delta_{\mathbf{G},\mathbf{k}_l-\mathbf{k}_i}$ selects out only those terms for which $\mathbf{k}_l = \mathbf{k}_i + \mathbf{G}$ for any of the set of $\mathbf{G}$ vectors for the lattice; all other terms are zero. Since we have a quasi-continuum of final states in the free-electron model, every $\mathbf{G}$ vector will give rise to some possible final state. We can restrict the sum in Eq. (5.214) to only those terms for which $\mathbf{k}_l = \mathbf{k}_i + \mathbf{G}$, which reduces the sum in effect to the following sum over all vectors $\mathbf{k}_i + \mathbf{G}$ for the lattice in question,

$$\mathscr{E}_i^{(2)} = \sum_{\mathscr{E}(\mathbf{k}_i+\mathbf{G})\neq\mathscr{E}(\mathbf{k}_i)} \frac{|\mathscr{U}_{\mathbf{G}}|^2}{\mathscr{E}(\mathbf{k}_i) - \mathscr{E}(\mathbf{k}_i+\mathbf{G})}. \tag{5.215}$$

Collecting the zero-, first-, and second-order contributions (5.199), (5.200), and (5.215), we thus obtain from Eq. (5.198) the expression

$$\mathscr{E}_i = \frac{\hbar^2k_i^2}{2m} + \lambda\mathscr{U}_0 + \lambda^2 \sum_{\mathscr{E}(\mathbf{k}_i+\mathbf{G})\neq\mathscr{E}(\mathbf{k}_i)} \frac{|\mathscr{U}_{\mathbf{G}}|^2}{\mathscr{E}(\mathbf{k}_i) - \mathscr{E}(\mathbf{k}_i+\mathbf{G})} \tag{5.216}$$

for the NFE model energy levels. This result shows that in general *the energy is no longer strictly a quadratic function of the wave vector k and the momentum* $\hbar k$*: The constant energy surfaces in* **k** *space can be nonspherical*!

EXERCISE Deduce the perturbed energy eigenfunctions corresponding to the perturbed energy eigenvalues given by Eq. (5.216). Can you draw therefrom any conclusions pertinent to the effect of the periodic lattice potential on electron propagation in solids?

One major requirement for validity of the perturbation result (5.216) can be seen immediately to be that

$$\mathscr{E}(\mathbf{k}_i) \neq \mathscr{E}(\mathbf{k}_i + \mathbf{G}) \tag{5.217}$$

for any nonzero $\mathscr{U}_\mathbf{G}$. This can be written

$$\hbar^2 k_i^2/2m \neq \hbar^2(\mathbf{k}_i + \mathbf{G})^2/2m, \tag{5.218}$$

which requires

$$|\mathbf{k}_i| \neq |\mathbf{k}_i + \mathbf{G}|. \tag{5.219}$$

This condition means that $\mathbf{k}_i$ cannot fall on the plane which bisects $\mathbf{G}$ and is perpendicular to it. (See diagram in Fig. 5.5.) We consider $\mathbf{k}_i$ to be measured from the origin in $\mathbf{k}$ space (i.e., *reciprocal space*). The planes which bisect the vectors $\mathbf{G}$ as measured from the origin in reciprocal space are called *Brillouin zone boundaries*, and the volume of $\mathbf{k}$ space delimited by these boundaries is called a *Brillouin zone*. (For details, refer to Chap. 6, §8.) Therefore it can be said that the *requirement for validity of the perturbation treatment is that the* $\mathbf{k}$ *vector of the state under consideration must not touch any of the Brillouin zone boundaries.* This excludes those states for which Bragg diffraction (§12) takes place, as can be seen from Eq. (5.189) and the fact that $|\mathbf{k}_f| = |\mathbf{k}_i|$ for elastic scattering. The perturbation treatment is therefore applicable only to those states inside the Brillouin zone; it breaks down for the electronic states on any portions of the Fermi surface which are in contact with the Brillouin zone boundaries. This is due to the fact that the energy is degenerate at the zone boundary, so that the correct set of zero-order basis states for a perturbation treatment must be properly chosen from suitable linear combinations of the individual plane-wave states corresponding to the zone boundaries. (The proper perturbation treatment of degenerate states has been discussed at length in §§ 1, 2, and 4 of this chapter.) When the perturbation operator is properly

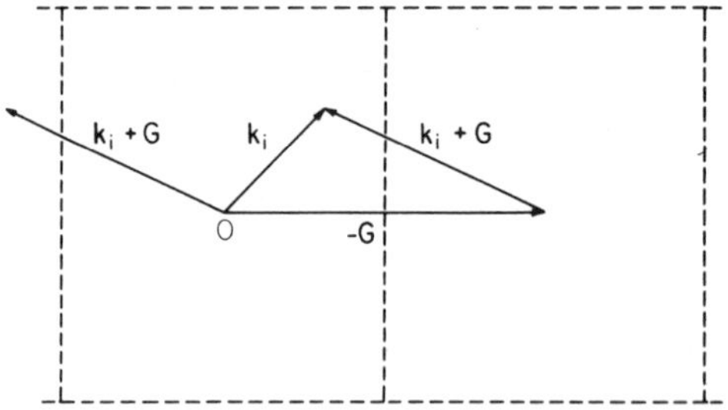

Fig. 5.5 Initial conduction electron states labeled by wave vectors $\mathbf{k}_i$ in reciprocal space. (If nondegenerate perturbation theory is to be used in evaluating the modifications in a given free-electron energy eigenvalue $\mathscr{E}_i$ and plane-wave eigenfunction ϕ_i introduced by the periodic lattice potential, the corresponding wave vector $\mathbf{k}_i$ cannot touch the Brillouin zone boundary indicated by the dashed lines. The Brillouin zone boundary in the figure is perpendicular to and bisects the reciprocal lattice vector $\mathbf{G}$. When the incident wave vector $\mathbf{k}_i$ touches a Brillouin zone boundary, the condition is met for diffraction of the incident beam, as illustrated in Fig. 5.2.)

diagonalized (as carried out explicitly in §13.2), it is found that the degeneracy is split, thereby opening up an *energy gap* in which there are *no* energies corresponding to allowed states for the conduction electrons! The energy range is thus divided into regions (or "bands") of *allowed energies* (called *energy bands*) over which there exists a quasi-continuum of electronic states meeting the condition of periodic boundary conditions, with these regions of allowed energies separated by regions of forbidden energies (called *energy gaps*) over which there exist no electronic states which satisfy periodic boundary conditions. One of the principal effects of the lattice potential is therefore to create the energy gaps which play such a central role in our understanding of the difference between metals and insulators. (Further details are given in the next section and in the discussion in Chap. 7, §9.)

13.2 Lattice Perturbation of States at Brillouin Zone Boundaries

The energy gaps discussed in the preceding section have such important implications for understanding the conductivity properties of solids that this topic deserves a more careful quantitative examination, using the first-order degenerate perturbation theory developed in §2. The starting point is the consideration of two plane-wave states

$$\xi_1 = V^{-1/2} e^{i\mathbf{k}_i \cdot \mathbf{r}}, \qquad \xi_2 = V^{-1/2} e^{i(\mathbf{k}_i + \mathbf{G}') \cdot \mathbf{r}},$$

where V is the volume of the metal crystal. These two states are degenerate whenever

$$\mathscr{E}_i^{(0)} = (\hbar^2/2m)\mathbf{k}_i \cdot \mathbf{k}_i = (\hbar^2/2m)(\mathbf{k}_i + \mathbf{G}') \cdot (\mathbf{k}_i + \mathbf{G}')$$

for any one of the reciprocal lattice vectors $\mathbf{G}'$. This constant energy condition can be written $|\mathbf{k}_i|^2 = |\mathbf{k}_i|^2 + 2\mathbf{k}_i \cdot \mathbf{G}' + |\mathbf{G}'|^2$, or equivalently, $\mathbf{k}_i \cdot \mathbf{G}' = -\frac{1}{2}|\mathbf{G}'|^2$. This vector equation maps out a plane in $\mathbf{k}$ space, somewhat *analogous* to the vector equation for a plane in real space. (Recall that the equation for a plane in real space can be written as $\mathbf{K} \cdot \mathbf{r} = \text{const}$, where $\mathbf{K}$ is some fixed vector which is perpendicular to the plane in question, and $\mathbf{r}$ is a variable vector in real space.)

Because $\mathbf{k}_i \cdot \mathbf{G}' = |\mathbf{k}_i||\mathbf{G}'| \cos \theta_i$, where θ_i is the angle between the vector $\mathbf{k}_i$ and $\mathbf{G}'$, the above equation can be written $|\mathbf{k}_i| \cos \theta_i = -\frac{1}{2}|\mathbf{G}'|$. This result can be interpreted as follows: Whenever the projection of $\mathbf{k}_i$ onto $\mathbf{G}'$ is exactly one-half the magnitude of $\mathbf{G}'$, and is oppositely directed to $\mathbf{G}'$, then the condition is satisfied for the above functions ξ_1 and ξ_2 to represent degenerate energies. For a given $\mathbf{G}'$, the vectors in $\mathbf{k}$ space which satisfy this condition are the subset of $\mathbf{k}_i$ vectors which extend from the origin and touch the plane which bisects the $\mathbf{G}'$ vector in question. (If we look ahead to Chap. 6, §§ 6 and 8, we can see that this represents a Brillouin zone boundary.)

To apply the theory developed in §2 to this situation, we need the matrix elements [cf. Eq. (5.49)] of the periodic lattice perturbation potential $\mathscr{U}_0(\mathbf{r})$ with respect to the above plane-wave states ξ_1 and ξ_2. Using the Fourier expansion

(5.187) for the periodic lattice potential, we obtain

$$\mathscr{U}_{ij} \equiv \langle i|\mathscr{U}_0(\mathbf{r})|j\rangle = \frac{1}{V}\int_{\Omega_\mathbf{r}} e^{-i\mathbf{k}_i\cdot\mathbf{r}}\left[\sum_{\mathbf{G}} \mathscr{U}_\mathbf{G} e^{i\mathbf{G}\cdot\mathbf{r}}\right] e^{i\mathbf{k}_j\cdot\mathbf{r}}\,d\mathbf{r}$$

$$= \frac{1}{V}\sum_{\mathbf{G}} \mathscr{U}_\mathbf{G}\int_{\Omega_\mathbf{r}} e^{i[(\mathbf{k}_j-\mathbf{k}_i)+\mathbf{G}]\cdot\mathbf{r}}\,d\mathbf{r} \simeq \sum_{\mathbf{G}} \mathscr{U}_\mathbf{G}\,\delta_{\mathbf{k}_i,\mathbf{G}+\mathbf{k}_j}$$

for plane-wave states represented by the vectors $\mathbf{k}_i$ and $\mathbf{k}_j$. (The reduction of the integral to the Kronecker delta function follows from the oscillatory nature of the integrand and the assumption that each dimension of the solid is a multiple of that wavelength of the oscillations.) Substituting $\mathbf{k}_i + \mathbf{G}'$ for $\mathbf{k}_j$ then gives $\mathscr{U}_{12}$, the matrix element of $\mathscr{U}(\mathbf{r})$ with respect to states ξ_1 and ξ_2,

$$\mathscr{U}_{12} = \sum_{\mathbf{G}} \mathscr{U}_\mathbf{G}\,\delta_{\mathbf{k}_i,\mathbf{G}+\mathbf{G}'+\mathbf{k}_i} = \mathscr{U}_{-\mathbf{G}'}.$$

Similarly,

$$\mathscr{U}_{21} = \sum_{\mathbf{G}} \mathscr{U}_\mathbf{G}\,\delta_{\mathbf{k}_i+\mathbf{G}',\mathbf{G}+\mathbf{k}_i} = \mathscr{U}_{\mathbf{G}'},$$

or with the aid of Eq. (5.211) and the above expression for $\mathscr{U}_{12}$, we obtain $\mathscr{U}_{21} = \mathscr{U}_{\mathbf{G}'} = \mathscr{U}^*_{-\mathbf{G}'} = \mathscr{U}^*_{12}$, where the asterisk denotes the complex conjugate. On the other hand, substituting $\mathbf{k}_i$ for $\mathbf{k}_j$ gives the diagonal matrix element of $\mathscr{U}(\mathbf{r})$ with respect to the state ξ_1,

$$\mathscr{U}_{11} = \sum_{\mathbf{G}} \mathscr{U}_\mathbf{G}\,\delta_{\mathbf{k}_i,\mathbf{G}+\mathbf{k}_i} = \mathscr{U}_0.$$

Similarly, the diagonal matrix element of $\mathscr{U}(\mathbf{r})$ with respect to the state ξ_2 is

$$\mathscr{U}_{22} = \sum_{\mathbf{G}} \mathscr{U}_\mathbf{G}\,\delta_{\mathbf{k}_i+\mathbf{G}',\mathbf{G}+\mathbf{G}'+\mathbf{k}_i} = \mathscr{U}_0.$$

Substituting these results for the matrix elements $\mathscr{U}_{ij}$ in place of $\mathscr{M}_{ij}$ in the secular determinant (5.53) [or (5.56)] and utilizing the symbol $\mathscr{E}_i^{(1)}$ for the first-order correction to the energy eigenvalue thus leads to the following determinantal equation,

$$\begin{vmatrix} \mathscr{U}_0 - \mathscr{E}_i^{(1)} & \mathscr{U}^*_{\mathbf{G}'} \\ \mathscr{U}_{\mathbf{G}'} & \mathscr{U}_0 - \mathscr{E}_i^{(1)} \end{vmatrix} = 0.$$

Expanding the determinant then leads to the *secular* eqution, $(\mathscr{U}_0 - \mathscr{E}_i^{(1)})^2 = \mathscr{U}_{\mathbf{G}'}\mathscr{U}^*_{\mathbf{G}'} = |\mathscr{U}_{\mathbf{G}'}|^2$, which gives $\mathscr{U}_0 - \mathscr{E}_i^{(1)} = \pm|\mathscr{U}_{\mathbf{G}'}|$, and thus leads to the two eigenvalues $\mathscr{E}_i^{(1)} = \mathscr{U}_0 \mp |\mathscr{U}_{\mathbf{G}'}|$ for the diagonalized representation. The component $\mathscr{U}_{\mathbf{G}'}$ of the periodic lattice potential thus *splits the degeneracy* of the degenerate plane-wave eigenstates of the unperturbed free-electron Hamiltonian. (This exercise in technical jargon is not meant to discourage the reader; on the contrary, it is meant to "gird the loins" for future royal battles in coping with the literature!) The two eigenvalues thus deduced differ from $\mathscr{U}_0$, the expectation value of a uniform (nonperiodic) potential. For definiteness, let us label the two roots $\mathscr{E}_{i1}^{(1)}$ and $\mathscr{E}_{i2}^{(1)}$,

$$\mathscr{E}_{i1}^{(1)} = \mathscr{U}_0 - |\mathscr{U}_{\mathbf{G}'}|, \qquad \mathscr{E}_{i2}^{(1)} = \mathscr{U}_0 + |\mathscr{U}_{\mathbf{G}'}|.$$

These constitute the first-order energy eigenvalue corrections produced by the periodic lattice potential. Note that the splitting introduced by the perturbation gives an energy separation $\Delta\mathscr{E}$ (i.e., an *energy gap* $\mathscr{E}_g$) between the states,

$$\mathscr{E}_g = \Delta\mathscr{E} = \mathscr{E}_{i2}^{(1)} - \mathscr{E}_{i1}^{(1)} = (\mathscr{U}_0 + |\mathscr{U}_{\mathbf{G}'}|) - (\mathscr{U}_0 - |\mathscr{U}_{\mathbf{G}'}|) = 2|\mathscr{U}_{\mathbf{G}'}|.$$

The energy gap is thus directly proportional to the relevant Fourier component of the periodic lattice potential. Figure 5.6 illustrates the introduction of the energy gap into the otherwise free-electron *dispersion* curve $\mathscr{E} \propto k^2$ characteristic of the free-electron model [see Eq. (3.11) and Fig. 3.12].

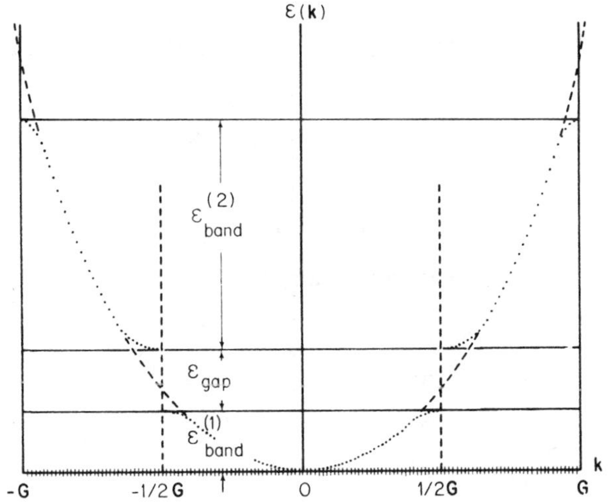

Fig. 5.6 The set of **k** vectors obtained by applying periodic boundary conditions to the rectangular parallelepiped domain of a free-electron metal fall on the parabola $\mathscr{E}(\mathbf{k}) = \hbar^2k^2/2m$, according to Eq. (3.11); the periodic lattice potential perturbs these closely spaced energies, especially in the neighborhood of the Brillouin zone boundaries where Bragg diffraction occurs, so that the various electronic states labeled by the **k** vectors (see dots) avoid certain ranges of energy (see $\mathscr{E}_{\text{gap}}$). The energy range is thus separated into energy bands and energy gaps by the periodic lattice potential. (The magic effect of the Brillouin zone boundaries can be traced to the fact that for each **k** vector touching the boundary, there exists a **k** vector $\mathbf{k} + \mathbf{G}$ of the same magnitude touching an opposite boundary, as can be visualized from Fig. 5.5; these two states have the same wavelength and the same energy $\mathscr{E} = \hbar^2k^2/2m$. Moreover, along the direction of **G** each state has oppositely directed components of **k** of equal magnitude which correspond to some integral multiple of the de Broglie wavelength between lattice sites parallel to **G**, so Bragg reflection by the lattice potential can serve to populate each of these states from the other. Thus two linearly independent standing waves can be created from the two oppositely directed traveling wave components, each standing wave being nonpropagating in directions parallel to **G**. The location of the peaks in the standing-wave electron probability density, in relation to the lattice sites, determines the interaction energy of the electron in question with the ionic lattice potential, as illustrated in Fig. 5.7.)

Next, let us deduce the corresponding eigenfunctions, again using the theory developed in §2. From Eqs. (5.52), (5.54), and the above results we obtain

$$a_{12} = a_{11}\,\mathscr{U}_{21}/(\mathscr{E}_{i1}^{(1)} - \mathscr{U}_{22}) = -a_{11}\,\mathscr{U}_{\mathbf{G}'}/|\mathscr{U}_{\mathbf{G}'}|,$$

$$a_{21} = a_{22}\,\mathscr{U}_{12}/(\mathscr{E}_{i2}^{(1)} - \mathscr{U}_{11}) = a_{22}\,\mathscr{U}_{\mathbf{G}'}^{*}/|\mathscr{U}_{\mathbf{G}'}|.$$

The perturbed eigenfunctions given by Eq. (5.38) thus take the form

$$\eta_1 = a_{11}\xi_1 + a_{12}\xi_2 = a_{11}\{\xi_1 - [\mathscr{U}_{\mathbf{G}'}/|\mathscr{U}_{\mathbf{G}'}|]\xi_2\},$$

$$\eta_2 = a_{21}\xi_1 + a_{22}\xi_2 = a_{22}\{[\mathscr{U}_{\mathbf{G}'}^{*}/|\mathscr{U}_{\mathbf{G}'}|]\xi_1 + \xi_2\},$$

where ξ_1 and ξ_2 are the plane-wave states $e^{i\mathbf{k}_i\cdot\mathbf{r}}$ and $e^{i(\mathbf{k}_i+\mathbf{G}')\cdot\mathbf{r}}$ with which we started our treatment. To proceed, it is convenient to write the complex Fourier component $\mathscr{U}_{\mathbf{G}'}$ in polar form

$$\mathscr{U}_{\mathbf{G}'} = |\mathscr{U}_{\mathbf{G}'}|\,e^{i\delta},$$

where δ is the phase angle and $|\mathscr{U}_{\mathbf{G}'}|$ is the modulus of the complex number. Thus we can write

$$\mathscr{U}_{\mathbf{G}'}/|\mathscr{U}_{\mathbf{G}'}| = e^{i\delta}, \qquad \mathscr{U}_{\mathbf{G}'}^{*}/|\mathscr{U}_{\mathbf{G}'}| = e^{-i\delta},$$

in which case

$$\eta_1 = a_{11}(\xi_1 - e^{i\delta}\xi_2), \qquad \eta_2 = a_{22}(e^{-i\delta}\xi_1 + \xi_2),$$

with a_{11} and a_{22} determined from the normalization conditions

$$\langle\eta_1|\eta_1\rangle = 1, \qquad \langle\eta_2|\eta_2\rangle = 1.$$

Substituting for ξ_1 and ξ_2 gives

$$\begin{aligned}
\eta_1 &= a_{11}V^{-1/2}[e^{i\mathbf{k}_i\cdot\mathbf{r}} - e^{i\delta}e^{i(\mathbf{k}_i+\mathbf{G}')\cdot\mathbf{r}}] \\
&= -a_{11}V^{-1/2}e^{i(\mathbf{k}_i\cdot\mathbf{r})}e^{i(\frac{1}{2}\delta)}e^{i(\frac{1}{2}\mathbf{G}'\cdot\mathbf{r})}\{e^{i[\frac{1}{2}(\mathbf{G}'\cdot\mathbf{r}+\delta)]} - e^{-i[\frac{1}{2}(\mathbf{G}'\cdot\mathbf{r}+\delta)]}\} \\
&= -2ia_{11}V^{-1/2}e^{i\mathbf{k}_i\cdot\mathbf{r}}e^{i[\frac{1}{2}(\mathbf{G}'\cdot\mathbf{r}+\delta)]}\sin[\tfrac{1}{2}(\mathbf{G}'\cdot\mathbf{r}+\delta)], \\
\eta_2 &= a_{22}V^{-1/2}[e^{-i\delta}e^{i\mathbf{k}_i\cdot\mathbf{r}} + e^{i(\mathbf{k}_i+\mathbf{G}')\cdot\mathbf{r}}] \\
&= a_{22}V^{-1/2}[e^{i(\mathbf{k}_i\cdot\mathbf{r})}e^{-i(\frac{1}{2}\delta)}e^{i(\frac{1}{2}\mathbf{G}'\cdot\mathbf{r})}(e^{i[\frac{1}{2}(\mathbf{G}'\cdot\mathbf{r}+\delta)]} + e^{-i[\frac{1}{2}(\mathbf{G}'\cdot\mathbf{r}+\delta)]}) \\
&= 2a_{22}V^{-1/2}e^{i\mathbf{k}_i\cdot\mathbf{r}}e^{i[\frac{1}{2}(\mathbf{G}'\cdot\mathbf{r}-\delta)]}\cos[\tfrac{1}{2}(\mathbf{G}'\cdot\mathbf{r}+\delta)].
\end{aligned}$$

The corresponding probability densities are thus given by

$$\eta_1^*\eta_1 = 4|a_{11}|^2V^{-1}\sin^2[\tfrac{1}{2}(\mathbf{G}'\cdot\mathbf{r}+\delta)],$$

$$\eta_2^*\eta_2 = 4|a_{22}|^2V^{-1}\cos^2[\tfrac{1}{2}(\mathbf{G}'\cdot\mathbf{r}+\delta)].$$

The normalization factors follow readily by integrating $\eta_1^*\eta_1$ and $\eta_2^*\eta_2$ over the volume of the metal and setting the result to unity. Taking into account that $\sin^2\alpha = \frac{1}{2} - \frac{1}{2}\cos 2\alpha$ and $\cos^2\alpha = \frac{1}{2} + \frac{1}{2}\cos 2\alpha$, and also the fact that the oscillatory part integrates to zero over any integer multiple of the lattice spacing (such as that corresponding to the length of the solid), we find

$$1 = 2|a_{11}|^2 V^{-1} V, \qquad 1 = 2|a_{22}|^2 V^{-1} V,$$

or equivalently,

$$|a_{11}| = |a_{22}| = 2^{-1/2}.$$

We note the following salient points regarding the perturbed energy eigenfunctions obtained by means of the above development:

(a) The two eigenfunctions are in the form of products of free-electron-like plane-wave factors $e^{i\mathbf{k}_i \cdot \mathbf{r}}$ and spatially periodic factors involving the sine and cosine of $\frac{1}{2}\mathbf{G}' \cdot \mathbf{r}$;
(b) The two eigenfunctions differ in phase by $\frac{1}{2}\pi$, with each being spatially periodic;
(c) The associated electron probability densities are spatially periodic.

If we are willing to look ahead to certain results proven in Chap. 7 which have already been referred to in §13.1, we can extend our understanding by noting the following additional facts:

(d) For every state $\mathbf{k}_i$ satisfying periodic boundary conditions, there exists a state $\mathbf{k}_i + \mathbf{G}$, where $\mathbf{G}$ is an *arbitrary* reciprocal lattice vector. Therefore, for the specific case in which $|\mathbf{k}_i + \mathbf{G}'| = |\mathbf{k}_i|$ is satisfied mathematically for some state $\xi_1 = V^{-1/2} e^{i\mathbf{k}_i \cdot \mathbf{r}}$ and some specific reciprocal lattice vector $\mathbf{G}'$, there indeed exists a free-electron state $\xi_2 = V^{-1/2} e^{i(\mathbf{k}_i + \mathbf{G}') \cdot \mathbf{r}}$. Thus our development has real physical content.
(e) The vector $\mathbf{G}'$ represents in reciprocal space a Fourier component of the lattice potential in real space having wavelength $\lambda' = 2\pi/|\mathbf{G}'|$, where $n'\lambda'$ is equal to the lattice spacing d. The electron propagation vector $\mathbf{k}_i$ also corresponds to some wavelength namely, $\lambda_i \equiv 2\pi/|\mathbf{k}_i|$. The condition $|\mathbf{k}_i + \mathbf{G}'| = |\mathbf{k}_i|$, as stated in the form $|\mathbf{k}_i| \cos\theta_i = -\frac{1}{2}|\mathbf{G}'|$ developed at the beginning of this section, thus leads to

$$|k_i \cos\theta_i| = \left|\frac{2\pi}{\lambda_i}\cos\theta_i\right| = \left|-\frac{1}{2}\mathbf{G}'\right| = \left|-\frac{1}{2}\frac{2\pi}{\lambda'}\right| = \left|-\frac{1}{2}\left(\frac{2\pi}{d/n'}\right)\right|,$$

or equivalently,

$$|\lambda_i/\cos\theta_i| = 2\lambda' = 2d/n'.$$

This in turn can be written in the form (Chap. 1, §4.4)

$$n'[\tfrac{1}{2}\lambda_i^{(\perp)}] = d,$$

where the quantity $\lambda_i^{(\perp)} \equiv \lambda_i/\cos\theta_i$ can be interpreted as the *component* of the wavelength $\lambda_i = 2\pi/k_i$ (corresponding to the electron propagation vector $\mathbf{k}_i$) as measured along the direction $\mathbf{G}'$ (which in turn is *perpendicular* to the Brillouin zone boundary). This result shows that the wavelength corresponding to the projection of $\mathbf{k}_i$ onto the $\mathbf{G}'$ vector in question is some *harmonic* with respect to twice the separation distance between *spatial lattice points*, or to put it alternatively, exactly an integral number of half wavelengths $\frac{1}{2}\lambda_i^{(\perp)}$ will fit into the lattice spacing along the $\mathbf{G}'$ direction.

(f) The real space periodicity of the electron probability density can now be deduced. It can be noted that $\eta_1^*\eta_1$ and $\eta_2^*\eta_2$ vary respectively as the *square* of the sine and cosine functions which contain the position-dependent argument $\frac{1}{2}\mathbf{G}'\cdot\mathbf{r}$. This argument itself has a real space periodicity determined by $\frac{1}{2}(2\pi/\lambda')$ with respect to the $\mathbf{G}'$ direction, which leads to the basic unit $2\lambda'$ for periodicity. However, the periodicity of the squared sine and squared cosine functions are exactly doubled over that of the sine and cosine functions, so that the basic unit of periodicity for the probability density along $\mathbf{G}'$ is λ'. Moreover, $\lambda' = d/n'$, so we reach the conclusion that there are n' maxima in the probability density for every lattice separation distance d along the $\mathbf{G}'$ direction. To look at this result somewhat more generally, it can be said that *the probability density is invariant under lattice translation*. These results are quite analogous to the properties of *Bloch functions* which are developed in Chap. 7. The particular eigenfunction corresponding to the *lower* bound-state energy is that which gives rise to a *high* probability density at the nucleus, as illustrated in Fig. 5.7a, since the attractive Coulomb potential between the positively charged ion core and the negatively charged electron is then a maximum. Conversely, the remaining eigenfunction,

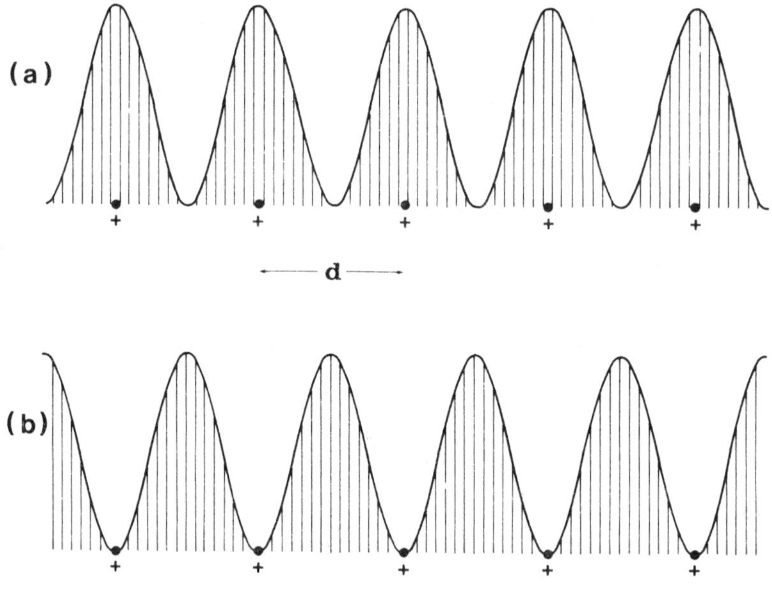

Fig. 5.7 Probability density distributions corresponding to standing-wave eigenfunctions. (a) Peaks in probability density occurring on lattice sites, thus allowing maximum interaction of the conduction electron with the positively charged ion cores, with a large attendant lowering in the electron potential energy. (b) Nodes in probability density occurring at lattice sites, thus minimizing the interaction of the conduction electrons with the positively charged ion cores, with a smaller attendant lowering in the electron potential energy.

differing in phase by $\frac{1}{2}\pi$, will have a *node* in the probability density at the nucleus as illustrated in Fig. 5.7b; the attractive Coulomb potential between nucleus and electron is in this case a minimum, so this particular eigenfunction corresponds to a *higher-lying* bound-state energy.

(g) The pre-exponential factors can be re-examined in light of the relation $|\mathbf{k}_i| \cos \theta_i = -\frac{1}{2}|\mathbf{G}'|$. It can be noted that both η_1 and η_2 contain the factor $e^{i(\mathbf{k}_i + \frac{1}{2}\mathbf{G}')\cdot\mathbf{r}}$. This factor reduces to unity when $\mathbf{k}_i$ is parallel to $-\mathbf{G}'$, in which case $\theta_i = 0$. In this case the wave functions have no free-electron propagation character at all; instead, they represent pure standing waves. Electrons occupying states described by wave vectors which touch the Brillouin zone boundaries are therefore nonpropagating along directions parallel to the reciprocal lattice vectors perpendicular to these boundaries. It was shown in §12 that electrons satisfying the conditions of the present problem, but having $\theta_i \neq 0$, undergo diffraction. Therefore such electrons propagate, but not rectilinearly in a direction parallel to $\mathbf{G}'$.

In the next section, we examine the effects of a *random* perturbing potential on the propagation of conduction electrons in solids. The contrast between the results deduced for the above-considered *perfectly periodic potential* and the results deduced for the *nonperiodic* (*or random*) potential are quite noteworthy and of great importance.

14 Differential Scattering Cross Section for Plane-Wave States and a Coulomb Potential

Equation (5.185) for the differential scattering cross section developed in §11 has been shown in §§ 12 and 13 to predict the diffraction of electrons by the periodic potential of a crystal. In addition to the periodic lattice potential, there are other potentials which modify the electron trajectory in a solid, and the effect of these potentials can also be considered with the aid of Eq. (5.185). As a specific example, let us consider the Coulomb potential of an ionized impurity atom. A free electron will be scattered by such a potential, and this often represents *a major contribution to the resistivity of solids at low temperatures* where there is very little lattice vibrational motion to produce conduction electron scattering.

The **Coulomb force** on an electronic charge $Z_e e$ exerted by an impurity of charge $Z_i e$ at a distance r is directed along the line joining the two particles and has a magnitude F (given in SI units) of

$$F = \frac{Z_i Z_e e^2}{4\pi\varepsilon} \frac{1}{r^2}. \tag{5.220}$$

The quantity ε is the appropriate dielectric constant. (In cgs units, the quantity $4\pi\varepsilon$ should be replaced by ε.) For electrons, $Z_e = -1$, while for electron holes, $Z_e = +1$. If the electronic charge and the impurity charge are of opposite sign, the force is attractive. Since $\mathbf{F} = -\nabla\mathcal{U}$, in the present situation of a central force $F = -d\mathcal{U}/dr$. Because $\mathcal{U}(\mathbf{r})$ is the potential energy at position r with respect to the vacuum level (i.e., at $r = \infty$), we obtain

$$\mathscr{U}(r) = -\int_{\infty}^{r} F\,dr = -\int_{\infty}^{r} \frac{Z_i Z_e e^2}{4\pi\varepsilon r^2} = \frac{Z_i Z_e e^2}{4\pi\varepsilon r} \quad \textbf{(Coulomb potential).} \tag{5.221}$$

This is the appropriate potential energy for our problem of scattering by ionized impurities.

Equation (5.185) for the differential scattering cross section $\sigma(\theta, \phi)$ shows that we must now evaluate the triple integral

$$\mathscr{I} \equiv \int \mathscr{U}(r')e^{i(\mathbf{k}_i - \mathbf{k}_f)\cdot\mathbf{r}'}\,d\mathbf{r}', \tag{5.222}$$

where $\mathscr{U}(r)$ is the spherically symmetric Coulomb scattering potential given above. Since $\mathscr{U}(r')$ is spherically symmetric and the range of integration covers all space, the direction of the resultant vector $\mathbf{k}_i - \mathbf{k}_f$ with respect to the coordinate system chosen for the dummy variable $\mathbf{r}'$ will not affect the value of the integral. The integration can thus be carried out in a coordinate system $\hat{\mathbf{x}}'$, $\hat{\mathbf{y}}'$, $\hat{\mathbf{z}}'$ such that $\hat{\mathbf{z}}'$ is parallel to $\mathbf{k}_i - \mathbf{k}_f$. In the corresponding spherical polar coordinate system r', θ', ϕ' we will thus have

$$d\mathbf{r}' = r'^2 \sin\theta'\,dr'\,d\theta'\,d\phi', \tag{5.223}$$

$$(\mathbf{k}_i - \mathbf{k}_f)\cdot\mathbf{r}' = |\mathbf{k}_i - \mathbf{k}_f| r' \cos\theta', \tag{5.224}$$

$$\mathscr{U}(\mathbf{r}') = Z_i Z_e e^2/4\pi\varepsilon r'. \tag{5.225}$$

Hence

$$\mathscr{I} = \frac{Z_i Z_e e^2}{4\pi\varepsilon}\int_0^{2\pi} d\phi' \int_0^{\pi} d\theta' \int_0^{\infty} dr' \exp[i|\mathbf{k}_i - \mathbf{k}_f| r' \cos\theta'] r' \sin\theta'. \tag{5.226}$$

The integrand is independent of ϕ', so this integration can be performed immediately to yield a multiplicative factor of 2π. The integration over θ' is easily performed by making the variable change

$$\eta = \cos\theta', \tag{5.227}$$

$$d\eta = -\sin\theta'\,d\theta' \tag{5.228}$$

to give

$$\begin{aligned}\mathscr{I} &= \left(\frac{Z_i Z_e e^2}{4\pi\varepsilon}\right) 2\pi \int_0^{\infty} dr' \int_{-1}^{+1} d\eta\, e^{[i|\mathbf{k}_i - \mathbf{k}_f| r'\eta] r'} \\ &= \left(\frac{Z_i Z_e e^2}{4\pi\varepsilon}\right) \frac{2\pi}{i|\mathbf{k}_i - \mathbf{k}_f|} \int_0^{\infty} dr' [e^{i|\mathbf{k}_i - \mathbf{k}_f| r'} - e^{-i|\mathbf{k}_i - \mathbf{k}_f| r'}] \\ &= \frac{Z_i Z_e e^2}{\varepsilon |\mathbf{k}_i - \mathbf{k}_f|} \int_0^{\infty} dr' \sin[|\mathbf{k}_i - \mathbf{k}_f| r'].\end{aligned} \tag{5.229}$$

The scalar $|\mathbf{k}_i - \mathbf{k}_f|$ can be denoted by β, if desired, and the integral then has the form

$$\mathscr{K} = \int_0^\infty dr' \sin \beta r'. \tag{5.230}$$

It is apparent that the integrand is an undamped oscillatory function of r'. Convergence of such an integral is weak, so that it is convenient to consider $\mathscr{K}$ to be the limiting value of a different integral,

$$\mathscr{K} = \lim_{\alpha \to 0} \mathscr{K}', \tag{5.231}$$

where

$$\mathscr{K}' \equiv \int_0^\infty e^{-\alpha r'} \sin \beta r' \, dr' = \beta/(\alpha^2 + \beta^2). \tag{5.232}$$

Thus

$$\mathscr{K} = \beta^{-1} = [|\mathbf{k}_i - \mathbf{k}_f|]^{-1} \tag{5.233}$$

and we obtain

$$\mathscr{I} = Z_i Z_e e^2/\varepsilon|\mathbf{k}_i - \mathbf{k}_f|^2. \tag{5.234}$$

Substituting this result into Eq. (5.185) for the differential scattering cross section yields

$$\sigma(\theta, \phi) = \left(\frac{m^*}{2\pi\hbar^2}\right)^2 \left[\frac{Z_i Z_e e^2}{\varepsilon|\mathbf{k}_i - \mathbf{k}_f|^2}\right]^2 = \frac{Z_i^2 Z_e^2 e^4 m^{*2}}{4\pi^2\hbar^4\varepsilon^2|\mathbf{k}_i - \mathbf{k}_f|^4}. \tag{5.235}$$

[In cgs units we would need to multiply the right-hand side of this result by $(4\pi)^2$.]

The above expression can be converted to a form which involves the scattering angle between the vectors $\mathbf{k}_i$ and $\mathbf{k}_f$. Let us choose a coordinate system with $\mathbf{k}_i$ parallel to the z axis and with $\mathbf{k}_f$ at an angle θ with respect to the z axis, so that θ is both the polar angle and the scattering angle. The magnitude $|\mathbf{k}_i - \mathbf{k}_f|$ can be obtained by analytical geometry: the distance d between two points located at (x_1, y_1, z_1) and (x_2, y_2, z_2) is

$$d^2 = (x_1 - x_2)^2 + (y_1 - y_2)^2 + (z_1 - z_2)^2, \tag{5.236}$$

so that

$$|\mathbf{k}_i - \mathbf{k}_f|^2 = (0 - k_f \sin\theta \cos\phi)^2 + (0 - k_f \sin\theta \sin\phi)^2 + (k_i - k_f \cos\theta)^2. \tag{5.237}$$

The quantities k_i and k_f are $|\mathbf{k}_i|$ and $|\mathbf{k}_f|$, respectively, which are equal since we consider elastic scattering and spherical energy surfaces in $\mathbf{k}$ space. Thus

$$|\mathbf{k}_i - \mathbf{k}_f|^2 = k_i^2 [\sin^2\theta \cos^2\phi + \sin^2\theta \sin^2\phi + 1 - 2\cos\theta + \cos^2\theta]. \tag{5.238}$$

Using the relation $\sin^2 \alpha + \cos^2 \alpha = 1$ for arbitrary α first for ϕ and then for θ reduces this relation to

$$|\mathbf{k}_i - \mathbf{k}_f|^2 = 2k_i^2(1 - \cos\theta). \tag{5.239}$$

However,

$$\begin{aligned}\cos\theta = \cos 2(\tfrac{1}{2}\theta) &= \cos^2(\tfrac{1}{2}\theta) - \sin^2(\tfrac{1}{2}\theta)\\ &= [1 - \sin^2(\tfrac{1}{2}\theta)] - \sin^2(\tfrac{1}{2}\theta) = 1 - 2\sin^2(\tfrac{1}{2}\theta),\end{aligned} \tag{5.240}$$

so that

$$|\mathbf{k}_i - \mathbf{k}_f|^2 = 4k_i^2 \sin^2(\tfrac{1}{2}\theta) = 4k_i^2 \csc^{-2}(\tfrac{1}{2}\theta). \tag{5.241}$$

Substituting this result into the above expression for $\sigma(\theta, \phi)$ yields

$$\sigma(\theta, \phi) = (Z_i^2 Z_e^2 e^4 m^{*2}/64\pi^2 \hbar^4 \varepsilon^2 k_i^4) \csc^4(\tfrac{1}{2}\theta). \tag{5.242}$$

In terms of the electron energy $\mathscr{E} = \hbar^2 k_i^2/2m^*$, this expression becomes

$$\sigma(\theta, \phi) = (Z_i^2 Z_e^2 e^4/256\pi^2 \varepsilon^2 \mathscr{E}^2) \csc^4(\tfrac{1}{2}\theta). \tag{5.243}$$

Note that *Coulomb scattering is anisotropic, varying as* (*cosecant*)^4 *of half the scattering angle*. Note further that *the differential scattering cross section is energy dependent, decreasing quadratically with increasing energy*. The square of the dielectric constant ε reflects the screening effect of the dielectric medium, and the product $Z_i^2 Z_e^2$ reflects the effect of the charge magnitudes. Of course, Z_e^2 can be considered to be unity for the scattering of electrons and electron holes. (In cgs units, the right-hand side of Eq. (5.243) should be multiplied by $16\pi^2$.)

The total scattering cross section σ_{tot} for Coulomb scattering given by integrating the above expression for $\sigma(\theta, \phi)$ over all solid angles diverges. The Coulomb force is a long-range force, and all scattering has been considered to contribute to σ_{tot} irrespective of how small the scattering angle may be. This suggests choosing some minimum scattering angle θ_{min} and disregarding all deflections smaller than this, in which case σ_{tot} is finite. This is easily shown by using the results above,

$$\begin{aligned}\sigma_{tot} &= \int_0^{2\pi} d\phi \int_{\theta_{min}}^{\pi} d\theta\,(\sin\theta)\,\sigma(\theta, \phi)\\ &= 2\pi(Z_i^2 Z_e^2 e^4/256\pi^2 \varepsilon^2 \mathscr{E}^2) \int_{\theta_{min}}^{\pi} d\theta \sin\theta \csc^4(\tfrac{1}{2}\theta).\end{aligned} \tag{5.244}$$

However,

$$\begin{aligned}\sin\theta \csc^4(\tfrac{1}{2}\theta)\, d\theta &= \frac{\sin 2(\tfrac{1}{2}\theta)}{\sin^4(\tfrac{1}{2}\theta)}\, d\theta = \frac{2\sin(\tfrac{1}{2}\theta)\cos(\tfrac{1}{2}\theta)}{\sin^4(\tfrac{1}{2}\theta)}\, d\theta\\ &= 4\sin^{-3}(\tfrac{1}{2}\theta)\, d(\sin\tfrac{1}{2}\theta),\end{aligned} \tag{5.245}$$

which integrates to $-2\sin^{-2}(\tfrac{1}{2}\theta)$. Inserting the upper and lower limits of π and

θ_{min} gives $-2[1 - \sin^{-2}(\frac{1}{2}\theta_{min})]$, so that

$$\sigma_{tot} = (Z_i^2 Z_e^2 e^4/64\pi\varepsilon^2\mathscr{E}^2)[\sin^{-2}(\tfrac{1}{2}\theta_{min}) - 1]. \tag{5.246}$$

Clearly if θ_{min} is allowed to approach zero the quantity σ_{tot} will diverge.

There is in addition another factor, namely, anisotropic scattering does not completely randomize the velocity, so that a factor of $(1 - \cos\theta)$ is necessary in the integral for computing an effective relaxation time from the collision rate. Such a factor is still not sufficient to give a convergent result, since $1 - \cos\theta = 2\sin^2(\frac{1}{2}\theta)$, and the product of this factor with $4\sin^{-3}(\frac{1}{2}\theta)\, d(\sin\frac{1}{2}\theta)$ of the integral gives $8\sin^{-1}(\frac{1}{2}\theta)\, d(\sin\frac{1}{2}\theta)$. This integrates to $8\ln(\sin\frac{1}{2}\theta)$ which diverges at $\theta = 0$. Again we see the need for restricting the small angle scattering if a finite value is to be obtained for the total scattering cross section for a Coulomb potential. This can be done in a physically realistic way for many randomly located scattering centers contributing to the resistivity of a solid by considering the maximum impact parameter for the electrons with respect to impurities to be given by half the mean distance between impurities. This is the procedure used [see R. A. Smith (1963), for example] in the derivation of the *Conwell–Weisskopf* formula for the mobility in a solid under conditions for which ionized impurity scattering predominates.

A shorter-range scattering force is given by the *screened* Coulomb potential

$$\mathscr{U}(\mathbf{r}) \propto r^{-1}\exp(-\lambda r), \tag{5.247}$$

where the redistribution of conduction electrons [cf. Kittel (1971)] effectively nullifies the Coulomb potential at distances much greater than λ^{-1}. The matrix element of this potential is readily evaluated by modifying slightly the above procedure used for the ordinary Coulomb potential. The results for the screened Coulomb potential are useful for computing the relaxation time for conduction electrons in a metal, whereas the ordinary Coulomb potential is more appropriate for a low-conductivity semiconductor or an insulator.

To summarize, the techniques of time-independent and time-dependent perturbation theory developed in this chapter have enabled us to study both the effects of periodic lattice potentials due to the orderly array of ion cores and the effects of randomly located charged impurity scattering centers on the motion of conduction electrons previously considered from the viewpoint of the free-electron model (Chap. 3). The random scattering will of course produce attenuation of the conduction current commensurate with Fermi–Dirac statistics; the coherent scattering resulting from the perfectly periodic lattice potential was found to perturb the energy eigenvalues and to introduce a splitting of the energy levels for the subset of propagation vectors which map out certain planar surfaces (called Brillouin zone boundaries) in wave vector (or *momentum*) space. The energy level splitting at the Brillouin zone boundaries leads to the extremely important concept of *energy gaps*; these gaps alternate with *energy bands* that contain closely spaced energy eigenvalues associated with unattenuated conduction electron propagation in the solid. The introduction of Fermi–Dirac statistics (Chap. 2) or the Pauli exclusion principle to the energy band picture then leads us to the remarkable conclusion that the electrons in a fully occupied (or "filled") band cannot be excited into adjacent unoccupied energy levels by an externally applied force. This in turn prohibits any displacement of the Fermi surface in wave vector (or momentum) space with the concomitant production of a conduction current previously deduced for the free-electron model (Chap. 3, §3). We therefore reach the conclusion that *an externally applied force cannot produce a conduction current in the normal way if the electronic levels in the energy bands are completely filled or completely empty*! The quantum treatment of the periodic lattice potential thus enables us to understand the difference between solids which are good electrical conductors and solids which are electrical insulators. In the case of electrical conductors such as metals, which involve at least one energy band which is only partly filled, the resistivity is due entirely to *deviations from* the perfectly periodic potential. Such deviations can be produced for example, by thermal vibrations of the ion cores, or a random array of charged impurity scattering centers.

The fundamentals of crystal lattices and the development of the associated *reciprocal space* for the Fourier expansion of functions having the lattice periodicity are necessary topics for developing a deeper understanding of energy bands. These topics are developed systematically in the next two chapters (Chaps. 6 and 7), culminating in the construction of *Bloch functions* which provide a complete set of basis states for theoretical treatments involving propagating electrons in crystalline solids.

PROJECT 5.12 Conduction Electron Scattering by Screened Coulomb Potential of Charged Impurities

Use the screened Coulomb potential for conduction electron scattering by charged impurities in a metal to derive an expression for the conduction electron relaxation time. [*Hint*: See Eq. (5.247).]

PROJECT 5.13 Conwell–Weisskopf Formula for Mobility in a Semiconductor

Derive the Conwell–Weisskopf formula for the electron mobility in a low-conductivity semiconductor under conditions for which the resistivity is dominated by ionized impurity scattering.

PROJECT 5.14 Magnetoresistance

Give a treatment of the magnetoresistance of a solid based on the Boltzmann transport equation. Explain your results in physical terms.

PROBLEMS

1. Consider an unperturbed Hamiltonian $\mathscr{H}_0$ based on an infinite one-dimensional square-well potential,

$$\mathscr{U}(x) = \begin{cases} 0 & (0 \leqslant x \leqslant L) \\ \infty & (x < 0 \quad \text{and} \quad x > L). \end{cases}$$

Find the first-order correction to the lowest energy level $\mathscr{E}_1$ and the corresponding wave function for the following perturbation,

$$\mathscr{V}(x) = \begin{cases} V_0 & (0 \leqslant x \leqslant \delta) \\ 0 & (x < 0 \quad \text{and} \quad x > \delta). \end{cases}$$

In the above, V_0 is a constant energy and δ is a constant distance.

2. Consider an unperturbed Hamiltonian $\mathscr{H}_0$ based on an infinite one-dimensional square-well potential,

$$\mathscr{U}(x) = \begin{cases} 0 & (-\frac{1}{2}L \leqslant x \leqslant \frac{1}{2}L) \\ \infty & (|x| > \frac{1}{2}L). \end{cases}$$

Find the first-order corrections to the three lowest energy levels $\mathscr{E}_1$, $\mathscr{E}_2$, and $\mathscr{E}_3$ and the corresponding wave functions for perturbations $\mathscr{V}(x)$ of the type

(a) $$\mathscr{V}(x) = \begin{cases} V_0 & (-\delta \leqslant x \leqslant \delta < \frac{1}{2}L) \\ 0 & (|x| > \delta); \end{cases}$$

(b) $$\mathscr{V}(x) = \begin{cases} V_0[1 + (x/\delta)] & (-\delta \leqslant x \leqslant \delta < \frac{1}{2}L) \\ 0 & (|x| > \delta); \end{cases}$$

(c) $$\mathscr{V}(x) = \begin{cases} V_0(x/\delta) & (-\delta \leqslant x \leqslant \delta < \frac{1}{2}L) \\ 0 & (|x| > \delta). \end{cases}$$

In all these cases, V_0 is a constant energy and δ is a constant distance. Be sure to interpret your results physically!

3. In Problem 2, choose $L = 4$ Å, $\delta = 1$ Å, and $V_0 = 0.1$ eV. Evaluate the resulting shift in the lowest three energy levels, both in electron-volts and in relative percentages.

4. In Problems 2 and 3, evaluate quantitatively the first-order changes in the wave functions for the three lowest energy levels. Plot the perturbed and unperturbed wave functions so as to illustrate the admixture of adjacent states produced by the perturbations. Can you observe any general trend in the effect of perturbations of various symmetry on the different wave functions?

5. Attempt to formulate some general rules for the effect of perturbations of various symmetry on the energy levels and corresponding wave functions of even parity and odd parity, using as a tool the eigenstates of the one-dimensional infinite square-well potential problem. [*Hint*: Consider the effect of making a variable change (x to $-x$) in the integrals for the matrix elements of the perturbation operator. Keep in mind that any number which is equal to its negative must be zero.]

6. Develop the third-order stationary-state perturbation equations which lead to the corresponding perturbed stationary-state energies and corresponding perturbed eigenfunctions.

7. The *Schiff* symbol S implies a summation over discrete states and an integration over continuum states. Carry through the evaluation of the coefficients in Eq. (5.12) for this general case. [*Hint*: The discrete set has weighting factors a_j analogous to Fourier series coefficients for the harmonics. The continuum set has weighting factors $\chi(k)$ analogous to Fourier integral distribution functions for the wavelength content of the superposition.]

8. Deduce the third-order perturbation coefficients for the specific example of a constant perturbation turned on suddenly at $t = 0$, assuming the initial state of the system to be an arbitrary superposition stationary state of the system.

9. Compute the third-order perturbation corrections to energy eigenvalues and energy eigenfunctions for a harmonic perturbation turned on at $t = 0$.

10. Use Fermi's Golden Rule to deduce as much information as you can for the physical situation in which a uniform time-dependent electric field is allowed to perturb a hydrogen atom which at time $t = 0$ is in the ground state. In particular, what is the minimum frequency of the electric field required to ionize the atom (i.e., promote a transition from the ground state to an unbound state characterized by a plane-wave eigenfunction appropriate in free space)? (For an in-depth treatment of this problem, see (7.73) in ter Haar (1975).)

PART IV

Energy Bands in Crystals

CHAPTER 6

THE PERIODICITY OF CRYSTALLINE SOLIDS

We shall regard the perfect solid as an aggregate of atoms arranged in unbroken lattice array. F. Seitz (1950)

1 Generalities

1.1 Prologue

The apex of our application of the discipline of quantum mechanics to gain a basic insight into the fundamental nature of the electronic properties of solid-state materials will be the development of Bloch functions and the proof of Bloch's theorem (Chap. 7). Before descending into the very bowels of solid-state theory, it is imperative to make sure that (a) we have a clear understanding of the essentials of crystal lattices, and (b) we understand how to make three-dimensional Fourier-series expansions of periodic functions that have the symmetry of the crystal lattice. It is to these ends that the present concise chapter is devoted.

1.2 Crystalline and Amorphous Solids

Solids are most likely to be found in the *crystalline* state, although amorphous forms are not uncommon. Crystalline solids are characterized by a regular three-dimensional pattern for the location of the atoms making up the solid; this three-dimensional pattern is made by the continual repetition (or stacking) of a small basic arrangement of atoms (i.e., the unit cell) to fill a three-dimensional space. The perfect crystalline solid (viz., the *single crystal* or *monocrystal*) has the property known as *translational invariance*, as illustrated by Fig. 6.1. This means that translation of the atoms comprising the solid by certain elementary distances (related to the size of the unit cell) in certain directions leads to the same arrangement of atoms in space throughout the solid, except of course at the surfaces which separate the solid from surrounding free space.

Figure 6.2 illustrates a typical arrangement of atoms in an amorphous material. The amorphous state is not completely without order; however, in

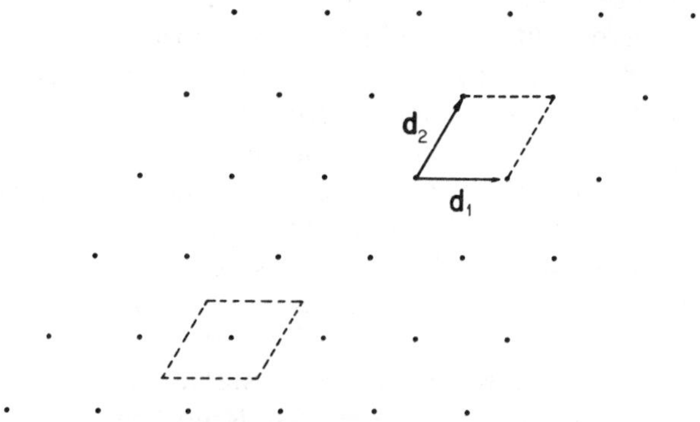

Fig. 6.1 Translational invariance of a two-dimensional lattice of points in space. [Note that translation of all points through the vector distance $\mathbf{d}_1$ or $\mathbf{d}_2$, or any integer multiple of such, unaffects the appearance of an array of such points extending to infinity in both directions. All points in the lattice can be mapped out by the set of vectors $\mathbf{R_m} = m_1\mathbf{d}_1 + m_2\mathbf{d}_2$ (m_1, m_2 integers). The two parallelograms indicate two choices for the elemental *unit cell* of the lattice; there is an average of one lattice point per unit cell. The entire two-dimensional space can be filled by the contiguous stacking of any one type of unit cell. These concepts are readily extended to a three-dimensional lattice of points, the corresponding unit cell being a parallelepiped.]

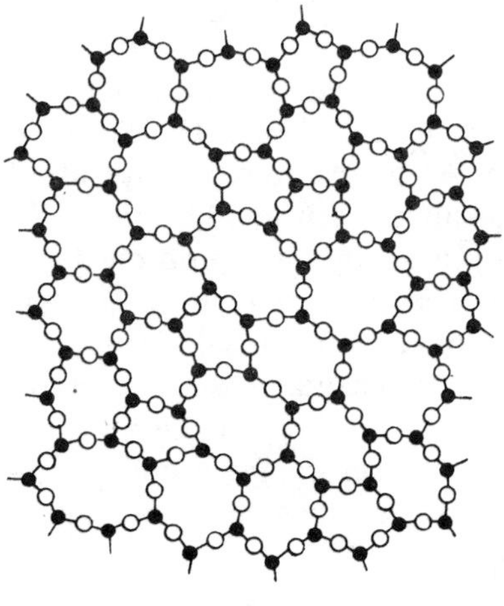

Fig. 6.2 Atom arrangement in an amorphous material lacks translational invariance, even though short-range ordering can be noted.

contrast to the crystalline state, the amorphous state does not have the property of translational invariance. This can be due to the fact that the basic unit cells are arranged somewhat randomly with respect to each other instead of being stacked in a contiguous regular array. The energy of the solid is generally higher for the amorphous state, so there is a thermodynamic tendency for the amorphous solid to change to the crystalline state. However, the viscosity of the amorphous solid at room temperature is generally so large that for all practical purposes this ordering process takes an infinitely long time. The amorphous form is thus frozen (or "quenched in") with respect to a *laboratory* time frame.

Somewhat intermediate to cases of monocrystalline and amorphous solids is the case of *polycrystalline* materials. A polycrystal consists of a large number of randomly oriented monocrystals, with thin intervening regions called *grain boundaries*. The polycrystalline state differs from the amorphous state in that there exists translational invariance within the confines of the macroscopic monocrystals comprising the polycrystalline substance, whereas in the amorphous material there is no translational invariance to be found on a macroscopic scale.

Examples of readily procured monocrystals are, for example, NaCl and LiF, and also pure copper and pure silver. The crystals must be prepared by means of an appropriate procedure from the melted material: Unless special care is given during formation of the sample, even these substances will be polycrystalline. For example, table salt and rolled copper sheets are polycrystalline. Examples of amorphous substances are various types of common glass, carbon when formed as a low-temperature decomposition product, and some of the oxides formed on certain metals.

The amorphous and crystalline forms of a solid often have quite different physical and electrical properties. One of the most spectacular examples is provided by carbon (cf. Table 1.4), which can vary in physical properties from those associated with powdery carbon black to those possessed by the hard brilliant diamond form of the same elemental material.

Even a monocrystal can depart somewhat from the condition of strict translational invariance. At temperatures above absolute zero, the random oscillations of the atoms about their equilibrium positions due to the kinetic energy of thermal motion causes the configuration at any instant of time to have a small deviation from perfect periodicity, even though the *time average* of the instantaneous atom positions may be spatially periodic. In addition, a variety of *point defects* are found experimentally, such as impurity atoms, misplaced atoms, and vacant positions. There are also extended defects known as *dislocations* which represent microscopic atomic configurations made up of atoms which have been misplaced in some appropriately regular manner. For further details, see Kittel (1971).

2 Unit Cells and Bravais Lattices

It is important to develop the mathematical consequences of the *periodicity* (i.e., the *translational invariance*) of perfect single crystals. The smallest distance over which a crystal structure is repetitive in the major symmetry directions gives us an intuitive understanding of the size of the basic *unit cell* (or elemental building block) for the crystal under consideration. The repetition distance along three different symmetry directions can be used to define *elementary translation vectors* $\mathbf{d}_1$, $\mathbf{d}_2$, $\mathbf{d}_3$ in these directions. Each unit cell in the crystal can be considered to be a parallelepiped with edges defined by the elementary

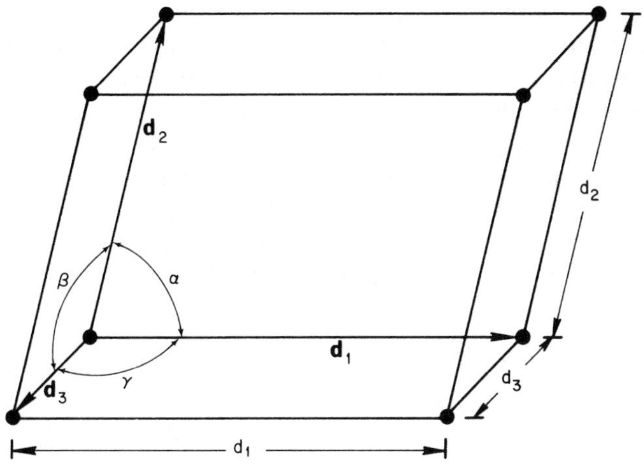

Fig. 6.3 Elemental parallelepiped unit cell characterized by the elementary translation vectors $\mathbf{d}_1$, $\mathbf{d}_2$, and $\mathbf{d}_3$ oriented at specific angles. [Translation of this unit cell through all distances $\mathbf{R}_\mathbf{m} = m_1\mathbf{d}_1 + m_2\mathbf{d}_2 + m_3\mathbf{d}_3$ (m_1, m_2, m_3 integers) maps out a three-dimensional space lattice.]

translation vectors (cf. Fig. 6.3). The crystal can then be considered to be built up by the adjacent stacking of such parallelepipeds until the volume of the crystal is filled. The volume of each unit cell v_c in the crystal is simply the parallelepiped volume

$$v_c = |\mathbf{d}_1 \cdot (\mathbf{d}_2 \times \mathbf{d}_3)|. \tag{6.1}$$

EXERCISE Prove relation (6.1) for v_c.

A *lattice* is a set of periodic points in space with coordinates determined by all integral multiples of the elementary translation vectors $\mathbf{d}_j$. For example,

$$\mathbf{R}_\mathbf{m} = m_1\mathbf{d}_1 + m_2\mathbf{d}_2 + m_3\mathbf{d}_3, \tag{6.2}$$

where $\mathbf{m} = (m_1, m_2, m_3)$ represents a triplet of arbitrary integers, can be considered to map out a lattice. Likewise, $\mathbf{R}_\mathbf{m} + \mathbf{R}'$, where $\mathbf{R}'$ is some fixed position vector in space, also maps out a lattice. (This second mapping is not so convenient because in this system of coordinates there is generally no lattice point at the origin.) Each unit cell with edges defined by $\mathbf{d}_1$, $\mathbf{d}_2$, $\mathbf{d}_3$ contains on the average a single lattice point; unit cells with this property are called *primitive*.

EXERCISE Construct a primitive unit cell for copper.

A *lattice translation operator* $T_\mathbf{j}$ can be defined which indicates a translation of the lattice by the vector $j_1\mathbf{d}_1 + j_2\mathbf{d}_2 + j_3\mathbf{d}_3$. This can be indicated by writing

$$T_\mathbf{j} \rightarrow j_1\mathbf{d}_1 + j_2\mathbf{d}_2 + j_3\mathbf{d}_3. \tag{6.3}$$

The lattice is *invariant* under this operation, which means that after translation the lattice appears to be exactly the same as before translation, even to the extent that the same spatial positions are occupied by lattice points.

There are other symmetry operators such as rotations and reflections which also leave the lattice invariant. These are designated *point operations*. Point symmetry operations provide an important tool for the classification of crystal structures. For example, a lattice which transforms into itself when rotated around some axis by an angle $(1/n)360°$ is said to have an *n-fold rotation axis*.

It is evident that a single primitive unit cell can be translated repeatedly throughout all space. In this way the complete lattice can be generated. This is evident in the two-dimensional lattice shown in the sketch (Fig. 6.1). It can also be noted that there is some arbitrariness in the choice of a primitive unit cell.

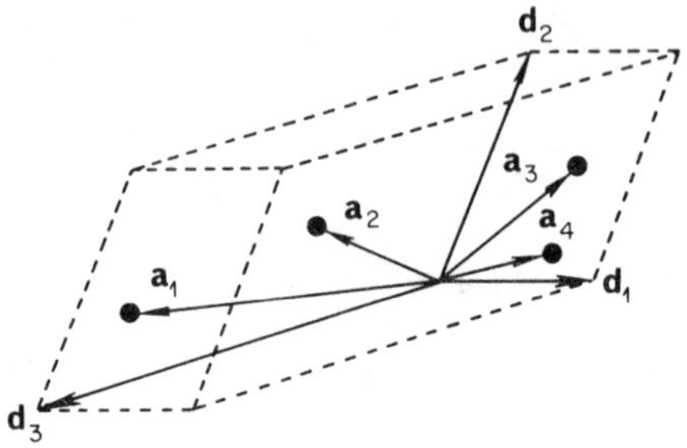

Fig. 6.4 Basis vectors $\mathbf{a}_j$ locate atom positions *within* each unit cell.

A *basis* is a set of vectors $\mathbf{a}_1, \mathbf{a}_2, \ldots$ (see Fig. 6.4) which locate the positions of the various atoms within each unit cell. The basis is of course the same for each unit cell in the crystal; otherwise, the cells considered would not be *unit* cells. The number of atoms in a unit cell of a crystal is equal to the number of atoms in the basis.

Bravais first introduced the mathematical concept of the lattice in 1848. He showed that in three dimensions there exists fourteen types. The classification is based on the elementary translation vectors $\mathbf{d}_1$, $\mathbf{d}_2$, $\mathbf{d}_3$ and the angles α, β, γ

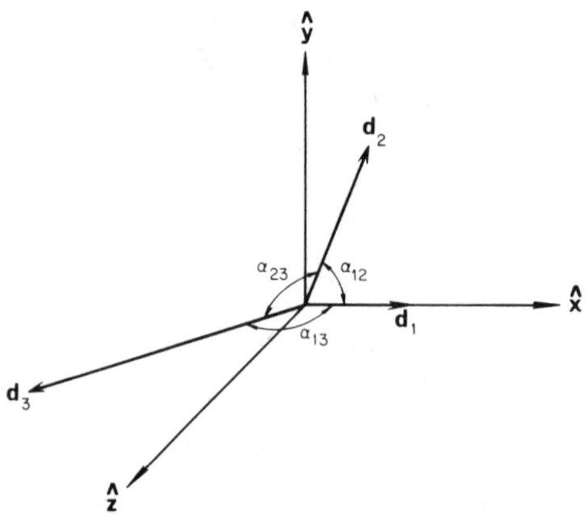

Fig. 6.5 Elementary translation vectors $\mathbf{d}_1, \mathbf{d}_2, \mathbf{d}_3$ and associated angles α_{12}, α_{13}, and α_{23} used in the classification of lattice types in Table 6.1, where $\alpha = \alpha_{12}$, $\beta = \alpha_{23}$, and $\gamma = \alpha_{13}$.

Table 6.1

The Seven Crystal Systems Incorporating Fourteen Bravais Lattices

Crystal system	Characteristic symmetry	Bravais lattice	Unit cell parameters
1. Triclinic	None	Simple	$d_1 \neq d_2 \neq d_3$ $\alpha \neq \beta \neq \gamma \neq 90^\circ$
2. Monoclinic	One 2-fold rotation axis	Simple Base centered	$d_1 \neq d_2 \neq d_3$ $\alpha = \beta = 90^\circ$ $\gamma \neq 90^\circ$
3. Tetragonal	One 4-fold rotation axis (or a 4-fold rotation-inversion axis)	Simple Body centered	$d_1 = d_2 \neq d_3$ $\alpha = \beta = \gamma = 90^\circ$
4. Trigonal (Rhombohedral)	One 3-fold rotation axis	Simple	$d_1 = d_2 = d_3$ $\alpha = \beta = \gamma \neq 90^\circ$
5. Hexagonal	One 6-fold rotation axis	Simple	$d_1 = d_2 \neq d_3$ $\alpha = \beta = 90^\circ$ $\gamma = 120^\circ$
6. Orthorhombic	Three mutually perpendicular 2-fold rotation axes	Simple Body centered Base centered Face centered	$d_1 \neq d_2 \neq d_3$ $\alpha = \beta = \gamma = 90^\circ$
7. Cubic	Four 3-fold rotation axes (along cube diagonals)	Simple Body centered Face centered	$d_1 = d_2 = d_3$ $\alpha = \beta = \gamma = 90^\circ$

between these vectors (cf. Table 6.1 and Fig. 6.5). If $d_1 \neq d_2 \neq d_3$ (where $d_j \equiv |\mathbf{d}_j|$), and $\alpha \neq \beta \neq \gamma$, the symmetry is *triclinic*. Whenever two of the angles are 90° but the third is not, with $d_1 \neq d_2 \neq d_3$, the symmetry is *monoclinic*. For all three angles equal to 90° with $d_1 \neq d_2 \neq d_3$, the symmetry is designated *orthorhombic*.

If there are lattice points only at the corners of the parallelepiped cells, they are called *simple*. There may be in addition lattice points located at the center of the cell (*body centered*), or located at the centers of two opposite faces (*base centered*), or located at the centers of all six faces (*face centered*). This leads to a *monoclinic base-centered* lattice in addition to the *monoclinic simple* lattice; similarly, there exist *orthorhombic simple*, *orthorhombic base-centered*, *orthorhombic body-centered*, and *orthorhombic face-centered* lattices, as can be noted in Fig. 6.6. This gives a subtotal of seven Bravais lattices for which $d_1 \neq d_2 \neq d_3$. The apparently missing ones (e.g., monoclinic body centered) can be represented by one of the basic types listed (e.g., the simple triclinic).

In addition to the subset of seven lattices thus far described, there exist seven others. The most complex of these is the *hexagonal*, which has one sixfold rotation axis. Consider a planar hexagon with the six edges, each having length *a*, with points at each of the six corners plus one point in the center; if one such

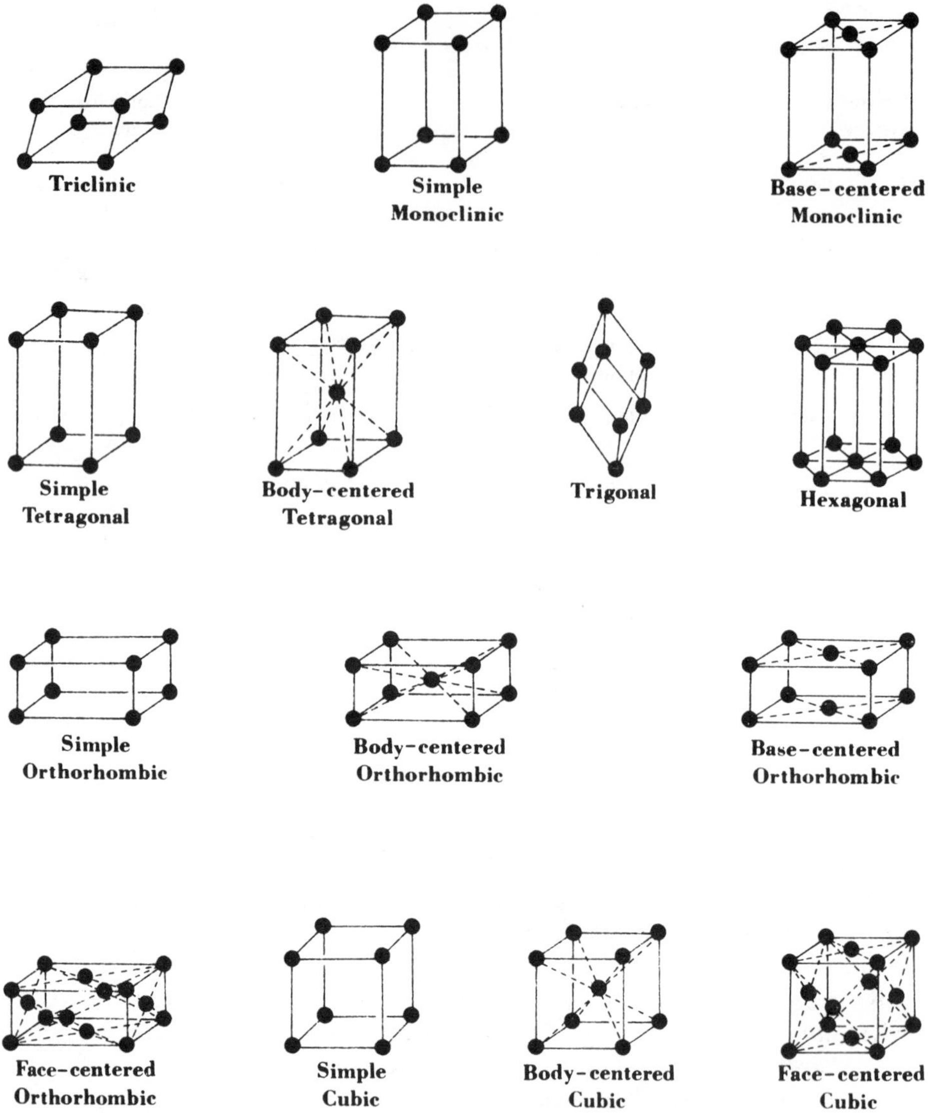

Fig. 6.6 The fourteen types of three-dimensional Bravais lattices. (Refer to Table 6.1 for specific relations with respect to the lengths of the elementary translation vectors and among the angles.)

hexagon is placed above another, with the separation distance being c (with $c \neq a$), the resulting cell is a hexagonal structure which represents a compound cell for the hexagonal lattice. This cell can be considered to be defined by three equal coplanar axes of length a (which make angles of 120° with respect to each other) and a fourth axis of length c which is perpendicular to the other three. The division of each hexagon into three equal parts obtained by drawing three radial lines from the center point to the corners to form parallelograms then yields three primitive cells for this lattice. (See the hexagonal unit cell in Fig. 6.6.)

The remaining Bravais lattices (see Fig. 6.6) have at least two equal values for the parameters d_1, d_2, d_3. The *tetragonal* lattice has $d_1 = d_2 \neq d_3$ and all three angles equal to 90°. (It differs from the orthorhombic lattice discussed above only insofar as it has two equal values for the d_j.) The tetragonal lattice has the *simple* and the *body-centered* forms. The *cubic* lattice has $d_1 = d_2 = d_3$ and all angles equal to 90°; it can be considered to have the *simple*, the *body-centered*, and the *face-centered* forms. The last of fourteen Bravais lattices is the *rhombohedral*, for which $d_1 = d_2 = d_3$ and $\alpha = \beta = \gamma \neq 90°$; it thus can be considered to be a cubic lattice skewed along the body diagonal.

EXERCISE Sketch the fourteen above-listed Bravais lattices and circle the lattice points. (*Hint*: See Fig. 6.6.)

EXERCISE Deduce the various n-fold rotation axes for the fourteen Bravais lattices. (*Hint*: It sometimes helps if you stand on your head inside the lattice!)

EXERCISE Make up your own table listing the pertinent information regarding relative angles, elementary lattice translation vector lengths, and the n-fold rotation axes for the Bravais lattices. (*Hint*: See Table 6.1.)

EXERCISE Is the diamond structure included as one of the fourteen Bravais lattices? If not, show how it can be obtained by adding an appropriate basis to one of the Bravais lattices. Extend your considerations to the crystal structures of silicon and germanium.

3 Miller Indices and Crystal Directions

Suppose it is observed that a plane can be passed through a lattice such that it intercepts certain groups of points in the lattice. Two such planes are illustrated in Fig. 6.7. It is not difficult to visualize such a plane, given any specific lattice. The intercepted points make a periodic geometric pattern in the plane due to the periodicity of the points in the lattice. It is a matter of convenience in discussing crystals to have a system for defining the location of such planes. For example, if points of intersection of a plane with the x, y, and z axes of a Cartesian coordinate system are given, the location of the plane in space is defined. For a crystal, however, it makes more sense to choose a coordinate system having the three axes along the major symmetry directions of the crystal, even if these axes are not orthogonal. Also it is convenient to measure distances in units of the elementary translation vector lengths d_1, d_2, d_3. The location of any given crystal plane can then be determined by giving the intercepts of the plane in such a coordinate system. The system designated as *Miller indices* is based on taking the *reciprocal* of the intercepts in this coordinate system and then reducing the result to the three lowest integers (hkl) having the same ratio. It is assumed that the origin of the coordinate system is chosen at one of the lattice sites.

EXERCISE Deduce Miller indices for several major planes in the three cubic Bravais lattices.

EXERCISE Show that parallel planes placed in *equivalent* positions in a crystal lattice can have the same set of Miller indices (hkl).

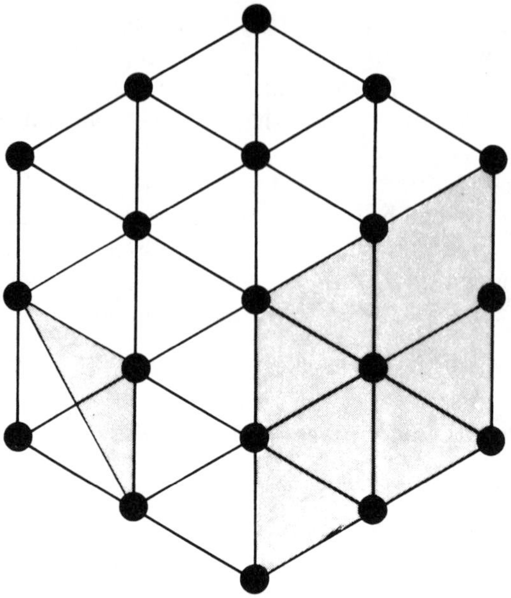

Fig. 6.7 A simple cubic array of lattice points. [The two shaded areas indicate two nonparallel planes which contain a high density of lattice points; in Miller index notation, the plane indicated by the rectangular area is called a (100) plane and the plane indicated by the triangular area is called a (111) plane. Each of these planes is a member of a set of parallel planes which contain all points of the lattice. Equivalent sets of parallel planes exist; for example, in the simple cubic lattice the (001) and the (100) planes are equivalent in the sense that each has the same density of lattice points and the same geometrical arrangement of such points.]

If one of the integers h, k, l is negative, it is conventional to place the minus sign above (instead of in front of) the integer. Due to the symmetry of the lattice, it is often the case that planes which are equivalent from a symmetry standpoint may have somewhat different Miller indices. Planes equivalent by symmetry are denoted by $\{hkl\}$, or simply hkl.

EXERCISE Show that the (100) and the (001) planes in a cubic crystal are equivalent in the sense of having the same lattice point density and geometrical arrangement.

The matter of defining crystal directions is much simpler than that of locating planes. Again it proves convenient to use a coordinate system with axes pointed along major crystal directions, with distances measured in units of the corresponding elementary translation vector lengths d_1, d_2, d_3. Direction is then determined by the triplet of numbers $[hkl]$ which are needed to orient the vector $h\mathbf{d}_1 + k\mathbf{d}_2 + l\mathbf{d}_3$ along the direction in question. If the origin of the coordinate system is chosen to be at a lattice point, then lines passing along directions for which h, k, l are integers intercept a regular linear array of points in the lattice. In this way any specific line of lattice points (or atoms in a crystal) can be indicated. Note that no reciprocals have been introduced in characterizing the crystal direction, in contrast to the Miller index system of locating crystal planes. A negative index for crystal direction is indicated by placing a bar over the integer.

EXERCISE Show that the [*hkl*] direction is perpendicular to the (*hkl*) plane in a cubic lattice.

4 Some Specific Crystal Structures

It is interesting to list crystal structures of some commonly found substances. Copper, silver, and gold are face-centered cubic (fcc) with lattice parameters of 3.61, 4.08, and 4.07 Å. [Note the positions of these elements in the Periodic Table (Table 1.4).] Magnesium has the hexagonal close-packed (hcp) structure with $a = 3.20$ and $c = 5.20$ Å. Iron is body-centered cubic (bcc) below 910°C with the lattice parameter being 2.86 Å. Between 910 and 1400°C, iron is fcc; above 1400°C, iron is bcc. Cobalt is hcp at temperatures below 100°C but fcc at higher temperatures. Tantalum has a bcc structure, with a lattice constant of 3.30 Å. Sodium metal is bcc with a lattice constant of 4.28 Å. The compound NaCl consists in two interpenetrating face-centered cubic lattices, one for sodium ions and one for chlorine ions. The resulting arrangement of lattice sites including both ionic types is simple cubic (sc). The size of each fcc unit cell is 5.63 Å. Cesium chloride has a different crystal structure from NaCl; it consists of a simple cubic lattice of cesium ions with a chlorine ion located in the center of each unit cell. The cesium chloride structure can be viewed as two interpenetrating simple cubic structures, each having a lattice constant of 4.11 Å.

5 Crystal Bonding

Electrical forces are responsible for the bonding of atoms in a solid. The *cohesive energy* is the difference in the total energy of a collection of neutral free atoms which are both stationary and greatly separated and the total energy of the solid obtained by the condensation of this ensemble of atoms. The cohesive energy is defined as the magnitude of the *free energy of formation* of the crystal. In the condensed state, some of the electrons have acquired a translational kinetic energy in addition to having a modified potential energy. It is convenient to examine crystal binding on the basis of

(a) the various types of electrical forces which can predominate in a solid (such as monopolar and dipolar forces);
(b) the distribution of electrons with respect to the ions within the solid; and
(c) the magnitude of the cohesive energy (which reflects the strength of the bonding).

The electric forces in *ionic crystals* are primarily due to the Coulomb interaction between electrical monopoles, the monopoles being the negative and positive ions making up the crystal. Sodium chloride, for example, can as a first approximation be considered to be a collection of individual Na^+ and Cl^- ions arranged in a periodic array. Ionic crystals have a relatively strong bond; they are characterized by a relatively high melting point, a low coefficient of thermal expansion, and a high degree of hardness.

The electrical forces in inert gas crystals, on the other hand, are primarily due to the Coulomb interaction between electrical dipoles. The dipoles are produced by the charge separation of the positive core (comprising nucleus plus inner shell electrons) from the negative outer shell electrons at any given time. Classically an outer shell electron can be viewed as a point negative charge traveling in a closed path around the inner positive core. This can be considered to be an electric dipole charge configuration which is continuously changing its spatial orientation. The force between such fluctuating electric dipoles is known as the *Van der Waals interaction.* This force decreases rapidly with increasing separation distance between atoms because the charge polarization itself depends upon the dipolar electric field of the adjacent polarized atom.

Electrical dipolar fields fall off with distance as r^{-3}, whereas the electrical monopolar fields fall off as r^{-2}. The bonding in ionic crystals is therefore much stronger than that in an inert gas crystal, and the cohesive energy in the ionic crystal is correspondingly larger. For example, the experimental cohesive energy for argon is approximately 1.85 kcal/mole; the corresponding energy for NaCl is approximately 185 kcal/mole! The physical properties of a crystal (such as melting point, hardness, and tensile strength) likewise depend upon the nature of the electrical bond. For example, the melting point of argon is 84°K whereas that for NaCl is 1074°K.

Implicitly we have assumed above that electrons are readily transferred from the metallic to the nonmetallic constituent in an ionic crystal such as NaCl (i.e., from Na to Cl), whereas we have assumed that a given outer shell electron is tightly held in its orbit around the parent core ion in rare gas crystals. Thus item (b) as well as items (a) and (c) in the above listing has been invoked in distinguishing ionic crystals from crystals made up of condensed rare gas atoms.

Metallic bonding differs from the types mentioned above. In a metal the outermost electrons are nearly free from the parent ions, and these *free electrons* (Chap. 3) for the most part determine the physical properties. Metals have high electrical and thermal conductivities.

Covalent bonds are characterized by the mutual sharing of a pair of electrons of opposite spin located in the region between two atomic constituents of a solid. The bond is highly directional, but apparently is not well characterized by any particular one of the ordinary physical properties such as hardness, melting point, or electrical conductivity. Examples of covalently bonded substances include the elemental solids C, Sn, Pb, Si, and Ge. [Refer to the Periodic Table (Table 1.4) to locate the positions of these elements. The electronic configurations of the free atom can be obtained from Table 1.3.]

6 The Reciprocal Lattice: Fourier Space for Arbitrary Functions That Have the Lattice Periodicity

The above discussion of electronic bonding leads to a consideration of periodic electron densities in the solid. We now attack the general theoretical problem of providing a suitable mathematical framework for describing arbitrary functions which have the lattice periodicity, such as indicated in Fig. 6.8. What we need is a suitable generalization of the three-dimensional Fourier series for orthogonal coordinates using the development in §4 of Chap. 1.

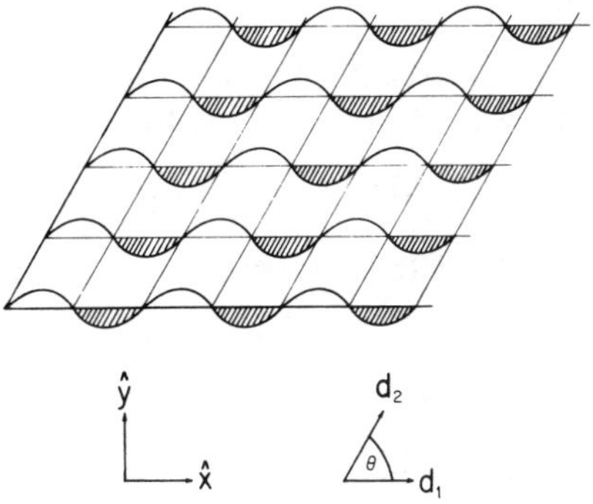

Fig. 6.8 Representation of a function which is periodic along two nonorthogonal directions denoted by $\mathbf{d}_1$ and $\mathbf{d}_2$.

Let us first define vectors $\mathbf{b}_1$, $\mathbf{b}_2$, $\mathbf{b}_3$ in terms of the elementary lattice translation vectors as follows,

$$\mathbf{b}_1 \equiv v_c^{-1}\, \mathbf{d}_2 \times \mathbf{d}_3, \tag{6.4}$$

$$\mathbf{b}_2 \equiv v_c^{-1}\, \mathbf{d}_3 \times \mathbf{d}_1, \tag{6.5}$$

$$\mathbf{b}_3 \equiv v_c^{-1}\, \mathbf{d}_1 \times \mathbf{d}_2, \tag{6.6}$$

where

$$v_c^{-1} = 1/(\mathbf{d}_1 \cdot \mathbf{d}_2 \times \mathbf{d}_3). \tag{6.7}$$

Note that these vectors $\mathbf{b}_j$ ($j = 1, 2, 3$) satisfy the conditions

$$\mathbf{b}_j \cdot \mathbf{d}_l = \delta_{jl} \qquad (l = 1, 2, 3), \tag{6.8}$$

where $\delta_{jl} = 1$ if $j = l$ but is zero otherwise. These relations hold even if $\mathbf{d}_1$, $\mathbf{d}_2$, $\mathbf{d}_3$ are nonorthogonal. The *reciprocal lattice* is then mapped out by the *reciprocal lattice vectors* $\mathbf{G}_l$ defined by

$$\mathbf{G}_l = 2\pi(l_1\mathbf{b}_1 + l_2\mathbf{b}_2 + l_3\mathbf{b}_3) \qquad \textbf{(reciprocal lattice vectors)}, \tag{6.9}$$

where $\boldsymbol{l}$ represents any arbitrary triplet of integers l_1, l_2, l_3. Often $\mathbf{G}_l$ will be found abbreviated simply as $\mathbf{G}$.

As in the case of the direct lattice, the reciprocal lattice is made up of contiguous primitive parallelepiped unit cells. Let us consider a coordinate system in reciprocal space with the origin located at one of the reciprocal lattice sites. (This is analogous to the coordinate system most frequently utilized for the direct lattice.) We then label the specific parallelepiped primitive unit cell within the first octant (i.e., all coordinates positive in sign) of reciprocal space and having one corner at the origin as the *initial parallelepiped reciprocal cell.* The entire reciprocal lattice can then be reproduced by translating the initial parallelepiped reciprocal cell by means of the reciprocal lattice vectors defined by Eq. (6.9).

EXERCISE Show that the reciprocal lattice of a fcc lattice is bcc, and vice versa.

Let us now examine some of the properties of the vectors $\mathbf{G}_l$. If $\mathbf{R}_\mathbf{m}$ represents some arbitrary vector between lattice points, then

$$\begin{aligned}\mathbf{G}_l \cdot \mathbf{R}_\mathbf{m} &= 2\pi(l_1\mathbf{b}_1 + l_2\mathbf{b}_2 + l_3\mathbf{b}_3)\cdot(m_1\mathbf{d}_1 + m_2\mathbf{d}_2 + m_3\mathbf{d}_3)\\ &= 2\pi(l_1m_1 + l_2m_2 + l_3m_3). \end{aligned} \tag{6.10}$$

Therefore

$$\exp(i\mathbf{G}_l \cdot \mathbf{R}_\mathbf{m}) = 1 \tag{6.11}$$

for arbitrary integer triplets $\boldsymbol{l}$ and $\mathbf{m}$. This also has significance from the standpoint of the lattice translation operator $T_\mathbf{j}$, since

$$T_\mathbf{j}(i\mathbf{G}_l \cdot \mathbf{r}) = i\mathbf{G}_l \cdot (\mathbf{r} + \mathbf{R}_\mathbf{j}) = i\mathbf{G}_l \cdot \mathbf{r} + i\mathbf{G}_l \cdot \mathbf{R}_\mathbf{j}, \tag{6.12}$$

which shows us that

$$T_\mathbf{j}\exp(i\mathbf{G}_l \cdot \mathbf{r}) = \exp(i\mathbf{G}_l \cdot \mathbf{r} + i\mathbf{G}_l \cdot \mathbf{R}_\mathbf{j}) = \exp(i\mathbf{G}_l \cdot \mathbf{r}). \tag{6.13}$$

That is, all functions $\exp(i\mathbf{G}_l \cdot \mathbf{r})$ are invariant under all possible lattice translations. The importance becomes apparent when one recognizes that this feature is in common with that in the complex basis states

$$\exp(i\mathbf{K}_\mathbf{n} \cdot \mathbf{r}) \equiv \exp\{i2\pi[(n_1x/\Lambda_1) + (n_2y/\Lambda_2) + (n_3z/\Lambda_3)]\} \tag{6.14}$$

for a three-dimensional Fourier expansion of an arbitrary periodic function with a fundamental periodicity Λ_1 in the x direction, Λ_2 in the y direction, and Λ_3 in the z direction. That is, for any set of integers n_1, n_2, n_3, an increase of x by Λ_1 leads to no change in the value of the function, and similarly for an increase in y by Λ_2 and an increase in z by Λ_3.

Let us now briefly consider the form of a three-dimensional Fourier series in an orthogonal coordinate system. In one dimension the complex Fourier series for an arbitrary periodic function $f(x)$ with periodicity Λ_1 is as follows [Chap. 1, §4.2, Eq. (1.40)],

$$f(x) = \sum_{n=-\infty}^{\infty} C_n \exp(i2\pi nx/\Lambda_1), \tag{6.15}$$

with

$$C_n = \frac{1}{\Lambda_1}\int_{x_0}^{x_0+\Lambda_1} f(x)\exp(-i2\pi nx/\Lambda_1)\,dx \qquad (n = 0, \pm 1, \pm 2, \ldots). \tag{6.16}$$

The distance x_0 is arbitrary and so can be chosen to be zero. The basis functions $\exp(i2\pi nx/\Lambda_1)$ represent a complete set of orthogonal functions. For a function

$f(\mathbf{r})$ which is periodic in three dimensions, either of two approaches can be used. The C_n in the above expansion can be considered to be periodic functions of y and thus Fourier expanded in a similar manner, and this process then repeated with all coefficients considered to be periodic functions of z. A somewhat different approach can be based on the fact that the product function $f_1(x)f_2(y)f_3(z) \equiv f(\mathbf{r})$ is periodic in the x, y, and z directions if $f_1(x)$ is periodic in x with periodicity Λ_1, $f_2(y)$ is periodic in y with periodicity Λ_2, and $f_3(z)$ is periodic in z with periodicity Λ_3. The resulting product Fourier series obtained simply by direct multiplication can then be written

$$f(\mathbf{r}) = \sum_{n_1=-\infty}^{\infty} \sum_{n_2=-\infty}^{\infty} \sum_{n_3=-\infty}^{\infty} C_{n_1n_2n_3} \exp\{i2\pi[(n_1x/\Lambda_1) + (n_2y/\Lambda_2) + (n_3z/\Lambda_3)]\}, \tag{6.17}$$

where

$$C_{n_1n_2n_3} \equiv C_{n_1}C_{n_2}C_{n_3}, \tag{6.18}$$

with

$$C_{n_j} = \frac{1}{\Lambda_j}\int_0^{\Lambda_j} f_j(\zeta) \exp(-i2\pi n_j\zeta/\Lambda_j)\, d\zeta \qquad (j = 1, 2, 3). \tag{6.19}$$

An apparent simplification can be effected by using the vector notation

$$\mathbf{K_n} \equiv 2\pi[(n_1/\Lambda_1)\hat{\mathbf{x}} + (n_2/\Lambda_2)\hat{\mathbf{y}} + (n_3/\Lambda_3)\hat{\mathbf{z}}], \tag{6.20}$$

since $f(\mathbf{r})$ can then be written as

$$f(\mathbf{r}) = \sum_{\mathbf{n}} C_{\mathbf{n}} \exp(i\mathbf{K_n}\cdot\mathbf{r}), \tag{6.21}$$

with

$$C_{\mathbf{n}} \equiv C_{n_1n_2n_3} = v^{-1}\int_{\text{cell}} f(\mathbf{r}) \exp(-i\mathbf{K_n}\cdot\mathbf{r})\, d\Omega, \tag{6.22}$$

where $d\Omega$ is the volume element

$$d\Omega \equiv dx\, dy\, dz \tag{6.23}$$

and v is the volume $\Lambda_1\Lambda_2\Lambda_3$ of the cell representing the basic unit of periodicity for the function $f(\mathbf{r})$. The functions $\exp(i\mathbf{K_n}\cdot\mathbf{r})$ can therefore be considered to represent a complete set of orthogonal basis functions for the three-dimensional Fourier series expansion of periodic functions in a Cartesian coordinate system.

For rectangular coordinates, however, the definitions for $\mathbf{b}_1$, $\mathbf{b}_2$, and $\mathbf{b}_3$ yield $\mathbf{b}_1 = \hat{\mathbf{d}}_1/d_1$, $\mathbf{b}_2 = \hat{\mathbf{d}}_2/d_2$, and $\mathbf{b}_3 = \hat{\mathbf{d}}_3/d_3$, where $\hat{\mathbf{d}}_j$ represents a unit vector in the $\mathbf{d}_j$ direction. For the special case of a rectangular lattice with $\mathbf{d}_1 = \hat{\mathbf{x}}d_1$, $\mathbf{d}_2 = \hat{\mathbf{y}}d_2$, and $\mathbf{d}_3 = \hat{\mathbf{z}}d_3$, then

$$\mathbf{G}_l = 2\pi[(l_1/d_1)\hat{\mathbf{x}} + (l_2/d_2)\hat{\mathbf{y}} + (l_3/d_3)\hat{\mathbf{z}}] = \mathbf{K}_l, \tag{6.24}$$

where d_1, d_2, and d_3 are the periodicities Λ_1, Λ_2, and Λ_3 along the three orthogonal directions. Thus, in the special case of a rectangular lattice, the functions $\exp(i\mathbf{G}_l \cdot \mathbf{r})$ provide a complete set of orthogonal basis states which can be used for Fourier series expansions of arbitrary functions which have the lattice periodicity. The role of the reciprocal lattice in providing a Fourier space for the expansion of functions within the lattice is therefore indicated, even though the use of the reciprocal lattice is rather superfluous for this case. It is in the more general case of nonorthogonal elementary lattice translation vectors that the reciprocal lattice becomes very powerful in its role in providing a Fourier space for functions having the lattice periodicity.

How do we know (or prove) that the functions $\exp(i\mathbf{G}_l \cdot \mathbf{r})$ constitute a complete set of orthogonal basis states for the expansion of arbitrary functions which have the lattice periodicity? The requirements which must be satisfied are simply those which can be expected in any Fourier series representation. From the standpoint of a given lattice, we must require that

(a) each of the functions appearing in such a Fourier representation must be invariant with respect to all lattice translations $T_\mathbf{j}$;

(b) the fundamental Fourier components based on the lattice spacings in the three principal directions plus all possible shorter wavelength harmonics must be included within the basis set; and

(c) the functions must be orthogonal in the sense that

$$\iiint \exp(i\mathbf{G}_l \cdot \mathbf{r}) \exp(-i\mathbf{G}_\mathbf{m} \cdot \mathbf{r}) \, dx_1 \, dx_2 \, dx_3 \propto \delta_{l\mathbf{m}}, \tag{6.25}$$

where $\delta_{l\mathbf{m}}$ is unity if the triplet of integers (l_1, l_2, l_3) corresponds exactly to the set (m_1, m_2, m_3), but is zero otherwise. (The proportionality is used instead of an equality since the normalization factor will depend upon the number of unit cells contained in the domain of integration and the volume per unit cell.)

The requirement listed as (a) is to be expected on physical grounds since any other function would destroy the required periodicity. This requirement is indeed met, since we have already shown that

$$T_\mathbf{j} \exp(i\mathbf{G}_l \cdot \mathbf{r}) = \exp(i\mathbf{G}_l \cdot \mathbf{r}) \tag{6.26}$$

for any $\mathbf{j}$ and any l.

The requirement (b) is intuitive, since the omission of any harmonic would mean that a function with that particular periodicity could not be Fourier expanded in terms of the basis set. (For a more complete discussion of basis sets, see Chap. 1, §2.2 and Chap. 5, §1.3.) In addition, all possible harmonics are needed in a Fourier expansion in order for the various continuous wiggles and spikes which are present in the original function to be mirrored precisely by the Fourier representation. Is the requirement (b) met by the $\mathbf{G}_l$? It is met in the rectangular lattice, since we have already shown that the $\mathbf{G}_l$ reduce to the set $\mathbf{K}_l$ in this case, and the set $\exp(i\mathbf{K}_l \cdot \mathbf{r})$ is certainly complete. In the more general case, suppose that we ask whether there exists a basis function for some arbitrary harmonic n for the $\mathbf{d}_1$ direction represented within the set $\exp(i\mathbf{G}_l \cdot \mathbf{r})$. If so, it meets the requirement that

$$\mathbf{G}_{l'} \cdot (\mathbf{r} + \mathbf{d}_1) - \mathbf{G}_{l'} \cdot \mathbf{r} = 2\pi n \tag{6.27}$$

for some triplet $\boldsymbol{l}'$. This in turn reduces to

$$\mathbf{G}_{l'} \cdot \mathbf{d}_1 = 2\pi n, \tag{6.28}$$

or

$$2\pi(l'_1\mathbf{b}_1 + l'_2\mathbf{b}_2 + l'_3\mathbf{b}_3) \cdot \mathbf{d}_1 = 2\pi n, \tag{6.29}$$

or

$$l'_1 = n. \tag{6.30}$$

Since n represents the nth harmonic, and is therefore an integer, the requirement is thus satisfied by every one of the subset of basis states $\exp(i\mathbf{G}_{nl_2l_3} \cdot \mathbf{r})$, with arbitrary l_2, l_3. Similarly, the basis state representing the harmonic n_1 in the $\mathbf{d}_1$ direction, n_2 in the $\mathbf{d}_2$ direction, and n_3 in the $\mathbf{d}_3$ direction is simply $\exp(i\mathbf{G}_\mathbf{n} \cdot \mathbf{r})$, where $\mathbf{n} \equiv (n_1, n_2, n_3)$.

The requirement listed under (c) will now be examined, namely, the orthogonality of the various basis states $\exp(i\mathbf{G}_l \cdot \mathbf{r})$. Why is orthogonality necessary? Let us first consider an analogy. An ordinary vector in a three-dimensional space can be resolved into components along three major axes, but the length of the vector will not be equal to the square root of the sum of the squares of the vector components along the three axes of a nonorthogonal coordinate system. (To visualize this, it may prove helpful to refer to Fig. 6.9.) In addition, the scalar product of two such vectors becomes quite complicated for nonorthogonal axes, and in general the component of an arbitrary vector along any given direction will not be equal to the ordinary scalar product of the

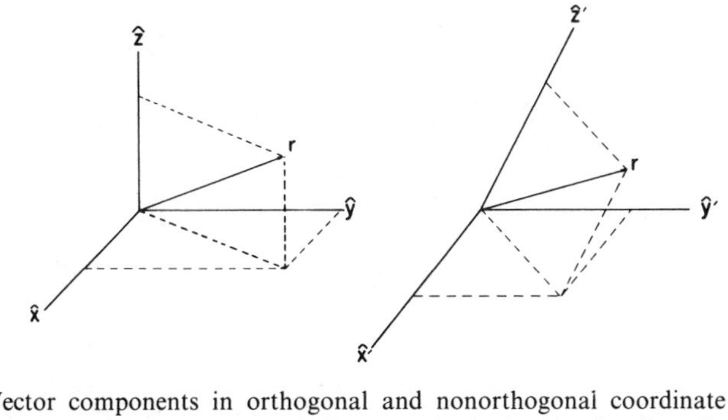

Fig. 6.9 Vector components in orthogonal and nonorthogonal coordinate systems. [In a nonorthogonal system, a given component of a vector is obtained by drawing a line from the tip of the vector parallel to the coordinate axis in question to intersect the plane defined by the remaining two axes; the length of this line thus differs in general from the length of the perpendicular projection. Other details relating to covariant and contravariant components of nonorthogonal coordinate systems can be found in Stratton (1941).]

vector with a unit vector along this specific direction (Fig. 6.9). It may be recalled that the Fourier coefficients of the basis states in Fourier space are the abstract vector space (Chap. 1, §2.2) analogs of the vector components in real space, and the integral in Eq. (1.41) for obtaining a given Fourier coefficient C_n in the one-dimensional Fourier series expansion [Eq. (1.40)] is the equivalent of taking the scalar product of the function with the basis state in question. Therefore the orthogonality of basis states is an important property which we will show allows an evaluation of the Fourier coefficients for an expansion in a lattice with nonorthogonal elementary translation vectors. The importance of orthogonal functions for use as basis states in series expansions can hardly be overemphasized. Let us therefore consider the integral of the product of one basis state with the complex conjugate of another,

$$\mathscr{I} = \iiint \exp(i\mathbf{G}_l \cdot \mathbf{r}) \exp(-i\mathbf{G}_\mathbf{m} \cdot \mathbf{r})\, dx_1\, dx_2\, dx_3, \tag{6.31}$$

where in units of d_1, d_2, and d_3,

$$\mathbf{r} = x_1\mathbf{d}_1 + x_2\mathbf{d}_2 + x_3\mathbf{d}_3 \tag{6.32}$$

in the nonorthogonal system. If $\mathbf{m} = \boldsymbol{l}$, the integrand is unity and the integral immediately yields the value unity, assuming the domain of integration to be over one unit cell in the real lattice. If the integration domain is over the entire crystal, we obtain N_c, which is the number of unit cells in the crystal. If $\mathbf{m} \neq \boldsymbol{l}$, then the triple integral becomes

$$\mathscr{I} = \iiint \exp\{i2\pi[(l_1 - m_1)x_1 + (l_2 - m_2)x_2 + (l_3 - m_3)x_3]\}\, dx_1\, dx_2\, dx_3, \tag{6.33}$$

which can in turn be separated into the product of three integrals of the form

$$\mathscr{I}_j \equiv \int_{x_j}^{x_j+1} \exp[i2\pi(l_j - m_j)x'_j]\, dx'_j. \tag{6.34}$$

Since $\mathbf{m} \neq \boldsymbol{l}$, at least one of the three integers m_j differs from the corresponding integer l_j. For such an integer, the integration yields

$$\mathscr{I}_j = [i2\pi(l_j - m_j)]^{-1} \exp[i2\pi(l_j - m_j)x'_j]\Big|_{x_j}^{x_j+1} = 0. \tag{6.35}$$

Combining the results for both cases, the orthonormality relation

$$\mathscr{I} = \delta_{l\mathbf{m}} \tag{6.36}$$

is obtained. The basis functions $\exp(i\mathbf{G}_l \cdot \mathbf{r})$ are therefore orthogonal.

It is now easy to obtain the Fourier coefficients $A_\mathbf{n}$ for the Fourier series

$$f(\mathbf{r}) = \sum_\mathbf{n} A_\mathbf{n} \exp(i\mathbf{G}_\mathbf{n} \cdot \mathbf{r}) \tag{6.37}$$

for the periodic function $f(\mathbf{r})$ as expanded in terms of the nonorthogonal elementary lattice translation vectors. [Equation (5.187) constitutes an application of this series.] Multiplying the above relation (6.37) by $\exp(-i\mathbf{G_m}\cdot\mathbf{r})$ and integrating over one lattice constant in each direction in the lattice gives

$$\iiint f(\mathbf{r}) \exp(-i\mathbf{G_m}\cdot\mathbf{r})\, dx_1\, dx_2\, dx_3$$

$$= \sum_{\mathbf{n}} A_{\mathbf{n}} \iiint \exp[i(\mathbf{G_n} - \mathbf{G_m})\cdot\mathbf{r}]\, dx_1\, dx_2\, dx_3$$

$$= \sum_{\mathbf{n}} A_{\mathbf{n}}\, \delta_{\mathbf{nm}} = A_{\mathbf{m}}, \tag{6.38}$$

so that

$$A_{\mathbf{m}} = \iiint f(\mathbf{r}) \exp(-i\mathbf{G_m}\cdot\mathbf{r})\, dx_1\, dx_2\, dx_3. \tag{6.39}$$

Aside from its role in Fourier space, the reciprocal lattice has certain other valuable usages. It may be recalled that in a cubic system the vector denoted by $[hkl]$ is perpendicular to a set of planes denoted by the Miller indices (hkl); this is not the case for nonorthogonal elementary lattice translation vectors. It can be shown, however, that *each vector of the reciprocal lattice is normal to a set of planes in the direct lattice*. Furthermore, if the components of a given reciprocal lattice vector **G** have no common factor, then **G** can be shown to have a magnitude which is inversely proportional to the spacing of the lattice planes perpendicular to **G**. Other interesting properties are that the direct lattice is the reciprocal lattice to its own reciprocal lattice, and the volume of a unit cell in the reciprocal lattice is inversely proportional to the volume of a unit cell of the direct lattice.

EXERCISE Prove that each vector of the reciprocal lattice is perpendicular to a set of planes in the direct lattice.

EXERCISE (a) Prove that a reciprocal lattice vector $\mathbf{G_m}$ has a magnitude which is inversely proportional to the spacing of the lattice planes perpendicular to $\mathbf{G_m}$ provided the integers m_1, m_2, m_3 have no common factor.

(b) What is the magnitude of $\mathbf{G_m}$ relative to the spacing of perpendicular lattice planes when m_1, m_2, m_3 possess a common factor?

EXERCISE Prove that the volumes of unit cells in the direct and reciprocal lattices vary reciprocally.

EXERCISE Prove that the direct lattice is reciprocal to its own reciprocal lattice.

The reciprocal lattice is very important in treating diffraction phenomena and electron motion in crystals (Chap. 5, §§ 12 and 13).

PROJECT 6.1 The Reciprocal Lattice

Interpret the reciprocal lattice geometrically with respect to the direct lattice. [*Hint*: First of all, consider the geometry of the triplet of vectors $\mathbf{b}_1, \mathbf{b}_2, \mathbf{b}_3$ relative to the orientations and magnitudes of the triplet of lattice translation vectors $\mathbf{d}_1$, $\mathbf{d}_2$, $\mathbf{d}_3$. Recall that the vector (or cross) product of two vectors defines a third vector perpendicular to the original two vectors; the magnitude of the new vector is the product of the magnitudes of the two initial vectors and the sine of the angle between them. For a nonrectangular coordinate system, the new triplet $\mathbf{b}_1$, $\mathbf{b}_2$, $\mathbf{b}_3$ will therefore be spatially rotated through some angles with respect to the initial triplet $\mathbf{d}_1$, $\mathbf{d}_2$, $\mathbf{d}_3$. Extend these considerations to the general reciprocal lattice itself, and then apply your results to several specific lattices including the face-centered cubic lattice and the hexagonal close-packed lattice.]

7 Wigner–Seitz Cell

It is not absolutely necessary that a primitive unit cell in a given lattice be chosen to be a parallelepiped, and in fact for some important theoretical developments it is much more useful to choose the unit cell to be of a different geometry. One alternate type, of especial interest, is the so-called *Wigner–Seitz primitive cell.* It is constructed by first drawing lines from a given lattice site to all nearby lattice sites, and then placing one plane at the midpoint and perpendicular to each line. The inner volume bounded by these planes is the desired cell. (See Fig. 6.10, for example.) Such cells can be stacked adjacent to one another to fill all space, as is evident from the method of construction. Furthermore, it is clear that each cell contains a single lattice point, so the cell is primitive.

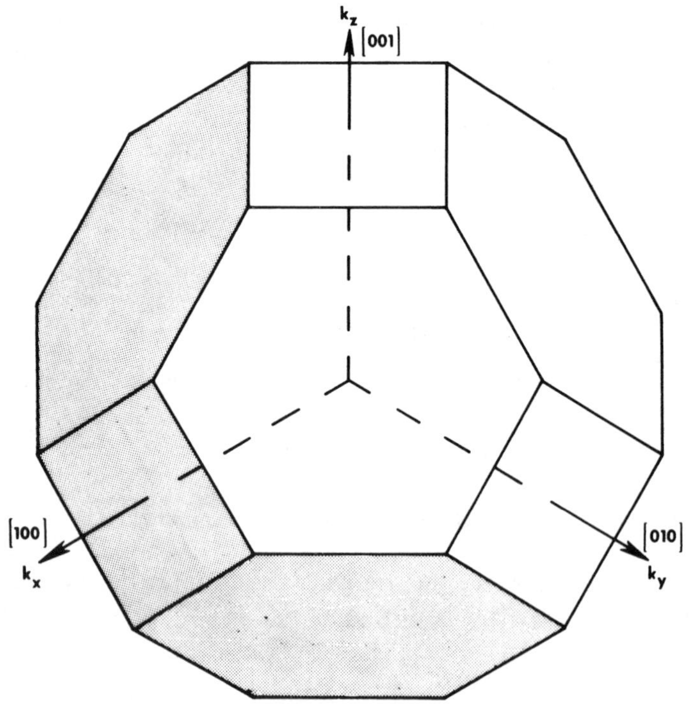

Fig. 6.10 The Wigner–Seitz primitive cell for the three-dimensional bcc lattice.

EXERCISE Construct the Wigner–Seitz cell for a two-dimensional rectangular lattice. [*Hint*: See Fig. 6.1 and also Ziman (1964).]

8 First Brillouin Zone

Consider an arbitrary direct lattice together with its corresponding reciprocal lattice. The Wigner–Seitz unit cell can of course be constructed for the direct lattice structure in the manner described in §7. In an analogous way, the Wigner–Seitz *method of construction* can be used to *delineate a primitive unit cell in the reciprocal lattice*; this cell in the reciprocal lattice is called the *Brillouin Zone* of the *direct lattice* in question. The Brillouin zone for a square lattice is illustrated by the central region in Fig. 6.11. The Brillouin zone is an extremely important concept in the theory of solids, as the reader will come to appreciate more and more as he carefully works through the essentials of energy band theory (Chap. 7).

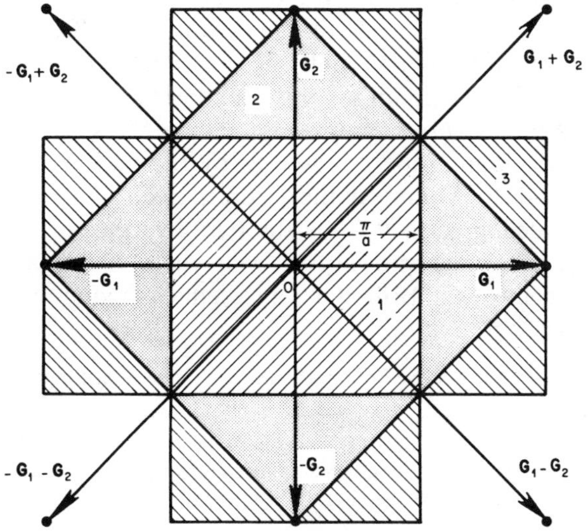

Fig. 6.11 Brillouin zone structure for the two-dimensional square lattice. [The first, second, and third Brillouin zones are shown and labeled; the second and third zones can be noted to be segmented. The construction of Brillouin zones in three dimensions is carried out geometrically as follows: Map out the reciprocal lattice points determined by the reciprocal lattice vectors $\mathbf{G}_l$ defined by Eq. (6.9), draw the reciprocal lattice vectors from the origin to all nearby reciprocal lattice points, and bisect these reciprocal lattice vectors with infinite planes. The innermost volume is the first Brillouin zone. Higher Brillouin zones, each having the same volume in reciprocal space as the first Brillouin zone, are obtained from a proper selection of adjacent volume segments. The two-dimensional construction for the square lattice can be viewed in the same way as the three-dimensional construction, but the conceptual difficulty is less because all **G** vectors then lie in the plane of the paper; the bisecting planes, being thus perpendicular to the plane of the paper, appear as lines, and the Brillouin zones appear as area segments instead of volume segments.]

EXERCISE Construct the Brillouin zone for the two-dimensional rectangular lattice. (*Hint*: First study the construction in Fig. 6.11 for the square lattice.)

9 Higher Brillouin Zones

The above-described construction giving rise to the "Brillouin zone" yields not only the inner central volume (usually referred to as the *first* Brillouin zone), but also an entire series of volume segments of various geometrical shape *outside of* the central volume. These are illustrated for the two-dimensional square lattice by Fig. 6.11. A detailed study shows that segments can be chosen from those directly in contact with the central zone which have a *net* volume which is exactly the same as the volume of the central zone, and moreover, these segments can be spatially rearranged so that they fit together like a three-dimensional puzzle to form a replica of the central zone. These segments are therefore referred to as the *second Brillouin zone*, and the central zone (which is the Wigner–Seitz cell for the reciprocal lattice) is then referred to as the *first Brillouin zone*. A *third Brillouin zone* likewise exists which is made up of volume segments adjacent to those making up the first and second Brillouin zones, and moreover, these volume segments can also be spatially rearranged so that they fit together like a three-dimensional puzzle to form a replica of the central zone. In fact, there is no limit to the number of Brillouin zones which can be delineated in this manner for any given direct lattice, and each zone can be spatially rearranged to form a replica of the first Brillouin zone. These constructs are quite useful for the visualization of *Fermi surfaces* of various real metals, since the energy gaps produced at the Brillouin zone boundaries by the periodic lattice potential (Chap. 5, §13.2) divide the volume of the occupied electronic states in **k** space (or *momentum* space) delimited at 0°K by the Fermi surface into zone segments which themselves rearrange into various geometrical shapes under the same spatial rearrangement of the Brillouin zone segments required to replicate the first Brillouin zone.

EXERCISE Construct the first five Brillouin zones for the two-dimensional rectangular lattice.

EXERCISE Show how the segments of the second, third, fourth, and fifth Brillouin zones of the two-dimensional rectangular lattice can be rearranged to form replicas of the first Brillouin zone. (*Hint*: Construct the Brillouin zones with pencil and ruler on two sheets of paper, number each segment, cut out the segments from one sheet, and attempt to rearrange them into the desired geometrical pattern.)

EXERCISE Use modeling clay to construct the segments of the second Brillouin zone for the three-dimensional simple-cubic lattice.

PROJECT 6.2 Brillouin Zones

Use wooden blocks and a saw to construct actual models of the three-dimensional geometrical segments making up the first, second, and third Brillouin zones of the sc lattice, and show how the segments of the second and the third Brillouin zones fit together to replicate the first Brillouin zone. Repeat for the bcc and the fcc lattices.

PROBLEMS

1. (a) Find the translation vectors for the primitive cell of the hexagonal space lattice.
(b) Find the volume of the primitive cell in Part (a).
(c) Find the volume of one hexagonal cell in zinc sulfide.

2. (a) Find the reciprocal lattice vectors for a tetragonal crystal having lattice parameters $d_1 = 2.06$ Å, $d_2 = 3.21$ Å, $d_3 = 4.06$ Å.

(b) How many unit cells are there in a macroscopic crystal of this material having a volume of 1 mm^3?

(c) If this macroscopic crystal has the same number of unit cell lengths in each of the three directions, what are the reciprocal crystal vectors? [*Hint*: See Eq. (7.8).]

3. If $\mathbf{d}_1 = \frac{1}{2}a(\hat{\mathbf{x}} + \hat{\mathbf{y}})$, $\mathbf{d}_2 = \frac{1}{2}a(\hat{\mathbf{y}} + \hat{\mathbf{z}})$, $\mathbf{d}_3 = \frac{1}{2}a(\hat{\mathbf{z}} + \hat{\mathbf{x}})$, then determine the reciprocal lattice vectors.

4. (a) How many lattice points are there in a fcc unit cell?

(b) Show that the fcc lattice can be viewed in terms of trigonal unit cells.

(c) What are the corresponding basis vectors?

(d) What are the corresponding reciprocal lattice vectors?

5. Lithium, which has a bcc structure, has an atomic weight of 6.9 g/mole and a density of 0.53 g/cm^3.

(a) Find the edge length a of the conventional unit cell.

(b) Using the information from Part (a), find d_1, d_2, d_3 for the primitive unit cell. Also find α_{12}, α_{13}, and α_{23}.

6. Show directly that $T_{\mathbf{R}} \exp(i\mathbf{G}_l \cdot \mathbf{r}_\mathbf{n}) = \exp[i\mathbf{G}_l \cdot (\mathbf{r}_\mathbf{n} + \mathbf{R})]$ for the special case where $\boldsymbol{l} = (1, 1, 1)$, $\mathbf{n} = (1, 1, 1)$, and $\mathbf{R} = (-\frac{7}{2}a\hat{\mathbf{x}} + \frac{1}{2}a\hat{\mathbf{y}} - a\hat{\mathbf{z}})$.

7. A series of identical pyramids are placed in a two-dimensional array on a horizontal surface. All edges of the base of each pyramid are in contact with the edges of bases of similar pyramids. Considering the array to extend from $-\infty$ to ∞ in both horizontal directions, expand the resulting top surface in a Fourier series representation.

8. Repeat Problem 7 for a single pyramid, with the Fourier integral representation replacing the Fourier series.

9. (a) Corresponding to the $\{l\ m\ n\}$ planes in a three-dimensional lattice, there are $\{l\ m\}$ planes (actually lines) in a two-dimensional lattice. Deduce and verify a general expression for the distance between adjacent parallel $\{l\ m\}$ planes in a simple two-dimensional square lattice.

(b) Deduce the vector form of the Bragg condition for reflection from the $\{l\,m\}$ planes in a simple two-dimensional square lattice.

(c) Describe elastic and inelastic neutron scattering by this lattice, and tell how such a lattice could be used as a neutron monochromator.

10. Construct the Wigner–Seitz cell for a two-dimensional rectangular lattice.

11. Construct the Brillouin zone for the two-dimensional rectangular lattice.

12. Deduce Miller indices for several major planes in the three cubic Bravais lattices.

13. Show that the (100) and the (001) planes in a cubic crystal are in some sense equivalent.

14. What is the geometry of the points in the {111} plane of a sc lattice? Is the geometry modified for the same plane in bcc and fcc lattices?

15. Show that parallel planes placed in *equivalent* positions in a crystal lattice can have the same set of Miller indices (hkl).

16. Show that the [hkl] direction is perpendicular to the (hkl) plane in a cubic lattice.

17. Prove that each vector in the reciprocal lattice is perpendicular to a set of planes in the direct lattice.

18. (a) Prove that a reciprocal lattice vector $\mathbf{G}_\mathbf{m}$ has a magnitude which is inversely proportional to the spacing of the lattice planes perpendicular to $\mathbf{G}_\mathbf{m}$ provided the integers m_1, m_2, m_3 have no common factor.

(b) What is the magnitude of $\mathbf{G}_\mathbf{m}$ relative to the spacing of perpendicular lattice planes when m_1, m_2, m_3 possess a common factor?

19. Explain the transformation needed to convert the fcc lattice to the hexagonal close-packed lattice.

20. Deduce the various n-fold rotation axes for the fourteen Bravais lattices.

21. Is the diamond structure included as one of the fourteen Bravais lattices? If not, show how it can be obtained by adding an appropriate basis to one of the Bravais lattices.

22. Explain the difference between the diamond, fluorite, and zincblende structures. [*Hint*: See R. A. Smith (1963).]

23. Prove that the unit cell volume v_c is given by $v_c = |\mathbf{d}_1 \cdot (\mathbf{d}_2 \times \mathbf{d}_3)|$.

24. Prove that the volumes of unit cells in the direct and reciprocal lattices vary reciprocally.
25. Show that the reciprocal lattice of a fcc lattice is bcc and vice versa.
26. Prove that the direct lattice is reciprocal to its own reciprocal lattice.
27. Construct a primitive unit cell for copper.
28. Prove that there are $V/8\pi^3$ allowed $\mathbf{k}$ vectors per unit volume of reciprocal space, where V is the volume of the crystal in real space.
29. Deduce a vector equation for the planes delineating the Brillouin zones.
30. Give one example where Brillouin zones are important. (*Hint*: See Chap. 5.)
31. Consider a simple cubic lattice with lattice spacing a.
(a) Write the direct lattice basis vectors.
(b) Calculate the reciprocal lattice basis vectors.
(c) Where are surfaces which enclose the first Brillouin zone? (Either write the equations for the planes or explain clearly with words.)
32. How can one ascertain whether the Fermi surface overlaps a certain Brillouin zone in the free-electron approximation?
33. Consider a linear crystal made up of four equally spaced ions of equal mass, the masses being connected by springs with the end ones being fixed in position. Find the equations of motion, the frequencies, and the relative amplitudes for the two ions which are capable of motion.

CHAPTER 7

BLOCH'S THEOREM AND ENERGY BANDS FOR A PERIODIC POTENTIAL

The present theories of metals seem enormously complicated, in contrast with the beautiful simplicity of Lorentz's theory. J. C. Slater (1934)

1 Fourier Series Expansions for Arbitrary Functions of Position within the Crystal

The *reciprocal lattice* was constructed in Chap. 6 to serve as a firm foundation for three-dimensional expansions having the *lattice* periodicity. That is, any arbitrary periodic function $f(\mathbf{r})$ which is a continuous function of position $\mathbf{r}$ that is commensurate with the lattice periodicity may be expanded in the particular Fourier series,

$$f(\mathbf{r}) = \sum_{\mathbf{n}} A_{\mathbf{n}} \exp(i\mathbf{G}_{\mathbf{n}} \cdot \mathbf{r}), \tag{7.1}$$

as shown in §6 of Chap. 6. The vectors $\mathbf{G}_{\mathbf{n}}$ are the *reciprocal lattice vectors*

$$\mathbf{G}_{\mathbf{n}} = 2\pi(n_1\mathbf{b}_1 + n_2\mathbf{b}_2 + n_3\mathbf{b}_3), \tag{7.2}$$

where the $\mathbf{b}_j$ are defined by Eqs. (6.4)–(6.7), and $\mathbf{r}$ is the position measured in units of the elemental translation vectors,

$$\mathbf{r} = x_1\mathbf{d}_1 + x_2\mathbf{d}_2 + x_3\mathbf{d}_3. \tag{7.3}$$

The *Fourier coefficients* for this expansion were shown to be determined by

$$A_{\mathbf{m}} = \iiint f(\mathbf{r}') \exp(-i\mathbf{G}_{\mathbf{m}} \cdot \mathbf{r}')\, dx'_1\, dx'_2\, dx'_3, \tag{7.4}$$

where the integration extends over *any* unit cell in the direct lattice. Since distance $\mathbf{r}$ is considered to be measured in units of the elemental translation vectors $\mathbf{d}_1$, $\mathbf{d}_2$, $\mathbf{d}_3$, the integrations extend from x_j to $x_j + 1$ with $j = 1, 2, 3$.

Let us now consider a crystal (Fig. 7.1) to be made up of N_c unit cells, with lengths L_1d_1, L_2d_2, and L_3d_3 in three major crystal directions. We specify on

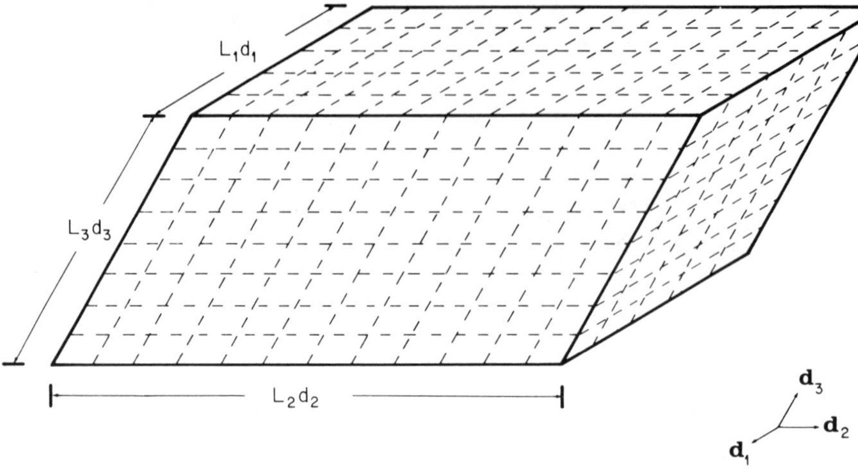

Fig. 7.1 Crystal consisting of $N_c = L_1L_2L_3$ unit cells, each unit cell delineated by the lattice translation vectors $\mathbf{d}_1$, $\mathbf{d}_2$, $\mathbf{d}_3$.

physical grounds that L_1, L_2, and L_3 are integers; the resulting crystal contains $N_c = L_1L_2L_3$ lattice points arranged in the same symmetry as the elemental unit cell defined by the elementary translation vectors $\mathbf{d}_1$, $\mathbf{d}_2$, $\mathbf{d}_3$. The crystal volume Ω will be given by

$$\Omega = N_c v_c, \tag{7.5}$$

where v_c is the volume of the elemental unit cell. Thus, $\Omega = L_1L_2L_3\,(\mathbf{d}_1 \cdot \mathbf{d}_2 \times \mathbf{d}_3)$.

Suppose that there exists some function $g(\mathbf{r})$ which is commensurate with the periodicity of this *bulk crystal* consisting of N_c unit cells; $g(\mathbf{r})$ does not necessarily possess the crystal *lattice* periodicity as defined by the *individual* unit cells. It is evident that a Fourier series can be constructed for this function $g(\mathbf{r})$; however, the reciprocal lattice vectors $\mathbf{G}_\mathbf{n}$ must be replaced by a different set of vectors (let us label them $\mathbf{k}_\mathbf{n}$) which reflect the *dimensions of the solid* instead of the unit cell (or *lattice*) dimensions. The key point to recognize in determining the $\mathbf{k}_\mathbf{n}$ is the fact that the basic *symmetry* of the periodic function is unchanged; hence, we can expect the $\mathbf{k}_\mathbf{n}$ vectors to have the same spatial *orientations* as the $\mathbf{G}_\mathbf{n}$ vectors, so that in practice only the magnitudes will differ.

Recall that the reciprocal lattice (Chap. 6, §6) for the unit cell is based upon the vectors

$$\mathbf{b}_j \equiv \mathbf{d}_k \times \mathbf{d}_l / (\mathbf{d}_j \cdot \mathbf{d}_k \times \mathbf{d}_l), \tag{7.6}$$

where (j, k, l) represents the triplets of integers (1, 2, 3), (2, 3, 1), and (3, 1, 2). A "reciprocal space" based upon the periodicity of the *solid* can therefore be constructed in a similar manner from three vectors defined as follows:

$$\mathscr{B}_j \equiv \frac{(L_k\mathbf{d}_k) \times (L_l\mathbf{d}_l)}{(L_j\mathbf{d}_j) \cdot (L_k\mathbf{d}_k) \times (L_l\mathbf{d}_l)} = \mathbf{b}_j / L_j \qquad (j = 1, 2, 3). \tag{7.7}$$

In terms of **reciprocal crystal vectors $\mathbf{k_n}$** defined by

$$\mathbf{k_n} \equiv 2\pi(n_1\mathscr{B}_1 + n_2\mathscr{B}_2 + n_3\mathscr{B}_3) = 2\pi[(n_1/L_1)\mathbf{b}_1 + (n_2/L_2)\mathbf{b}_2 + (n_3/L_3)\mathbf{b}_3], \tag{7.8}$$

the set of basis states for the *crystalline solid* will be $\exp(i\mathbf{k_n}\cdot\mathbf{r})$. It is easily shown that each of these basis states has the periodicity of the solid. For example, a translation by $L_1\mathbf{d}_1$ gives

$$\begin{aligned}\exp(i\mathbf{k_n}\cdot\mathbf{r}) \to \exp[i\mathbf{k_n}\cdot(\mathbf{r} + L_1\mathbf{d}_1)] &= \exp(i\mathbf{k_n}\cdot\mathbf{r})\exp(iL_1\mathbf{k_n}\cdot\mathbf{d}_1)\\ &= \exp(i\mathbf{k_n}\cdot\mathbf{r})\exp(in_1 2\pi) = \exp(i\mathbf{k_n}\cdot\mathbf{r}),\end{aligned} \tag{7.9}$$

since n_1 is an integer.

All harmonics having periodicity commensurate with the solid are contained within the set of vectors $\mathbf{k_n}$. For example, a wavelength equal to $\frac{1}{3}$ the length of the solid in one particular direction $\mathbf{d}_j$ is found in all of the vectors $\mathbf{k_n}$ having $n_j = 3$. The set is therefore *complete.*

EXERCISE Show that there exists among the set of $\mathbf{k_n}$ vectors given by Eq. (7.8) a harmonic which has wavelength (i.e., a periodicity) $L_2d_2/5$ along the $\mathbf{d}_2$ direction.

The *orthogonality* of the functions $\exp(i\mathbf{k_n}\cdot\mathbf{r})$ over the domain of the solid is readily established; this can be done in a manner analogous to that used to deduce the properties of the functions $\exp(i\mathbf{G_n}\cdot\mathbf{r})$ in Chap. 6, §6. That is,

$$\iiint (e^{i\mathbf{k_n}\cdot\mathbf{r}})^*(e^{i\mathbf{k_m}\cdot\mathbf{r}})\,dx_1\,dx_2\,dx_3 = \delta_{\mathbf{nm}}N_c \tag{7.10}$$

whenever the domains of integration are chosen to be from x_1 to $x_1 + L_1$, from x_2 to $x_2 + L_2$, and from x_3 to $x_3 + L_3$. The set of functions $N_c^{-1/2}\exp(i\mathbf{k_n}\cdot\mathbf{r})$ are normalized *with respect to the domain of the entire solid.*

EXERCISE Prove the result stated in Eq. (7.10).

One Fourier series for a function $g(\mathbf{r})$ having the periodicity of the *bulk crystal* is given by

$$g(\mathbf{r}) = \sum_{\mathbf{n}} A_{\mathbf{n}}e^{i\mathbf{k_n}\cdot\mathbf{r}}, \tag{7.11}$$

with

$$A_{\mathbf{m}} = N_c^{-1}\int_{x_1}^{x_1+L_1} dx_1' \int_{x_2}^{x_2+L_2} dx_2' \int_{x_3}^{x_3+L_3} dx_3'\,g(\mathbf{r}')e^{-i\mathbf{k_m}\cdot\mathbf{r}'}. \tag{7.12}$$

An alternate form is given by

$$g(\mathbf{r}) = N_c^{-1/2}\sum_{\mathbf{n}} C_{\mathbf{n}}\exp(i\mathbf{k_n}\cdot\mathbf{r}), \tag{7.13}$$

with

$$C_{\mathbf{m}} = N_c^{-1/2}\iiint g(\mathbf{r}')\exp(-i\mathbf{k_m}\cdot\mathbf{r}')\,dx_1'\,dx_2'\,dx_3', \tag{7.14}$$

where the integration is over the domain of the crystal in the three principal crystal directions, respectively. Yet a third form is given by

$$g(\mathbf{r}) = N_c^{-1} \sum_{\mathbf{n}} \mathscr{C}'_{\mathbf{n}} \exp(i\mathbf{k_n}\cdot\mathbf{r}) \tag{7.15}$$

where

$$\mathscr{C}'_{\mathbf{m}} = N_c^{1/2} C_{\mathbf{m}}, \tag{7.16}$$

with $C_{\mathbf{m}}$ defined by Eq. (7.14).

For all of the above forms of the Fourier series,

$$\mathbf{k_n} = 2\pi[(n_1/L_1)\mathbf{b}_1 + (n_2/L_2)\mathbf{b}_2 + (n_3/L_3)\mathbf{b}_3], \tag{7.17}$$

where n_1, n_2, and n_3 are integers; the above forms therefore differ only with respect to normalization of the basis functions (Chap. 1, §4). It is apparent from Eq. (7.17) that the points which are mapped out by the set of $\mathbf{k_n}$ vectors are equally spaced along the directions of $\mathbf{b}_1$, $\mathbf{b}_2$, $\mathbf{b}_3$. Therefore it can be inferred geometrically (Fig. 7.2) that the sum of any two $\mathbf{k_n}$ vectors represents yet a third $\mathbf{k_n}$ vector; this conclusion can be verified very simply by algebraically adding the components of the two vectors.

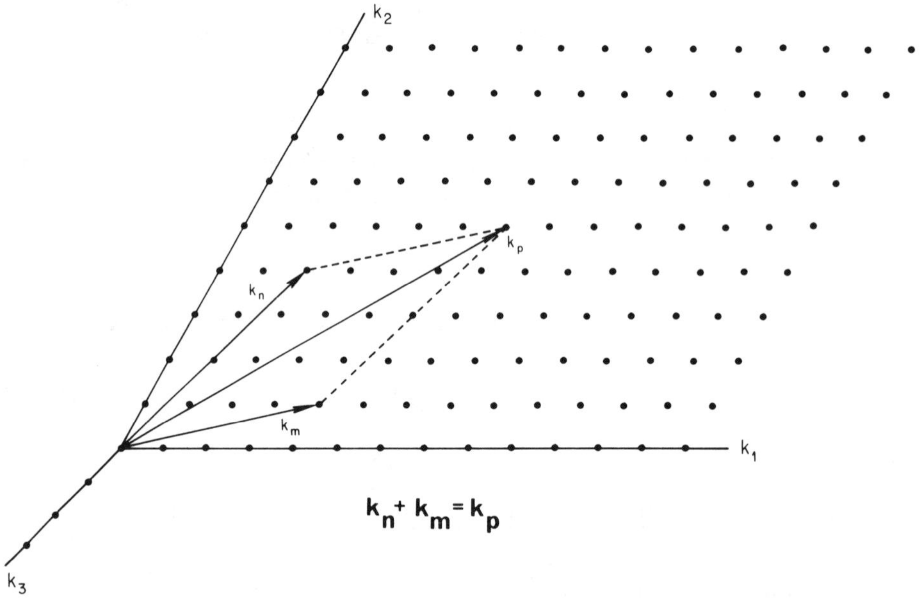

Fig. 7.2 The vector sum of any two **k** vectors (such as $\mathbf{k_m}$ and $\mathbf{k_n}$) yields a third vector (represented by $\mathbf{k_p}$) which is likewise a member of the set of **k** vectors.

EXERCISE Show algebraically that $\mathbf{k_n} + \mathbf{k_{n'}} = \mathbf{k_{n''}}$ where $\mathbf{k_n}$ and $\mathbf{k_{n'}}$ are two arbitrary reciprocal crystal vectors and $\mathbf{k_{n''}}$ represents a third reciprocal crystal vector. What is the relationship among **n**, **n**′, and **n**″? Given the triplets **n** and **n**′ to be specifically (2, 7, 3) and (4, 6, 12), what is the triplet **n**″?

Any arbitrary function which satisfies the *Dirichlet conditions* (Chap. 1, §4) and which also has the periodicity of the solid can thus be represented by a linear combination of the *complete set of functions* $\exp(i\mathbf{k_n}\cdot\mathbf{r})$. The functional dependence *within* a given period (such as the domain of the solid) can therefore, apart from certain constraints, be completely arbitrary. For all practical purposes, then, nearly any arbitrary function of position in the solid, whether or not it has any periodicity *within* the solid, can be represented *over the domain of the solid* by a *Fourier series* involving only linear combinations of the discrete set of basis states $\exp(i\mathbf{k_n}\cdot\mathbf{r})$. It is true that such a linear combination will also have nonzero values outside of the solid, which in most cases will be physically meaningless for the problem at hand, but this does not negate the fact that the three-dimensional Fourier *series* can provide an exact representation of an arbitrary continuous function *within the spatial region occupied by the solid.* The behavior of the Fourier series *outside of the solid,* being nonphysical insofar as the description of the physical properties of the solid itself is concerned, can simply be ignored. The advantage of using the Fourier *series* representation instead of the corresponding Fourier *integral* representation (for which the linear combination *can* be chosen in such a way as to have a zero value outside of the solid), is simply that a smaller set of **k** vectors is required. In contrast to a *discrete* set of **k** vectors, a *continuous* set is required for a Fourier *integral* representation of an arbitrary function of position within the solid (refer to Chap. 1, §4.3).

EXERCISE Set up the Fourier integral representation of aperiodic functions having the dimensions of the solid. (*Hint*: Refer to Chap. 1, §4.3.)

Let us now turn our attention to the relationship between the **k** vectors and the **G** vectors. It can be seen that whenever the three integers in the triplet (m_1, m_2, m_3) in $\mathbf{k_m}$

$$\mathbf{k_m} = 2\pi[(m_1/L_1)\mathbf{b}_1 + (m_2/L_2)\mathbf{b}_2 + (m_3/L_3)\mathbf{b}_3] \tag{7.18}$$

are integral multiples of L_1, L_2, L_3, respectively, such as $m_1/L_1 = n_1$, $m_2/L_2 = n_2$, $m_3/L_3 = n_3$, then $\mathbf{k_m}$ becomes equal to one of the **G** vectors, namely,

$$\mathbf{G_n} = 2\pi(n_1\mathbf{b}_1 + n_2\mathbf{b}_2 + n_3\mathbf{b}_3) \qquad \textbf{(reciprocal lattice vectors)}. \tag{7.19}$$

Therefore we reach the following conclusion: *The* **G** *vectors constitute a subset of the* **k** *vectors.* It is thus evident that the **k** vectors are a discrete set of vectors in the reciprocal lattice containing the subset **G** which delineate the reciprocal lattice sites. This is illustrated in Fig. 7.3.

A continuation of this examination of the properties of **G** vectors and **k** vectors leads to another result: *Any* **k** *vector extending beyond one unit cell in the reciprocal lattice can be resolved into the sum of one of the* **G** *vectors and a* **k** *vector which does not extend beyond one unit cell.* (Later in this section we show that we can be even more specific than this in resolving the **k** vectors.) The above statement can be proved quite easily in two different ways, geometrically and

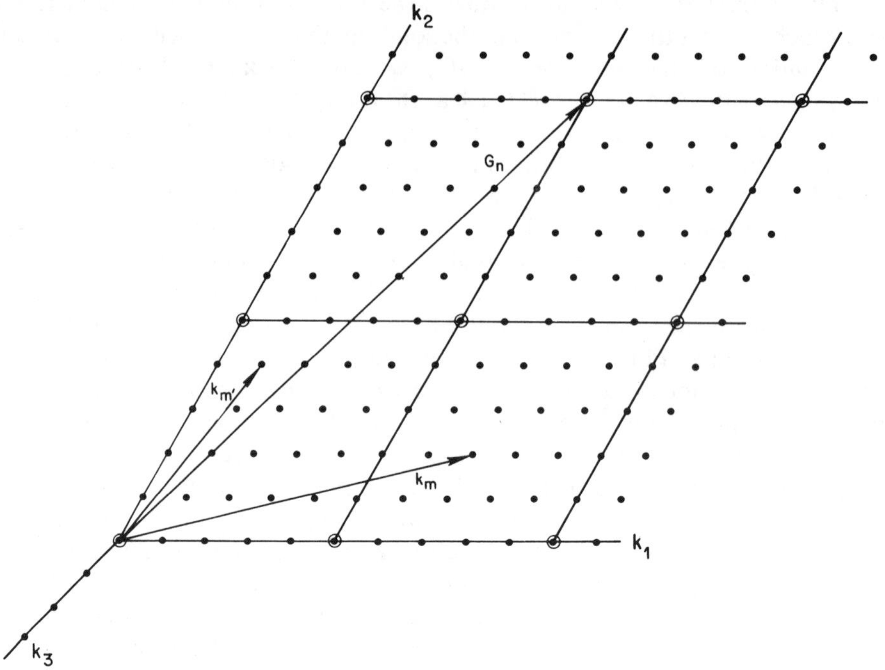

Fig. 7.3 The *reciprocal crystal vectors* $\mathbf{k_m} = 2\pi[(m_1/L_1)\mathbf{b}_1 + (m_2/L_2)\mathbf{b}_2 + (m_3/L_3)\mathbf{b}_3]$ map out points denoted by the solid circles; the *reciprocal lattice vectors* $\mathbf{G_n} = 2\pi[n_1\mathbf{b}_1 + n_2\mathbf{b}_2 + n_3\mathbf{b}_3]$ map out points denoted by the open circles. It can thus be seen that the **G** vectors constitute a subset of the **k** vectors. (The elemental parallelepiped unit cells in *reciprocal space* are indicated by the solid lines drawn between the reciprocal lattice points.)

algebraically. The geometric proof is based on the above conclusion that the **G** vectors represent a subset of the **k** vectors.

EXERCISE Prove geometrically the above statement regarding resolution of **k** vectors. (*Hint*: A **k** vector reaching to a point represented by any of the allowed **k** values within an arbitrary cell in the reciprocal lattice can be resolved into a **G** vector to a corner of the arbitrary cell plus a vector from the corner of the cell to the point in question within the cell. This can be noted from Fig. 7.4. The vector from the corner to the point in question within the cell is equivalent by symmetry to a vector from the origin to the corresponding point in a cell having one corner at the origin. This follows because of the symmetry of the reciprocal lattice, the fact that the points delineating the **k** values are uniformly spaced throughout the reciprocal lattice, and the convention that parallel vectors of equal length are equivalent.)

The algebraic proof to be given now is perhaps easier to understand mathematically, though it lacks the intuitive value of the geometric proof. Let

$$\mathbf{k_m} = 2\pi[(m_1/L_1)\mathbf{b}_1 + (m_2/L_2)\mathbf{b}_2 + (m_3/L_3)\mathbf{b}_3] \tag{7.20}$$

represent any vector extending from the origin of the reciprocal lattice to one of the possible points allowed by the requirement that m_1, m_2, m_3 be integers, and assume that at least one of the three quantities m_1/L_1, m_2/L_2, m_3/L_3 has a

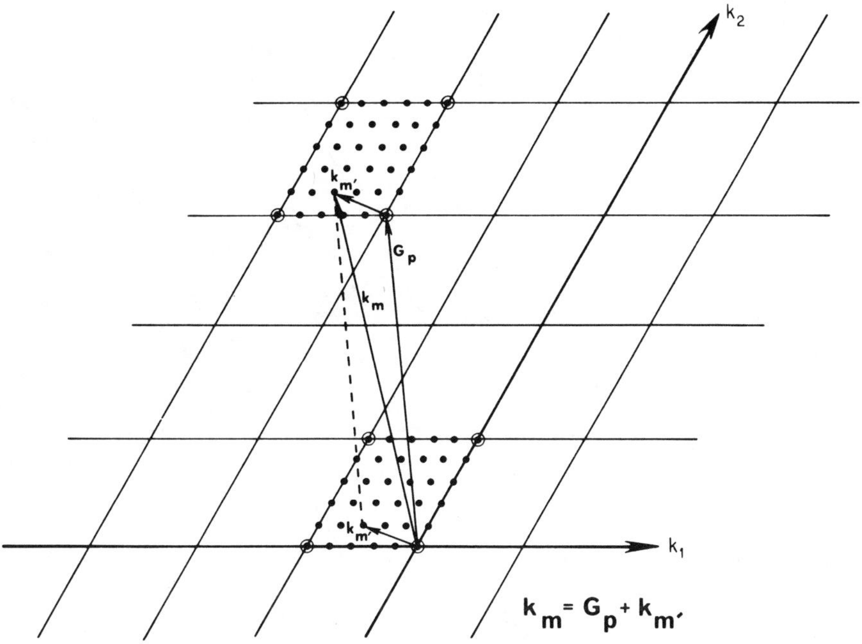

Fig. 7.4 Any **k** vector $\mathbf{k_m}$ extending beyond the parallelepiped unit cells adjacent to the origin in reciprocal space can be represented by the vector sum of the **G** vector $\mathbf{G_p}$ extending to the nearest corner of the parallelepiped unit cell containing the point delineated by $\mathbf{k_m}$ and a **k** vector $\mathbf{k_{m'}}$ within one of the parallelepiped unit cells adjacent to the origin. (Thus the entire set of vectors $\mathbf{k_m}$ can be mapped into vectors in the set $\mathbf{k_{m'}}$ which do not extend beyond one parallelepiped unit cell in reciprocal space. Conversely, the restricted set of **k** vectors denoted by $\mathbf{k_{m'}}$ can be used to map out the entire set of **k** vectors $\mathbf{k_m}$ by suitable addition of $\mathbf{k_{m'}}$ vectors with vectors from the reciprocal lattice set $\mathbf{G_p}$.)

magnitude exceeding unity so the $\mathbf{k_m}$ lies outside of the domain of the cells immediately at the origin. The ratio (or ratios) which exceed unit magnitude can be written as the sum of an integer plus a fraction with magnitude less than unity. Thus

$$m_1/L_1 = p_1 + (m_1'/L_1), \tag{7.21}$$

$$m_2/L_2 = p_2 + (m_2'/L_2); \tag{7.22}$$

$$m_3/L_3 = p_3 + (m_3'/L_3), \tag{7.23}$$

where p_1, p_2, p_3 are integers (positive, negative, or zero), and m_1', m_2', m_3' represent integers which are less than the integers L_1, L_2, L_3, respectively. There is no restriction on this transformation. Thus

$$\begin{aligned}\mathbf{k_m} &= 2\pi\{[p_1 + (m_1'/L_1)]\mathbf{b}_1 + [p_2 + (m_2'/L_2)]\mathbf{b}_2 + [p_3 + (m_3'/L_3)]\mathbf{b}_3\}\\ &= 2\pi[p_1\mathbf{b}_1 + p_2\mathbf{b}_2 + p_3\mathbf{b}_3] + 2\pi[(m_1'/L_1)\mathbf{b}_1 + (m_2'/L_2)\mathbf{b}_2 + (m_3'/L_3)\mathbf{b}_3]\\ &= \mathbf{G_p} + \mathbf{k_{m'}}.\end{aligned} \tag{7.24}$$

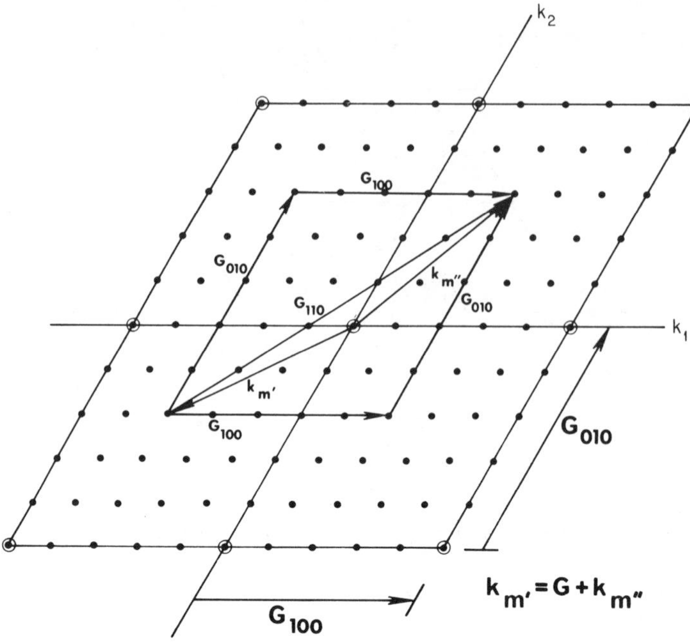

Fig. 7.5 The $\mathbf{k}_{m'}$ vectors can be mapped between adjacent parallelepiped unit cells which are in contact with the origin in reciprocal space by means of appropriate **G** vectors.

The physical interpretation of this resolution of the **k** vectors is quite interesting, especially since it has far-reaching physical consequences: Because $|\mathbf{k}| = 2\pi/\lambda$, the large magnitude **k** vectors (corresponding to small wavelengths λ) are resolved into the sum of a vector corresponding to some harmonic having the *lattice* periodicity and a vector having a λ value *greater* than the fundamental *lattice* periodicity (but *less* than the *bulk crystal* periodicity).

Let us now consider a further resolution of the **k** vectors. It is now possible to show that by means of reciprocal lattice vectors *all* **k** vectors can be reduced to **k** vectors lying in any *single* unit cell at the origin in reciprocal space, such as the initial parallelepiped reciprocal cell. Thus far we have already shown that any **k** vector $\mathbf{k}_m$ reaching beyond one unit cell in the reciprocal lattice can be resolved into one of the **G** vectors $\mathbf{G}_p$ and a **k** vector $\mathbf{k}_{m'}$ which does not extend beyond one unit cell,

$$\mathbf{k}_m = \mathbf{k}_{m'} + \mathbf{G}_p. \tag{7.25}$$

It is an evident extension of the geometrical proof, however, that any vector $\mathbf{k}_{m'}$ reaching to a point in any one of the unit cells in the reciprocal lattice which has the origin of the reciprocal lattice at one corner (Fig. 7.5) can be resolved into some **k** vector $\mathbf{k}_{m''}$ lying in any one of the other adjacent cells plus a suitable **G** vector $\mathbf{G}_l$,

$$\mathbf{k}_{m'} = \mathbf{k}_{m''} + \mathbf{G}_l. \tag{7.26}$$

Thus

$$\mathbf{k_m} = \mathbf{k_{m'}} + \mathbf{G_p} = \mathbf{k_{m''}} + \mathbf{G}_l + \mathbf{G_p}, \tag{7.27}$$

where $\mathbf{k_{m''}}$ lies in any cell desired. However,

$$\mathbf{G}_l + \mathbf{G_p} = \mathbf{G_{p'}}, \tag{7.28}$$

where $\mathbf{G_{p'}}$ itself is a reciprocal lattice vector; therefore

$$\mathbf{k_m} = \mathbf{k_{m''}} + \mathbf{G_{p'}}, \tag{7.29}$$

where $\mathbf{k_{m''}}$ lies in the desired cell. The corresponding algebraic justification readily follows from the preceding algebraic proof by recognizing that all integers may be either positive or negative. Thus it is allowable for the reduced **k** vector $\mathbf{k_{m'}}$ to lie in a different octant of the coordinate system than the unreduced **k** vector $\mathbf{k_m}$. The physical significance of this further reduction is that the *reduced* **k** vectors (i.e., the new **k** vectors obtained by resolving the original **k** vectors) may be chosen to lie in one unit cell in any chosen octant of reciprocal space.

EXERCISE Prove Eqs. (7.26)–(7.29), both geometrically and algebraically.

The arguments based on the geometry of the reciprocal lattice can be extended somewhat further. It has been shown that the $\mathbf{k_m}$ vectors are equally spaced throughout the reciprocal lattice in such a way that each $\mathbf{k_m}$ can be reduced to the sum of a reciprocal lattice vector and a corresponding $\mathbf{k_m}$ vector $\mathbf{k_{m'}}$ lying within an elemental parallelepiped unit cell at the origin. Once such a reduction has been effected, it is clear geometrically that *all* such reduced vectors $\mathbf{k_{m'}}$ can be translated to *any arbitrary* parallelepiped primitive cell in the reciprocal lattice by means of some specific reciprocal lattice vector. Therefore we could in effect "reduce" all $\mathbf{k_m}$ vectors into a single parallelepiped primitive cell lying *anywhere* in the reciprocal lattice, so the reduction need not be into a cell in contact with the origin. This fact is useful in examining the periodicity of electronic wave functions in a crystal with respect to momentum space.

EXERCISE Carry out the indicated translation of the **k** vectors, choosing some specific $\mathbf{G_p}$, and examine the range of wavelengths associated with the new set of **k** vectors.

Likewise, the **k** vectors may all be reduced to the Wigner–Seitz cell (Chap. 6, §§ 7, 8) of the reciprocal lattice (i.e., to the first Brillouin zone). This can be seen in essence by means of the geometrical technique.

EXERCISE Show geometrically that the **k** vectors can be reduced to the first Brillouin zone. [*Hint*: The Wigner–Seitz cell is made up of geometric portions of the parallelepiped cells immediately surrounding the origin; since it is also primitive, it has the same volume as a parallelepiped primitive cell. We have already seen that a **k** vector can be reduced to *any* given parallelepiped primitive cell in reciprocal space, including those with one corner at the origin. For parallelepiped primitive cells with one corner at the origin, portions immediately in the neighborhood of the origin are already contained within the corresponding Wigner–Seitz cell in reciprocal space. It follows geometrically that any **k** vector in these cells reaching a point *outside* of the Wigner–Seitz cell in reciprocal space can be reduced to a **k** vector *within* the Wigner–Seitz cell in reciprocal space simply by reducing it to an appropriate one of the adjacent primitive parallelepiped cells. This is further clarified by Figs. 7.6 and 7.7.]

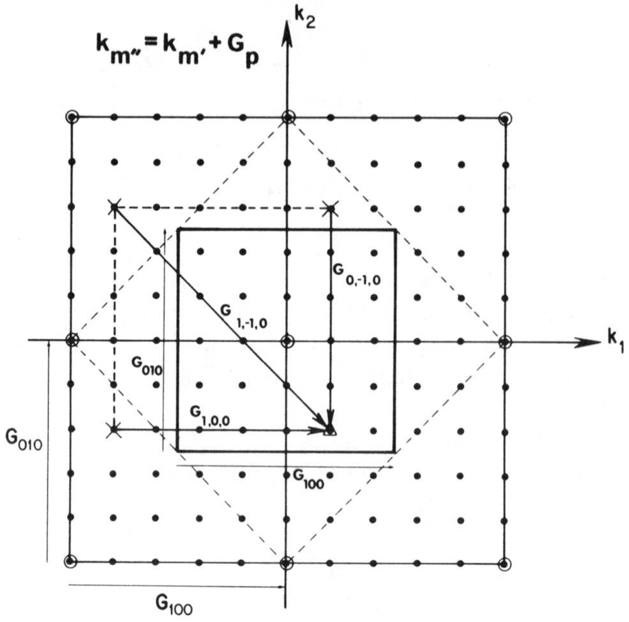

Fig. 7.6 Mapping of **k** vectors $\mathbf{k}_{m'}$ from square unit cells in contact with the origin in reciprocal space into the first Brillouin zone for the case of the two-dimensional square lattice. (The three points marked × locate the tips of **k** vectors $\mathbf{k}_{m'}$ which lie within adjacent square unit cells in contact with the origin of reciprocal space but lie outside of the first Brillouin zone; these three points map into the single point denoted by the triangle which locates the tip of the **k** vector $\mathbf{k}_{m''}$ which lies both within an adjacent square unit cell in contact with the origin in reciprocal space and within the central square area delineating the first Brillouin zone.)

Thus the following statement can be made: *Any* **k** *vector extending beyond the first Brillouin zone* (viz., *the Wigner–Seitz primitive cell of the reciprocal lattice*) *can be resolved into the sum of some* **G** *vector and a* **k** *vector within the first Brillouin zone* (Chap. 6, §8). The physical significance of the choice of the *first Brillouin zone* for **k** vector reduction is that the corresponding series of wavelengths $\lambda = 2\pi/|\mathbf{k}|$ may be restricted to values equal to or greater than *twice* the lattice spacing. This choice includes **k** vectors directed in the negative as well as in the positive directions in reciprocal space. Since traveling waves are constructed by taking the product of a basis function $\exp(i\mathbf{k}_n \cdot \mathbf{r})$ with the time factor $\exp(i\omega t)$, where ω is an angular frequency and t is the time, this means that both forward and reverse traveling waves in the direct lattice are included. (For a review of traveling waves, see Chap. 1, §§ 2.2 and 4.4.)

It may be recalled that there is a sequence of Brillouin zones; the one containing the origin of reciprocal space is designated as the first. Higher Brillouin zones were discussed in Chap. 6, §9. Since the Brillouin zone segments are contiguous in the reciprocal lattice, the points mapped out by the **k** vectors can be translated from one Brillouin zone to another by a **G** vector mapping procedure somewhat analogous to that described above for the primitive

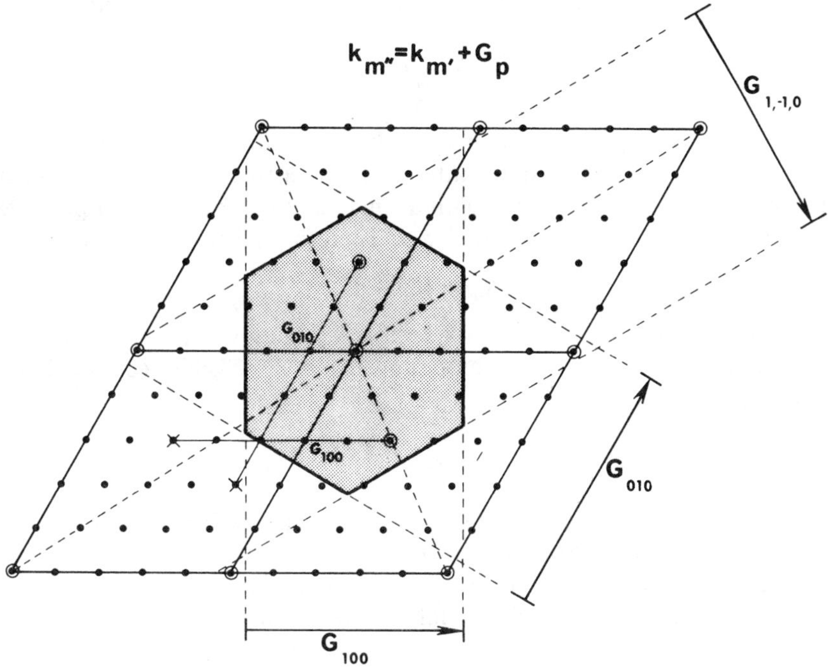

Fig. 7.7 Mapping of **k** vectors $\mathbf{k}_{\mathbf{m}'}$ from parallelepiped unit cells in contact with the origin in reciprocal space into the first Brillouin zone. (The two points marked × locate the tips of **k** vectors $\mathbf{k}_{\mathbf{m}'}$ which lie within adjacent parallelepiped unit cells in contact with the origin of reciprocal space but lie outside of the first Brillouin zone; these points map respectively into the open circles which locate the tips of **k** vectors $\mathbf{k}_{\mathbf{m}''}$ which lie both within an adjacent parallelepiped unit cell in contact with the origin in reciprocal space and within the central area delineating the first Brillouin zone. It is interesting that different **G** vectors are sometimes required for the mapping of two nearby points in the parallelepiped cell, which leads to a large separation between the points after mapping into the first Brillouin zone.)

parallelepiped unit cells. Because the volumes of the various Brillouin zones in reciprocal space are equal, the mapping procedure can be used to translate entire sections of Brillouin zones into one another.

EXERCISE Show how one can map the second and third Brillouin zones into the first Brillouin zone for the case of a rectangular lattice. Point out in particular the different **G** vectors which are required for the mapping, and tell how this differs from the mapping procedure for primitive parallelepiped unit cells. (*Hint*: See exercises in Chap. 6, §9. Also refer to Fig. 7.6.)

In order to show the usefulness of the above mathematical reductions of the set of **k** vectors, let us refer to the Fourier series

$$g(\mathbf{r}) = \sum_{\mathbf{m}} A_{\mathbf{m}} e^{i\mathbf{k}_{\mathbf{m}}\cdot\mathbf{r}} \tag{7.30}$$

for any arbitrary function $g(\mathbf{r})$ having the periodicity of the solid. This series can be formally written as a sum over the different *zones* (e.g., the *elemental primitive cells in the reciprocal lattice*) wherein the **k** vectors are to be found,

$$g(\mathbf{r}) = \sum_{\text{zones}} \sum_{\substack{\mathbf{m}\text{ within}\\ \text{a zone}}} A_{\mathbf{m}} e^{i\mathbf{k}_{\mathbf{m}}\cdot\mathbf{r}}. \tag{7.31}$$

However, we have shown above that the elemental primitive cells of the reciprocal lattice can be translated into one another by the **G** vectors; the entire reciprocal lattice can be mapped with a single unit cell by using all possible **G** vectors for translation of this initial cell. Therefore all possible **k** vectors $\mathbf{k}_{\mathbf{m}}$

$$\mathbf{k}_{\mathbf{m}} = \mathbf{k}_{\mathbf{m}'} + \mathbf{G}_{\mathbf{p}} \tag{7.32}$$

can be obtained from the subset of **k** vectors $\mathbf{k}_{\mathbf{m}'}$ lying within a given primitive cell in the reciprocal lattice by vector sums using all possible reciprocal lattice vectors. Thus we can write

$$g(\mathbf{r}) = \sum_{\mathbf{p}} \sum_{\mathbf{m}'} A_{\mathbf{m}',\mathbf{p}} e^{i(\mathbf{k}_{\mathbf{m}'} + \mathbf{G}_{\mathbf{p}})\cdot\mathbf{r}}, \tag{7.33}$$

where the sum over $\mathbf{m}'$ represents a sum over all **k** vectors in a given primitive cell in the reciprocal lattice, and the sum over **p** constitutes a sum over all possible reciprocal lattice vectors $\mathbf{G}_{\mathbf{p}}$. The sum over reciprocal lattice vectors represents in essence a summation over all possible zones. (Unless stated otherwise, the domain of $\mathbf{m}'$ is often considered to be the first Brillouin zone.)

Let us now attempt to interpret the meaning of this form of the Fourier series. The terms can be grouped in the following way,

$$g(\mathbf{r}) = \sum_{\mathbf{m}'} e^{i\mathbf{k}_{\mathbf{m}'}\cdot\mathbf{r}} \sum_{\mathbf{p}} A_{\mathbf{m}',\mathbf{p}} e^{i\mathbf{G}_{\mathbf{p}}\cdot\mathbf{r}}. \tag{7.34}$$

Since each basis state $\exp(i\mathbf{G}_{\mathbf{p}}\cdot\mathbf{r})$ is translationally invariant,

$$T_{\mathbf{j}} e^{i\mathbf{G}_{\mathbf{p}}\cdot\mathbf{r}} = e^{i\mathbf{G}_{\mathbf{p}}\cdot\mathbf{R}_{\mathbf{j}}} e^{i\mathbf{G}_{\mathbf{p}}\cdot\mathbf{r}} = e^{i\mathbf{G}_{\mathbf{p}}\cdot\mathbf{r}}, \tag{7.35}$$

the sums

$$w_{\mathbf{m}'}(\mathbf{r}) \equiv \sum_{\mathbf{p}} A_{\mathbf{m}',\mathbf{p}} e^{i\mathbf{G}_{\mathbf{p}}\cdot\mathbf{r}} \tag{7.36}$$

made up of linear combinations of these basis states are translationally invariant,

$$T_{\mathbf{j}} w_{\mathbf{m}'}(\mathbf{r}) = w_{\mathbf{m}'}(\mathbf{r} + \mathbf{R}_{\mathbf{j}}) = \sum_{\mathbf{p}} A_{\mathbf{m}',\mathbf{p}} e^{i\mathbf{G}_{\mathbf{p}}\cdot\mathbf{R}_{\mathbf{j}}} e^{i\mathbf{G}_{\mathbf{p}}\cdot\mathbf{r}} = \sum_{\mathbf{p}} A_{\mathbf{m}',\mathbf{p}} e^{i\mathbf{G}_{\mathbf{p}}\cdot\mathbf{r}} = w_{\mathbf{m}'}(\mathbf{r}). \tag{7.37}$$

Therefore any arbitrary function of position *over the domain of the solid* can be written

$$g(\mathbf{r}) = \sum_{\mathbf{m}'} w_{\mathbf{m}'}(\mathbf{r}) e^{i\mathbf{k}_{\mathbf{m}'}\cdot\mathbf{r}}, \tag{7.38}$$

where $\mathbf{m}'$ is restricted to either the **k** vectors within a single primitive parallelepiped cell in the reciprocal lattice, or to the **k** vectors within a single

Brillouin zone (commonly, the first Brillouin zone), and $w_{\mathbf{m}'}(\mathbf{r})$ is a function of position having the lattice periodicity. This represents a very important conclusion. The physical interpretation of this result is as follows: *Any position-dependent physical quantity in the solid can be represented by a linear combination of terms, each of which is the product of a coefficient consisting of some function having the lattice periodicity and a phase factor of unit modulus*; *the phase factors represent some subset of the complete set of complex basis states* $\exp(i\mathbf{k}\cdot\mathbf{r})$ *for a Fourier series having the periodicity of the solid.* The domain of $\mathbf{m}'$ can be *any* unit cell in the reciprocal lattice, as already mentioned. Whenever the domain is chosen to be one of the parallelepiped unit cells at the origin, the subset $\exp(i\mathbf{k}_{\mathbf{m}'}\cdot\mathbf{r})$ have wave vectors $\mathbf{k}_{\mathbf{m}'}$ corresponding to wavelengths in each principal direction in the crystal greater than the fundamental periodicity of the crystal lattice in that direction. Whenever the subset of $\mathbf{k}$ vectors are chosen to lie within the first Brillouin zone, the subset of $\mathbf{k}$ vectors correspond to wavelengths in each principal direction greater than *twice* the fundamental periodicity of the lattice in that direction. These results are quite surprising: it has been shown that a linear combination of terms, each representing the product of a factor having the lattice periodicity with a phase factor representing a wavelength greater than the fundamental lattice periodicity, can in fact represent an arbitrary function of position over the domain of the solid, even including functions with sharply varying nonperiodic (relative to the lattice) segments containing Fourier components with wavelengths much less than the fundamental lattice periodicity. Therefore nothing definite can be said regarding the functional dependence of a linear combination of such terms from the standpoint of any restriction on the possible shape or sharpness of the functional variation relative to the lattice spacing.

2 The Periodic Potential Characteristic of the Perfect Monocrystal

A *lattice* has been defined in Chap. 6, §2 as a set of periodic points in space with coordinates determined by all integral multiples of the elementary translation vectors $\mathbf{d}_j$. If identical point charges were placed at every lattice site, the result would be a periodic Coulomb potential. This is illustrated in Fig. 7.8. Likewise, if a set of basis vectors $\mathbf{a}_1, \mathbf{a}_2, \ldots,$ (Fig. 6.4) was used to locate an array of charges of various magnitude (and sign) within each unit cell of the lattice, the result would again be a periodic Coulomb potential, although the functional dependence would be more complex than the case with charges located only at lattice sites.

To develop our understanding further, consider also the fact that a Coulomb potential exists in the neighborhood of and within any isolated neutral atom, because any given atom is composed of a positively charged parent nucleus surrounded by a characteristic number of rapidly orbiting satellite electrons. The electron motion may indeed cause rapid temporal fluctuations in the Coulomb potential at any given spatial point, but the time-average of the potential can in principle be deduced by solving the Schrödinger equation [using

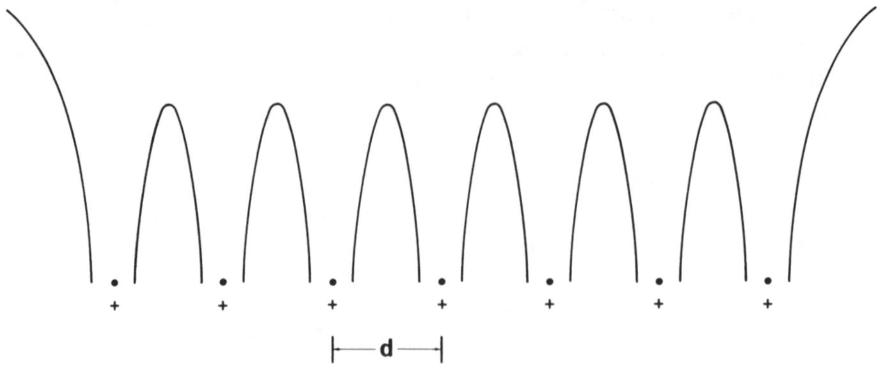

Fig. 7.8 An electron undergoes a periodic change in potential energy as it propagates through an array of charge arranged with a symmetry determined by the lattice.

Eq. (2.13)] for the system comprised of the positively charged nucleus and the negatively charged electrons. With the exception of the hydrogen atom, which has a single electron, the solution requires a consideration of a number of indistinguishable electrons, and thus is a many-body problem which is generally difficult to solve exactly. An alternative approach is to consider the simultaneous solution of a number of Schrödinger equations, namely one for each electron in the atom. This set of Schrödinger equations must be solved self-consistently, with the potential energy for a given electron being considered to be the Coulomb interaction energy of that electron with the positively charged nucleus and with all other electrons in the atom. This interaction with all other electrons in the atom is considered quantum mechanically in the sense that the Coulomb energy is that obtained from a point charge interaction of the electron with the "smeared-out" (time-averaged) charge of each of the other electrons obtained from the wave functions of the occupied quantum states. In addition, other quantum effects, such as the electron exchange interaction (Chap. 2, §1.4) must be included. The population of the various quantum energy levels must be considered to be in accordance with the Pauli exclusion principle (Chap. 2, §1.3), namely, each state represented by a complete set of quantum numbers (spin included) can contain at most one electron. Much research has been directed toward obtaining solutions to this difficult quantum-mechanical problem for various atoms. The greater the number of electrons in the atom, the greater is the difficulty in solving the problem. In general, numerical techniques and computer solutions are required, and even then a number of approximations are employed to make the problem tractable.

As mentioned, the solutions obtained from such an approach are the stationary-state solutions which yield the time-averaged probability densities for the individual electrons. The charge density distribution resulting from superimposing the charge densities of all of the individual electrons then can be used to calculate the Coulomb potential in the neighborhood of and within the atom.

Suppose now that one atom is placed at each site in some given lattice to form a solid, or even further, assume that a lattice with a basis is filled with identical groups of atoms, each group being composed of several atoms, some of which may be identical. The exact quantum-mechanical problem is then vastly more complex than that for a single atom, since in principle each electron interacts with every other electron and with every nucleus in the solid. The problem of computing electron energies and electronic wave functions in solids is a broad field in itself, and a variety of approaches has been used to obtain approximate solutions. Our present interest is directed toward the question of whether or not some general properties may be expected because of the lattice periodicity, irrespective of the type and arrangement of atoms in any one unit cell. In the limiting case in which wave functions of individual atoms filling the lattice (viz., the *atomic* wave functions) are not much perturbed by the interaction between atoms, as perhaps might be expected in a hypothetical solid having *extremely large* translation vectors relative to the atomic diameter, the repeating *atomic* Coulomb potential would theoretically constitute a periodic potential throughout all space.

If the lattice translation vectors of this hypothetical solid are considered to decrease gradually in magnitude from some very large value towards smaller values more characteristic of the atomic diameter, which is the usual case for a real solid, there must then be an interaction between various neighboring atoms. This is due to the fact that the Coulomb energies of one of the charged particles in a given atom with the charged particles of an adjacent atom will then be significant with respect to the Coulomb energies between charged particles in a given atom. Although such interaction will most certainly lead to a modification in the atomic wave functions and energy levels, there is no obvious reason to expect such interaction to cause departures from *periodicity* in the Coulomb potential. Any foreign (or "test") electron which is injected into a hypothetically perfect crystal with a given velocity should sense a nearly periodic potential as it travels through the solid. The one-electron approach to determining energy levels and wave functions is based on this physical picture of a single electron traveling through a perfectly periodic potential. If the potential is considered to be due to an array of singly charged ions (Fig. 7.8) which are arranged regularly throughout a given lattice, as contrasted with a corresponding array of atoms, the wave functions and energy levels so deduced may be identified with those corresponding to the outermost shells of the atom, since these higher energy states are responsible for the mobile conduction electrons in the solid. If on the average there is one conduction electron per atom in the solid, then the one-electron approach of solving the Schrödinger equation for a single electron in a periodic potential involves the assumption that the other conduction electrons themselves, each perhaps being in a state of rapid translational motion, constitute a periodic contribution to the overall lattice potential when temporally averaged. The question which we wish to address ourselves to is therefore whether or not any general properties can be deduced for the electronic wave functions characteristic of such a one-electron picture of the solid.

3 The Hamiltonian for an Electron in a Periodic Potential

The usual one-electron approach for deducing the quantum levels and wave functions for individual electrons in a solid is to write the time-independent Schrödinger equation (1.130),

$$\mathscr{H}\phi = \mathscr{E}\phi, \tag{7.39}$$

where some *spatially periodic* function is chosen for the potential energy, and then solve this eigenvalue equation subject to physically reasonable boundary conditions to obtain the energy eigenfunctions and energy eigenvalues for the electron in the crystal. The time-dependent eigenfunction is then obtained by taking the product of the time-independent function $\phi(\mathbf{r})$ with the usual time-dependent function [Eq. (1.129)] which gives

$$\psi(\mathbf{r}, t) = \phi(\mathbf{r})e^{-i\omega t} \tag{7.40}$$

with $\omega \equiv \mathscr{E}/\hbar$.

The periodic potential energy $V(\mathbf{r})$ satisfies the condition

$$V(\mathbf{r}) = V(\mathbf{r} + \mathbf{R_j}) \tag{7.41}$$

for all lattice translation vectors

$$\mathbf{R_j} = j_1\mathbf{d}_1 + j_2\mathbf{d}_2 + j_3\mathbf{d}_3, \tag{7.42}$$

where (j_1, j_2, j_3) represents arbitrary integer triplets and $\mathbf{d}_1$, $\mathbf{d}_2$, $\mathbf{d}_3$ are the elemental lattice translation vectors in the three principal directions in the lattice. The potential energy *operator* $\mathscr{V}(\mathbf{r})$ in the position representation is simply the potential energy $V(\mathbf{r})$; the corresponding kinetic energy operator is $-(\hbar^2/2m)\nabla^2$, with m denoting the electron mass and ∇^2 denoting the Laplacian differential operator. (For a review of these fundamentals, see Chap. 1.) Thus the Schrödinger equation for an electron in a periodic potential is

$$-(\hbar^2/2m)\nabla^2\phi^{(s)} + \mathscr{V}(\mathbf{r})\phi^{(s)} = \mathscr{E}^{(s)}\phi^{(s)}, \tag{7.43}$$

where the superscript (s) allows for the possibility that there may be a *set* of eigenfunctions (with corresponding eigenvalues) which satisfy this time-independent Schrödinger equation. The solutions $\phi^{(s)}$ to this equation then yield the time-dependent solutions

$$\psi^{(s)}(\mathbf{r}, t) = \phi^{(s)}(\mathbf{r})e^{-i\omega_s t}, \tag{7.44}$$

with

$$\omega_s \equiv \mathscr{E}^{(s)}/\hbar, \tag{7.45}$$

to the time-dependent Schrödinger equation

$$-(\hbar^2/2m)\nabla^2\psi^{(s)} + \mathscr{V}(\mathbf{r})\psi^{(s)} = i\hbar\, \partial\psi^{(s)}/\partial t. \tag{7.46}$$

4 Fourier Series Derivation of Bloch's Theorem

4.1 Expansion of the Energy Eigenfunctions in a Series of Momentum Eigenfunctions

It was shown in §1 that any arbitrary function of position meeting the Dirichlet conditions can be represented over the domain of the solid by a Fourier series of the form (7.33). The requirement that the physical probability density be continuous requires continuity of the wave functions, so the solutions to the time-independent Schrödinger equation (7.43) must indeed be required to satisfy the Dirichlet conditions (Chap. 1, §4.1). Thus one approach to solving the Schrödinger equation for an electron in a solid is to use a Fourier series expansion for the energy eigenfunctions,

$$\phi^{(s)} = \sum_{l} \sum_{\mathbf{m}'} B^{(s)}_{\mathbf{m}',l} \exp[i(\mathbf{k}_{\mathbf{m}'} + \mathbf{G}_l) \cdot \mathbf{r}]. \tag{7.47}$$

This wave function in the Fourier series representation is in effect a linear combination of momentum eigenfunctions. That is, each plane-wave basis state $\exp[i(\mathbf{k}_{\mathbf{m}'} + \mathbf{G}_l) \cdot \mathbf{r}]$ satisfies the eigenvalue equation for the momentum operator $\mathbf{p}^{\text{op}} = -i\hbar\nabla = -i\hbar(\hat{\mathbf{x}}\ \partial/\partial x + \hat{\mathbf{y}}\ \partial/\partial y + \hat{\mathbf{z}}\ \partial/\partial z)$,

$$\mathbf{p}^{\text{op}} \exp[i(\mathbf{k}_{\mathbf{m}'} + \mathbf{G}_l) \cdot \mathbf{r}] = -i\hbar\nabla \exp[i(\mathbf{k}_{\mathbf{m}'} + \mathbf{G}_l) \cdot \mathbf{r}] = \mathbf{p}_{\mathbf{m}',l} \exp[i(\mathbf{k}_{\mathbf{m}'} + \mathbf{G}_l) \cdot \mathbf{r}], \tag{7.48}$$

where the eigenvalues $\mathbf{p}_{\mathbf{m}',l}$ are the constant vectors

$$\mathbf{p}_{\mathbf{m}',l} \equiv \hbar\nabla[(\mathbf{k}_{\mathbf{m}'} + \mathbf{G}_l) \cdot \mathbf{r}]. \tag{7.49}$$

EXERCISE (a) Show that for an orthogonal system of elemental translation vectors $\mathbf{d}_1$, $\mathbf{d}_2$, $\mathbf{d}_3$, the right-hand side of Eq. (7.49) reduces to $\hbar\mathbf{k}_{\mathbf{m}'} + \hbar\mathbf{G}_l$. (b) Show that for a nonorthogonal system, the right-hand side of Eq. (7.49) reduces to

$$2\pi\hbar\{[(m'_1/L_1) + l_1]\mathbf{d}_1/d_1^2 + [(m'_2/L_2) + l_2]\mathbf{d}_2/d_2^2 + [(m'_3/L_3) + l_3]\mathbf{d}_3/d_3^2\}.$$

These same basis states are likewise kinetic energy eigenfunctions; the kinetic energy operator $\mathscr{T}^{\text{op}}$ is

$$-(\hbar^2/2m)\nabla^2 = (2m)^{-1}(-i\hbar\nabla) \cdot (-i\hbar\nabla) = (2m)^{-1}\mathbf{p}^{\text{op}} \cdot \mathbf{p}^{\text{op}}, \tag{7.50}$$

so that

$$\begin{aligned} \mathscr{T}^{\text{op}} \exp[i(\mathbf{k}_{\mathbf{m}'} + \mathbf{G}_l) \cdot \mathbf{r}] &= (2m)^{-1}(-i\hbar\nabla) \cdot (\mathbf{p}_{\mathbf{m}',l} \exp[i(\mathbf{k}_{\mathbf{m}'} + \mathbf{G}_l) \cdot \mathbf{r}]) \\ &= (2m)^{-1}\, \mathbf{p}_{\mathbf{m}',l} \cdot (\mathbf{p}^{\text{op}} \exp[i(\mathbf{k}_{\mathbf{m}'} + \mathbf{G}_l) \cdot \mathbf{r}]) \\ &= (2m)^{-1}\mathbf{p}_{\mathbf{m}',l} \cdot \mathbf{p}_{\mathbf{m}',l} \exp[i(\mathbf{k}_{\mathbf{m}'} + \mathbf{G}_l) \cdot \mathbf{r}]. \end{aligned} \tag{7.51}$$

Thus the constant kinetic energy eigenvalues $\mathscr{T}_{\mathbf{m}',l}$ are given by

$$\mathscr{T}_{\mathbf{m}',l} = (2m)^{-1}\mathbf{p}_{\mathbf{m}',l} \cdot \mathbf{p}_{\mathbf{m}',l}, \tag{7.52}$$

with the constant vectors $\mathbf{p}_{\mathbf{m}',l}$ defined by Eq. (7.49). On the other hand, these plane-wave basis states are not generally eigenfunctions of the Hamiltonian

$\mathscr{H} = \mathscr{T}^{\text{op}} + \mathscr{V}(\mathbf{r})$, where $\mathscr{V}(\mathbf{r})$ is the potential energy operator corresponding to the periodic lattice potential $V(\mathbf{r})$. That is,

$$\mathscr{H} \exp[i(\mathbf{k}_{\mathbf{m}'} + \mathbf{G}_l) \cdot \mathbf{r}] = (\mathscr{T}_{\mathbf{m}',l} + \mathscr{V}(\mathbf{r})) \exp[i(\mathbf{k}_{\mathbf{m}'} + \mathbf{G}_l) \cdot \mathbf{r}], \tag{7.53}$$

which represents an eigenvalue equation only in the special case in which $V(\mathbf{r})$ is a constant; otherwise, the quantity $[\mathscr{T}_{\mathbf{m}',l} + \mathscr{V}(\mathbf{r})]$ is position dependent and thus cannot be an eigenvalue.

EXERCISE Prove that the plane-wave basis states individually are not in general energy eigenfunctions.

Some linear combination of these basis states, such as Eq. (7.47), is therefore required to represent an energy eigenfunction $\phi^{(s)}$. The Fourier coefficients $B^{(s)}_{\mathbf{m}',l}$ in Eq. (7.47) must be chosen so that the Schrödinger equation is satisfied for the particular periodic potential in question. A linear combination of the complete set of energy eigenfunctions $\phi^{(s)}$ will then be needed to represent an arbitrary function, including that corresponding to the most general electron wave function Φ for the system.

4.2 Generalized Solutions of the Schrödinger Equation for a Periodic Potential; Energy Bands

It is convenient to use an arbitrary Fourier series expansion such as Eq. (7.1) for the periodic lattice potential for use in the Schrödinger equation. Thus we write

$$V(\mathbf{r}) = \sum_{\mathbf{n}} V_{\mathbf{n}} e^{i\mathbf{G}_{\mathbf{n}} \cdot \mathbf{r}}, \tag{7.54}$$

where the Fourier coefficients $V_{\mathbf{n}}$ are characteristic of (and determined by) the particular functional form of the periodic potential $V(\mathbf{r})$ in accordance with the general relation (7.4),

$$V_{\mathbf{n}} = \iiint V(\mathbf{r}')e^{-i\mathbf{G}_{\mathbf{n}} \cdot \mathbf{r}'} \, dx_1' \, dx_2' \, dx_3'. \tag{7.55}$$

This is the same approach which was used in Chap. 5, §12. Substituting Eqs. (7.47) and (7.54) into the time-independent Schrödinger equation (7.43) then gives

$$\begin{aligned} &-\frac{\hbar^2}{2m} \sum_{l} \sum_{\mathbf{m}'} B^{(s)}_{\mathbf{m}',l} \nabla^2 \exp[i(\mathbf{k}_{\mathbf{m}'} + \mathbf{G}_l) \cdot \mathbf{r}] \\ &+ \sum_{l} \sum_{\mathbf{m}'} \sum_{\mathbf{n}} B^{(s)}_{\mathbf{m}',l} V_{\mathbf{n}} \exp[i(\mathbf{k}_{\mathbf{m}'} + \mathbf{G}_{\mathbf{n}} + \mathbf{G}_l) \cdot \mathbf{r}] \\ &= \sum_{l} \sum_{\mathbf{m}'} \mathscr{E}^{(s)} B^{(s)}_{\mathbf{m}',l} \exp[i(\mathbf{k}_{\mathbf{m}'} + \mathbf{G}_l) \cdot \mathbf{r}]. \end{aligned} \tag{7.56}$$

However, each term in the first sum involving $-(\hbar^2/2m)\nabla^2 = \mathscr{T}^{\text{op}}$ has already been evaluated in Eq. (7.51), thus giving

$$-\frac{\hbar^2}{2m}\sum_{l}\sum_{\mathbf{m}'} B^{(s)}_{\mathbf{m}',l}\nabla^2 \exp[i(\mathbf{k}_{\mathbf{m}'} + \mathbf{G}_l)\cdot\mathbf{r}]$$
$$= \sum_{l}\sum_{\mathbf{m}'} B^{(s)}_{\mathbf{m}',l}\mathscr{T}_{\mathbf{m}',l} \exp[i(\mathbf{k}_{\mathbf{m}'} + \mathbf{G}_l)\cdot\mathbf{r}], \tag{7.57}$$

where the values of the constants $\mathscr{T}_{\mathbf{m}',l}$ are given by Eq. (7.52). The vector sum $\mathbf{G}_\mathbf{n} + \mathbf{G}_l$ appears in the arguments of the second group of terms in Eq. (7.56); however,

$$\mathbf{G}_\mathbf{n} + \mathbf{G}_l = \mathbf{G}_\mathbf{j}, \tag{7.58}$$

where $\mathbf{j} = (j_1, j_2, j_3)$ is determined by $j_1 = n_1 + l_1$, $j_2 = n_2 + l_2$, $j_3 = n_3 + l_3$. Thus $\mathbf{j} = \mathbf{n} + \boldsymbol{l}$, so we write

$$\sum_{l}\sum_{\mathbf{m}'}\sum_{\mathbf{n}} B^{(s)}_{\mathbf{m}',l}V_\mathbf{n} \exp[i(\mathbf{k}_{\mathbf{m}'} + \mathbf{G}_\mathbf{n} + \mathbf{G}_l)\cdot\mathbf{r}]$$
$$= \sum_{l}\sum_{\mathbf{m}'}\sum_{\mathbf{j}-l} B^{(s)}_{\mathbf{m}',l}V_{\mathbf{j}-l} \exp[i(\mathbf{k}_{\mathbf{m}'} + \mathbf{G}_\mathbf{j})\cdot\mathbf{r}]. \tag{7.59}$$

The integer triplet $\mathbf{j} - \boldsymbol{l} = \mathbf{n}$ takes on all possible values in the sum for each $\boldsymbol{l}$; thus a replacement of the summation index $\mathbf{j} - \boldsymbol{l}$ by the alternate summation index $\mathbf{j}$, although representing a reordering of the terms in the sum, does not change the set of terms appearing in the sum. Since convergence is not a problem with Fourier series representations of continuous functions, the reordering of terms does not change the value of the sum. Reordering the terms in this manner converts Eq. (7.59) to

$$\sum_{l}\sum_{\mathbf{m}'}\sum_{\mathbf{n}} B^{(s)}_{\mathbf{m}',l}V_\mathbf{n} \exp[i(\mathbf{k}_{\mathbf{m}'} + \mathbf{G}_\mathbf{n} + \mathbf{G}_l)\cdot\mathbf{r}]$$
$$= \sum_{l}\sum_{\mathbf{m}'}\sum_{\mathbf{j}} B^{(s)}_{\mathbf{m}',l}V_{\mathbf{j}-l} \exp[i(\mathbf{k}_{\mathbf{m}'} + \mathbf{G}_\mathbf{j})\cdot\mathbf{r}]. \tag{7.60}$$

Substituting Eqs. (7.57) and (7.60) into Eq. (7.56) gives

$$\sum_{l}\sum_{\mathbf{m}'}\sum_{\mathbf{j}} B^{(s)}_{\mathbf{m}',l}V_{\mathbf{j}-l} \exp[i(\mathbf{k}_{\mathbf{m}'} + \mathbf{G}_\mathbf{j})\cdot\mathbf{r}]$$
$$- \sum_{l}\sum_{\mathbf{m}'} B^{(s)}_{\mathbf{m}',l}[\mathscr{E}^{(s)} - \mathscr{T}_{\mathbf{m}',l}] \exp[i(\mathbf{k}_{\mathbf{m}'} + \mathbf{G}_l)\cdot\mathbf{r}] = 0. \tag{7.61}$$

If the Kronecker delta function $\delta_{\mathbf{j}l} = \delta_{l\mathbf{j}}$ is introduced, where $\delta_{\mathbf{j}l} =$ unity if $j_1 = l_1$, $j_2 = l_2$, and $j_3 = l_3$ but is zero otherwise, the right-hand side can be written as a triple sum involving $\exp[i(\mathbf{k}_{\mathbf{m}'} + \mathbf{G}_\mathbf{j})\cdot\mathbf{r}]$, so that Eq. (7.61) becomes

$$0 = \sum_{l}\sum_{\mathbf{m}'}\sum_{\mathbf{j}} B^{(s)}_{\mathbf{m}',l}\{V_{\mathbf{j}-l} \exp[i(\mathbf{k}_{\mathbf{m}'} + \mathbf{G}_\mathbf{j})\cdot\mathbf{r}]$$
$$- \delta_{\mathbf{j}l}[\mathscr{E}^{(s)} - \mathscr{T}_{\mathbf{m}',l}] \exp[i(\mathbf{k}_{\mathbf{m}'} + \mathbf{G}_\mathbf{j})\cdot\mathbf{r}]\}. \tag{7.62}$$

The order of the summations is unimportant for a convergent series of terms, so this equation can be written in the form

$$0 = \sum_{\mathbf{m}'} e^{i\mathbf{k}_{\mathbf{m}'}\cdot\mathbf{r}} \left[\sum_{\mathbf{j}} e^{i\mathbf{G}_{\mathbf{j}}\cdot\mathbf{r}} \left\{ \sum_{\boldsymbol{l}} B^{(s)}_{\mathbf{m}',\boldsymbol{l}} [V_{\mathbf{j}-\boldsymbol{l}} - \delta_{\mathbf{j}\boldsymbol{l}}(\mathscr{E}^{(s)} - \mathscr{T}_{\mathbf{m}',\boldsymbol{l}})] \right\} \right]. \tag{7.63}$$

It is now helpful to recall the fact that basis vectors $e^{i\mathbf{k}_{\mathbf{m}}\cdot\mathbf{r}}$ for the complex Fourier series representation of an arbitrary function over the domain of the solid are linearly independent functions, where $\mathbf{k}_{\mathbf{m}} = 2\pi[(m_1/L_1)\mathbf{b}_1 + (m_2/L_2)\mathbf{b}_2 + (m_3/L_3)\mathbf{b}_3]$. This has been discussed in some detail in §1. Both the $\mathbf{k}_{\mathbf{m}'}$ vectors and the $\mathbf{G}_{\mathbf{j}}$ vectors represent subsets of the $\mathbf{k}_{\mathbf{m}}$ vectors, the $\mathbf{k}_{\mathbf{m}'}$ vectors being those which lie within some given primitive unit cell of the reciprocal lattice and the $\mathbf{G}_{\mathbf{j}}$ vectors being those which map out the reciprocal lattice points. Therefore the set of functions $\exp[i\mathbf{k}_{\mathbf{m}'}\cdot\mathbf{r}]$, as well as the set of functions $\exp[i\mathbf{G}_{\mathbf{j}}\cdot\mathbf{r}]$, are linearly independent. From the definition of linear independence, each coefficient of $\exp[i\mathbf{k}_{\mathbf{m}'}\cdot\mathbf{r}]$ must be zero in order that the linear combination represented by Eq. (7.63) be zero. Therefore the Schrödinger equation in the form (7.63) can be satisfied only if

$$\sum_{\mathbf{j}} e^{i\mathbf{G}_{\mathbf{j}}\cdot\mathbf{r}} \left\{ \sum_{\boldsymbol{l}} B^{(s)}_{\mathbf{m}',\boldsymbol{l}} [V_{\mathbf{j}-\boldsymbol{l}} - \delta_{\boldsymbol{l}\mathbf{j}}(\mathscr{E}^{(s)} - \mathscr{T}_{\mathbf{m}',\boldsymbol{l}})] \right\} = 0 \tag{7.64}$$

for each allowable triplet $\mathbf{m}'$, the allowable triplets being those for which $\mathbf{k}_{\mathbf{m}'}$ is confined to lie within some given primitive unit cell or Brillouin zone of the reciprocal lattice. The left-hand side of Eq. (7.64) itself represents a linear combination of the linearly independent functions $\exp[i\mathbf{G}_{\mathbf{j}}\cdot\mathbf{r}]$; therefore, it can be zero only if

$$\sum_{\boldsymbol{l}} B^{(s)}_{\mathbf{m}',\boldsymbol{l}} [V_{\mathbf{j}-\boldsymbol{l}} - \delta_{\boldsymbol{l}\mathbf{j}}(\mathscr{E}^{(s)} - \mathscr{T}_{\mathbf{m}',\boldsymbol{l}})] = 0 \qquad (j_1, j_2, j_3 \quad \text{arbitrary}) \tag{7.65}$$

for each triplet $\mathbf{j}$. [Every possible triplet (j_1, j_2, j_3) represents an allowable $\mathbf{j}$.] Next, Eq. (7.65), with $\mathbf{j}$ variable, can be recognized as a set of linear homogeneous algebraic equations for the quantities $B^{(s)}_{\mathbf{m}',\boldsymbol{l}}$. The domain of the triplet $\boldsymbol{l}$ is the same as that of the triplet $\mathbf{j}$, namely, all possible integer triplet values. Nonzero values for the coefficients $B^{(s)}_{\mathbf{m}',\boldsymbol{l}}$ follow from the solution of these equations by means of Cramer's Rule [Wylie (1951)] only if the determinant of the coefficients $a_{\mathbf{j}\boldsymbol{l}}$,

$$a_{\mathbf{j}\boldsymbol{l}} \equiv V_{\mathbf{j}-\boldsymbol{l}} - \delta_{\boldsymbol{l}\mathbf{j}}[\mathscr{E}^{(s)} - \mathscr{T}_{\mathbf{m}',\boldsymbol{l}}], \tag{7.66}$$

is equal to zero:

$$\det(a_{\mathbf{j}\boldsymbol{l}}) = 0. \tag{7.67}$$

The symbol det, meaning *determinant*, can be represented alternatively by double bars, in which case Eqs. (7.66) and (7.67) give

$$\| V_{\mathbf{j}-\boldsymbol{l}} - \delta_{\boldsymbol{l}\mathbf{j}}(\mathscr{E}^{(s)} - \mathscr{T}_{\mathbf{m}',\boldsymbol{l}}) \| = 0 \qquad \textbf{(secular equation)}. \tag{7.68}$$

The utility of a determinant of such a high order ($j_1, j_2, j_3, l_1, l_2, l_3$ each take on *all* possible integer values) may be questioned, so let us hasten to state that our objective in this development is primarily to deduce some of the *general* properties of the eigenfunctions and eigenvalues of the Schrödinger equation for a periodic potential rather than develop a practical method to carry out actual numerical computations for various specific systems. In principle, at least, the above determinantal equation represents an algebraic equation which determines the eigenvalues $\mathscr{E}^{(s)}$. The determinant itself is called the *secular determinant*, and the algebraic equation resulting from evaluating the determinant and setting the result equal to zero is called the *secular equation*. The degree of this algebraic equation will be equal to the number of rows (or columns) in the determinant, which is designated the *order* of the determinant. Since $\mathbf{j}$ and $\boldsymbol{l}$ represent dummy indices in the determinantal equation which run over all possible triplets, it can be seen that the order of the determinant is very high. It is readily noted from the development that the number of rows and the number of columns are equal, since each number is derived from a one-to-one correspondence with the reciprocal lattice vectors. It is thus evident that the number of roots $\mathscr{E}^{(s)}(\mathbf{k}_{\mathbf{m}'})$ of the secular equation for a given $\mathbf{k}_{\mathbf{m}'}$ is equal to the number of reciprocal lattice vectors, since the number of reciprocal lattice vectors defines the order of the secular determinant. Furthermore, each of the roots of the corresponding secular equation will involve all Fourier coefficients $V_{\mathbf{n}} = V_{\mathbf{j}-\boldsymbol{l}}$ of the lattice potential and all reciprocal lattice vectors $\mathbf{G}_{\boldsymbol{l}}$ through the $\mathscr{T}_{\mathbf{m}',\boldsymbol{l}}$ quantities defined by Eq. (7.52). In addition, there will be an explicit dependence of each root on the vector $\mathbf{k}_{\mathbf{m}'}$ which occurs explicitly in the $\mathscr{T}_{\mathbf{m}',\boldsymbol{l}}$. For each different vector $\mathbf{k}_{\mathbf{m}'}$, the determinantal equation will therefore be different, so the set of roots of the equation will be characteristic of the $\mathbf{k}_{\mathbf{m}'}$ vector in question. For any specific $\mathbf{k}_{\mathbf{m}'}$ the determinantal equation has multiple roots, the actual number being given by the order of the determinant, as mentioned already. The order of the determinant is equal to the number of reciprocal lattice vectors $\mathbf{G}_{\boldsymbol{l}}$ and so is in principle unlimited. Thus there are an unlimited number of solutions $\mathscr{E}^{(s)}$ for a given $\mathbf{k}_{\mathbf{m}'}$, and the index s will serve to label the solution in question. The index s will be designated the *band index*. The explicit dependence of a given $\mathscr{E}^{(s)}$ on the vector $\mathbf{k}_{\mathbf{m}'}$ can be noted by adding $\mathbf{k}_{\mathbf{m}'}$ as the argument of $\mathscr{E}^{(s)}$ to give $\mathscr{E}^{(s)}(\mathbf{k}_{\mathbf{m}'})$. The allowed values of $\mathbf{k}_{\mathbf{m}'}$ are very closely spaced in the reciprocal lattice; the discrete set of energy levels $\mathscr{E}^{(s)}(\mathbf{k}_{\mathbf{m}'})$ for a fixed *band index s* with $\mathbf{k}_{\mathbf{m}'}$ ranging over its domain of one unit cell in the reciprocal lattice is called an *energy band*. Since $\mathscr{E} = \hbar\omega$, the function $\mathscr{E}^{(s)}(\mathbf{k}_{\mathbf{m}'})$ versus $\mathbf{k}_{\mathbf{m}'}$ represents what is generally known as a *dispersion relation* in wave propagation. A schematic illustration of $\mathscr{E}^{(s)}(\mathbf{k}_{\mathbf{m}'})$ versus $|\mathbf{k}_{\mathbf{m}'}|$ for a given direction $\mathbf{k}$ in $\mathbf{k}$ space is illustrated in Fig. 7.9 for three values of the band index s. The results illustrated in this figure are deduced by a reduction of the $\mathbf{k}$ vectors (and corresponding energy eigenvalues) of Fig. 5.6 for the perturbation treatment of a periodic lattice potential to the first Brillouin zone in accordance with the procedure described in §1.

For a given $\mathbf{k}_{\mathbf{m}'}$, each of the energies $\mathscr{E}^{(s)}(\mathbf{k}_{\mathbf{m}'})$ labeled by the band index s can be used in the set of linear homogeneous algebraic equations (7.65) to determine the Fourier coefficients $B^{(s)}_{\mathbf{m}',\boldsymbol{l}}$ corresponding to all possible integer triplets $\boldsymbol{l}$. It is well

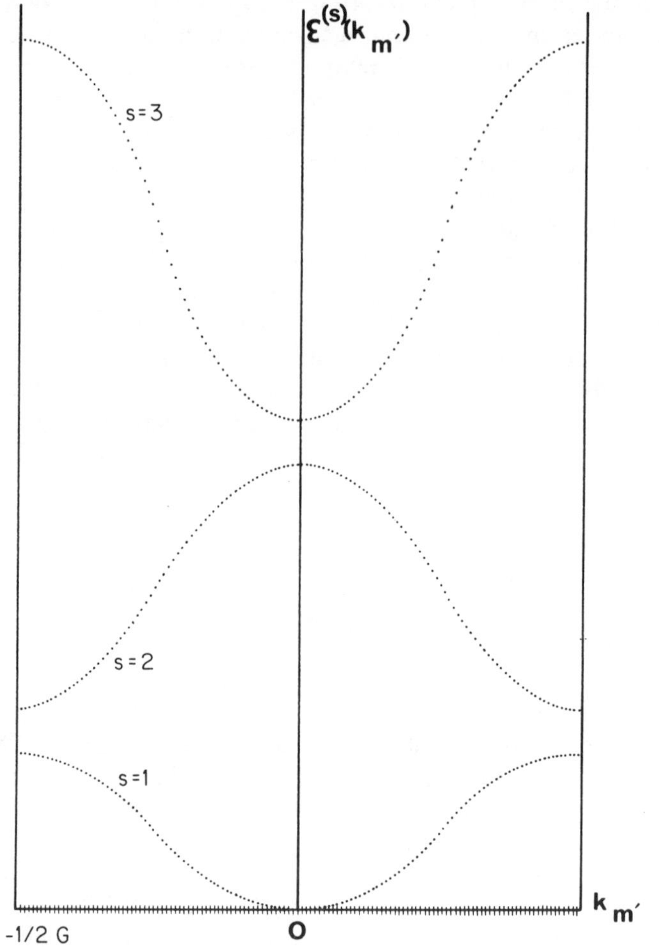

Fig. 7.9 Energy eigenvalues for Bloch functions. [These discrete energies are obtained by solving the algebraic *secular equation* (7.68) obtained upon expansion of the *secular determinant*. It can be seen that for each given value of the *crystal momentum* $\hbar\mathbf{k}_{\mathbf{m}'}$ there are a number of energy eigenvalues $\mathscr{E}^{(s)}(\mathbf{k}_{\mathbf{m}'})$ labeled by the energy band index s. The quasi-continuum of discrete energies noted within a given band is broken by the energy gaps, namely, the regions of energy between bands containing no energy eigenvalues; these gaps occur for $\mathbf{k}_{\mathbf{m}'}$ vectors which reach the boundaries of the Brillouin zone.]

known that the solution of a set of N independent simultaneous linear homogeneous algebraic equations in N unknowns yields $N - 1$ of the unknowns in terms of the remaining unknown, the remaining unknown being arbitrary. Later we see that this flexibility is useful for *normalizing* the resulting wave functions. The dependence of the Fourier coefficients upon the vector $\mathbf{k}_{\mathbf{m}'}$ is explicitly denoted by the subscript $\mathbf{m}'$, and the dependence upon the band index is explicitly denoted by the superscript s.

For a given band index s, the linear combination given by Eq. (7.47) for $\phi^{(s)}$ involves all possible values of $\mathbf{k}_{\mathbf{m}'}$. However, as shown above, the energy values $\mathscr{E}^{(s)}(\mathbf{k}_{\mathbf{m}'})$ obtained from the secular equation depend in general upon the specific vector $\mathbf{k}_{\mathbf{m}'}$. Although there will undoubtedly be many degenerate states which occur in any particular solution consistent with the above approach, there is certainly no reason to believe that all energies $\mathscr{E}^{(s)}(\mathbf{k}_{\mathbf{m}'})$ for fixed s and variable $\mathbf{k}_{\mathbf{m}'}$ will be degenerate. Therefore we are forced to restrict the linear combination (7.47) to those terms which when grouped together constitute an eigenfunction $\phi^{(s)}$ corresponding to a specific eigenvalue; otherwise the time-independent Schrödinger eigenvalue equation (7.43) will not be satisfied. To ensure that we do have a grouping of terms corresponding to a single energy $\mathscr{E}^{(s)}(\mathbf{k}_{\mathbf{m}''})$, each coefficient $B^{(s)}_{\mathbf{m}',l}$ in Eq. (7.47) can be set equal to zero for $\mathbf{m}' \neq \mathbf{m}''$. Two points should be made regarding this procedure. First, we must show that the functions resulting from this modified procedure indeed constitute a complete set of energy eigenfunctions. Second, we must recognize that each resulting function will probably not represent the only possible eigenfunction corresponding to this particular energy, since there is most likely degeneracy for specific subsets of the different $\mathbf{k}_{\mathbf{m}'}$ states, especially for a three-dimensional system.

4.3 Solutions to the Schrödinger Equation Characterized by a Wave Vector $\mathbf{k}_{\mathbf{m}}$; Bloch Functions

Let us now address ourselves to the above question: Can energy eigenfunctions $\phi^{(s)}$ be constructed from individual subsets of the set of terms appearing in the general expansion (7.47), each subset corresponding to a single value of $\mathbf{k}_{\mathbf{m}'}$? Our approach is to choose one of the $\mathbf{k}_{\mathbf{m}'}$ vectors, $\mathbf{k}_{\mathbf{m}''}$ for example, set all other coefficients $B^{(s)}_{\mathbf{m}',l}$ equal to zero,

$$\begin{aligned}\phi^{(s)}_{\mathbf{m}''} &\equiv \sum_{l}\sum_{\mathbf{m}'} \delta_{\mathbf{m}',\mathbf{m}''}\, B^{(s)}_{\mathbf{m}',l} \exp[i(\mathbf{k}_{\mathbf{m}'} + \mathbf{G}_l)\cdot\mathbf{r}] \\ &= \sum_{l} B^{(s)}_{\mathbf{m}'',l} \exp[i(\mathbf{k}_{\mathbf{m}''} + \mathbf{G}_l)\cdot\mathbf{r}] \\ &= \exp[i(\mathbf{k}_{\mathbf{m}''}\cdot\mathbf{r})] \sum_{l} B^{(s)}_{\mathbf{m}'',l} \exp(i\mathbf{G}_l\cdot\mathbf{r}), \end{aligned} \tag{7.69}$$

and then ascertain whether the resulting linear combination can provide a solution to the Schrödinger equation (7.43) by substituting Eq. (7.69) into Eq. (7.43). Since the only difference in the new trial solution (7.69) and the previous trial solution (7.47) is the addition of the Kronecker delta function $\delta_{\mathbf{m}',\mathbf{m}''}$ as a factor multiplying every coefficient $B^{(s)}_{\mathbf{m}',l}$, the development is the same with this exception. For example, Eq. (7.56) is obtained with the modification that $B^{(s)}_{\mathbf{m}',l}$ is replaced everywhere by $\delta_{\mathbf{m}',\mathbf{m}''}B^{(s)}_{\mathbf{m}',l}$; this can be noted to reduce the sums over $\mathbf{m}'$ to a single term involving $\mathbf{m}''$, if desired, although such a reduction is not mandatory. Likewise, Eq. (7.61) follows, but again with $\delta_{\mathbf{m}',\mathbf{m}''}B^{(s)}_{\mathbf{m}',l}$ replacing $B^{(s)}_{\mathbf{m}',l}$ everywhere. This then leads to the similarly modified form of Eq. (7.63), namely,

$$0 = \sum_{\mathbf{m}'} e^{i\mathbf{k}_{\mathbf{m}'}\cdot\mathbf{r}} \left[\sum_{\mathbf{j}} e^{i\mathbf{G}_{\mathbf{j}}\cdot\mathbf{r}} \left\{ \sum_{l} \delta_{\mathbf{m}',\mathbf{m}''} B^{(s)}_{\mathbf{m}',l} [V_{\mathbf{j}-l} - \delta_{l\mathbf{j}}(\mathscr{E}^{(s)} - \mathscr{T}_{\mathbf{m}',l})] \right\} \right]. \quad (7.70)$$

Using the Kronecker delta function property to reduce each sum over $\mathbf{m}'$ to a single term gives

$$0 = e^{i\mathbf{k}_{\mathbf{m}''}\cdot\mathbf{r}} \left[\sum_{\mathbf{j}} e^{i\mathbf{G}_{\mathbf{j}}\cdot\mathbf{r}} \left\{ \sum_{l} B^{(s)}_{\mathbf{m}'',l} [V_{\mathbf{j}-l} - \delta_{l\mathbf{j}}(\mathscr{E}^{(s)} - \mathscr{T}_{\mathbf{m}'',l})] \right\} \right]. \quad (7.71)$$

The factor $\exp(i\mathbf{k}_{\mathbf{m}''}\cdot\mathbf{r})$ is never zero; dividing Eq. (7.71) by this factor gives

$$0 = \sum_{\mathbf{j}} e^{i\mathbf{G}_{\mathbf{j}}\cdot\mathbf{r}} \left\{ \sum_{l} B^{(s)}_{\mathbf{m}'',l} [V_{\mathbf{j}-l} - \delta_{l\mathbf{j}}(\mathscr{E}^{(s)} - \mathscr{T}_{\mathbf{m}'',l})] \right\}. \quad (7.72)$$

This represents a linear combination of linearly independent functions $\exp(i\mathbf{G}_{\mathbf{j}}\cdot\mathbf{r})$, and hence it can be zero only if the coefficient is zero for each triplet $\mathbf{j}$,

$$\sum_{l} B^{(s)}_{\mathbf{m}'',l} [V_{\mathbf{j}-l} - \delta_{l\mathbf{j}}(\mathscr{E}^{(s)} - \mathscr{T}_{\mathbf{m}'',l})] = 0 \qquad (j_1, j_2, j_3 \ \text{arbitrary}). \quad (7.73)$$

This can be recognized as a set of linear homogeneous equations ($\mathbf{j}$ variable) for the quantities $B^{(s)}_{\mathbf{m}'',l}$, so nonzero values for each of these quantities requires that the determinant of the coefficients be zero,

$$\| V_{\mathbf{j}-l} - \delta_{l\mathbf{j}}(\mathscr{E}^{(s)} - \mathscr{T}_{\mathbf{m}'',l}) \| = 0. \quad (7.74)$$

A comparison of Eq. (7.74) with our former secular determinant (7.68) reveals that the secular equation for $\mathscr{E}^{(s)}$ will be modified only by the substitution of the subscript $\mathbf{m}''$ for $\mathbf{m}'$, the subscript $\mathbf{m}''$ representing the specific vector $\mathbf{k}_{\mathbf{m}''}$. It is thus seen that each eigenvalue $\mathscr{E}^{(s)}(\mathbf{k}_{\mathbf{m}''})$ presently determined by using the restricted expansion (7.69) must be identical to one of the eigenvalues $\mathscr{E}^{(s)}(\mathbf{k}_{\mathbf{m}'})$ previously determined by means of the general expansion (7.47), namely, it is equal to the one in which $\mathbf{k}_{\mathbf{m}'}$ has the value $\mathbf{k}_{\mathbf{m}''}$. The set of energy eigenvalues $\mathscr{E}^{(s)}(\mathbf{k}_{\mathbf{m}''})$ (s denoting the root) obtained by solving the algebraic equation can then be used in the set of linear algebraic equations (7.73) to determine the coefficients $B^{(s)}_{\mathbf{m}'',l}$. It can be noted that the set of equations (7.73) is the same as the previous set (7.65) with the exception that $\mathbf{m}''$ appears in place of $\mathbf{m}'$. Therefore the coefficients $B^{(s)}_{\mathbf{m}'',l}$ are identical to a subset of the coefficients $B^{(s)}_{\mathbf{m}',l}$ previously determined by means of the general expansion (7.47). Specifically, this subset is delimited by $\mathbf{k}_{\mathbf{m}'}$ having the specific value $\mathbf{k}_{\mathbf{m}''}$. It is clear that by choosing successively each of the various $\mathbf{k}_{\mathbf{m}'}$ vectors in the first Brillouin zone in reciprocal space for $\mathbf{k}_{\mathbf{m}''}$ in the restricted expansion (7.69), a set of energy eigenfunctions $\phi^{(s)}_{\mathbf{m}''}$ can be generated. By considering all energies and all coefficients obtained in this way from these restricted expansions, a one-to-one correspondence is noted with the set of energy eigenvalues and Fourier coefficients previously deduced by means of the general expansion. Thus the superposition of the complete set of restricted expansions yields the general expansion. Our conclusions are therefore the following:

1. Each restricted expansion of the type (7.69) provides a solution to the Schrödinger equation corresponding to a single energy eigenvalue.

2. The energy eigenvalues obtained from the restricted expansion (7.69) are identical to those obtained by means of the general Fourier series expansion.

3. The coefficients obtained by means of all restricted expansions of the type (7.69), for $\mathbf{k}_{\mathbf{m}''}$ *within the first Brillouin zone, are identical to those obtained by means of the general Fourier series expansion (7.47).*

On the basis of the first conclusion, it is thus *possible* to find energy eigenfunctions for a periodic potential which have the form

$$\phi_{\mathbf{m}''}^{(s)}(\mathbf{r}) = u_{\mathbf{m}''}^{(s)}(\mathbf{r}) e^{i\mathbf{k}_{\mathbf{m}''}\cdot\mathbf{r}} \qquad \textbf{(Bloch functions)}, \tag{7.75}$$

with the energy eigenvalue $\mathscr{E}^{(s)}(\mathbf{k}_{\mathbf{m}''})$ corresponding to the wave vector $\mathbf{k}_{\mathbf{m}''}$ and the function $u_{\mathbf{m}''}^{(s)}(\mathbf{r})$ defined to be the linear combination

$$u_{\mathbf{m}''}^{(s)}(\mathbf{r}) = \sum_{l} B_{\mathbf{m}'',l}^{(s)} \exp(i\mathbf{G}_l\cdot\mathbf{r}), \tag{7.76}$$

where the allowable energy eigenvalues and the corresponding coefficients $B_{\mathbf{m}'',l}^{(s)}$ are determined by the Schrödinger equation in accordance with the procedure which we have developed above. It is readily seen that the functions $u_{\mathbf{m}''}^{(s)}(\mathbf{r})$ have the lattice periodicity,

$$\begin{aligned} u_{\mathbf{m}''}^{(s)}(\mathbf{r} + \mathbf{R_j}) &= \sum_{l} B_{\mathbf{m}'',l}^{(s)} \exp[i\mathbf{G}_l\cdot(\mathbf{r} + \mathbf{R_j})] \\ &= \sum_{l} B_{\mathbf{m}'',l}^{(s)} \exp(i\mathbf{G}_l\cdot\mathbf{r}) \exp[i2\pi(l_1 j_1 + l_2 j_2 + l_3 j_3)] \\ &= \sum_{l} B_{\mathbf{m}'',l}^{(s)} \exp(i\mathbf{G}_l\cdot\mathbf{r}) = u_{\mathbf{m}''}^{(s)}(\mathbf{r}). \end{aligned} \tag{7.77}$$

Thus the energy eigenfunction (7.75) represents the product of a function $u(\mathbf{r})$ which has the lattice periodicity and a phase factor $e^{i\mathbf{k}\cdot\mathbf{r}}$. Eigenfunctions of this form are generally known as *Bloch functions*. The phase factor corresponds to a sinusoidal spatial modulation of the real and imaginary parts of the wave function in accordance with the identity

$$e^{i\mathbf{k}_{\mathbf{m}'}\cdot\mathbf{r}} = \cos(\mathbf{k}_{\mathbf{m}'}\cdot\mathbf{r}) + i\sin(\mathbf{k}_{\mathbf{m}'}\cdot\mathbf{r}). \tag{7.78}$$

The wavelength $\lambda_{\mathbf{m}'} = 2\pi/|\mathbf{k}_{\mathbf{m}'}|$ of the modulation is greater than twice the lattice spacing provided we assume that we have chosen $\mathbf{k}_{\mathbf{m}'}$ to lie within the first Brillouin zone. These results are indicated in Fig. 7.10.

On the basis of conclusions 2 and 3 above, we see that the set of functions $\phi_{\mathbf{m}''}^{(s)}(\mathbf{r})$ of the type (7.75) which are obtained by permitting $\mathbf{m}''$ to range over all values within the first Brillouin zone involve all $\mathbf{k}_{\mathbf{m}''}$ values and all $\mathbf{G}_l$ values. The question is whether or not this new set of functions $\phi_{\mathbf{m}''}^{(s)}(\mathbf{r})$ is complete. If this additional point can be demonstrated, then we will have proven quite generally

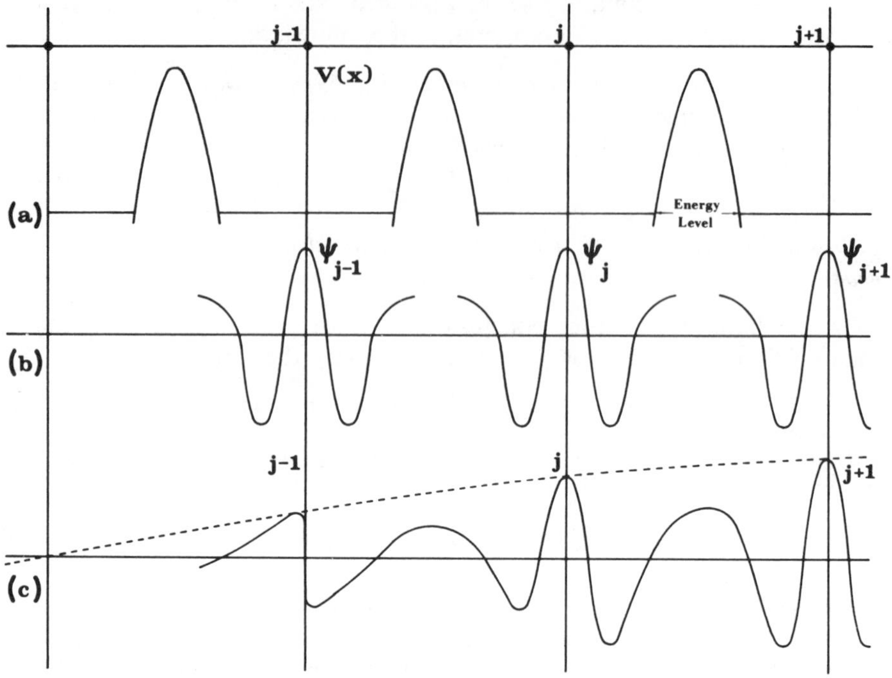

Fig. 7.10 Energy eigenfunctions for a periodic lattice potential as sketched in (a) can be chosen to have the Bloch form $\phi^{(s)}_{\mathbf{m}'}(\mathbf{r}) = u^{(s)}_{\mathbf{m}'}(\mathbf{r})\, e^{i\mathbf{k}_{\mathbf{m}'}\cdot\mathbf{r}}$, where the functions $u^{(s)}_{\mathbf{m}'}(\mathbf{r})$ illustrated in (b) and distinguished by the band index s and the crystal momentum wave vector $\mathbf{k}_{\mathbf{m}'}$ are spatially periodic with the lattice periodicity, and the real and imaginary parts of the factor $e^{i\mathbf{k}_{\mathbf{m}'}\cdot\mathbf{r}}$ lead to a modulation in the functions $u^{(s)}_{\mathbf{m}'}(\mathbf{r})$ as indicated in (c) with a wavelength $\lambda_{\mathbf{m}'} = 2\pi/|\mathbf{k}_{\mathbf{m}'}|$ greater than twice the lattice translation distance d.

that *a complete set of energy eigenfunctions for a periodic potential can always be found which have the Bloch form* (7.75). This in essence is *Bloch's theorem.*

5 Properties of Bloch Functions

5.1 Completeness of the Set

The approach which we use to prove completeness of the set of Bloch functions $\phi^{(s)}_{\mathbf{m}''}(\mathbf{r})$ given by Eq. (7.75), assuming that all $\mathbf{k}_{\mathbf{m}''}$ in the first Brillouin zone and all roots s of the secular determinant are considered, is to demonstrate that any arbitrary function $g(\mathbf{r})$ which can be represented by the complex Fourier series expansion (7.34) can also be represented by a linear combination of the Bloch functions. Considering *any specific function* $g(\mathbf{r})$, then from Eqs. (7.11) and (7.12) as expressed in terms of Eqs. (7.32) and (7.34) we can write

$$g(\mathbf{r}) = \sum_{\mathbf{m}'} e^{i\mathbf{k}_{\mathbf{m}'}\cdot\mathbf{r}} \sum_{\mathbf{p}} A_{\mathbf{m}',\,\mathbf{p}} e^{i\mathbf{G}_{\mathbf{p}}\cdot\mathbf{r}}, \tag{7.79}$$

with the coefficients $A_{\mathbf{m}',\mathbf{p}}$ determined by

$$A_{\mathbf{m}',\mathbf{p}} = N^{-1}\int_{x_1}^{x_1+L_1} dx_1' \int_{x_2}^{x_2+L_2} dx_2' \int_{x_3}^{x_3+L_3} dx_3'\, g(\mathbf{r}')e^{-i(\mathbf{k}_{\mathbf{m}'}+\mathbf{G}_{\mathbf{p}})\cdot\mathbf{r}'}. \quad (7.80)$$

Let us *attempt* to represent $g(\mathbf{r})$ in addition by a linear combination of all of the Bloch functions which we can generate,

$$g(\mathbf{r}) = \sum_s \sum_{\mathbf{m}'} \Gamma^{(s)}_{\mathbf{m}'}\phi^{(s)}_{\mathbf{m}'}(\mathbf{r}), \quad (7.81)$$

where the $\Gamma^{(s)}_{\mathbf{m}'}$ denote arbitrary expansion coefficients to be determined by requiring self-consistency between Eqs. (7.81) and (7.79). (The subscript $\mathbf{m}'$ is used instead of $\mathbf{m}''$ for simplicity; both represent arbitrary triplets corresponding to $\mathbf{k}_{\mathbf{m}}$ vectors in the first Brillouin zone.) Substituting Eqs. (7.75) and (7.76) into Eq. (7.81) gives

$$\begin{aligned} g(\mathbf{r}) &= \sum_s \sum_{\mathbf{m}'} \Gamma^{(s)}_{\mathbf{m}'} e^{i\mathbf{k}_{\mathbf{m}'}\cdot\mathbf{r}} \sum_{\mathbf{p}} B^{(s)}_{\mathbf{m}',\mathbf{p}} e^{i\mathbf{G}_{\mathbf{p}}\cdot\mathbf{r}} \\ &= \sum_{\mathbf{m}'} e^{i\mathbf{k}_{\mathbf{m}'}\cdot\mathbf{r}} \sum_{\mathbf{p}} e^{i\mathbf{G}_{\mathbf{p}}\cdot\mathbf{r}} \sum_s \Gamma^{(s)}_{\mathbf{m}'} B^{(s)}_{\mathbf{m}',\mathbf{p}}, \end{aligned} \quad (7.82)$$

where $\mathbf{p}$ has been used for a dummy index in place of $\boldsymbol{l}$ and the rearrangement of the summations represents only a reordering of the terms. Equating the right-hand sides of Eqs. (7.79) and (7.82) then requires

$$0 = \sum_{\mathbf{m}'} [w_{\mathbf{m}'}(\mathbf{r}) - \mathscr{S}_{\mathbf{m}'}]e^{i\mathbf{k}_{\mathbf{m}'}\cdot\mathbf{r}}, \quad (7.83)$$

where

$$w_{\mathbf{m}'}(\mathbf{r}) \equiv \sum_{\mathbf{p}} A_{\mathbf{m}',\mathbf{p}} e^{i\mathbf{G}_{\mathbf{p}}\cdot\mathbf{r}} \quad (7.84)$$

is the grouping of terms defined by Eq. (7.36) appearing in the complex Fourier series and

$$\mathscr{S}_{\mathbf{m}'} \equiv \sum_{\mathbf{p}} \mathscr{S}_{\mathbf{m}',\mathbf{p}} e^{i\mathbf{G}_{\mathbf{p}}\cdot\mathbf{r}}, \quad (7.85)$$

with

$$\mathscr{S}_{\mathbf{m}',\mathbf{p}} \equiv \sum_s \Gamma^{(s)}_{\mathbf{m}'} B^{(s)}_{\mathbf{m}',\mathbf{p}}, \quad (7.86)$$

is the grouping of terms appearing in the linear combination (7.82) of Bloch functions. Linear independence of the basis states $\exp(i\mathbf{k}_{\mathbf{m}'}\cdot\mathbf{r})$ appearing in (7.83) requires that

$$\mathscr{S}_{\mathbf{m}'} = w_{\mathbf{m}'}(\mathbf{r}) \qquad \text{(all allowed triplets } \mathbf{m}') \quad (7.87)$$

for each allowed triplet $\mathbf{m}'$ in order for Eq. (7.83) to be satisfied. However, use of the definitions (7.84) and (7.85) in Eq. (7.87) then gives

$$0 = \sum_{\mathbf{p}} e^{i\mathbf{G}_{\mathbf{p}}\cdot\mathbf{r}}(\mathscr{S}_{\mathbf{m}',\mathbf{p}} - A_{\mathbf{m}',\mathbf{p}}). \quad (7.88)$$

Linear independence of the basis states $\exp(i\mathbf{G}_\mathbf{p}\cdot\mathbf{r})$ appearing in Eq. (7.88) requires that

$$\mathscr{S}_{\mathbf{m}',\mathbf{p}} = A_{\mathbf{m}',\mathbf{p}} \qquad \text{(all triplets } \mathbf{p}) \tag{7.89}$$

for all triplets $\mathbf{p}$. Substitution of the definition (7.86) into Eq. (7.89) then gives a set of nonhomogeneous algebraic equations

$$\sum_s \Gamma^{(s)}_{\mathbf{m}'} B^{(s)}_{\mathbf{m}',\mathbf{p}} = A_{\mathbf{m}',\mathbf{p}} \qquad \text{(all triplets } \mathbf{p}\text{; all allowed triplets } \mathbf{m}') \tag{7.90}$$

for the expansion coefficients $\Gamma^{(s)}_{\mathbf{m}'}$ in terms of the known coefficients $B^{(s)}_{\mathbf{m}',\mathbf{p}}$ and $A_{\mathbf{m}',\mathbf{p}}$. It has been shown that the number of roots of the secular determinant is equal to the number of reciprocal lattice vectors, since the number of reciprocal lattice vectors defines the order of the secular determinant. Thus for each given triplet $\mathbf{m}'$, the left-hand side of Eq. (7.90) involving the sum over the band index s contains exactly as many terms as there are independent equations obtained from Eq. (7.90) by varying $\mathbf{p}$ over all possible triplets. Thus, for each $\mathbf{m}'$, the set of linear nonhomogeneous algebraic equations (7.90) obtained by varying $\mathbf{p}$ can be written in matrix form

$$\mathsf{B}\mathbf{\Gamma} = \mathbf{A}. \tag{7.91}$$

In this matrix equation, B is a square matrix with elements $B^{(s)}_{\mathbf{m}',\mathbf{p}}$, where s is the column index and $\mathbf{p}$ is the row index; $\mathbf{\Gamma}$ is a column matrix with elements $\Gamma^{(s)}_{\mathbf{m}'}$, where s is the row index, and $\mathbf{A}$ is a column matrix with $\mathbf{p}$ denoting the row index. This set of equations can in principle be solved by Cramer's Rule, or equivalently in the terminology of matrix theory, the equations can be solved by finding the inverse matrix B^{-1} for B, where by definition,

$$\mathsf{B}^{-1}\mathsf{B} = \mathsf{I}, \tag{7.92}$$

with I representing the corresponding square matrix with the diagonal elements having value unity and the off-diagonal elements having value zero. Operating on Eq. (7.91) with B^{-1} gives

$$\mathsf{B}^{-1}\mathsf{B}\mathbf{\Gamma} = \mathsf{B}^{-1}\mathbf{A}, \tag{7.93}$$

from which

$$\mathbf{\Gamma} = \mathsf{B}^{-1}\mathbf{A}. \tag{7.94}$$

The coefficients $\Gamma^{(s)}_{\mathbf{m}'}$ appearing as the elements of $\mathbf{\Gamma}$ are thus in principle determined uniquely. This procedure is of course repeated for each triplet $\mathbf{m}'$. In this way we find that there are exactly enough equations to determine all coefficients $\Gamma^{(s)}_{\mathbf{m}'}$ in the Bloch function expansion (7.81) representing the arbitrary function $g(\mathbf{r})$. Therefore we have proved that the set of Bloch functions is complete.

To summarize the results of our complete development based on Fourier series expansions, we have proven the following statement of *Bloch's theorem*: *There exists a complete set of energy eigenfunctions for any periodic potential which has the property that any eigenfunction is the product of a function having the lattice periodicity and a phase factor having unit magnitude which represents a modulation wavelength in each principal direction which can be chosen to be equal to or greater than twice the lattice periodicity in that direction.*

5.2 Alternatives to the Bloch Form

Bloch's theorem is very important in the theory of crystalline solids. The question naturally arises as to whether or not *all* energy eigenfunctions are required to have the Bloch form. It is readily demonstrated that for the usual situation in three-dimensional systems wherein the energy eigenvalues $\mathscr{E}^{(s)}(\mathbf{k}_{\mathbf{m}'})$ are at least partially degenerate, alternate complete sets of energy eigenfunctions can be constructed which do not have the Bloch form. It is necessary simply to choose a linear combination of two degenerate Bloch functions having different vectors $\mathbf{k}_{\mathbf{m}'}$,

$$\eta_1(\mathbf{r}) \equiv \alpha_{11}\phi^{(s)}_{\mathbf{m}'}(\mathbf{r}) + \alpha_{12}\phi^{(s')}_{\mathbf{m}''}(\mathbf{r}), \tag{7.95}$$

where α_{11} and α_{12} are arbitrary constants in the complex plane subject only to the possible requirement that $\eta_1(\mathbf{r})$ be normalized. For $\alpha_{11} \neq 0$ and $\alpha_{12} \neq 0$, $\eta_1(\mathbf{r})$ does *not* have the Bloch form (7.75). In principle, the band indices s and s' could be the same or they could be different. One naturally occurring case of degeneracy for a given band index is found in the isotropic free-electron model system whenever $|\mathbf{k}_{\mathbf{m}'}| = |\mathbf{k}_{\mathbf{m}''}|$ with the directions of $\mathbf{k}_{\mathbf{m}'}$ and $\mathbf{k}_{\mathbf{m}''}$ being different. By the definition of degeneracy, $\mathscr{E}^{(s)}(\mathbf{k}_{\mathbf{m}'}) = \mathscr{E}^{(s')}(\mathbf{k}_{\mathbf{m}''})$ for the two Bloch functions, so that

$$\mathscr{H}\eta_1(\mathbf{r}) = \alpha_{11}\mathscr{H}\phi^{(s)}_{\mathbf{m}'}(\mathbf{r}) + \alpha_{12}\mathscr{H}\phi^{(s')}_{\mathbf{m}''}(\mathbf{r}) = \mathscr{E}^{(s)}(\mathbf{k}_{\mathbf{m}'})\eta_1(\mathbf{r}). \tag{7.96}$$

That is, the new function $\eta_1(\mathbf{r})$ is also an eigenfunction of the Hamiltonian corresponding to the same energy eigenvalue. Next, a second function $\eta_2(\mathbf{r})$ is constructed,

$$\eta_2(\mathbf{r}) \equiv \alpha_{21}\phi^{(s)}_{\mathbf{m}'}(\mathbf{r}) + \alpha_{22}\phi^{(s')}_{\mathbf{m}''}(\mathbf{r}), \tag{7.97}$$

subject to the possible conditions that it be normalized and be orthogonal to $\eta_1(\mathbf{r})$ over the domain of the solid. Note that $\eta_2(\mathbf{r})$ does not in general have the Bloch form either. Nevertheless, it follows from the above development that

$$\mathscr{H}\eta_2(\mathbf{r}) = \mathscr{E}^{(s)}(\mathbf{k}_{\mathbf{m}'})\eta_2(\mathbf{r}), \tag{7.98}$$

so that $\eta_2(\mathbf{r})$ is also an eigenfunction of $\mathscr{H}$ corresponding to the same degenerate manifold. Since it is evident that the functions $\eta_1(\mathbf{r})$ and $\eta_2(\mathbf{r})$ will generally be linearly independent whenever $\phi^{(s)}_{\mathbf{m}'}(\mathbf{r})$ and $\phi^{(s)}_{\mathbf{m}''}(\mathbf{r})$ are linearly independent, it follows that these new functions represent equally valid energy eigenfunctions which can be used in place of the Bloch functions $\phi^{(s)}_{\mathbf{m}'}(\mathbf{r})$ and $\phi^{(s')}_{\mathbf{m}''}(\mathbf{r})$ as basis functions in the complete set of energy eigenfunctions. In this way we obtain an alternate complete set of energy eigenfunctions, only a portion of which have the Bloch form. Thus Bloch's theorem must *not* be interpreted as a requirement on the functional form of the energy eigenfunctions for a periodic potential; rather, it tells us that we can always choose the energy eigenfunctions for a periodic potential to have the Bloch form whenever this may be desirable.

5.3 Normalization of Bloch Functions

One question which arises for complete sets of functions is whether or not the functions are *orthonormal*. That is, are the functions *orthogonal* and *normalized*? First let us examine the normalization. The coefficients $B^{(s)}_{\mathbf{m}',l}$ for a given $\mathbf{m}'$ and a given s and variable l occur in a given Bloch function $\phi^{(s)}_{\mathbf{m}'}(\mathbf{r})$, and all of these coefficients are determined by using a single root $\mathscr{E}^{(s)}(\mathbf{k}_{\mathbf{m}'})$ of the secular determinant in solving a set of linear homogeneous algebraic equations for these coefficients. The solution to a set of homogeneous equations yields all of the variables in terms of any specific one of them, and this specific one can be chosen arbitrarily. (This is readily seen to be the case since multiplication of each variable in a *homogeneous* algebraic equation by any constant factor results in the same equation. This of course is not the case for a nonhomogeneous algebraic equation.) Thus by choosing the single arbitrary coefficient in each Bloch function to have the appropriate magnitude, we effect normalization of each Bloch function, viz.,

$$1 = \langle \mathbf{m}', s|\mathbf{m}', s\rangle \equiv \iiint \phi^{(s)}_{\mathbf{m}'}(\mathbf{r})^* \phi^{(s)}_{\mathbf{m}'}(\mathbf{r})\, dx_1\, dx_2\, dx_3, \tag{7.99}$$

where the integrations are carried out over the lengths of the crystal in the three principal directions.

5.4 Number of Bloch Functions in Each Energy Band

It is easy to determine the number of Bloch functions in any given energy band when we recall the one-to-one correspondence between the Bloch functions and the set of $\mathbf{k}_{\mathbf{m}'}$ vectors, as indicated by Eq. (7.75). In the first parallelepiped unit cell in reciprocal space, there are

$$\mathscr{N}_{\mathbf{k}_{\mathbf{m}'}} = L_1 L_2 L_3 = N_c \tag{7.100}$$

different values of $\mathbf{k}_{\mathbf{m}'}$, as one can readily determine from Eqs. (7.20)–(7.23). Therefore, *an energy band has the same number of Bloch functions as there are lattice sites in the crystal.*

Recalling that all points mapped out in reciprocal space by the $\mathbf{k}_{\mathbf{m}}$ vectors are equally spaced, as also are the reciprocal lattice points determined by the $\mathbf{G}_{\mathbf{p}}$ vectors, it can be concluded that the number $\mathscr{N}_{\mathbf{k}_{\mathbf{m}'}}$ of points mapped out by the $\mathbf{k}_{\mathbf{m}}$ vectors is the same for *any* parallelepiped unit cell. This is likewise the number of points in the first Brillouin zone in reciprocal space, because the first Brillouin zone is a primitive unit cell which therefore must have the same volume as any other primitive cell in reciprocal space. All Brillouin zones have the same volume in $\mathbf{k}$ space, so that by once again using the property that the points mapped out by the $\mathbf{k}_{\mathbf{m}}$ vectors are equally spaced, we can conclude that each Brillouin zone has associated with it a number $\mathscr{N}_{\mathbf{k}_{\mathbf{m}'}} = L_1 L_2 L_3$ of Bloch functions equal to the number of lattice points in the solid.

The number of Bloch functions, when applied to electrons in crystals to determine electronic states, can be considered to be doubled because of the two allowable values of electron spin. The number of allowable values of the magnetic quantum number consistent with the orbital angular momentum of the electron state in question (Chap. 1, §12.4.2) is also an important factor for energy bands in real metals (§8).

EXERCISE Show that there are $L_1 L_2$ distinct values of $\mathbf{k}_{\mathbf{m}'}$ in the first, second, and third Brillouin zones for the two-dimensional rectangular lattice.

5.5 Linear Independence of Bloch Functions

At this point let us examine the dimensionality of the complete manifold of Bloch functions and compare it with the dimensionality of the manifold of Fourier series basis states for the solid. The number $\mathcal{N}_{\mathrm{BF}}$ of Bloch functions is equal to

$$\mathcal{N}_{\mathrm{BF}} = (\mathcal{N}_{\mathbf{k}_{\mathbf{m}'}})(\mathcal{N}_s), \tag{7.101}$$

which is the product of the number $\mathcal{N}_{\mathbf{k}_{\mathbf{m}'}}$ of functions in each subset corresponding to a given band index s (viz., the number in a given band) with the number $\mathcal{N}_s$ of bands. The number $\mathcal{N}_{\mathbf{k}_{\mathbf{m}'}}$ in each band is equal to the number $N_c = L_1 L_2 L_3$ of $\mathbf{k}_{\mathbf{m}'}$ vectors in the first Brillouin zone, as argued in §5.4, and the number $\mathcal{N}_s$ of bands is equal to the number $\mathcal{N}_{\mathbf{G}}$ of reciprocal lattice vectors, as shown in §4.2. From the geometry of the reciprocal lattice, it can also be seen that the product $(\mathcal{N}_{\mathbf{k}_{\mathbf{m}'}})(\mathcal{N}_{\mathbf{G}})$ is equal to the total number $\mathcal{N}_{\mathbf{k}_{\mathbf{m}}}$ of allowed $\mathbf{k}_{\mathbf{m}}$ vectors in the Fourier series basis set used in the expansion of arbitrary functions of position within the solid. There is a one-to-one correspondence between the complex Fourier series basis states $\exp(i\mathbf{k}_{\mathbf{m}} \cdot \mathbf{r})$ and the $\mathbf{k}_{\mathbf{m}}$ vectors, so $\mathcal{N}_{\mathbf{k}_{\mathbf{m}}}$ is equal to the number $\mathcal{N}_{\mathrm{FS}}$ of basis states in the Fourier series. Thus

$$\mathcal{N}_{\mathrm{BF}} = (\mathcal{N}_{\mathbf{k}_{\mathbf{m}'}})(\mathcal{N}_s) = (\mathcal{N}_{\mathbf{k}_{\mathbf{m}'}})(\mathcal{N}_{\mathbf{G}}) = \mathcal{N}_{\mathbf{k}_{\mathbf{m}}} = \mathcal{N}_{\mathrm{FS}}, \tag{7.102}$$

which shows us that the dimensionality $\mathcal{N}_{\mathrm{BF}}$ of the set of Bloch functions is equal to the dimensionality of the complete set of linearly independent complex Fourier series basis states. When this fact is coupled with the fact that the set of Bloch functions is *complete*, as already proved, we can conclude quite rigorously that *the Bloch functions must be linearly independent*.

Thus far we have proven that the Bloch functions can be normalized, that they constitute a complete set, and that they are linearly independent. We have not yet proved that the Bloch functions are orthogonal.

5.6 Orthogonality of Bloch Functions over the Domain of the Crystal

First of all let us consider two Bloch functions [$\phi_{\mathbf{m}'}^{(s_1)}$ and $\phi_{\mathbf{m}''}^{(s_2)}$] corresponding to different wave vectors $\mathbf{k}_{\mathbf{m}'}$ and $\mathbf{k}_{\mathbf{m}''}$ lying within the first Brillouin zone,

$$\phi_{\mathbf{m}'}^{(s_1)}(\mathbf{r}) = e^{i\mathbf{k}_{\mathbf{m}'} \cdot \mathbf{r}} \sum_{\mathbf{p}'} A_{\mathbf{m}',\mathbf{p}'}^{(s_1)} e^{i\mathbf{G}_{\mathbf{p}'} \cdot \mathbf{r}}, \tag{7.103}$$

$$\phi_{\mathbf{m}''}^{(s_2)}(\mathbf{r}) = e^{i\mathbf{k}_{\mathbf{m}''} \cdot \mathbf{r}} \sum_{\mathbf{p}''} A_{\mathbf{m}'',\mathbf{p}''}^{(s_2)} e^{i\mathbf{G}_{\mathbf{p}''} \cdot \mathbf{r}}. \tag{7.104}$$

The band indices s_1 and s_2 may be the same or they may be different. The spatial domain Ω_c over which we wish to prove these functions orthogonal is simply the volume occupied by the crystal. The triple integral involved in the orthogonality relation will have limits for each of the three principal directions which reflect the boundaries of the crystal. The domains of integration can therefore be taken as x_1 to $x_1 + L_1$, x_2 to $x_2 + L_2$, and x_3 to $x_3 + L_3$. The integral can be written

$$\begin{aligned}\langle \mathbf{m}', s_1 | \mathbf{m}'', s_2 \rangle &= \iiint_{\Omega_c} \phi_{\mathbf{m}'}^{(s_1)}(\mathbf{r})^* \phi_{\mathbf{m}''}^{(s_2)}(\mathbf{r})\, dx_1\, dx_2\, dx_3 \\ &= \iiint_{\Omega_c} \sum_{\mathbf{p}'} \sum_{\mathbf{p}''} A_{\mathbf{m}',\mathbf{p}'}^{(s_1)*} A_{\mathbf{m}'',\mathbf{p}''}^{(s_2)} \exp\{i[(\mathbf{k}_{\mathbf{m}''} - \mathbf{k}_{\mathbf{m}'}) \\ &\quad + (\mathbf{G}_{\mathbf{p}''} - \mathbf{G}_{\mathbf{p}'})] \cdot \mathbf{r}\}\, dx_1\, dx_2\, dx_3. \end{aligned} \tag{7.105}$$

In terms of the quantity $\mathscr{I}_{\mathbf{m}'', \mathbf{m}', \mathbf{p}'', \mathbf{p}'}$ defined as

$$\mathscr{I}_{\mathbf{m}'', \mathbf{m}', \mathbf{p}'', \mathbf{p}'} = \iiint_{\Omega_c} \exp\{i[(\mathbf{k}_{\mathbf{m}''} - \mathbf{k}_{\mathbf{m}'}) + (\mathbf{G}_{\mathbf{p}''} - \mathbf{G}_{\mathbf{p}'})] \cdot \mathbf{r}\}\, dx_1\, dx_2\, dx_3, \tag{7.106}$$

the orthogonality integral takes the form

$$\langle \mathbf{m}', s_1 | \mathbf{m}'', s_2 \rangle = \sum_{\mathbf{p}'} \sum_{\mathbf{p}''} A_{\mathbf{m}',\mathbf{p}'}^{(s_1)*} A_{\mathbf{m}'',\mathbf{p}''}^{(s_2)} \mathscr{I}_{\mathbf{m}'', \mathbf{m}', \mathbf{p}'', \mathbf{p}'}. \tag{7.107}$$

However,

$$\mathbf{k}_{\mathbf{m}'} + \mathbf{G}_{\mathbf{p}'} = \mathbf{k}_{\mathbf{m}_1}, \tag{7.108}$$

$$\mathbf{k}_{\mathbf{m}''} + \mathbf{G}_{\mathbf{p}''} = \mathbf{k}_{\mathbf{m}_2}, \tag{7.109}$$

where $\mathbf{k}_{\mathbf{m}_1}$ and $\mathbf{k}_{\mathbf{m}_2}$ are also members of the set of $\mathbf{k}$ vectors satisfying periodic boundary conditions for the crystalline solid. Thus

$$\mathscr{I}_{\mathbf{m}'', \mathbf{m}', \mathbf{p}'', \mathbf{p}'} = \iiint_{\Omega_c} e^{i(\mathbf{k}_{\mathbf{m}_2} - \mathbf{k}_{\mathbf{m}_1}) \cdot \mathbf{r}}\, dx_1\, dx_2\, dx_3. \tag{7.110}$$

In addition,

$$\mathbf{k}_{\mathbf{m}_2} - \mathbf{k}_{\mathbf{m}_1} = \mathbf{k}_{\mathbf{m}}, \tag{7.111}$$

where $\mathbf{k}_{\mathbf{m}}$ likewise is a member of the set of $\mathbf{k}$ vectors. Therefore the integrand is simply one of the functions $\exp(i\mathbf{k}_{\mathbf{m}} \cdot \mathbf{r})$. Since these functions are periodic in each principal direction with the periodicity of the solid, the integral over the domain of the solid is zero unless $\mathbf{m}$ is equal to the triplet of integers (0, 0, 0). If $\mathbf{m} =$ (0, 0, 0), then it is readily seen that the triple integral has the value $L_1 L_2 L_3 \equiv N_c$, the number of unit cells in the crystal. Thus in terms of the Kronecker delta function, we can write

$$\mathscr{I}_{\mathbf{m}'', \mathbf{m}', \mathbf{p}'', \mathbf{p}'} = N_c \, \delta_{\mathbf{k}_{\mathbf{m}_1}, \mathbf{k}_{\mathbf{m}_2}}. \tag{7.112}$$

The further argument can be made that the two vectors $\mathbf{k}_{\mathbf{m}_1} \equiv \mathbf{k}_{\mathbf{m}'} + \mathbf{G}_{\mathbf{p}'}$ and $\mathbf{k}_{\mathbf{m}_2} \equiv \mathbf{k}_{\mathbf{m}''} + \mathbf{G}_{\mathbf{p}''}$ cannot be equal unless $\mathbf{k}_{\mathbf{m}'} = \mathbf{k}_{\mathbf{m}''}$ and $\mathbf{G}_{\mathbf{p}'} = \mathbf{G}_{\mathbf{p}''}$. This follows from the fact that $\mathbf{k}_{\mathbf{m}'}$ and $\mathbf{k}_{\mathbf{m}''}$ lie within the first Brillouin zone in reciprocal space, whereas the reciprocal lattice vectors $\mathbf{G}_{\mathbf{p}'}$ and $\mathbf{G}_{\mathbf{p}''}$ must of necessity extend across at least one unit cell in reciprocal space. Thus the above delta function can be represented by the product of two delta functions,

$$\delta_{\mathbf{k}_{\mathbf{m}_1}, \mathbf{k}_{\mathbf{m}_2}} = \delta_{\mathbf{k}_{\mathbf{m}'}, \mathbf{k}_{\mathbf{m}''}} \, \delta_{\mathbf{G}_{\mathbf{p}'}, \mathbf{G}_{\mathbf{p}''}}. \tag{7.113}$$

The delta functions on the right-hand side can be written also in terms of the triplets of integers $\mathbf{m}'$, $\mathbf{m}''$, $\mathbf{p}'$, and $\mathbf{p}''$. Thus

$$\delta_{\mathbf{k}_{\mathbf{m}_1}, \mathbf{k}_{\mathbf{m}_2}} = \delta_{\mathbf{m}', \mathbf{m}''} \, \delta_{\mathbf{p}', \mathbf{p}''}, \tag{7.114}$$

and so

$$\mathscr{I}_{\mathbf{m}'', \mathbf{m}', \mathbf{p}'', \mathbf{p}'} = N_c \, \delta_{\mathbf{m}', \mathbf{m}''} \, \delta_{\mathbf{p}', \mathbf{p}''}. \tag{7.115}$$

The orthogonality integral therefore takes the form

$$\langle \mathbf{m}', s_1 | \mathbf{m}'', s_2 \rangle = N_c \sum_{\mathbf{p}'} \sum_{\mathbf{p}''} A_{\mathbf{m}', \mathbf{p}'}^{(s_1)*} A_{\mathbf{m}'', \mathbf{p}''}^{(s_2)} \, \delta_{\mathbf{m}', \mathbf{m}''} \, \delta_{\mathbf{p}', \mathbf{p}''}. \tag{7.116}$$

The delta function $\delta_{\mathbf{p}', \mathbf{p}''}$ reduces the sum over $\mathbf{p}''$ to a single term, so that

$$\langle \mathbf{m}', s_1 | \mathbf{m}'', s_2 \rangle = N_c \, \delta_{\mathbf{m}', \mathbf{m}''} \sum_{\mathbf{p}'} A_{\mathbf{m}', \mathbf{p}'}^{(s_1)*} A_{\mathbf{m}'', \mathbf{p}'}^{(s_2)}. \tag{7.117}$$

Recalling that the triplet $\mathbf{m}'$ serves to distinguish between different Bloch functions within a given band, it can be seen that whenever $s_1 = s_2$ the derived result proves that all of the Bloch functions within a given energy band are mutually orthogonal. Since the result holds in addition for arbitrary band indices s_1 and s_2, the derived result also proves that members of the set of Bloch functions from different bands are also mutually orthogonal, provided only that $\mathbf{m}' \neq \mathbf{m}''$.

The remaining pairs of Bloch functions to be considered from the standpoint of orthogonality are restricted in number due to the requirements $\mathbf{m}' = \mathbf{m}''$ and $s_1 \neq s_2$. These further subdivide into two parts, namely, the situation in which the members of a pair are nondegenerate and the situation in which the members of a pair are degenerate. For the nondegenerate case we can immediately invoke the well-known orthogonality theorem proven in §8 of Chap. 1 that eigenfunctions belonging to different eigenvalues are orthogonal. Applying this theorem to the present situation, we can say that whenever pairs of Bloch functions $\phi_{\mathbf{m}'}^{(s_1)}$ and $\phi_{\mathbf{m}'}^{(s_2)}$ having the *same* wave vector $\mathbf{k}_{\mathbf{m}'}$ but lying in different energy bands $(s_1 \neq s_2)$ correspond to energy eigenvalues $\mathscr{E}^{(s_1)}(\mathbf{k}_{\mathbf{m}'})$ and $\mathscr{E}^{(s_2)}(\mathbf{k}_{\mathbf{m}'})$ which are different, then without a doubt the eigenfunctions must be orthogonal in accordance with the statement of the orthogonality theorem. The only situation then remaining is the particular degenerate case in which $\mathbf{k}_{\mathbf{m}'} = \mathbf{k}_{\mathbf{m}''}$, i.e., $\mathscr{E}^{(s_1)}(\mathbf{k}_{\mathbf{m}'}) = \mathscr{E}^{(s_2)}(\mathbf{k}_{\mathbf{m}'})$ for some $\mathbf{k}_{\mathbf{m}'}$ but with $s_1 \neq s_2$. This would require an overlapping of the energy bands s_1 and s_2 in the same spatial direction (viz, parallel to $\mathbf{k}_{\mathbf{m}'}$). It would represent the situation of a multiple root of the secular equation for some given $\mathbf{k}_{\mathbf{m}'}$, and thus any linear combination of the solutions would represent an energy eigenfunction having the Bloch form. The Gram–Schmidt orthogonalization process can then be invoked to construct an orthogonal set within this subspace of functions. The new Bloch functions thus obtained are then orthogonal to each other and to all remaining Bloch functions. Therefore we can state with confidence that we have proven the orthogonality of the Bloch functions over the domain of the crystal for all cases with the possible exception of the very specialized degenerate case $\mathscr{E}^{(s_1)}(\mathbf{k}_{\mathbf{m}'}) = \mathscr{E}^{(s_2)}(\mathbf{k}_{\mathbf{m}'})$, $(s_1 \neq s_2)$, and even for this case we have argued that it is possible to construct orthogonal Bloch functions.

This completes our rather ambitious program of proving the properties of completeness and orthogonality and illustrating the feasibility of normalization for the set of Bloch functions. It can now be generally assumed without further ado that *the Bloch functions represent a complete orthonormal set over the domain of the crystal.*

The questions to which we now address our attention are those of the meaning of the reduced zone solutions and the relationship between eigenfunctions and energy eigenvalues obtained when different reduced zones are utilized. Since the arbitrary zone chosen for our earlier development yielded a complete set of Bloch functions, it is intuitive that the solutions obtained using any other reduced zone will be either the same as or at least linearly dependent upon the solutions already obtained. To develop our understanding of this matter, let us consider the free-electron case.

6 Correspondence with the Free-Electron Model

6.1 Reduced Zone Scheme

6.1.1 Mapping Energy Eigenvalues. Before proceeding further with our development, it is quite important that we first distinguish apparent changes due to the mathematical approach from actual physical consequences of a periodic lattice potential. This distinction can be made quite effectively by examining the results of §5 in the limiting case in which the periodic lattice potential reduces to zero. This represents the situation in which each of the Fourier coefficients $V_{\mathbf{n}}$ in Eq. (7.54) is zero. The reader will find it very instructive at this stage to review the perturbation treatments given in §§ 13.1 and 13.2 of Chap. 5, where it is shown that the Fourier components of the periodic lattice potential perturb the free electron energy eigenvalues and introduce energy gaps at the Brillouin zone boundaries. The free electron dispersion curve is thus modified from a pure parabolic form to the shape illustrated in Fig. 5.6. The energy gap at the zone boundary was shown to decrease linearly with decreasing value of the relevant

Fourier coefficient of the lattice potential. In the limit in which all Fourier components of the lattice potential are reduced to zero, the energy eigenfunctions and energy eigenvalues for the conduction electrons in the crystal must necessarily reduce to those which we have already obtained by means of the free electron model of Chap. 3. This limit in the present development gives us insight into the meaning of a separation of wave vectors $\mathbf{k_m}$ into categories for which $\mathbf{k_m}$ lies either within the initial parallelepiped reciprocal cell or the first Brillouin zone (viz., $\mathbf{k_m} = \mathbf{k_{m'}}$), or lies outside of these regions (i.e., $\mathbf{k_m} = \mathbf{k_{m'}} + \mathbf{G_p}$).

The energy eigenfunctions satisfying periodic boundary conditions in the three-dimensional free-electron model can be written in the form

$$\phi_{\mathbf{k_m}}(\mathbf{r}) = N_c^{-1/2} e^{i\mathbf{k_m}\cdot\mathbf{r}}, \tag{7.118}$$

where N_c is the number of unit cells in the crystal. For a rectangular Cartesian coordinate system the components of the $\mathbf{k}$ vector are given by Eq. (3.10); for our presently considered general approach including nonorthogonal coordinate systems,

$$\mathbf{k_m} = 2\pi[(m_1/L_1)\mathbf{b}_1 + (m_2/L_2)\mathbf{b}_2 + (m_3/L_3)\mathbf{b}_3], \tag{7.119}$$

and the position vector is

$$\mathbf{r} = x_1\mathbf{d}_1 + x_2\mathbf{d}_2 + x_3\mathbf{d}_3. \tag{7.120}$$

The corresponding energy eigenvalues are given by

$$\mathscr{E}(\mathbf{k_m}) = (\hbar^2/2m)\mathbf{k_m}\cdot\mathbf{k_m}. \tag{7.121}$$

A plot of $\mathscr{E}(\mathbf{k_m})$ versus $k_\mathbf{m} \equiv |\mathbf{k_m}|$ in the direction of the vector $\mathbf{k_m}$ is shown in Fig. 7.11. The curve is parabolic in shape, and it extends monotonically to arbitrarily large values of the total energy. There seems to be little reason on the basis of this diagram to restrict our consideration to the limited region encompassed by the initial reciprocal cell or the first Brillouin zone, and to be sure, the reduced zone scheme does not have a great deal of physical content whenever there is no perturbing potential having the lattice periodicity acting to alter the wave functions or energy eigenvalues. The results in this limit are therefore expected to be purely mathematical. From a mathematical standpoint, however, the reduced zone scheme still exists, provided only that a lattice of points is defined throughout the region of the solid. The reciprocal lattice, defined in terms of the usual reciprocal lattice vectors $\mathbf{G}_l$, remains the same. All $\mathbf{k_m}$ vectors can thus be mapped into the initial reciprocal cell or the first Brillouin zone in accordance with the prescription already given by Eq. (7.29),

$$\mathbf{k_m} = \mathbf{k_{m'}} + \mathbf{G}_l. \tag{7.122}$$

(It may be recalled from §1 that this reduction could be justified rigorously by an algebraic proof and also by an equivalent geometrical proof.) As was already mentioned, any consequences of this mapping will be purely mathematical in nature, and thus more apparent than real, since the free-electron limit must indeed follow whenever the perturbing periodic potential is reduced to zero.

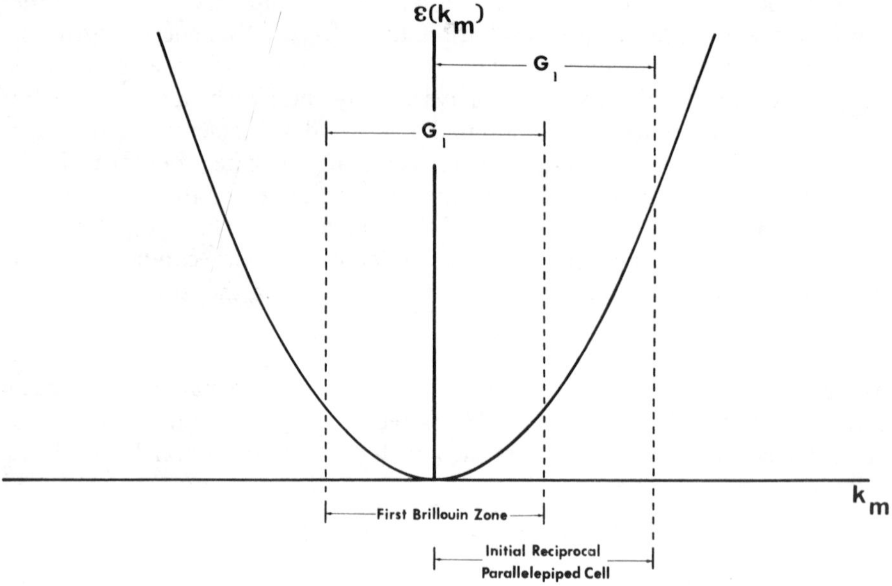

Fig. 7.11 Extended zone parabolic dispersion curve for energy $\mathscr{E}(\mathbf{k_m})$ versus $|\mathbf{k_m}|$ for the free-electron model, with the boundaries of the first Brillouin zone and those of the initial parallelepiped unit cell in reciprocal space indicated for a particular direction $\mathbf{k_m}$ in $\mathbf{k}$ space. [The discrete set of electron energy eigenvalues $\mathscr{E}(\mathbf{k_m})$ corresponding to the set of $\mathbf{k_m}$ values obtained by applying periodic boundary conditions to a crystal such as the one illustrated in Fig. 7.1 lie on the parabola; the periodic lattice potential is considered to be zero for the present case, so the energy gaps at the Brillouin zone boundary are zero, in contrast to the situation illustrated in Fig. 5.6.]

That is, in the limit where the perturbing periodic potential is reduced to zero, the energy eigenvalues are of course entirely the same as those of the free-electron model, so that

$$\mathscr{E}(\mathbf{k_m}) = (\hbar^2/2m)\mathbf{k_m} \cdot \mathbf{k_m} = (\hbar^2/2m)[(\mathbf{k_{m'}} + \mathbf{G}_l) \cdot (\mathbf{k_{m'}} + \mathbf{G}_l)]. \tag{7.123}$$

These energy eigenvalues are generally not the same as those given by

$$\mathscr{E}(\mathbf{k_{m'}}) = (\hbar^2/2m)\mathbf{k_{m'}} \cdot \mathbf{k_{m'}}. \tag{7.124}$$

The condition required for $\mathscr{E}(\mathbf{k_m})$ to equal $\mathscr{E}(\mathbf{k_{m'}})$ is

$$|\mathbf{k_{m'}} + \mathbf{G}_l| = |\mathbf{k_{m'}}|, \tag{7.125}$$

which is a very special case. Figure 7.12 illustrates schematically the two possible situations to be considered, namely, that in which $\mathbf{k_{m'}}$ lies inside of the reduced zone and that in which $\mathbf{k_{m'}}$ touches the boundary of the reduced zone. It can be concluded from these diagrams that $|\mathbf{k_{m'}} + \mathbf{G}_l| = |\mathbf{k_{m'}}|$ if (and only if) the wave vector $\mathbf{k_{m'}}$ touches the reduced zone boundary. This occurs for only a very small subset of the $\mathbf{k_m}$ vectors; the condition is not met for the great majority of electronic states. Thus, that two electronic states happen to be characterized by the same wave vector $\mathbf{k_{m'}}$ in the reduced zone does not imply that the energies of

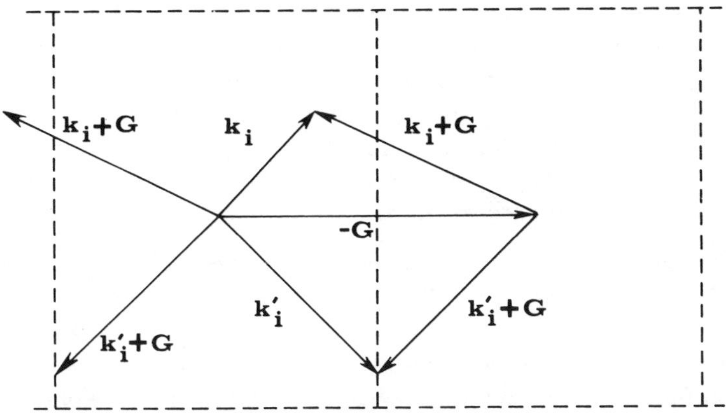

Fig. 7.12 Illustration that $|\mathbf{k}_i + \mathbf{G}|$ is not equal to $|\mathbf{k}_i|$ unless $\mathbf{k}_i$ is a member of the subset $\mathbf{k}'_i$ of wave vectors which touch the Brillouin zone boundary.

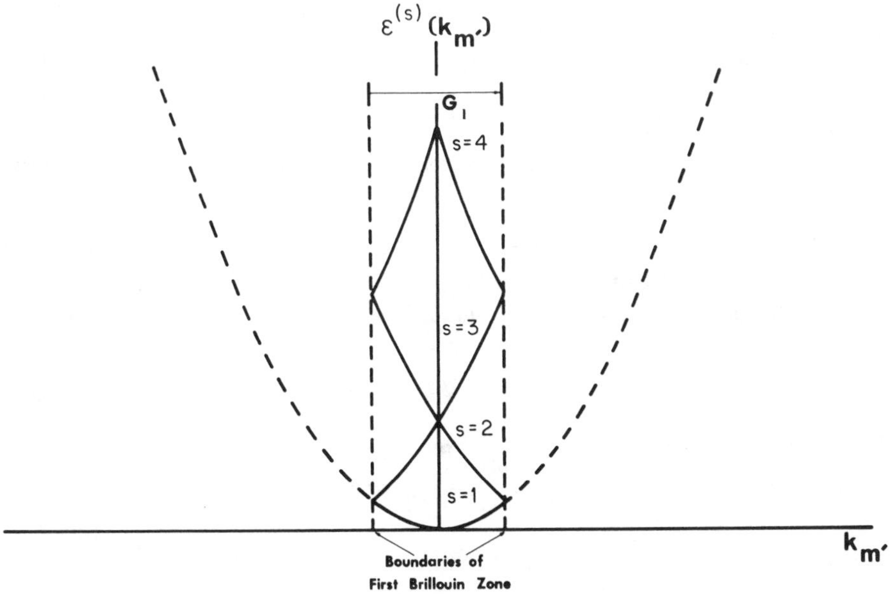

Fig. 7.13 Reduced zone scheme for labeling energy eigenvalues which are obtained from the extended zone parabolic dispersion curve $\mathscr{E}(\mathbf{k}_m)$ versus $\mathbf{k}_m$ illustrated in Fig. 7.11 for the free-electron model. (The various branches $1, 2, 3, \ldots$, labeled by the band index s, are translated from the parabola into the first Brillouin zone by means of appropriate reciprocal lattice vectors $\mathbf{G}_p$. The parabola in Fig. 7.11 is actually the cross section of a three-dimensional paraboloid of revolution, so the energy bands denoted by $s = 1, 2, 3, \ldots$ in the present figure represent cross-sectional lines obtained by cutting three-dimensional sheets which have been translated piecewise from the paraboloid of revolution into the first Brillouin zone by appropriate reciprocal lattice vectors.)

the two states are the same. On the contrary, the energies are generally different. We therefore adopt the rule that the free-electron energy eigenvalue is conserved when the $\mathbf{k_m}$ vectors are mapped into the reduced zone, so that there will appear many energy eigenvalues for each $\mathbf{k_{m'}}$ vector in the reduced zone.

It is easily seen that all vectors $\mathbf{k_m}$ within any given primitive parallelepiped cell in reciprocal space will require the same specific reciprocal lattice vector $\mathbf{G}_l$ for mapping into the initial parallelepiped reciprocal cell.

EXERCISE Show explicitly that the situation is somewhat different for the mapping of the higher Brillouin zones into the first Brillouin zone. (*Hint*: Several **G** vectors are required for the mapping.)

The set of closely spaced energy eigenvalues $\mathscr{E}(\mathbf{k_m})$ constituting the portion of the *energy dispersion surface* contained within any given primitive parallelepiped cell in reciprocal space will thus map as one sheet of the dispersion surface within the reduced zone. An entirely similar type of mapping can be carried out for each primitive parallelepiped cell in reciprocal space so that in the end many sheets are mapped into the reduced zone. These energy sheets represent energy as a function of the three components of the **k** vector, and so are difficult to illustrate in textbook diagrams. Therefore what is generally illustrated is energy as a function of the magnitude of **k** for specific directions in **k** space. Diagrams of this type represent cross-sectional views of the energy sheets. It is convenient to label the individual branches in these diagrams by means of the previously introduced band index s. The resulting reduced zone picture illustrated in Fig. 7.13 contains the same information as the *extended zone* free electron dispersion curve shown in Fig. 7.11.

6.1.2 Mapping Energy Eigenfunctions. Next, we develop further insight by examining the effect of the mapping process on the free-electron energy eigenfunctions. It is conventional to label Bloch states by the vector $\mathbf{k_{m'}}$ and the band index s,

$$\phi_{\mathbf{m}'}^{(s)}(\mathbf{r}) = e^{i\mathbf{k_{m'}}\cdot\mathbf{r}}u_{\mathbf{m}'}^{(s)}(\mathbf{r}), \tag{7.126}$$

where in general

$$u_{\mathbf{m}'}^{(s)}(\mathbf{r}) = \sum_l B_{\mathbf{m}',l}^{(s)} \exp(i\mathbf{G}_l\cdot\mathbf{r}). \tag{7.127}$$

It is clear that free-electron eigenfunctions of the plane-wave form

$$\phi_{\mathbf{k_m}}(\mathbf{r}) = N_c^{-1/2}e^{i\mathbf{k_m}\cdot\mathbf{r}} = e^{i\mathbf{k_{m'}}\cdot\mathbf{r}}[N_c^{-1/2}e^{i\mathbf{G_p}\cdot\mathbf{r}}] \tag{7.128}$$

fall into groups distinguished by whether or not $\mathbf{k_m} = \mathbf{k_{m'}} + \mathbf{G_p}$ lies inside of the initial parallelepiped reciprocal cell. If $\mathbf{k_m}$ does lie within the initial parallelepiped reciprocal cell (i.e., $\mathbf{G_p} = 0$), then $\mathbf{k_m} = \mathbf{k_{m'}}$, and the lattice periodic portion of the corresponding Bloch function is simply a constant,

$$u_{\mathbf{m}'}^{(1)}(\mathbf{r}) = N_c^{-1/2} \qquad \text{(free-electron limit; } s = 1). \tag{7.129}$$

If $\mathbf{k_m}$ lies outside of the initial parallelepiped reciprocal cell, then $\mathbf{k_m} = \mathbf{k_{m'}} + \mathbf{G_p}$

with $\mathbf{G_p} \neq 0$, so the lattice periodic portion of the corresponding Bloch function for this group is position dependent,

$$u_{\mathbf{m}'}^{(s)}(\mathbf{r}) = N_c^{-1/2} e^{i\mathbf{G_p}\cdot\mathbf{r}} \qquad \text{(free-electron limit; } s \neq 1\text{).} \tag{7.130}$$

It is readily seen that the Bloch functions $u_{\mathbf{m}'}^{(s)}(\mathbf{r})e^{i\mathbf{k_{m'}}\cdot\mathbf{r}}$ in both situations are entirely equivalent to the corresponding free-electron plane-wave form. Entirely similar considerations hold if the reduced zone is chosen to be the first Brillouin zone or any one of the primitive parallelepiped cells in reciprocal space.

Note that in the free-electron limit, the lattice periodic portion of the Bloch function happens to be independent of the reduced wave vector $\mathbf{k_{m'}}$ over a given parallelepiped unit cell and over large regions of a given Brillouin zone. This is not generally the case for a *nonzero* lattice potential. However, in practice it is sometimes assumed to be a reasonable first approximation in constructing Bloch functions.

It can also be noted above that the band index s is correlated uniquely (at least in the free-electron limit) with the vector $\mathbf{G_p}$ required in the mapping of the $\mathbf{k_m}$ vectors from any given primitive parallelepiped cell into the initial parallelepiped reciprocal cell. Thus a one-to-one correspondence exists between the band index s and the reciprocal lattice vectors $\mathbf{G_p}$. That is, one (and only one) reciprocal lattice vector is required in constructing each complete sheet $\mathscr{E}^{(s)}(\mathbf{k_{m'}})$ corresponding to any specific band index. (The situation is somewhat different when the reduced zone is chosen to be the first Brillouin zone, since more than one $\mathbf{G_p}$ is involved in mapping any given energy sheet.)

EXERCISE Consider which **G** vectors are involved in mapping the energy eigenvalues from the second to the first Brillouin zone for the two- and three-dimensional rectangular lattices.

Equation (7.128) leads to another very interesting and significant conclusion: The vector $\mathbf{k_{m'}}$ labeling the energy eigenfunction $\phi_{\mathbf{m}'}^{(s)}(\mathbf{r})$ does *not* give directly the electron momentum eigenvalue $\mathbf{p_m}$, since in general $\mathbf{p_m} \neq \hbar\mathbf{k_{m'}}$. This result, so easily seen in this limit of the free-electron correspondence where of necessity

$$\mathbf{p_m} = \hbar\mathbf{k_m} = \hbar(\mathbf{k_{m'}} + \mathbf{G_p}),$$

has far-reaching consequences for electrons propagating in nonzero lattice potentials. In this more complicated case, the lattice periodic portion $u_{\mathbf{m}'}^{(s)}(\mathbf{r})$ of the Bloch function contains a mixture of many plane waves of different $\mathbf{G_p}$ value, and in some sense these contribute to the true momentum of the electron. Thus $\hbar\mathbf{k_{m'}}$ is referred to merely as the *crystal momentum* or the *quasi-momentum* of the electron.

PROJECT 7.1 Energy Band Theory for a Linear Array of Uniformly Spaced Delta Function Potentials

1. (a) Reduce the problem of the motion of an electron in a series of very narrow deep potential wells to one of periodic discontinuous boundary conditions.

(b) Express these periodic discontinuous boundary conditions in matrix form, and obtain the dispersion relation.

2. (a) Deduce the effective mass m^* for the first band from the dispersion relation derived above. Qualitatively describe the variation of m^* with lattice parameter and energy of the bound level, and relate this to the qualitative dependence of the corresponding $\mathscr{E}$ versus k curve on these parameters.

(b) Define the crystal momentum P and quantitatively relate it to the electron velocity, momentum, and effective mass, and also to the force experienced by an electron when an external electric field is applied to the crystal.

3. (a) Give an analytical expression and a graphical representation for the eigenstate ψ_{40} as deduced using the above approach.

(b) Discuss the physical significance of the arbitrariness in **k** vector for a given band, and give your concept of "space harmonics." Discuss the physical significance of the multivaluedness in energy for a given **k** vector in the reduced zone scheme, and relate this to space harmonics. [*Hint*: See R. A. Smith (1963).]

PROJECT 7.2 Energy Bands Deduced by Means of the Kronig–Penney Model

Solve the Schrödinger equation for a particle of mass m under the influence of a one-dimensional periodic potential energy of the form

$$U(x) = U_0 \quad (-D \leqslant x \leqslant 0), \qquad U(x) = 0 \quad (0 \leqslant x \leqslant L - D), \qquad U(x + L) = U(x) \quad (-\infty < x < \infty).$$

Numerically evaluate the relationship for the energy eigenvalues to obtain the energy ranges for which there exist a quasi-continuum of allowed energy values (i.e., the *energy bands*), and likewise delineate the energy ranges over which there exist no allowed energy values (i.e., the *energy gaps*). Use for your computations some reasonable set of numerical values for the parameters, such as the electronic mass, $U_0 = 1$–5 eV, $D = 0.1$–1 Å, and $L = 2$–5 Å.

PROJECT 7.3 Effective Masses of Electrons and Electron Holes

1. What is the "effective mass" of an electron?
2. Provide theoretical justification for the fact that the carriers in a nearly full band can be considered to be of positive sign.
3. Justify the concept of an electron hole in terms of energy band theory.

PROJECT 7.4 Effective Mass Approximation at Semiconductor Band Edges

Justify the fact that the free-electron-like dispersion relation $\mathscr{E} = \hbar^2k^2/2m^*$ can be applied to the valence and conduction bands for semiconductors. (That is, explain why the regions of the Brillouin zone constituting the top of the valence band and the bottom of the conduction band are the most important as far as the thermal production of carriers is concerned, and then explain how a Taylor series expansion of $\mathscr{E}(\mathbf{k})$ versus **k** about these extremum points yields results which can be described by a quadratic form in the **k**-vector components, very similar to the $\mathscr{E}(\mathbf{k})$ versus **k** dispersion relation for the free-electron model.)

PROJECT 7.5 Carrier Statistics in Semiconductors

Semiconductors are extremely important technologically because of the need for such components in sophisticated instrumentation.

1. Formulate the statistics of the electron and electron–hole carriers in terms of the chemical potential and the gap energy.
2. State and prove the law of mass action for this system, carefully delineating any necessary assumptions for the derivation, and obtain

$$n_e = n_h = 2(k_BT/2\pi\hbar^2)^{3/2}(m_e^*m_h^*)^{3/4}\exp(-\mathscr{E}_{gap}/2k_BT)$$

for the electron and electron hole densities n_e and n_h as a function of the energy $\mathscr{E}_{gap}$ and temperature and the electron and electron hole effective masses m_e^* and m_h^*.

3. Design a useful electrical circuit (or similar device) in which a semiconducting material plays some clever role in the functional operation.

PROJECT 7.6 Effective Mass

Consider the application of an external force **F** to the conduction electrons in a solid having an energy band structure $\mathscr{E}(\mathbf{k})$. Show that the force **F** produces a momentum change in each electron in accordance with Newton's second law $\mathbf{F} = m\mathbf{a}$, provided the electron is assumed to respond in the solid as if it had inertia properties which can be described by

$$\left(\frac{1}{m_{\text{eff}}}\right)_{ij} = \frac{1}{\hbar^2}\frac{\partial^2 \mathscr{E}(\mathbf{k})}{\partial k_i\,\partial k_j},$$

where the left-hand side represents the *ij*th component of a tensor quantity known as the *inverse effective mass*.

6.2 Periodic Zone Scheme

6.2.1 Mapping into All Primitive Parallelepiped Cells in Reciprocal Space. There exist mathematical transformations which lead to yet another zone scheme, namely, the *periodic zone scheme*. It is fundamental that the Bloch functions can be indexed with wave vectors from any of the various primitive parallelepipeds in reciprocal space, corresponding merely to different choices of the reduced zone for obtaining the general solution to the Schrödinger equation for a periodic potential. (It is thus evident that any additional solutions found for the Schrödinger equation will be redundant, since we already have obtained the most general possible solution with the Fourier series approach.) The reduction to an arbitrary primitive parallelepiped cell other than the initial parallelepiped reciprocal cell can be readily effected by adding to $\mathbf{k}_{\mathbf{m}'}$ the appropriate reciprocal lattice vector $\mathbf{G}_{\mathbf{p}}$ which is required for translation of any particular wave vector from the initial parallelepiped reciprocal cell to the new reduced zone, so that the usual Bloch functions

$$\phi_{\mathbf{m}'}^{(s)}(\mathbf{r}) = e^{i\mathbf{k}_{\mathbf{m}'}\cdot\mathbf{r}}u_{\mathbf{m}'}^{(s)}(\mathbf{r}) \tag{7.131}$$

take the form

$$\phi_{\mathbf{m}'}^{(s)}(\mathbf{r}) = e^{i(\mathbf{k}_{\mathbf{m}'}+\mathbf{G}_{\mathbf{p}})\cdot\mathbf{r}}[u_{\mathbf{m}'}^{(s)}(\mathbf{r})e^{-i\mathbf{G}_{\mathbf{p}}\cdot\mathbf{r}}]. \tag{7.132}$$

By designating the new reduced vectors as $\mathbf{k}_{\mathbf{p}'}$, where

$$\mathbf{k}_{\mathbf{p}'} \equiv \mathbf{k}_{\mathbf{m}'} + \mathbf{G}_{\mathbf{p}}, \tag{7.133}$$

then the same Bloch function takes the new form

$$\phi_{\mathbf{p}'}^{(s)}(\mathbf{r}) = e^{i\mathbf{k}_{\mathbf{p}'}\cdot\mathbf{r}}u_{\mathbf{p}'}^{(s)}(\mathbf{r}), \tag{7.134}$$

where

$$u_{\mathbf{p}'}^{(s)}(\mathbf{r}) = u_{\mathbf{m}'}^{(s)}(\mathbf{r})e^{-i\mathbf{G}_{\mathbf{p}}\cdot\mathbf{r}}. \tag{7.135}$$

(The subscript $\mathbf{p}'$ labels the state within the new reduced zone.) The function $u_{\mathbf{p}'}^{(s)}(\mathbf{r})$ also possesses the lattice periodicity, so the new form $\phi_{\mathbf{p}'}^{(s)}(\mathbf{r})$ likewise is of the Bloch form. The corresponding energy eigenvalue cannot be changed by these essentially mathematical manipulations, but a new label such as $\mathscr{E}^{(s)}(\mathbf{k}_{\mathbf{p}'})$ is helpful in denoting that the energy sheets are for a new unit cell in reciprocal space. As with our earlier reduction to the initial parallelepiped reciprocal cell (or to the first Brillouin zone), the new reduction must likewise be carried out without a change in the energy eigenvalues. The easiest way to visualize the reduction to the new zone is first to consider all $\mathbf{k}_{\mathbf{m}}$ vectors reduced to the initial parallelepiped cell in reciprocal space, with the resultant energy sheets then reduced simultaneously to the new zone. This is equivalent to a translation of the energy sheets in $\mathbf{k}$ space from the initial parallelepiped reciprocal cell to the new parallelepiped primitive cell in question. If we consider a sequence of such translations so that each and every primitive parallelepiped cell in reciprocal space is sequentially considered as the new reduced zone, then it is evident that every primitive parallelepiped cell in reciprocal space will consequently take on an appearance very similar to the original reduced zone. Thus all zones in reciprocal space become filled with energy sheets! [Evidently this is an overabundance; however, this scheme does provide a very useful framework in describing open and closed orbits of electrons on the Fermi surface in reciprocal space. See, for example, Ziman (1964).]

6.2.2 Free-Electron Limit. Considering specifically the free-electron limit, the dispersion curves in the periodic zone scheme are illustrated for one spatial direction in Fig. 7.14. Note the periodicity of the dispersion curves in this diagram. (This periodicity remains a general characteristic of the dispersion curves even when the lattice potential is nonzero; it is thus the basis for the label "periodic zone scheme." The primary effects of a perturbing periodic potential is to separate the different branches of the dispersion sheets $\mathscr{E}^{(s)}(\mathbf{k}_{\mathbf{p}'})$ at the boundaries of the unit cells in reciprocal space with the introduction of energy gaps, as was proven by the perturbation treatment given in Chap. 5, §13.2.) The appearance of the curves in Fig. 7.14 over the domain of the first Brillouin zone is of interest, since this most often is the choice for the *reduced* zone. It is also interesting to compare the free-electron results deduced within the framework of the periodic zone scheme (Fig. 7.14) with the corresponding results expressed in terms of the extended zone scheme (Fig. 7.11). Since the physical content must be exactly the same, the apparent differences are in fact entirely mathematical in nature.

EXERCISE The periodic zone scheme dispersion curves illustrated in Fig. 7.14 for the free-electron model are modified by the perturbations due to the periodic lattice potential, as shown analytically in Chap. 5, §13.2. Sketch the dispersion curves in the periodic zone scheme for a nonzero lattice potential. [*Hint*: See Ziman (1964), and also refer to Figs. 7.15 and 7.16, which illustrate the corresponding results for the extended zone scheme and the reduced zone scheme.]

The Fermi surface for free electrons is spherical, and in the periodic zone scheme it appears in each unit cell of reciprocal space, as indicated in Fig. 7.17. The periodic lattice potential perturbs the energy bands most in the neighborhood of the Brillouin zone boundaries where the energy gap is introduced;

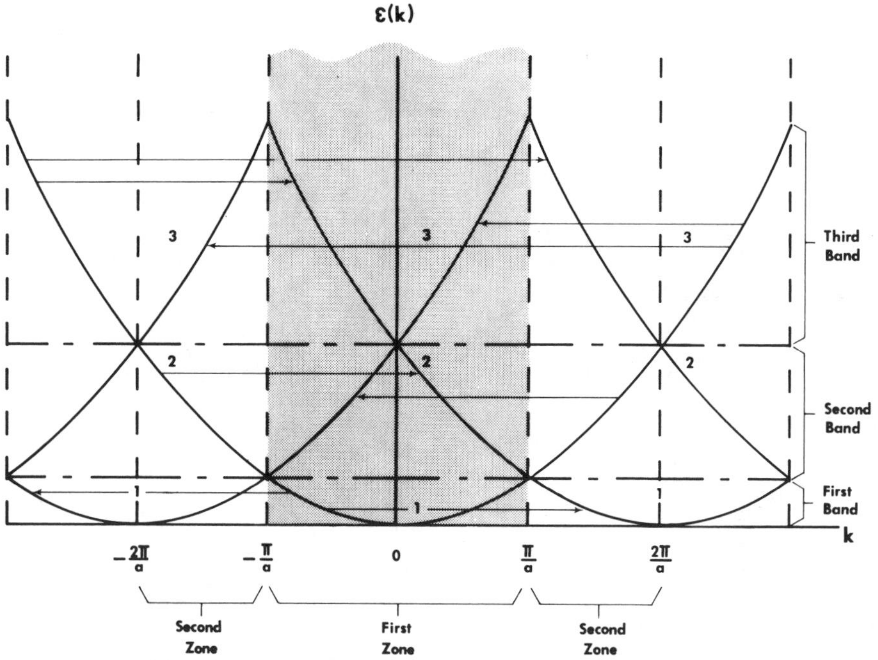

Fig. 7.14 Periodic (or "repeated") zone scheme for the energy dispersion curves $\mathscr{E}(\mathbf{k})$ versus $|\mathbf{k}|$ for the free-electron model. (The sheets represented by the cross-sectional lines labeled $s = 1, 2, 3, \ldots$ in the reduced zone scheme illustrated by Fig. 7.13 have been translated into every unit cell in reciprocal space by means of the set of reciprocal lattice vectors $\mathbf{G_p}$.)

this has a marked effect on the shape of the Fermi surface if it is large enough to reach the boundary of the first Brillouin zone. In the periodic zone scheme, the Fermi surface for some metals is connected from zone to zone, as indicated in Fig. 7.18. This leads to what is known as "open orbits in **k** space" in addition to normally closed orbits [cf. Ziman (1964)].

The free-electron energy eigenfunctions in the periodic zone scheme are readily deduced from the usual plane-wave form. The specific reciprocal lattice vector required for translation to the zone in question can be expressed in the form

$$\mathbf{G_{p''}} = \mathbf{G_p} - \mathbf{G}_{l'}, \tag{7.136}$$

where $\mathbf{G}_{l'}$ is the reciprocal lattice vector required to reduce the wave vector $\mathbf{k_m}$ to the initial parallelepiped reciprocal cell. Thus

$$\mathbf{k_m} = \mathbf{k_{m'}} + \mathbf{G}_{l'}, \tag{7.137}$$

and $\mathbf{G_p}$ is the reciprocal lattice vector required to translate $\mathbf{k_{m'}}$ from the initial parallelepiped reciprocal cell to the new reduced zone. The wave vector $\mathbf{k_{p'}}$ in the new reduced zone is thus

$$\mathbf{k_{p'}} = \mathbf{k_{m'}} + \mathbf{G_p}, \tag{7.138}$$

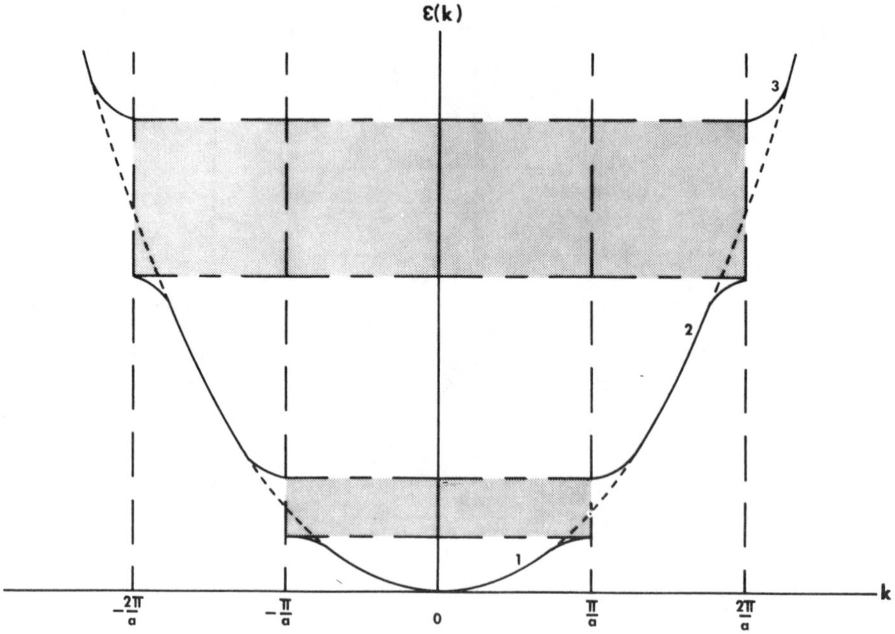

Fig. 7.15 The introduction of energy gaps due to the perturbing periodic lattice potential into the extended zone scheme for the energy dispersion curves $\mathscr{E}(\mathbf{k})$ versus $|\mathbf{k}|$ for Bloch functions.

and we obtain the result

$$\mathbf{k}_{\mathbf{p}'} = (\mathbf{k}_{\mathbf{m}} - \mathbf{G}_{l'}) + \mathbf{G}_{\mathbf{p}} = \mathbf{k}_{\mathbf{m}} + \mathbf{G}_{\mathbf{p}''}. \tag{7.139}$$

Therefore the usual free-electron plane-wave eigenfunction

$$\phi_{\mathbf{m}}(\mathbf{r}) = N^{-1/2} e^{i\mathbf{k}_{\mathbf{m}} \cdot \mathbf{r}} \tag{7.140}$$

takes the following Bloch form,

$$\phi_{\mathbf{p}'}^{(s)}(\mathbf{r}) = e^{i\mathbf{k}_{\mathbf{p}'} \cdot \mathbf{r}} u_{\mathbf{p}'}^{(s)}(\mathbf{r}), \tag{7.141}$$

with the requirement $\phi_{\mathbf{p}'}^{(s)}(\mathbf{r}) = \phi_{\mathbf{m}}(\mathbf{r})$ yielding

$$u_{\mathbf{p}'}^{(s)}(\mathbf{r}) = N^{-1/2} e^{i(\mathbf{k}_{\mathbf{m}} - \mathbf{k}_{\mathbf{p}'}) \cdot \mathbf{r}} = N^{-1/2} e^{-i\mathbf{G}_{\mathbf{p}''} \cdot \mathbf{r}}. \tag{7.142}$$

Although any given energy eigenfunction in the periodic zone scheme is the same as in the reduced zone scheme, the spatial dependence of the plane-wave factor $e^{i\mathbf{k}_{\mathbf{p}'} \cdot \mathbf{r}}$ and the lattice periodic factor $u_{\mathbf{p}'}^{(s)}(\mathbf{r})$ are thus noted to be individually different for the two schemes. Again we note that for the free electron model the lattice periodic factor $u_{\mathbf{p}'}^{(s)}(\mathbf{r})$ of the Bloch function is actually independent of the particular wave vector $\mathbf{k}_{\mathbf{p}'}$ within any specific zone, although $u_{\mathbf{p}'}^{(s)}(\mathbf{r})$ does change from zone to zone. The description of the free electron energy eigenfunctions and eigenvalues in terms of the periodic zone scheme emphasizes the fact that the results as expressed in this scheme are quite redundant. The source of this

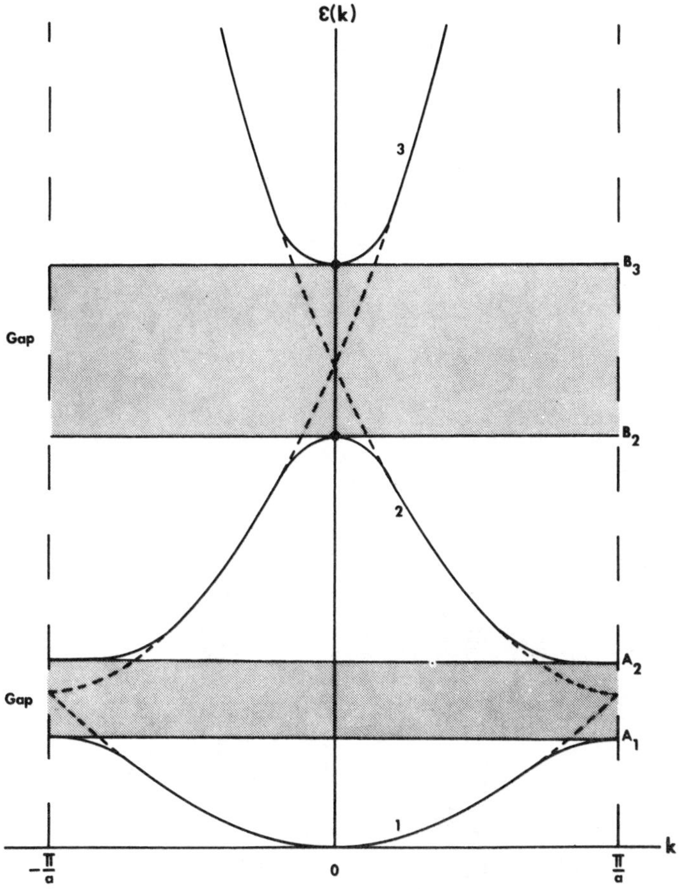

Fig. 7.16 The introduction of energy gaps due to the perturbing periodic lattice potential into the reduced zone scheme for the energy dispersion curves $\mathscr{E}(\mathbf{k})$ versus $|\mathbf{k}|$ for Bloch functions.

redundancy is actually in the mathematical approach where more than one reduced zone is considered; only one reduced zone is required for the general Fourier series solution developed previously for the Schrödinger equation. Therefore it is not surprising that we obtain a redundancy in the Bloch functions and energy eigenvalues in the periodic zone scheme.

To a large extent the insight given by the free-electron correspondence enables us to understand the results to be expected whenever the periodic lattice potential is nonzero. For free electrons, the energy eigenvalues $\mathscr{E}(\mathbf{k}_m)$ are unbounded, with the larger values of $|\mathbf{k}_m|$ corresponding to higher electron kinetic energies. Furthermore, there is a high degree of degeneracy of the energy eigenvalues for free electrons. Degeneracy occurs in the free-electron case for different $\mathbf{k}_m$ vectors which have the same magnitude, corresponding to electrons traveling with a given speed but in different directions. In the case of nonzero

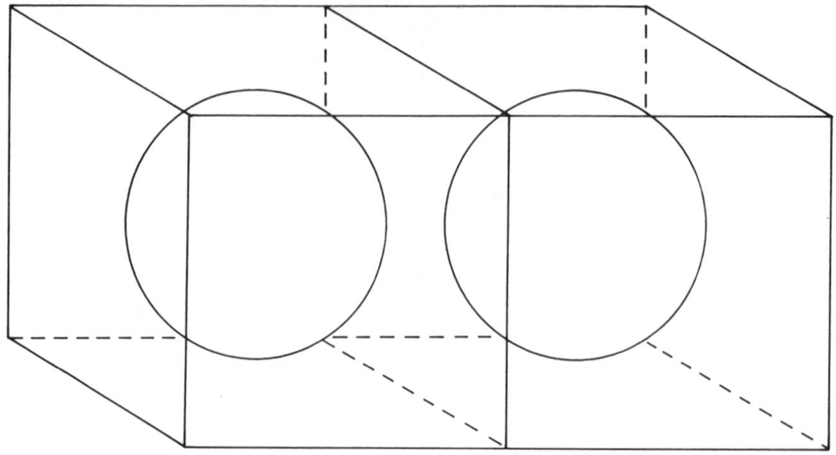

Fig. 7.17 Spherical Fermi surface which is characteristic of the free-electron model in the periodic zone scheme.

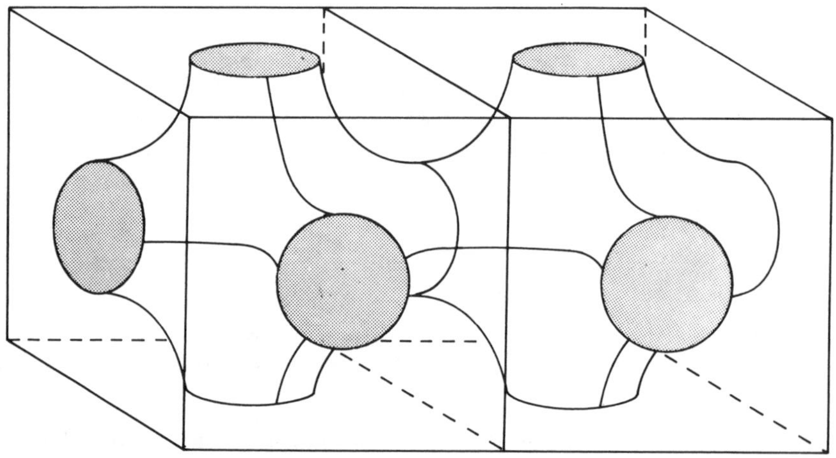

Fig. 7.18 Nonspherical Fermi surface which touches the Brillouin zone boundary appears to be connected between adjacent zones in the periodic zone scheme. (The energy gap at the Brillouin zone boundary prevents an immediate filling of states in the second Brillouin zone even when enough electrons are added to the system to expand the Fermi surface sufficiently for it to touch the boundary of the first Brillouin zone; thus, states parallel to the Brillouin zone boundary are successively populated first until energies lying above the energy gap are reached, and only then do electrons begin to populate the second Brillouin zone, corresponding to population of the next higher energy band. This factor, together with the modifications in $\mathscr{E}(\mathbf{k})$ versus $\mathbf{k}$ due to the lattice potential, accounts for the peculiar topology of the Fermi surface as indicated in the figure. It may also be noted that the cross-sectional area of the Fermi surface is a minimum at the "neck" located at the zone boundary; such "extremal" areas of the Fermi surface lead to quantum oscillatory effects in various physical properties such as the electrical conductivity, magnetic susceptibility, and ultrasonic absorption in the presence of applied magnetic fields.)

lattice potentials, these features are changed only by the fact that the magnitudes of $\mathbf{k}_{\mathbf{m}'}$ vectors pointing in different directions (but corresponding to some given energy) can be different. This reflects the fact that *the constant energy surfaces in* **k** *space can deviate from the spherical shape* $\{\mathscr{E}(\mathbf{k}) = (\hbar^2/2m)[k_x^2 + k_y^2 + k_z^2]\}$ characteristic of perfectly free electrons. The features of degeneracy and an infinity of energy eigenvalues are for the most part unchanged by the lattice potential. Whereas the extended zone scheme was found to provide a natural framework for the free electron approach, the reduced zone scheme was found to provide a more natural framework for interpreting the eigenvalues proceeding from the secular determinant as obtained by assuming a reduced zone Fourier series solution for the time-independent Schrödinger equation. The periodic zone scheme was found to be a natural result of the reduced zone Fourier series solution whenever all possible primitive parallelepiped cells in reciprocal space were sequentially chosen for the reduced zone. Whereas the reduced zone scheme provides exactly the same information as the extended zone scheme (only the format being different), the periodic zone scheme was shown to be highly redundant due to the fact that every zone in reciprocal space then contains exactly the same information as is contained in the single zone appearing in the reduced zone scheme.

An additional question which we address in this section is whether it is meaningful to attempt to deduce a unique correspondence between the Fourier series reduced zone dispersion sheets for the case of a nonzero periodic potential and some specific extended zone solution. One of the easiest ways to approach this question is to examine the behavior of our general formalism in the limit in which the Fourier coefficients of the lattice potential approach zero. In this limit the coefficients $V_{\mathbf{j}-\boldsymbol{l}}$ in the secular determinant (7.68) go to zero, and all off-diagonal terms are then zero. The secular equation then reduces to

$$\prod_{\boldsymbol{l}} [\mathscr{E}^{(s)} - \mathscr{T}_{\mathbf{m}',\boldsymbol{l}}] = 0. \tag{7.143}$$

This yields n energy eigenvalues $\mathscr{E}^{(s)} = \mathscr{T}_{\mathbf{m}',\boldsymbol{l}}$, where n represents a number equal to $(2l_1)(2l_2)(2l_3)$ in the limit where each of these integers approaches infinity. (Each of the integers l_1, l_2, and l_3 in the triplet $\boldsymbol{l}$ ranges from $-\infty$ to $+\infty$.) Thus for *each* value of $\mathbf{k}_{\mathbf{m}'}$ and *each* value of $\mathbf{G}_{\boldsymbol{l}}$ there exists a root

$$\mathscr{E}^{(s)}(k_{\mathbf{m}'}) = \mathscr{T}_{\mathbf{m}',\boldsymbol{l}} \equiv (\hbar^2/2m)[(\mathbf{k}_{\mathbf{m}'} + \mathbf{G}_{\boldsymbol{l}}) \cdot (\mathbf{k}_{\mathbf{m}'} + \mathbf{G}_{\boldsymbol{l}})]. \tag{7.144}$$

These roots are readily identified as the energy eigenvalues for the free-electron model, as expressed in reduced zone notation. In this scheme, there are $N_c = L_1 L_2 L_3$ independent values of $\mathbf{k}_{\mathbf{m}'}$; the dispersion surface $\mathscr{E}^{(s)}(\mathbf{k}_{\mathbf{m}'})$ for fixed $\mathbf{G}_{\boldsymbol{l}}$ and all possible values of $\mathbf{k}_{\mathbf{m}'}$ represents a given energy band labeled by the index s. Each band contains N_c energy eigenvalues corresponding to the N_c independent values of $\mathbf{k}_{\mathbf{m}'}$ which meet periodic boundary conditions for a crystal consisting of N_c unit cells. The order of the secular determinant is determined by the number of independent values of $\mathbf{G}_{\boldsymbol{l}}$, so that there is one energy band for every $\mathbf{G}_{\boldsymbol{l}}$ vector. All n of these energy bands clutter the reduced zone used in obtaining the general Fourier series solution. Any attempt to correlate the band index s with the $\mathbf{G_p}$ vectors for this reduced zone approach would be arbitrary. Since $\mathscr{E}^{(s)}(\mathbf{k_{m'}}) = \mathscr{T}_{\mathbf{m'},l}$ for the limit of free electrons, the band index could be readily chosen as l if desired. Translation of each dispersion sheet labeled with the variable index l by the corresponding vector $\mathbf{G}_l$ would then place it in the same parallelepiped unit cell as would be obtained by means of the extended zone scheme. It is to be anticipated, however, that the correlation between band index and reciprocal lattice vectors $\mathbf{G_p}$ will no longer be so straightforward in the more general case of a nonzero $V(\mathbf{r})$. This is due to the fact that each Bloch function and each energy eigenvalue will depend upon a number of different reciprocal lattice vectors, so that the correspondence will be unique only in the limit where $V(\mathbf{r}) \to 0$.

Let us again consider the periodic zone scheme. From our previous arguments based on writing the same Bloch function in forms appropriate to various zones, we would expect the new secular equation to yield the same form for the dispersion curves in the new reduced zone as was obtained in the original reduced zone. To be a bit more quantitative, it can be noted that the addition of the factor $e^{i\mathbf{G_p}\cdot\mathbf{r}}$ ($\mathbf{G_p}$ fixed) to the right-hand side of the general Fourier expansion (7.47) for the wave function does not change the wave function at all since the sum is over all possible reciprocal lattice vectors. Thus, considering $\mathbf{k_{m'}} + \mathbf{G_p} = \mathbf{k_{p'}}$, where $\mathbf{k_{p'}}$ is a wave vector in a new reduced zone, the index $\mathbf{m}'$ can be changed throughout in (7.47) to the index $\mathbf{p}'$. The remainder of the development leading from Eq. (7.47) to the secular determinant (7.68) remains the same, so the results obtained pertain to the new reduced zone. The set of energy eigenvalues obtained is identical to the original set, however, since the net result is simply a permutation in the row and column indices. The band index should perhaps be relabeled because of the shift in the reduced zone, but otherwise no new information has been obtained. This provides some additional mathematical justification for the periodicity already deduced for the periodic zone scheme. This redundancy from zone to zone is not surprising in view of the fact that our solution of the Schrödinger equation assuming a single reduced zone is perfectly general and therefore should yield all possible energy eigenvalues as well as a complete set of linearly independent eigenfunctions.

PROJECT 7.7 Electron–Electron Interactions

Develop the theory of electron–electron interactions. Derive Lindhard's expression, and apply it to obtain worthwhile results for certain specific limiting cases. [*Hint*: See Ziman (1964).]

PROJECT 7.8 Plasma Oscillations

Develop the theory of plasma oscillations, using the approach of a frequency-dependent dielectric constant. [*Hint*: See Ziman (1964).]

PROJECT 7.9 Cohesive Energy of a Metal

Derive expressions for the cohesive energy and the interatomic spacing of an idealized metal.

PROJECT 7.10 Fermi Surface for a Two-Dimensional Square Lattice

Construct the Fermi surface in the free-electron approximation for a two-dimensional square lattice having four valence electrons per atom, and reduce the Fermi surface obtained to the first Brillouin zone. [*Hint*: See Goldsmid (1968).]

PROJECT 7.11 Fermi Surfaces: The Nearly-Free-Electron Approach and Experimental Measurements

1. Tell how to construct Fermi surfaces using the NFE model.
2. List five experimental methods for Fermi surface studies.
3. Explain how cyclotron resonance is used to map out the Fermi surface of metals.
4. Develop equations for the anomalous skin effect using the intuitive approach of Pippard.
5. Explain how magnetoresistance can be used to study Fermi surfaces.
6. Explain how ultrasonic attenuation is employed to study Fermi surfaces.
7. What happens to a Fermi surface when the metal is subjected to a large hydrostatic pressure? [*Hint*: See Harrison (1970).]

PROJECT 7.12 Quantum Oscillatory Effects

Prove that extremal cross-sectional areas of the Fermi surface such as the "neck" orbit occurring at the Brillouin zone boundary in Fig. 7.18 give rise to oscillations in the density of states which are periodic in $(1/B)$, where B is the applied magnetic field. [*Hint*: See Harrison and Webb (1960).]

7 Additional Properties of Bloch Functions

7.1 Alternative Statements of Bloch's Theorem

It follows from the functional form of the Bloch functions

$$\phi_{\mathbf{m}'}^{(s)}(\mathbf{r}) = u_{\mathbf{k}_{\mathbf{m}'}}^{(s)}(\mathbf{r})e^{i\mathbf{k}_{\mathbf{m}'}\cdot\mathbf{r}} \tag{7.145}$$

that translating by an arbitrary direct lattice vector

$$\mathbf{R}_{\mathbf{j}} = j_1\mathbf{d}_1 + j_2\mathbf{d}_2 + j_3\mathbf{d}_3 \tag{7.146}$$

yields

$$\begin{aligned} T_{\mathbf{j}}\phi_{\mathbf{m}'}^{(s)}(\mathbf{r}) &= \phi_{\mathbf{m}'}^{(s)}(\mathbf{r} + \mathbf{R}_{\mathbf{j}}) = u_{\mathbf{k}_{\mathbf{m}'}}^{(s)}(\mathbf{r} + \mathbf{R}_{\mathbf{j}})e^{i\mathbf{k}_{\mathbf{m}'}\cdot(\mathbf{r}+\mathbf{R}_{\mathbf{j}})} \\ &= u_{\mathbf{k}_{\mathbf{m}'}}^{(s)}(\mathbf{r})e^{i\mathbf{k}_{\mathbf{m}'}\cdot\mathbf{r}}e^{i\mathbf{k}_{\mathbf{m}'}\cdot\mathbf{R}_{\mathbf{j}}} = e^{i\mathbf{k}_{\mathbf{m}'}\cdot\mathbf{R}_{\mathbf{j}}}\phi_{\mathbf{m}'}^{(s)}(\mathbf{r}). \end{aligned} \tag{7.147}$$

That is, the only effect of lattice translation on energy eigenfunctions having the Bloch form is to introduce a constant phase factor of unit modulus into the energy eigenfunction. Since we have already shown that energy eigenfunctions in a periodic potential can be chosen to have the Bloch form, it can be stated loosely that *the only effect of lattice translation on the energy eigenfunctions in a crystal is to introduce a phase factor*. This may be taken as an *alternate* (though nonrigorous) *statement of Bloch's theorem*.

Alternatively, it can be stated (again nonrigorously) on the basis of Eq. (7.147) that Bloch's theorem means simply that the energy eigenfunctions in a perfect crystal satisfy the mathematical condition that

$$\phi_{\mathbf{m}'}^{(s)}(\mathbf{r} + \mathbf{R}_\mathbf{j}) = e^{i\mathbf{k}_{\mathbf{m}'}\cdot\mathbf{R}_\mathbf{j}}\phi_{\mathbf{m}'}^{(s)}(\mathbf{r}) \tag{7.148}$$

for some fixed vector $\mathbf{k}_{\mathbf{m}'}$. Of course, from our detailed work above we are aware that $\mathbf{k}_{\mathbf{m}'}$ represents one of the set of wave vectors which satisfy periodic boundary conditions for the crystal.

7.2 Interpretation of the Fourier Series Solution Obtained in Reduced-Zone Notation

The trial wave function (7.47) used to solve the time-independent Schrödinger equation was written in terms of reduced zone notation in the sense that all Fourier components $e^{i\mathbf{k}_\mathbf{m}\cdot\mathbf{r}}$ were replaced by corresponding components $\exp[i(\mathbf{k}_{\mathbf{m}'} + \mathbf{G}_l)\cdot\mathbf{r}]$. All energy eigenvalues were found to be dependent upon the vectors $\mathbf{k}_{\mathbf{m}'} + \mathbf{G}_l$. Since in the reduced zone scheme we only consider the $\mathbf{k}$ vectors to span the range of the reduced zone in question, with the varying band index s giving all possible sheets of the dispersion surface, the energy eigenvalues and eigenfunctions thus obtained must be considered to be confined to the reduced zone. It is only by repeating the process of solution in a sequential manner, assuming a different reduced zone for each different solution, that the so-called *repeated* (or *periodic*) zone results discussed in §6.2 can be obtained.

7.3 Time Dependence of Wave Functions Having the Bloch Form

The time factor $\exp[-(i/\hbar)\mathscr{E}^{(s)}(\mathbf{k}_{\mathbf{m}'})t]$ can be added to the corresponding stationary-state Bloch function $\phi_{\mathbf{k}_{\mathbf{m}'}}^{(s)}(\mathbf{r})$ to give the time-dependent energy eigenfunctions $\psi_{\mathbf{k}_{\mathbf{m}'}}^{(s)}(\mathbf{r}, t)$ for a periodic potential,

$$\psi_{\mathbf{k}_{\mathbf{m}'}}^{(s)}(\mathbf{r}, t) = \phi_{\mathbf{k}_{\mathbf{m}'}}^{(s)}(\mathbf{r})e^{-(i/\hbar)\mathscr{E}^{(s)}(\mathbf{k}_{\mathbf{m}'})t}. \tag{7.149}$$

This particular form for the time factor is quite general, since it follows directly from a separation of variables in the time-dependent Schrödinger equation, as shown in §6 of Chap. 1. In this way we obtain

$$\begin{aligned}\psi_{\mathbf{k}_{\mathbf{m}'}}^{(s)}(\mathbf{r}, t) &= \exp\{i[\mathbf{k}_{\mathbf{m}'}\cdot\mathbf{r} - \hbar^{-1}\mathscr{E}^{(s)}(\mathbf{k}_{\mathbf{m}'})t]\}\sum_l B_{\mathbf{m}',l}^{(s)}\exp(i\mathbf{G}_l\cdot\mathbf{r})\\ &= u_{\mathbf{m}'}^{(s)}(\mathbf{r})\exp\{i[\mathbf{k}_{\mathbf{m}'}\cdot\mathbf{r} - \hbar^{-1}\mathscr{E}^{(s)}(\mathbf{k}_{\mathbf{m}'})t]\}.\end{aligned} \tag{7.150}$$

It can be seen immediately that the electron probability density has the lattice periodicity, namely,

$$\psi_{\mathbf{k}_{\mathbf{m}'}}^{(s)}(\mathbf{r}, t)^*\psi_{\mathbf{k}_{\mathbf{m}'}}^{(s)}(\mathbf{r}, t) = u_{\mathbf{m}'}^{(s)}(\mathbf{r})^*u_{\mathbf{m}'}^{(s)}(\mathbf{r}). \tag{7.151}$$

7.4 Bloch Functions and Charge Transport in Crystals

The significance of the Bloch form for wave functions in a crystal is twofold. First of all, it is of interest to be able to compute the probability amplitude $\psi^*\psi$ of a given energy eigenstate, and second, it can be shown that electrons in Bloch states can propagate through the crystal. Propagation is of course the all important factor when considering charge transport under the action of externally applied potential differences. The proof that electrons in Bloch states can propagate in crystals is rather involved [Jones and Zener (1934); R. A. Smith (1963)]. In essence it involves a computation of the current density for Bloch states in accordance with the prescription given by Eq. (1.264). It is found that the average velocity $\langle \mathbf{v} \rangle$ over a unit cell for an electron moving in a perfectly periodic potential is given by the *constant* value

$$\langle \mathbf{v} \rangle = \hbar^{-1} \nabla_{\mathbf{k}} \mathscr{E}(\mathbf{k}), \tag{7.152}$$

quite analogous to the group velocity of a wave packet as given by Eq. (1.187). Thus an electron in a Bloch state is not scattered and thereby slowed down by the atoms making up the periodic crystal lattice. The perfect lattice therefore offers zero electrical resistance to charge transport. From this standpoint the free-electron model of a metal is a rather good approximation. (This quantum-mechanical result of zero resistance due to the ions in the lattice is significantly different from the scattering which one might expect from the viewpoint of classical physics. Refer to the discussion in Chap. 3, §1.6.) That the electron velocity depends critically upon the dispersion relation $\mathscr{E}(\mathbf{k})$, according to the above expression, provides justification for efforts to deduce the various energy bands for a periodic potential.

One interesting point follows immediately from Eq. (7.152): At the Brillouin zone boundaries where Bragg reflection prevents electron propagation, such that the eigenfunctions become nonpropagating standing waves for which $\langle \mathbf{v} \rangle = 0$, the $\mathscr{E}(\mathbf{k})$ relation has zero slope. This is markedly distinct from the parabolic dispersion relation $\mathscr{E}(\mathbf{k}) = \hbar^2 k^2/2m$ characteristic of free electrons.

In addition to the importance of the energy band dispersion relation $\mathscr{E}(\mathbf{k})$ for determining the electron velocity, it can also be shown that the *inertial properties* of the conduction electrons can be deduced from the electronic energy bands. The so-called "effective mass" m^* is analogous to the free-electron mass m in limiting the acceleration $\mathbf{a} = \mathbf{F}/m$ which can be produced by an externally applied electrical force $\mathbf{F}$. Its value is characteristic of the interaction of the electron with the periodic lattice potential, and it can be determined by taking appropriate derivatives of $\mathscr{E}(\mathbf{k})$ [Project 7.6] with respect to the components of $\mathbf{k}$.

PROJECT 7.13 Are Bloch Functions Momentum Eigenfunctions?

Prove or disprove the following theorem: "Bloch functions are not necessarily momentum eigenfunctions."

PROJECT 7.14 Charge Transport by Electrons in Bloch States

Prove that electrons characterized by Bloch functions have a nonzero linear momentum. Deduce an expression for this momentum.

PROJECT 7.15 Crystal Momentum

Prove that the time rate of change of the crystal momentum of an electron in a solid is equal to the externally applied force.

PROJECT 7.16 Band Theory of Electron Acceleration

Describe the acceleration of an electron through an energy band, including a discussion of its crystal momentum, velocity, and periods in time and position.

PROJECT 7.17 Group Velocity as a Unit Cell Average of the Instantaneous Velocity

Prove that the average of the instantaneous velocity of an electron over one unit cell of a one-dimensional crystal is equal to the group velocity of the electron in the crystal. [*Hint*: See R. A. Smith (1963) and Ziman (1964).]

PROJECT 7.18 Group Velocity of Electrons in Bloch States

Prove that Bloch states represent electrons in translational motion in a crystal with group velocity $\mathbf{v} = \hbar^{-1} \nabla_{\mathbf{k}}\mathscr{E}(\mathbf{k})$.

PROJECT 7.19 Quasi-Classical Dynamics of Quantum Particles

Develop the equations $\hbar\, d\mathbf{k}/dt = -\nabla U(\mathbf{r})$ and $\mathbf{v} = \hbar^{-1} \nabla_{\mathbf{k}}\mathscr{E}(\mathbf{k})$, where $U(\mathbf{r})$ is the scalar potential which yields the externally applied forces $\mathbf{F} = -\nabla U(\mathbf{r})$ acting on the particles in the solid, and $\mathscr{E}(\mathbf{k})$ is the energy-band dispersion relation for the solid in the absence of the external forces. [*Hint*: See Ziman (1964).]

8 Energy Bands from the Viewpoint of the One-Electron Atomic Levels

The periodic lattice potential, being derived from an array of local one-electron potentials (§2), includes the possibility of orbital motion as well as translational motion of the conduction electrons. From an alternate viewpoint [viz, *the tight-binding approach*, as described by Ziman (1964), for example], the energy bands can be deduced from the overlap of the atomic wave functions, in which case the bound electronic states of the one-electron atom give rise to the energy bands in the solid. The degeneracy of the atomic energy levels with respect to the electron spin and magnetic quantum numbers generally leads to overlapping bands for those states derived from atomic levels having the same principal and orbital quantum numbers, so that all electrons from the nl level form a band, n being the principal quantum number and l being the orbital angular momentum quantum number. (It may be helpful for the reader to refer back to Chap. 1, §12.4. The degeneracy of states with respect to the orbital quantum number found for the case of the one-electron atom is removed by the noncentral periodic potential.) Since the magnetic quantum number m ranges in value from $-l$ to $+l$ in unit increments, there are $2l + 1$ values, and so, considering also electron spin, there will be $2(2l + 1)$ electron states per atom within the nl band. Thus for each atom in the solid any filled s band contains two electrons, any filled p band contains six electrons, any filled d band contains 10 electrons, etc., as indicated in Table 7.1 (refer also to Chap. 1, §12.4). Adjacent bands sometimes overlap in energy, in which case the electronic states take on the character of *both* atomic levels in question *as well as* the translational character of the Bloch functions. Band structure calculations in general are extremely complex, but with the advent of high speed digital computers and the

Table 7.1

Energy Bands from the Viewpoint of Overlapping Atomic Levels

Principal quantum number	Orbital quantum number	Magnetic quantum number	Spin quantum number	Energy band	Number of electrons
$n = 1$	$l = 0$	$m = 0$	$m_s = \pm\frac{1}{2}$	1s	2
$n = 2$	$l = 1$	$m = 1, 0, -1$	$m_s = \pm\frac{1}{2}$	2p	6
	$l = 0$	$m = 0$	$m_s = \pm\frac{1}{2}$	2s	2
$n = 3$	$l = 2$	$m = 2, 1, 0, -1, -2$	$m_s = \pm\frac{1}{2}$	3d	10
	$l = 1$	$m = 1, 0, -1$	$m_s = \pm\frac{1}{2}$	3p	6
	$l = 0$	$m = 0$	$m_s = \pm\frac{1}{2}$	3s	2

use of experimentally measured parameters, such calculations have become feasible for a large number of real metals.

PROJECT 7.20 Energy Band Computations

Qualitatively describe the following methods for the computation of electronic energy bands:

1. Cellular method.
2. Augmented plane-wave (APW) method.
3. Orthogonalized plane-wave (OPW) method.
4. Green's function method.
5. Perturbation method based on plane-wave expansions (NFE method).
6. Wigner–Seitz method.

PROJECT 7.21 Tight-Binding Approach to Energy Band Theory

1. How is the electronic wave function constructed in the tight-binding approach to energy band theory?
2. List the three-dimensional version of the following: (a) interaction integrals, (b) overlap integrals, (c) crystal field integrals.
3. Develop the dispersion relation $\mathscr{E}(\mathbf{k})$ versus $\mathbf{k}$ in terms of the above-listed integrals and other necessary quantities.
4. Use the above dispersion relation to set up as far as possible a computation of the lowest electronic energy bands for a square array of hydrogen atoms physically adsorbed on a chemically inert planar nonconducting surface.
5. Develop the theory of energy bands in the tight-binding limit for the degenerate case (i.e., the case in which the noninteracting bands would cross).
6. State qualitatively the variation in the character of the wave functions with crystal momentum.

PROJECT 7.22 Wannier Functions

Describe Wannier functions and their usefulness for representing electrons in crystals. Are they energy eigenfunctions? How are they similar and how do they differ from the complete set of Bloch functions? [*Hint*: See R. A. Smith (1963).]

9 Energy Gaps and Energy Bands: Insulators, Semiconductors, and Metals

As shown in §13 of Chap. 5, the periodic potential of the lattice modifies the plane-wave eigenfunctions characteristic of the free-electron model and perturbs the corresponding eigenvalues. These perturbations were evaluated (Chap. 5, §13.1) to arrive at the nearly-free-electron model. It was found, however, that the perturbation treatment failed for states at the Brillouin zone boundary because of degeneracy. A proper perturbation treatment (Chap. 5, §13.2) required diagonalization of the perturbation operator (viz, in this case the periodic potential of the lattice) in the manifold of planewave eigenstates. As is commonly found in such cases (Chap. 5, §2), the perturbation split the degeneracy of the energy levels. In the case of the periodic potential of a metal, the free-electron dispersion curve (see Fig. 7.11) becomes drastically modified in the neighborhood of the Brillouin zone boundaries; the quasi-continuum of allowable quantized energies for the free-electron eigenstates is ruptured and a forbidden energy range is introduced at the zone boundaries equal to the splitting of the energy levels introduced by the perturbation. (See Fig. 7.15.) The dispersion curve (or in actuality, the contoured *surface* or *sheet* from the three-dimensional viewpoint) becomes segmented, with the segments appearing in the different Brillouin zones being designated as energy *bands*. Thus in the presence of a periodic potential we have: (a) *energy bands* for which there exists a quasi-continuum of allowed quantized energies for which the conduction electrons propagate without attenuation by the ion cores, and (b) *energy gaps* wherein there exist no allowed (quantized or unquantized) energies for conduction electrons. There are a countable number of states in the quasi-continuum in any given band, which in conjunction with the Pauli exclusion principle results in a limitation in the number of electrons which can be accommodated in any band. If a given band is full, then the electrons in that band have a net zero momentum because for every momentum state $\mathbf{p}_\mathbf{k}$ there is also an allowed (and filled) momentum state $-\mathbf{p}_\mathbf{k}$. Therefore we arrive at the remarkable fact that *a filled energy band carries no electrical current.* This in fact is the "payoff" from our detailed (and at times laborious!) development of a solution to the Schrödinger equation for the case of a periodic potential. To elaborate on this point a bit, we have shown that the periodic potential leads to a situation in which a very large number of electrons (namely, the number required to fill an energy band) may give rise to absolutely no contribution to the electric current, whereas in the free-electron model these same electrons could give rise to arbitrarily large currents. In this way we have developed a basis for understanding the phenomenal differences between metals, semiconductors, and insulators. These differences are not due to the relative differences between the concentrations of electrons in the materials, since electron concentrations may differ by less than a factor of 10 in two materials for which the electrical conductivities differ by many orders of magnitude.

To extend our discussion, suppose we consider a material in which all electrons are accommodated in filled energy bands, with all of the energy bands at higher energies being empty. This situation would in actuality occur only if the temperature were near 0°K because otherwise thermal excitation (Chap. 3, §4) could excite some of the electrons in the uppermost filled band into the next higher empty band. If we consider the band gap separating the uppermost filled and next higher empty band to be quite large compared to $k_B T$, however, then at room temperature the number of electrons excited into the empty band could be almost negligible. For example, the probability for excitation of an electron across a 1 eV gap at room temperature is of the order of exp(− 1.0/0.025), which is only of the order of 4.25×10^{-18}. The application of an electric field to such a system imposes a potential energy difference across the substance, but even for electric fields as large as 10^5 V/cm, corresponding to 10,000 V placed across a 1-mm-long sample, the change in energy from atom to atom in the sample is only of the order of $(4 \times 10^{-8}$ cm$) \times (10^5$ V/cm$) \times$ (1 electronic charge) = 0.004 eV, which is far less than the band gap so that an electron on one atom would not be transferred by the electric field into the next higher band while undergoing rectilinear and orbital motion in the neighborhood of a given atomic site. Thus the filled band would still give no contribution to the electric current, and we would have what is commonly known as an *electrical insulator*. On the other hand, if we had a substance with a band gap small enough so that a reasonable number of electrons could be excited across the gap thermally, we would have what is known as an *intrinsic semiconductor*. An intrinsic semiconductor would have zero conductivity at 0°K but would have an increasing conductivity with increasing temperature due to the increased thermal excitation of electrons from the filled to the empty band. Empty states in the quasi-continuum of the empty band are readily available for filling with electrons by electric field excitation, thus leading to conduction in the same way as in the free electron model. A substance with a *partly filled band* can generally be conceived of as a *metal*, since empty states are available to the uppermost electrons for electric field excitation, thereby allowing a shifting of the Fermi surface in momentum space which leads to conduction (cf. Chap. 3), even in the absence of any thermal excitation. Thus in its own splendid way, quantum mechanics provides a firm theoretical structure for explaining simultaneously the dichotomous properties of a perfect metallic conductor (as brilliantly evidenced in the *free-electron model* of Chap. 3 and the Bloch function extension of this property to the real lattice) and a perfect electrical insulator (as spectacularly provided by the concepts of energy bands created by the periodic lattice potential which can be completely filled with electrons congruent with Pauli's atom *aufbau* concept of the exclusion principle).

PROBLEMS

1. Derive the major perturbing effects of a small periodic potential $V(x) = -\varepsilon h^2 g(x)/2m$ on the free-electron dispersion curve, where ε is a constant which can be chosen to be as small as desired. Describe the role played by the various Fourier components of the periodic potential. (*Hint*: See Chap. 5.)

2. Derive the energy-band dispersion relation $\mathscr{E}(\mathbf{k})$ in the nearly-free-electron approximation.

3. (a) Discuss the energy bands in a one-dimensional lattice with period d, where the potential energy is of the form $V = V_0\ (-b \leqslant x \leqslant 0)$; $V = 0\ (0 \leqslant x \leqslant d - b)$; $V(x) = V(x + d)$.

(b) Determine the energy values for the top of the first band and bottom of the second band at the zone boundary when V_0 has a value 0.1 eV, $d = 8$, $b = 3$ atomic units.

4. (a) Show that the free-electron wave functions in a one-dimensional periodic lattice of period d are degenerate for states at the Brillouin zone boundary.

(b) If a small perturbing potential is introduced at each atomic site, to first order in the perturbation show that the wave functions at the zone boundary are proportional to $\sin(n\pi x/d)$ and $\cos(n\pi x/d)$, where n is an integer.

5. Consider the behavior of an electron in a solid to be described by the plane wave $\exp[i(\mathbf{k}\cdot\mathbf{r} - \omega t)]$. Show that the quantity $\hbar\mathbf{k}$ corresponds to the momentum.

6. Prove that Bloch functions can represent electrons in translational motion in a crystal.

7. Verify that the function below satisfies Bloch's theorem (i.e., it is a Bloch function),

$$\phi_{\mathrm{B}}(\mathbf{r}) = \sum_{l} \exp(i\mathbf{k}_j \cdot \mathbf{R}_l)\, \phi(\mathbf{r} - \mathbf{R}_l),$$

where $\mathbf{R}_l$ is a direct lattice vector and $\phi(\mathbf{r})$ is an atomic wave function.

8. (a) Derive an expression for the electron effective mass in a one-dimensional lattice. You may use the relation $v = (1/\hbar)\, d\mathscr{E}/dk$.

(b) What could a negative effective mass possibly mean physically?

9. If an external electric field is applied to a solid, show that the time rate of change of the conduction-electron momentum is such that the electron behaves as if it has an inverse effective mass which is a tensor quantity with components $(1/m^*)_{ij} = (1/\hbar^2)\, \partial^2\mathscr{E}/\partial k_i \partial k_j$.

10. (a) Prove that the number of distinct allowed **k** vectors in each energy band is N_c, where N_c is the actual number of unit cells in the real crystal.

(b) Considering that there are two allowable values for the electronic spin for each allowed **k** vector, how many *electronic* states are there in each energy band?

(c) How is this number increased when we consider the fact that Bloch bands overlap for cases where the orbital angular momentum quantum number is unity or greater?

(d) Resolve any conflict in the following two statements:

(i) Each band of Bloch functions contains $2N_c$ electronic states.

(ii) The 3d band in copper can accommodate 10 electrons.

APPENDIX

PHYSICAL CONSTANTS: SYMBOLS, UNITS, AND VALUES

Symbol	Name	SI Unit	Value
h	Planck's constant	joule-second	6.6262×10^{-34}
$\hbar$	$h/2\pi$	joule-second	1.0546×10^{-34}
e	electronic charge magnitude	coulomb	1.6022×10^{-19}
m_e	electron rest mass	kilogram	9.1096×10^{-31}
m_p	proton rest mass	kilogram	1.6726×10^{-27}
m_n	neutron rest mass	kilogram	1.6749×10^{-27}
G	gravitational constant	newton-meter^2/kilogram^2	6.673×10^{-11}
a	Bohr radius ($4\pi\varepsilon_0\hbar^2/m_e e^2$)	meter	5.2918×10^{-11}
R_∞	Rydberg constant ($m_e e^4/64\pi^3\varepsilon_0^2\hbar^3 c$)	meter^{-1}	1.097×10^{7}
Ry	Rydberg energy ($m_e e^4/32\pi^2\hbar^2\varepsilon_0^2$)	joule	2.180×10^{-18}
N_{Avog}	Avogadro number	number/mole	6.0222×10^{23}
k_B	Boltzmann constant	joule/°K	1.3806×10^{-23}
ε_0	electric permittivity of free space	farad/meter	8.854×10^{-12}
μ_0	magnetic permeability of free space	henry/meter	$4\pi \times 10^{-7}$
c	velocity of light in free space	meter/second	2.997925×10^{8}

REFERENCES

Bloch, F. (1976). Heisenberg and the Early Days of Quantum Mechanics, *Phys. Today* **29**, 23.

Bohm, D. (1951). "Quantum Theory." Prentice-Hall, Englewood Cliffs, New Jersey.

Born, M. (1957). "Atomic Physics." Hafner, New York.

Bragg, Sir William (1931). "The Universe of Light," 1st ed. Macmillan, New York.

Bragg, Sir William (1959). "The Universe of Light." Reprinted by Dover, New York.

Brillouin, L. (1953). "Wave Propagation in Periodic Structures." Dover, New York.

Burington, R. S. (1956). "Handbook of Mathematical Tables and Formulas." Handbook Publ., Sandusky, Ohio.

Burstein, E., and Lundqvist, S. (1967). "Tunneling Phenomena in Solids." Plenum, New York.

Callaway, J. (1976). "Quantum Theory of the Solid State." Academic Press, New York.

Conley, J. W., Duke, C. B., Mahan, G. D., and Tiemann, J. J. (1966). Electron Tunneling in Metal-Semiconductor Barriers, *Phys. Rev.* **150**, 466.

Dirac, P. A. M. (1930). "The Principles of Quantum Mechanics," 1st ed. Oxford Univ. Press, London and New York.

Dirac, P. A. M. (1962). "The Principles of Quantum Mechanics," 4th ed. Oxford Univ. Press, London and New York.

Duke, C. B. (1969). "Tunneling in Solids" (*Solid State Phys.* Suppl. **10**). Academic Press, New York.

Einstein, A. (1917). On the Quantum Theory of Radiation, *Phys. Z.* **18**, 21. [As quoted from van der Waerden (1967).]

Eisberg, R. M. (1967). "Fundamentals of Modern Physics." Wiley, New York.

Eisberg, R. M. (1976). "Applied Mathematical Physics with Programmable Pocket Calculators." McGraw-Hill, New York.

Esaki, L. (1967). Tunneling, *in* "Tunneling Phenomena in Solids" (E. Burstein and S. Lundqvist, eds.). Plenum, New York.

Fromhold, A. T., Jr. (1976). "Theory of Metal Oxidation", Vol. I. North-Holland Publ., Amsterdam.

Furry, W. H. (1947). Two Notes on Phase-Integral Methods, *Phys. Rev.* **71**, 360.

Gamow, G. (1966). "Thirty Years That Shook Physics." Doubleday, Garden City, New York.

Giaever, I. (1960). Electron Tunneling between Two Superconductors, *Phys. Rev. Lett.* **5**, 464. (See also: Energy Gap in Superconductors Measured by Electron Tunneling, *Phys. Rev. Lett.* **5**, 147).

Goldsmid, H. J., ed. (1968). "Problems in Solid State Physics." Academic Press, New York.

Goldstein, H. (1956). "Classical Mechanics." Addison-Wesley, Cambridge, Massachusetts.

Good, R. H., Jr., and Müller, E. W. (1956). Field Emission, *in* "Handbuch der Physik," Vol. XXI (S. Flügge, ed.), p. 176. Springer-Verlag, Berlin and New York.

Gundlach, K. H., and Simmons, J. G. (1969). Range of Validity of the WKB Tunnel Probability, and Comparison of Experimental Data and Theory, *Thin Solid Films* **4**, 61.

Halliday, D., and Resnick, R. (1974). "Fundamentals of Physics." Wiley, New York.

Harrison, W. A. (1970). "Solid State Theory." McGraw-Hill, New York.

Harrison, W. A., and Webb, M. B. (1960). "The Fermi Surface." Wiley, New York.

Hartman, T. E. (1964). Tunneling through Asymmetric Barriers, *J. Appl. Phys.* **35**, 3283.
Heisenberg, W. (1925). Quantum Theoretical Re-interpretation of Kinematic and Mechanical Relations, *Z. Phys.* **33**, 879. [As quoted from van der Waerden (1967).]
Heisenberg, W. (1930). "The Physical Principles of the Quantum Theory" (Carl Eckart and Frank C. Hoyt, translators). Dover, New York.
Houston, W. V. (1948). "Principles of Mathematical Physics." McGraw-Hill, New York.
Ikenberry, E. (1962). "Quantum Mechanics for Mathematicians and Physicists." Oxford Univ. Press, New York.
Jackson, J. D. (1962). "Classical Electrodynamics." Wiley, New York.
Jaklevic, R. C., and Lambe, J. (1966). Molecular Vibration Spectra by Electron Tunneling, *Phys. Rev. Lett.* **17**, 1139.
Jones, H., and Zener, C. (1934). A General Proof of Certain Fundamental Equations in the Theory of Metallic Conduction, *Proc. R. Soc. London Ser. A* **144**, 107.
Josephson, B. D. (1962). Possible New Effects in Superconductive Tunneling, *Phys. Lett.* **1**, 251.
Kittel, C. (1971). "Introduction to Solid State Physics," 4th ed. Wiley, New York.
Langenberg, D. N., Scalapino, D. J., and Taylor, B. N. (1966). The Josephson Effects, *Sci. Am.* **214**, 30.
Langer, R. E. (1937). On the Connection Formulas and the Solutions of the Wave Equation, *Phys. Rev.* **51**, 669.
Leighton, R. B. (1959). "Principles of Modern Physics." McGraw-Hill, New York.
McGervey, J. D. (1971). "Introduction to Modern Physics." Academic Press, New York.
McKelvey, J. P. (1966). "Solid-State and Semiconductor Physics." Harper, New York.
Mandl, F. (1957). "Quantum Mechanics." Butterworth, London.
March, N. H., Young, W. H., and Sampanthar, S. (1967). "The Many-Body Problem in Quantum Mechanics." Cambridge Univ. Press, London and New York.
Merzbacher, E. (1970). "Quantum Mechanics," 2nd ed. Wiley, New York.
Messiah, A. (1965). "Quantum Mechanics." North-Holland Publ., Amsterdam.
Pauling, L., and Wilson, E. B., Jr. (1935). "Introduction to Quantum Mechanics." McGraw-Hill, New York.
Rojansky, V. (1938). "Introductory Quantum Mechanics." Prentice-Hall, Englewood Cliffs, New Jersey.
Rowell, J. M. (1963). Magnetic Field Dependence of the Josephson Tunnel Current, *Phys. Rev. Lett.* **11**, 200.
Sachs, M. (1963). "Solid State Theory." McGraw-Hill, New York.
Schiff, L. I. (1968). "Quantum Mechanics," 3rd ed. McGraw-Hill, New York.
Seitz, F. (1950). Imperfections in Nearly Perfect Crystals: A Synthesis. From Shockley *et al.* (1952).
Shockley, W., Hollomon, J. H., Maurer, R., and Seitz, F. eds. (1952). "Imperfections in Nearly Perfect Crystals." Wiley, New York.
Simmons, J. G. (1963a). Generalized Formula for the Electric Tunnel Effect between Similar Electrodes Separated by a Thin Insulating Film, *J. Appl. Phys.* **34**, 1793.
Simmons, J. G. (1963b). Electric Tunnel Effect between Dissimilar Electrodes Separated by a Thin Insulating Film, *J. Appl. Phys.* **34**, 2581.
Slater, J. C. (1934). The Electronic Structure of Metals, *Rev. Mod. Phys.* **6**, 210.
Smith, R. A. (1963). "Wave Mechanics of Crystalline Solids." Chapman & Hall, London.
Stratton, J. A. (1941). "Electromagnetic Theory." McGraw-Hill, New York.
Stratton, R. (1962). Volt-Current Characteristics for Tunneling Through Insulating Films, *J. Phys. Chem. Solids* **23**, 1177.
Taylor, P. L. (1970). "A Quantum Approach to the Solid State." Prentice-Hall, Englewood Cliffs, New Jersey.
ter Haar, D., ed. (1975). "Problems in Quantum Mechanics." Pion, London.
Tinkham, M. (1975). "Introduction to Superconductivity." McGraw-Hill, New York.
van der Waerden, B. L., ed. (1967). "Sources of Quantum Mechanics." North-Holland Publ., Amsterdam.
Wylie, C. R. Jr., (1951). "Advanced Engineering Mathematics." McGraw-Hill, New York.
Ziman, J. M. (1964). "Principles of the Theory of Solids." Cambridge Univ. Press, London.

INDEX

Boldface numbers are given when it proves helpful to identify for the reader the primary reference page(s) in a given listing that involves a string of numbers.

C

D

E

P

T

U